Applied Plate Theory for the Engineer

Applied Plate Theory for the Engineer

C. P. Heins
University of Maryland

Lexington Books
D.C. Heath and Company
Lexington, Massachusetts
Toronto

Library of Congress Cataloging in Publication Data

Heins, Conrad P
Applied plate theory for the engineer.

Includes index.
1. Plates (Engineering) I. Title.
TA660.P6H36 624'.1776 75-42951
ISBN 0-669-00569-x

Published simultaneously in Canada

Printed in the United States of America

International Standard Book Number: 0-669-00569-x

Library of Congress Catalog Card Number: 75-42951

To My Parents

Contents

List of Figures

List of Tables

Notation

A_0, An	Fourier Series Coefficient
be	Effective Plate Width
B, B^T	Displacement Matrix
Bn	Fourier Series Coefficient
C	Half Width between Ribs
D	Isotropic Plate Stiffness
D_x	Orthotropic Plate Stiffness—with respect to a section along the y-axis
D_y	Orthotropic Plate Stiffness—with respect to a section along the x-axis
D_θ	Orthotropic Plate Stiffners—with respect to a section along the r-axis
D_r	Orthotropic Plate Stiffners—with respect to a section along the θ-axis
E	Modulus of Elasticity
F	Stress Function
G	Shear Modulus
h	Plate Thickness
H	Torsional Plate Stiffness
I	Moment of Inertia
K_T	Torsional Constant
K	Stiffness Matrix
L	Span Length
M	Bending Moment
M_x	Bending Moment about x-axis
M_y	Bending Moment about y-axis
M_{xy}	Torsional Moment
M_r	Bending Moment about θ axis
M_θ	Bending Moment about r-axis
$M_{r\theta}$	Torsional Moment
n	Integer
N_x	Normal Force with respect to y Section
N_y	Normal Force with respect to x Section
N_{xy}	Shear Force

N_θ	Normal Force with respect to r Section
N_r	Normal Force with respect to θ Section
P	In-Plane Load per Unit of Length
q	Lateral Load
Q_x	Summation of Boundary Force with respect to x-axis
Q_y	Summation of Boundary Force with respect to y-axis
r	Coordinate Axis
R	Radius to Plate Boundary
R_x	Plate Reaction with respect to x Section
R_y	Plate Reaction with respect to y Section
S	Surface Length of Plate, Section Modulus
S_w	Torsional Warping Shear Function
t	Plate Thickness
u	Lateral Displacement with respect to x-axis
v	Vertical Displacement with respect to y-axis
V_x	Shear Force with respect to y Section
V_y	Shear Force with respect to x Section
W	Vertical Displacement
W_n	Torsional Warping Normal Function
x	Coordinate Axis
X	Body Force
$\bar{\bar{X}}$	Boundary Force
y	Coordinate Axis
Y	Body Force
$\bar{\bar{Y}}$	Boundary Force
z	Coordinate Axis
α	Plate Stiffness Ratio $H/\sqrt{D_xD_y}$, D_θ/D_r
β	Plate Stiffness Ratio H/D_y, H/D_r
ε	Normal Strain
θ	Arc Dimension
γ	Shear Strain
λ	Coefficient $n\pi/L$, $n\pi/\theta_T$
μ	Poisson's Ratio
σ	Normal Stress
τ	Shear Stress
ϕ	Stress Function

Applied Plate Theory for the Engineer

1 Introduction

In the design and analysis of many engineering structures, the interaction of the various supporting members and the floor system must be considered. This situation thus creates a highly indeterminate structure, requiring some knowledge relative to the floor (plate) action and the supporting members. If the floor has no main supporting members, and the floor or slab acts as an integral unit of the entire system, then plate action consideration is again required. In order to overcome the difficulty in applying such considerations to design, many simplified procedures have been documented in the various specifications and handbooks. Such factors as "Load Distribution," "Moment Coefficients," "Effective Slab Widths," "Equivalent Stiffness," etc., have been incorporated into design concepts to account for plate action.

These factors, however, may only be applicable under certain conditions, which must be understood by the engineer. If the problem under design does not meet these conditions, then further studies may be required using plate theory. Thus it is important for the engineer to have some knowledge of plate action and how this action may interact in the structural system. When the system action is understood, then appropriate solutions may be obtained by various schemes. Possibly subdividing the plate into equivalent beam strips, as used in retaining wall design, may be considered. Use of solutions representing part of the plate, with certain boundaries, may also be used. Developing design tables, after the general equations have been developed, may also be an appropriate solution. The development of such general equations usually involves determination of deformations as the unknown variables. Thus the prescribed boundaries are also stipulated relative to deformations, which are known. This approach is generally the procedure followed for many indeterminate structures, be it either a frame or a plate or combination thereof.

The solution of the general plate equations can be found in several manners. In some instances an "exact" solution can be found if the boundaries of the plate are known. If the plate has unusual boundaries or interacting elements, then approximate techniques must be employed.

The intention, therefore, of this text is to present a comprehensive review of classical plate theory, relative to in-plane forces and plate bending. The application of such theories and procedures used to find solutions will be given in detail, through problem solutions and computer techniques.

The text will be divided into four basic areas:

Theory of Elasticity
Plates Subjected to Forces in Their Planes
Plane Bending of Plates
Finite Difference and Finite Element Solutions

The above listed chapters pertain to the general theories and their solution. The application of these solutions to practical problems and associated computer programs are next described in:

Bridge and Building System Solutions
Computer Programs

Associated with plate theory is the orthotropic deck bridge, where application of the required theories has resulted in design data, which is given in the next chapter:

Bridge Data and Examples

The last chapter describes the combined action of bending and in-plane forces and is entitled:

Large Deflection Theory and Plate Buckling

Throughout the development contained herein, certain assumptions are made. The assumptions, relative to thin plate bending behavior, come under the broad category of small deflection theory. Such a theory is based on the following:

(1) There is no deformation in the middle plane of the plate, i. e., it is unstrained.
(2) Plane sections remain plane.
(3) Normal stresses in a direction perpendicular to the plane of the plate are disregarded.

These assumptions then permit stress relationships to be stated in terms of displacement and coordinate axes.

Exception to these assumptions pertain to large deflections of plates as discussed in chapter 9.

1.1 History of Plate Theory Development

Although the behavior of any complicated structural system may involve effects of static and dynamic loads, generally new solutions and techniques are first concerned with only static load effects. In the study of plate behavior, however, the first concentrated study involved free vibration of plates by L. Euler in 1766. J. Bernoulli, Euler's student, extended this work to consider gridworks. Bernoulli's equation, however, when solved and compared to experiments was not fruitful.

E.F.F. Chladni, a German physicist, performed vibration experiments on plates which permitted determination of the various modes of free vibration. Because of these experiments, he was invited by Napoleon and the French Academy of Science in 1809 to show his tests. His tests impressed the French, and thus a Research Competition on plate vibrations was initiated. Only one applicant entered, S. Germain, a French mathematician. The results of her studies yielded a plate differential equation that lacked the warping term. This inconsistency was noted by one of the judges of the competition, Lagrange, who was therefore the first person to present the general plate equation properly.

S.D. Poisson in 1829 successfully expanded the Germain-Lagrange plate equation to the solution of a plate under static loading. In this solution, however, the plate flexural rigidity D was set equal to a constant term.

The first satisfactory theory of bending of plates is attributed to C.L. Navier, who considered the plate thickness in the general plate equation as a function of rigidity D. He also introduced an "exact" method which transforms the differential equation into algebraic expressions by use of Fourier trigonometric series.

In 1850 G.R. Kirchoff published an important thesis on the theory of thin plates. In this thesis, Kirchoff stated two basic assumptions, now accepted and given previously as:

(1) Inside the plate, those lines which are initially perpendicular to the middle plane remain as straight lines during the bending of the plate and are perpendicular to the middle plane of the bent plate.
(2) When the plate exhibits a small deflection under transverse load, the middle plane of the plate does not undergo stretching.

Using these assumptions, Kirchoff developed the correct strain energy plate bending equation. He also pointed out that there exist only two boundary conditions on a plate edge.

Lord Kelvin provided additional insight relative to the condition of boundary equations by converting torsional moments along the edge of a plate into shearing forces. Thus the edges are subject to only two forces, shear and moment.

The solution of rectangular plates, with two parallel simple supports and the other two supports arbitrary, was successfully solved by M. Levy in the late nineteenth century.

Since that time classical plate solutions and developments have continued through the efforts of Westergaard, Nadai, Huber, Schleicher, and many others.

These classical solutions have now been augmented by numerical solutions through the advent of the digital computer. Solutions of very complicated problems can now be solved using finite difference methods or finite element techniques.

The advancement of the classical techniques have thus permitted new insight and new tcchniques. This text will present information to make use of these ideas.

Theory of Elasticity

2.1 General

In order to develop and apply the general plate equations to engineering problems, it is necessary to consider the general state of stress at a point or element. These stress equations will be developed relative to an in-plane state of stress and can then be applied to in-plane force problems and, with appropriate modifications, to bending problems.

2.2 Stress

2.2.1 State of Stress at a Point

Consider the infinitesimal cubic element dx, dy, dz, shown in Figure 2-1, subjected to stress vectors $\bar{S}_x$, $\bar{S}_y$ and $\bar{S}_z$. Each of the vectors have stress components, as shown by vector $\bar{S}_x$, which are noted by the following equation.

$$\left.\begin{aligned} \bar{S}_x &= (\sigma_x, \tau_{xy}, \tau_{xz}) \\ \bar{S}_y &= (\tau_{yx}, \sigma_y, \tau_{yz}) \\ \bar{S}_z &= (\tau_{zx}, \tau_{zy}, \sigma_z) \end{aligned}\right\} \tag{2.1}$$

If the state of stress at a point is required, the three vectors ($\bar{S}_x$, $\bar{S}_y$, $\bar{S}_z$) or their six independent components must be determined. The state of stress at a point, or stress tensor, is given by the following expression:

$$\begin{vmatrix} \sigma_x & \tau_{xy} & \tau_{xz} \\ \tau_{yx} & \sigma_y & \tau_{yz} \\ \tau_{zx} & \tau_{zy} & \sigma_z \end{vmatrix} \tag{2.2}$$

2.2.2 Equations of Equilibrium—Two-Dimensional

The general cubic element shown in Figure 2-1 subjected to the stress vectors, is in a state of equilibrium. If a top view of the element is consid-

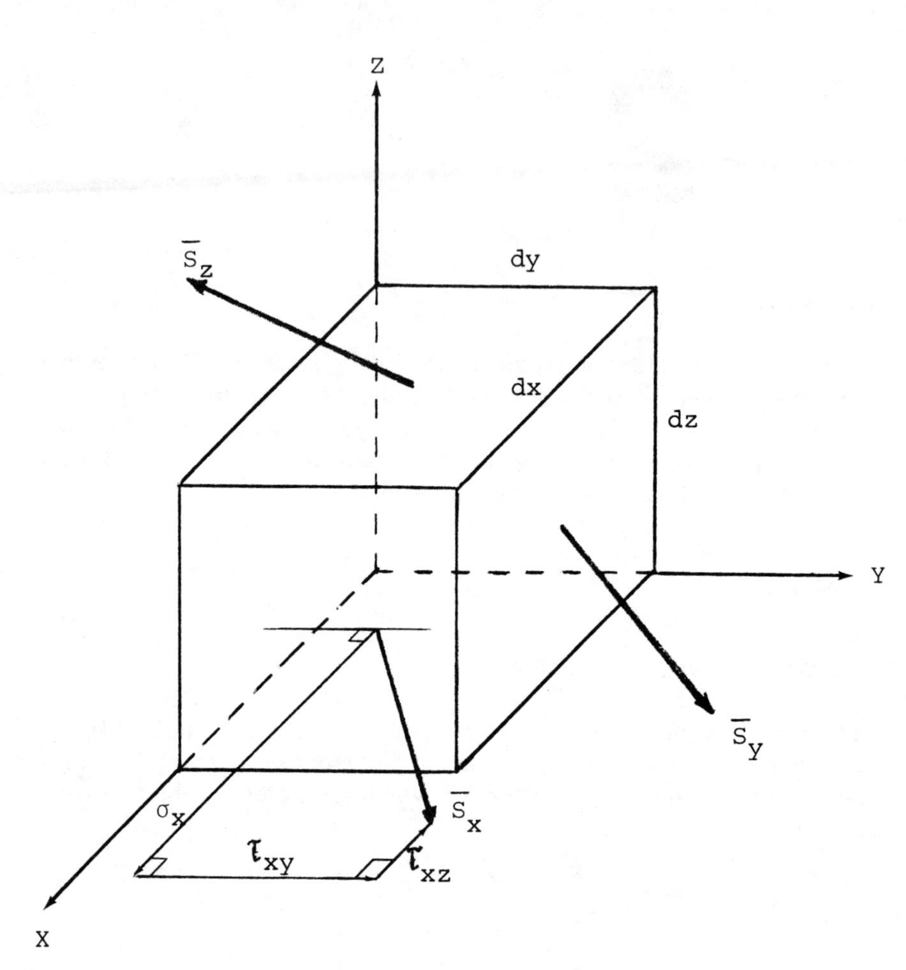

Figure 2-1. Body Stresses.

ered, and the stresses associated with vector $\bar{S}_z$ are neglected, we have a two-dimensional state of stress, as shown in Figure 2-2. The coordinate axis Z, is perpendicular to this element and is normal to the paper. The dpeth of thc element will be assumed unity. Initial stresses σ_x, σ_y, τ_{xy}, and τ_{yx} act on the positive faces; a distance dx, dy from these faces are incremental changes resulting in the forces shown. The incremental or differential stress can be written as a partial derivative by use of Taylor's series [1]. In general the series is written as follows:

$$f(x) = f(a) + f'(a)(x - a) + \frac{f''(a)}{2!}(x - a)^2 + \ldots \tag{2.3}$$

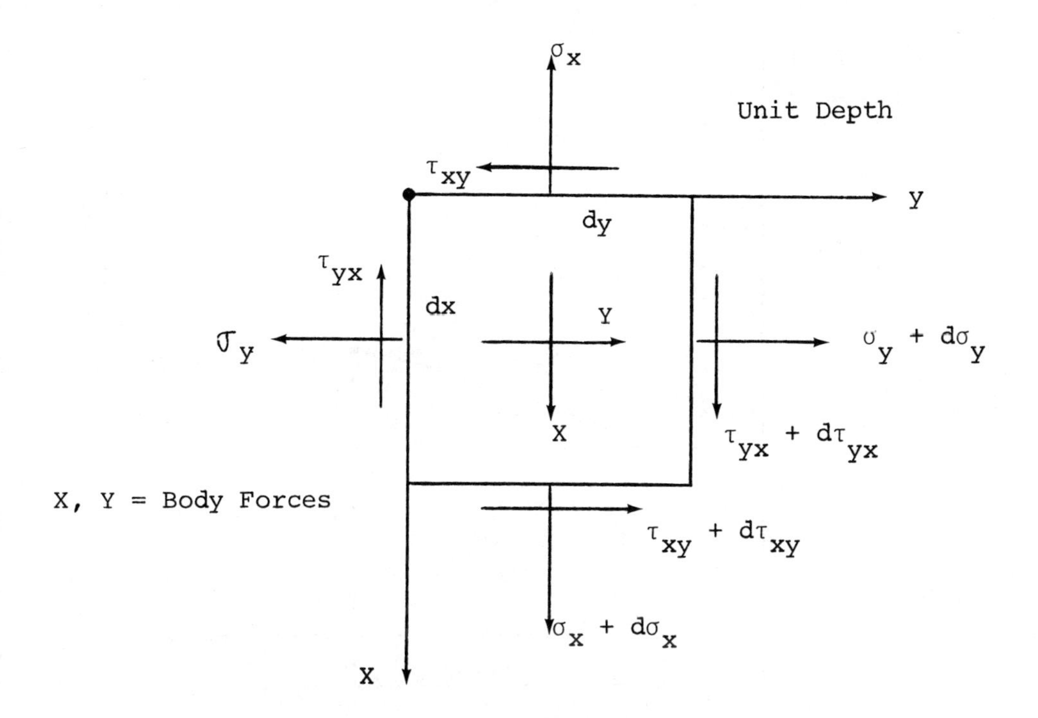

Figure 2-2. Plane Stress Element.

where: $f(x)$ = the function required

$f(a)$ = the initial function

a = distance to the initial function $f(a)$

In this instance it is required to evaluate the expression $\sigma_x + d\sigma_x$ or $f(x) = f(\sigma_x + d\sigma_x)$, and the initial variable is $f(a) = f(\sigma_x)$, as shown in Figure 2-3. The distance a, from the x coordinate is zero, as the value of $f(a) = f(\sigma_x)$ is at the origin of the element as shown in Figure 2-2. The derivatives of $f(a)$ are:

$$f'(a) = f'(\sigma_x) = \partial\sigma_x/\partial x$$

$$f''(a) = f''(\sigma_x) = \partial^2\sigma_x/\partial x^2$$

etc.

Substituting these variables into equation (2.3) gives:

$$f(\sigma_x + d\sigma_x) = \sigma_x + \frac{\partial\sigma_x}{\partial x}\,dx + \frac{\partial^2\sigma_x}{\partial x^2} \cdot \frac{dx^2}{2} + \ldots$$

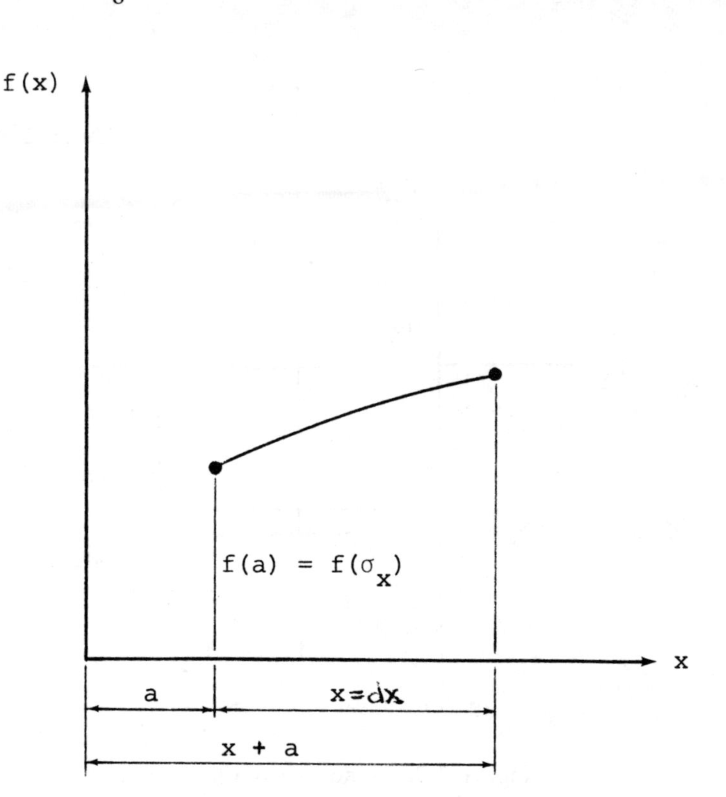

Figure 2-3. Variation of $f(\sigma_x)$.

where the higher order terms, i.e., $(\partial^2\sigma_x/\partial x^2)(dx^2/2)$, . . ., are neglected, thus the stress $\sigma_x + d\sigma_x$ is equal to $\sigma_x + (\partial\sigma_x/\partial x)\,dx$. Similar relationships can be developed for the other stresses σ_y, τ_{yx}, and τ_{xy}, resulting in the stresses as shown in Figure 2-4. Also applied to the element are known body forces X, Y, representing weight, inertia, etc. of the element.

With this set of stresses and forces, equilibrium of the element can now be evaluated. Considering the summation of the forces in the x direction gives

$$\sum F_x = 0$$

$$\left(\sigma_x + \frac{\partial\sigma_x}{\partial x}\,dx\right)dy \cdot 1 - \sigma_x\,dy \cdot 1 + \left(\tau_{yx} + \frac{\partial\tau_{yx}}{\partial y}\,dy\right) \cdot dx \cdot 1$$

$$- \tau_{yx}\,dx \cdot 1 + X\,dx \cdot dy \cdot 1 = 0 \tag{2.4}$$

Expanding this equation (2.4), and neglecting higher order terms gives

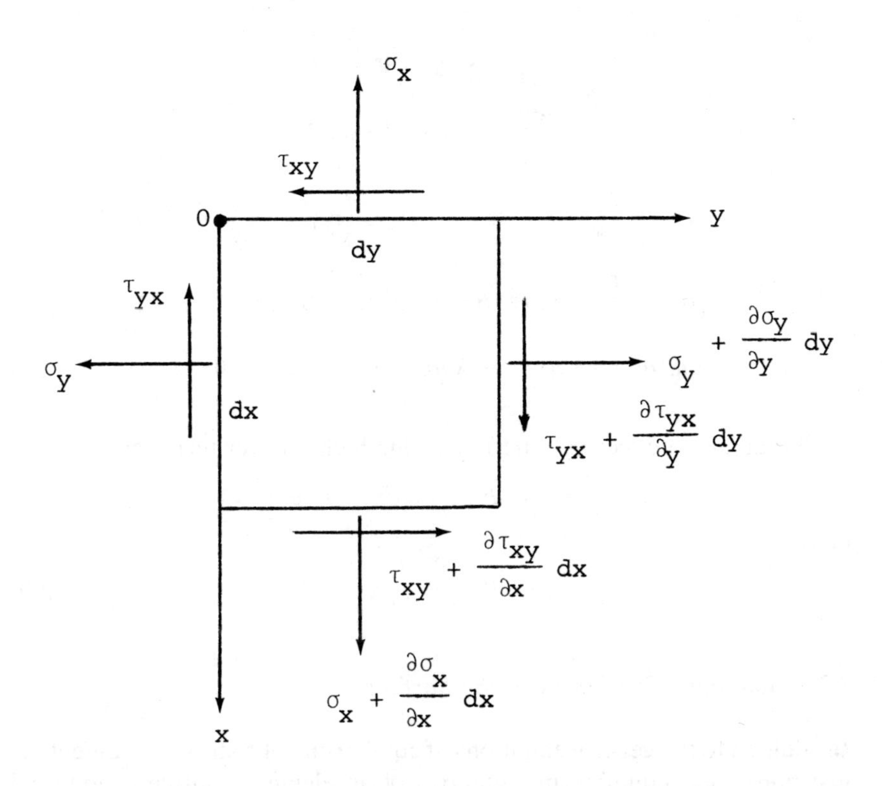

Figure 2-4. Plane Stress Differential Element.

$$\frac{\partial \sigma_x}{\partial x} + \frac{\partial \tau_{yx}}{\partial y} + X = 0 \tag{2.5}$$

Similarly, the sum of forces in the y direction gives

$$\sum F_y = 0$$

$$\left(\sigma_y + \frac{\partial \sigma_y}{\partial y}\, dy\right) dx \cdot 1 - \sigma_y \cdot dx \cdot 1 + \left(\tau_{xy} + \frac{\partial \tau_{xy}}{\partial x}\, dx\right) dy \cdot 1$$

$$- \tau_{xy} \cdot dy \cdot 1 + Y\, dx \cdot dy \cdot 1 = 0 \tag{2.6}$$

Expanding this equation (2.6), and neglecting higher order terms gives

$$\frac{\partial \tau_{xy}}{\partial x} + \frac{\partial \sigma_y}{\partial y} + Y = 0 \tag{2.7}$$

One additional equation of equilibrium can be obtained by considering summation of moment about the z-axis at point 0, shown in Figure 2-4. This gives

$$\sum M_0 = 0$$

$$\sigma_x dy \cdot 1 \cdot dy/2 + \left(\sigma_y + \frac{\partial \sigma_y}{\partial y} dy\right) dx \cdot 1 \cdot dx/2$$

$$+ \left(\tau_{xy} + \frac{\partial \tau_{xy}}{\partial x} dx\right) dy \cdot 1 \cdot dx - \left(\tau_{yx} + \frac{\partial \tau_{yx}}{\partial y} dy\right) dx \cdot 1 \cdot dy$$

$$- \left(\sigma_x + \frac{\partial \sigma_x}{\partial x} dx\right) \cdot dy \cdot 1 \cdot dy/2 - \sigma_y dx \cdot 1 \cdot dx/2$$

$$+ Y dx\, dy \cdot 1 \cdot dx/2 - X dx\, dy \cdot 1 \cdot dy/2 = 0 \tag{2.8}$$

Expanding, collecting, and neglecting higher order terms gives

$$\tau_{xy} dx\, dy - \tau_{yx} dx\, dy = 0$$

or

$$\tau_{xy} = \tau_{yx} \tag{2.9}$$

2.2.3 Boundary Conditions at the Surface

In addition to the general equations of equilibrium of a stressed element, as just presented, equilibrium equations of an element relative to external surface forces can be obtained and will be necessary in defining boundary conditions. A general element on the surface of a plate of unit thickness is shown in Figure 2-5. This element is subjected to a surface stress of magnitude R which has components $\bar{X}$ and $\bar{Y}$. The internal stresses on the element are shown in Figure 2-6, in addition to the external stresses. The summation of forces in the x and y direction on this element gives

$$\sum F_x = 0$$

$$(R \cdot ds \cdot 1) \cos \alpha = \sigma_x \cdot dy \cdot 1 + \tau_{yx} \cdot dx \cdot 1 \tag{2.10}$$

however $\bar{X} = R \cos \alpha$, therefore

$$\bar{X} = \sigma_x dy/ds + \tau_{yx} dx/ds \tag{2.11}$$

and $\sin \alpha = dx/ds$, $\cos \alpha = dy/ds$, therefore

$$\bar{X} = \sigma_x \cos \alpha + \tau_{yx} \sin \alpha \tag{2.12}$$

Similarly, $\Sigma F_y = 0$, gives

$$\bar{Y} = \sigma_y \sin \alpha + \tau_{yx} \cos \alpha \tag{2.13}$$

We now have a set of equations representing internal and external

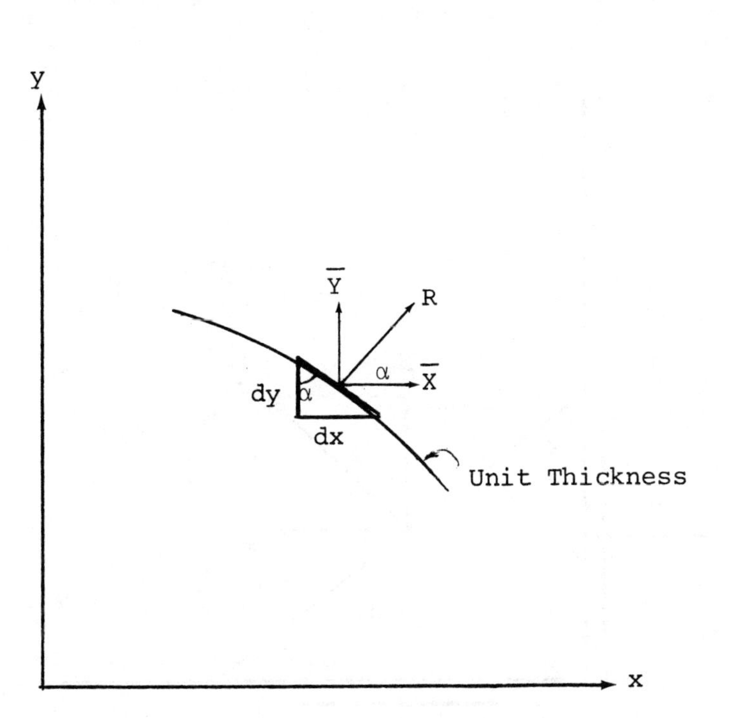

Figure 2-5. Forces on a Boundary.

equilibrium. Any stress field which fulfills equations (2.5) and (2.7), and (2.12) and (2.13) is statically admissible. In order to decide which of all possible conditions correspond to the solution, deformations and hence strains must be considered. This is very necessary as there are three unknowns (σ_x, σ_y, and τ_{xy}) and two equations of equilibrium (2.5) and (2.7), hence one more equation is required, which will be strain or stress related.

2.3 Strain

As in the case of stress at a point, a small element (dx, dy) will be examined, and the deformations of such an element determined. Such an element is shown in Figure 2-7, in which only two sides are shown—AC and AB. Consider a deformation of such an element of amount u and v, with the final position of the sides being $A'C'$ and $A'B'$. If we determine the change in length per unit length of each of these sides, then we will have defined the strain, i.e., $\epsilon = \Delta L/L$. Note that the changes in the element lengths $A'B'$ and $A'C'$ constitute component changes of magnitudes $B - \bar{B}$, $\bar{B} - B'$, and $C - \bar{C}$, $\bar{C} - C'$.

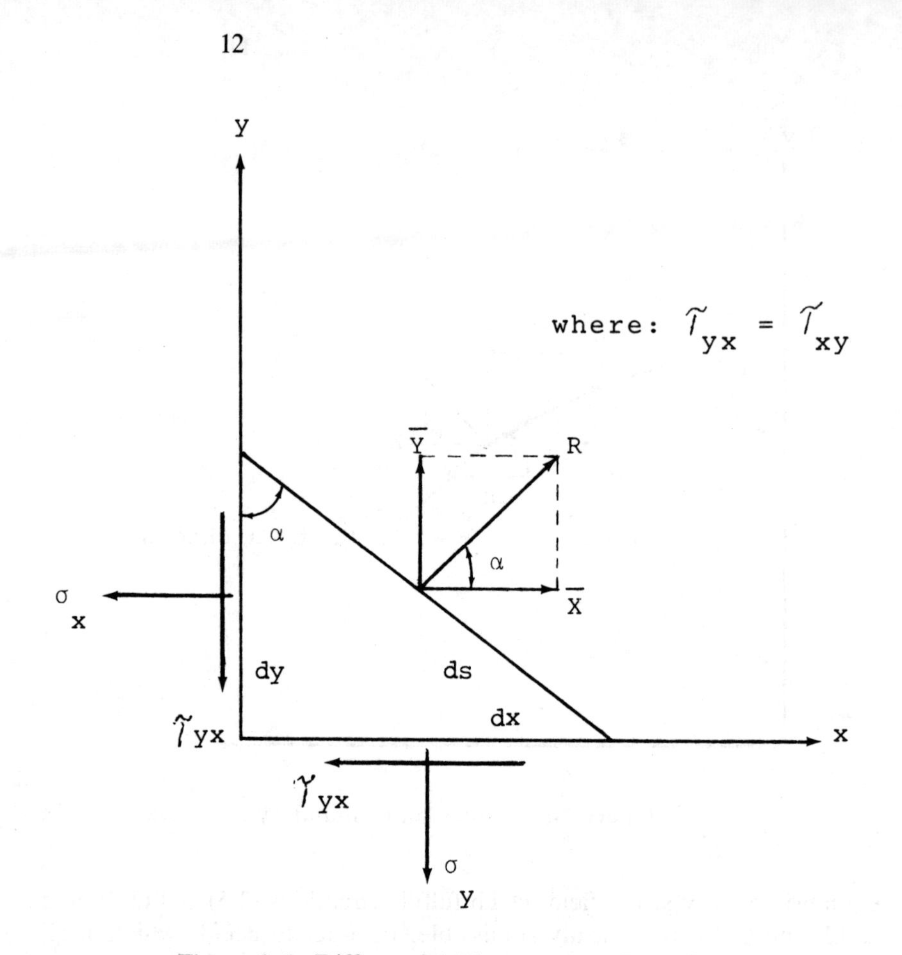

Figure 2-6. Differential Element on a Boundary.

In order to evaluate the induced strains in this element dx by dy, the Taylor's series equation can be applied. As will be recalled, the Taylor's series can yield the solution to a function $f(B'C')$ if the function is specified at $f(A')$, as illustrated in Figure 2-8. Previously, we used Taylor's series for a single variable problem, as shown in Figure 2-9, where

$$f(x) = f(a) + f'(a)(x - a) + \frac{f''(a)}{2!}(x - a)^2 + \ldots$$

The present problem, however, involves two variables x and y, therefore the Taylor's series can be written [1] as

$$f(x, y) = f(a, b) + \left[(x - a)\frac{\partial}{\partial x} + (y - b)\frac{\partial}{\partial y}\right] f(x, y)\big|_{ab}$$

$$+ \frac{1}{2!}\left[(x - a)\frac{\partial}{\partial x} + (y - b)\frac{\partial}{\partial y}\right]^2 f(x, y)\big|_{ab} + \ldots \qquad (2.14)$$

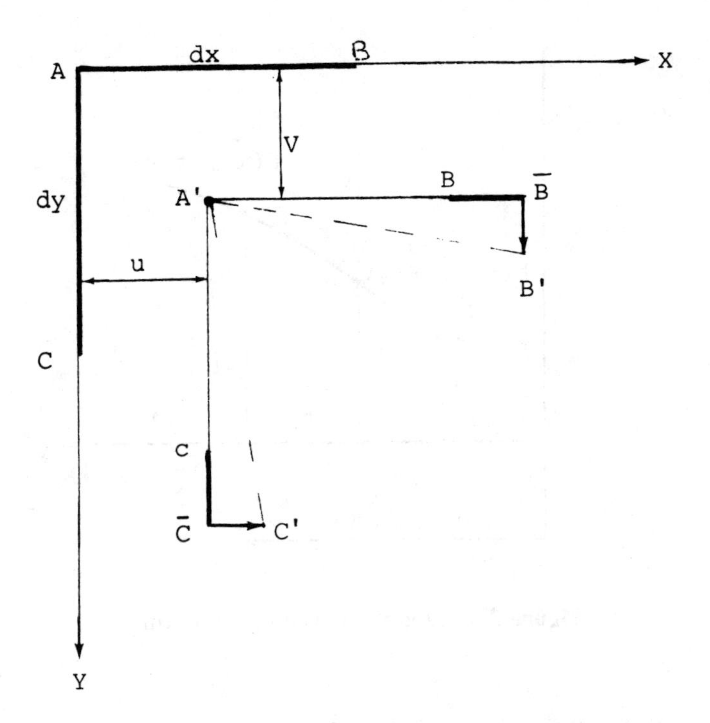

Figure 2-7. Displacement of a Differential Element.

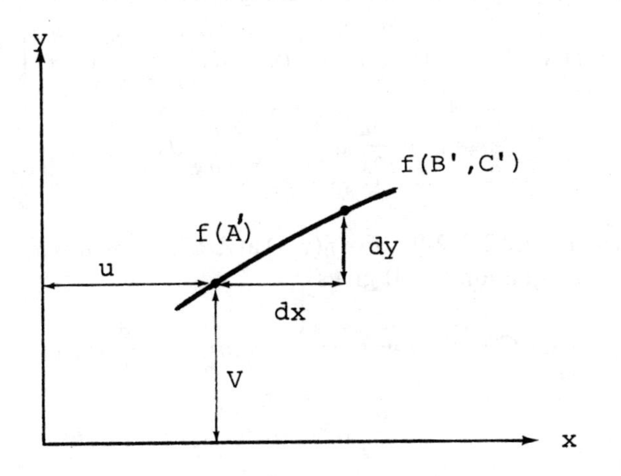

Figure 2-8. Variation of a Function.

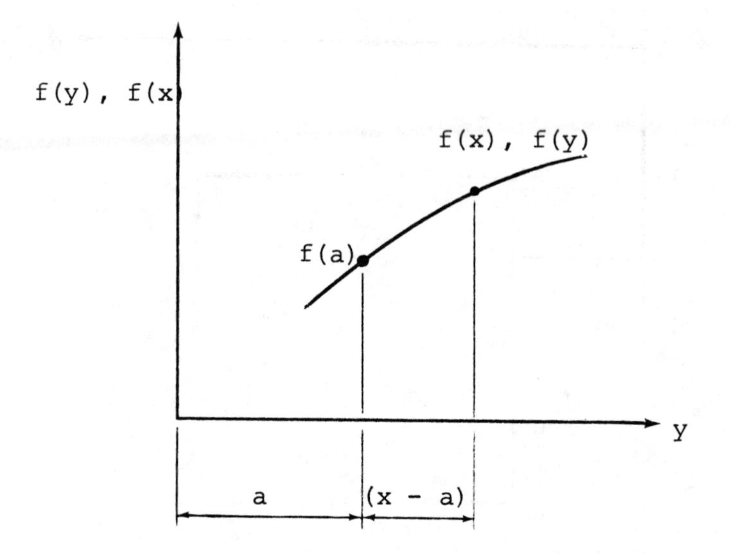

Figure 2-9. Double Variable Function.

This equation will first be applied for $f(x,y)$ at $A'B'$ and then for $f(x,y)$ at $A'C'$.

$A'B'$ change (Figure 2-7, 2-9): For $f(x, y)$ at $A'B'$, $x = u + dx$, $a = u$, $y = v$, and $b = v$, applying equation (2.14) using only two terms gives

$$f(A', B') = (u + v) + \left[(u + dx - u)\frac{\partial}{\partial x}(u, v)\right]$$

$$= u + \frac{\partial u}{\partial x} dx + v + \frac{\partial v}{\partial x} dx \tag{2.15}$$

$A'C'$ change (Figure 2-7, 2-9): For $f(x, y)$ at $A'C'$, $x = u$, $a = u$, $y = v + dy$, $b = v$, applying equation (2.14) gives

$$f(A'C') = (u + v) + \left[(v + dy - v)\frac{\partial}{\partial y}(u, v)\right]$$

$$= u + \frac{\partial u}{\partial y} dy + v + \frac{\partial v}{\partial y} dy \tag{2.16}$$

These two equations, (2.15) and (2.16), describe the respective changes in the position of the element and can be described as shown in Figure 2-10.

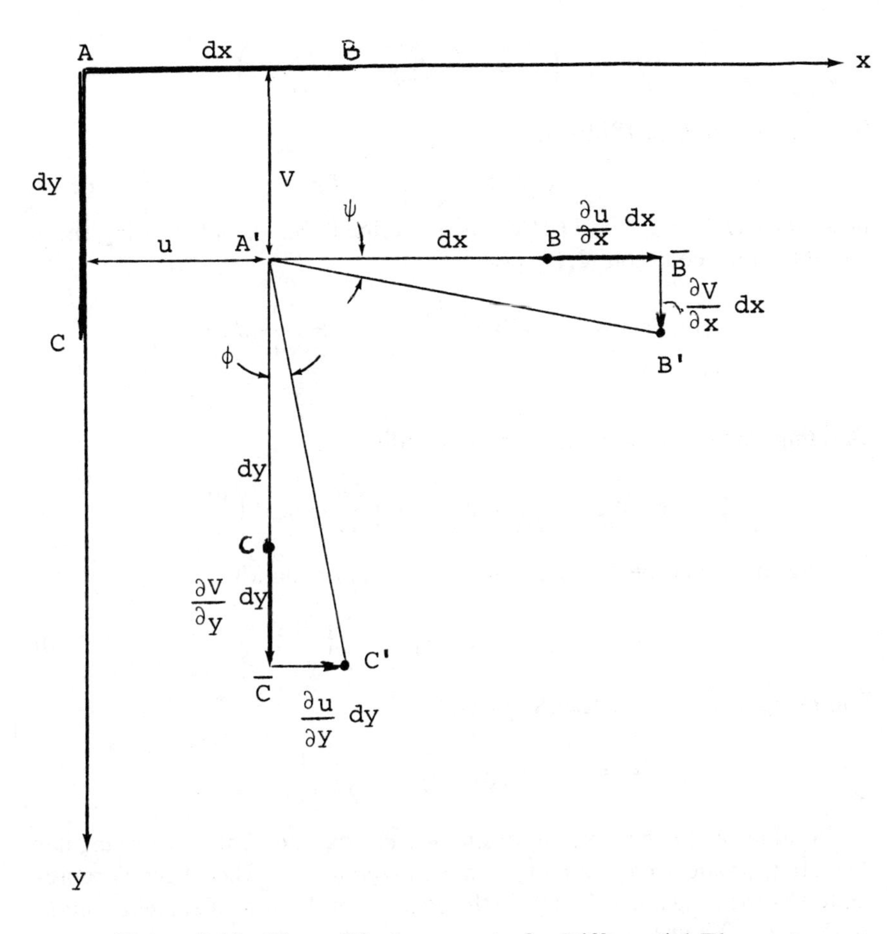

Figure 2-10. Shear Displacement of a Differential Element.

Thus by applying Taylor's series [1], all the respective components of the distorted elements are found.

The strains in the element are now evaluated by use of the general strain equation [2]:

$$\epsilon = \Delta L/L \tag{2.17}$$

The strain ϵ_x by definition is

$$\epsilon_x = \frac{A'B' - AB}{AB} \tag{2.18}$$

noting that $A'B'$ is the final length and AB is the original length. Examination of Figure 2-10, shows:

$$(A'B')^2 = \left(dx + \frac{\partial u}{\partial x}\,dx\right)^2 + \left(\frac{\partial v}{\partial x}\,dx\right)^2$$

Solving equation (2.18) for $A'B'$ gives

$$(A'B')^2 = (1 + \epsilon_x)^2 (AB)^2 \tag{2.19}$$

however $(AB)^2 = dx^2$ and $(A'B')^2$ is as defined above. Substituting these variables into equation (2.19) gives:

$$(1 + \epsilon_x)^2\,dx^2 = \left(dx + \frac{\partial u}{\partial x}\,dx\right)^2 + \left(\frac{\partial v}{\partial x}\,dx\right)^2$$

Dividing this equation by dx^2 and expanding yields:

$$1 + 2\epsilon_x + \epsilon_x^2 = \left[1 + 2\frac{\partial u}{\partial x} + \left(\frac{\partial u}{\partial x}\right)^2\right] + \left(\frac{\partial v}{\partial x}\right)^2$$

Solving for ϵ_x and neglecting ϵ_x^2 gives the final equation:

$$\epsilon_x = \frac{\partial u}{\partial x} + \frac{1}{2}\left(\frac{\partial u}{\partial x}\right)^2 + \frac{1}{2}\left(\frac{\partial v}{\partial x}\right)^2 \tag{2.20}$$

Following a similar procedure gives:

$$\epsilon_y = \frac{\partial v}{\partial y} + \frac{1}{2}\left(\frac{\partial v}{\partial y}\right)^2 + \frac{1}{2}\left(\frac{\partial u}{\partial y}\right)^2 \tag{2.21}$$

In addition to the normal strains ϵ_x, ϵ_y, the element has an angular distortion or shearing strain effect designated as γ_{xy}. This strain is represented by the total angle change in the element or the sum of angles ϕ and ψ, as shown in Figure 2-10.

The tangents of these angles are defined as follows:

$$\tan\psi = \psi = \frac{\partial v}{\partial x}\,dx \Big/ \left(dx + \frac{\partial u}{\partial x}\,dx\right)$$

$$\tan\phi = \phi = \frac{\partial u}{\partial y}\,dy \Big/ \left(dy + \frac{\partial v}{\partial y}\,dy\right)$$

The shearing angle is therefore;

$$\gamma_{xy} = \psi + \phi = \frac{\dfrac{\partial v}{\partial x}\,dx}{\left(1 + \dfrac{\partial u}{\partial x}\right)dx} + \frac{\dfrac{\partial u}{\partial y}\,dy}{\left(1 + \dfrac{\partial v}{\partial y}\right)dy}$$

or

$$\gamma_{xy} = \frac{\dfrac{\partial v}{\partial x} + \dfrac{\partial v}{\partial x} \cdot \dfrac{\partial v}{\partial y} + \dfrac{\partial u}{\partial y} + \dfrac{\partial u}{\partial y} \cdot \dfrac{\partial u}{\partial x}}{\left(1 + \dfrac{\partial u}{\partial x}\right)\left(1 + \dfrac{\partial v}{\partial y}\right)}$$

Assuming that $1 \gg (\partial u/\partial x)(\partial v/\partial y)$, the denominator can be assumed equal to one, therefore:

$$\gamma_{xy} = \frac{\partial v}{\partial x} + \frac{\partial u}{\partial y} + \frac{\partial v}{\partial x}\frac{\partial v}{\partial y} + \frac{\partial u}{\partial y}\frac{\partial u}{\partial x} \tag{2.22}$$

Neglecting second order terms, i.e., small strains, small displacements, gives

$$\epsilon_x = \frac{\partial u}{\partial x} \tag{2.23}$$

$$\epsilon_y = \frac{\partial v}{\partial y} \tag{2.24}$$

$$\gamma_{xy} = \frac{\partial v}{\partial x} + \frac{\partial u}{\partial y} \tag{2.25}$$

Equations (2.23), (2.24), and (2.25) are those which are used in the ordinary "Theory of Elasticity" [2, 3, 4].

Equations (2.23), (2.24), and (2.25) can be reduced to one equation by taking derivatives of the three equations and adding, which gives

$$\epsilon_x = \frac{\partial u}{\partial x}; \qquad \frac{\partial^2 \epsilon_x}{\partial y^2} = \frac{\partial^3 u}{\partial x\, \partial y^2}$$

$$\epsilon_y = \frac{\partial v}{\partial y}; \qquad \frac{\partial^2 \epsilon_y}{\partial x^2} = \frac{\partial^3 v}{\partial y\, \partial x^2}$$

$$\gamma_{xy} = \frac{\partial v}{\partial x} + \frac{\partial u}{\partial y}; \qquad -\frac{\partial^2 \gamma_{xy}}{\partial x\, \partial y} = -\frac{\partial^3 v}{\partial x^2\, \partial y} - \frac{\partial^3 u}{\partial y^2\, \partial x}$$

$$\frac{\partial^2 \epsilon_x}{\partial y^2} - \frac{\partial^2 \gamma_{xy}}{\partial x\, \partial y} + \frac{\partial^2 \epsilon_y}{\partial x^2} = 0 \tag{2.26}$$

Equation (2.26), is known as the compatability equation and shows that strains must be compatible and cannot be arbitrary, as the displacements u and v must be piecewise continuous. A strain field which fulfills equation (2.26) is geometrically admissible for a solution.

We now have three equations and three unknowns (σ_x, σ_y, τ_{xy}). The strain equation (2.26) can be related to the stresses, which will now be done.

2.4 Stress-Strain Relationships

The relationship between stress and strain for linearly elastic materials which follow Hooke's Law is given by the following equations:

2.4.1 One-Dimensional

$$\epsilon_x = \sigma_x/E \tag{2.27}$$

$$\epsilon_y = -\mu\sigma_x/E \tag{2.28}$$

2.4.2 Two-Dimensional

$$\epsilon_x = \sigma_x/E - \mu\sigma_y/E \tag{2.29}$$

$$\epsilon_y = \sigma_y/E - \mu\sigma_x/E \tag{2.30}$$

$$\gamma_{xy} = \tau_{xy}/G \tag{2.31}$$

$$G = E/2(1 + \mu) \tag{2.32}$$

where: E = Young's Modulus

G = Modulus of Rigidity

μ = Poisson's Ratio

2.4.3 Summary of Equations

We can now summarize our equations, in order to illustrate the total unknown variables and equations available to solve for these quantities: Equations of Equilibrium:

$$\frac{\partial \sigma_x}{\partial x} + \frac{\partial \tau_{xy}}{\partial y} + X = 0 \tag{2.5}$$

$$\frac{\partial \tau_{yx}}{\partial x} + \frac{\partial \sigma_y}{\partial y} + Y = 0 \tag{2.7}$$

Strain Displacements:

$$\epsilon_x = \frac{\partial u}{\partial x} \tag{2.23}$$

$$\epsilon_y = \frac{\partial v}{\partial y} \tag{2.24}$$

$$\gamma_{xy} = \frac{\partial u}{\partial y} + \frac{\partial v}{\partial x} \tag{2.25}$$

Stress-Strain Equations:

$$\epsilon_x = \frac{1}{E}(\sigma_x - \mu\sigma_y) \tag{2.29}$$

$$\epsilon_y = \frac{1}{E}(\sigma_y - \mu\sigma_x) \tag{2.30}$$

$$\gamma_{xy} = \frac{\tau_{xy}}{G} = \frac{2(1 + \mu)}{E}\tau_{xy} \tag{2.31}$$

$$G = E/2(1 + \mu) \tag{2.32}$$

If we now consider the total unknown variables, we have eight:

$$\sigma_x,\ \sigma_y,\ \tau_{xy},\ \epsilon_x,\ \epsilon_y,\ \gamma_{xy},\ u,\ v$$

There are now, however, eight equations, as given above, therefore all the unknowns can be determined. Originally, three unknowns were considered, σ_x, σ_y, τ_{xy}, and only two equations were available, equations (2.5) and (2.7). The additional equation was obtained by relating equations (2.23), (2.24), and (2.25), which resulted in the strain compatability equation (2.26):

$$\frac{\partial^2\epsilon_x}{\partial y^2} - \frac{\partial^2\gamma_{xy}}{\partial y\,\partial x} + \frac{\partial^2\epsilon_y}{\partial x^2} = 0 \tag{2.26}$$

Substituting the stress-strain equations (2.29) through (2.31) into this equation (2.26) gives:

$$\frac{\partial^2}{\partial y^2}(\sigma_x - \mu\sigma_y) - 2(1 + \mu)\frac{\partial^2\tau_{xy}}{\partial x\,\partial y} + \frac{\partial^2}{\partial x^2}(\sigma_y - \mu\sigma_x) = 0 \tag{2.33}$$

Plates Subjected to Force in their Plane-Plane Stress

3.1 Airy's Stress Function

In the past chapter, the general equations relating to a plate of unit thickness subjected to in-plane forces or stresses were developed. A plate subjected to these forces is shown in Figure 3-1, and the three general equations of interest as given previously are:

$$\frac{\partial \sigma_x}{\partial x} + \frac{\partial \tau_{xy}}{\partial y} = 0 \tag{3.1}$$

$$\frac{\partial \tau_{yx}}{\partial x} + \frac{\partial \sigma_y}{\partial y} = 0 \tag{3.2}$$

$$\frac{\partial^2}{\partial y^2}(\sigma_x - \mu\sigma_y) - 2(1 + \mu)\frac{\partial^2 \tau_{xy}}{\partial x\,\partial y} + \frac{\partial^2}{\partial x^2}(\sigma_y - \mu\sigma_x) = 0 \tag{3.3}$$

where the body forces X and Y are assumed equal to zero.

Consider now the condition that:

$$\sigma_x = \frac{\partial R(x, y)}{\partial y}$$

$$\tau_{xy} = -\frac{\partial R(x, y)}{\partial x} \tag{3.4}$$

where R is a variable function of the coordinates x and y, which is the condition for a plane stress problem as shown in Figure 3-1. Also note that the relationships given in (3.4) will be satisfied if substituted into equation (3.1).

Also consider the relationship:

$$\sigma_y = \frac{\partial S(x, y)}{\partial x}$$

$$\tau_{yx} = -\frac{\partial S(x, y)}{\partial y} \tag{3.5}$$

where S is a variable function of x and y; and if these relationships are substituted in equation (3.2), that equation will be satisfied.

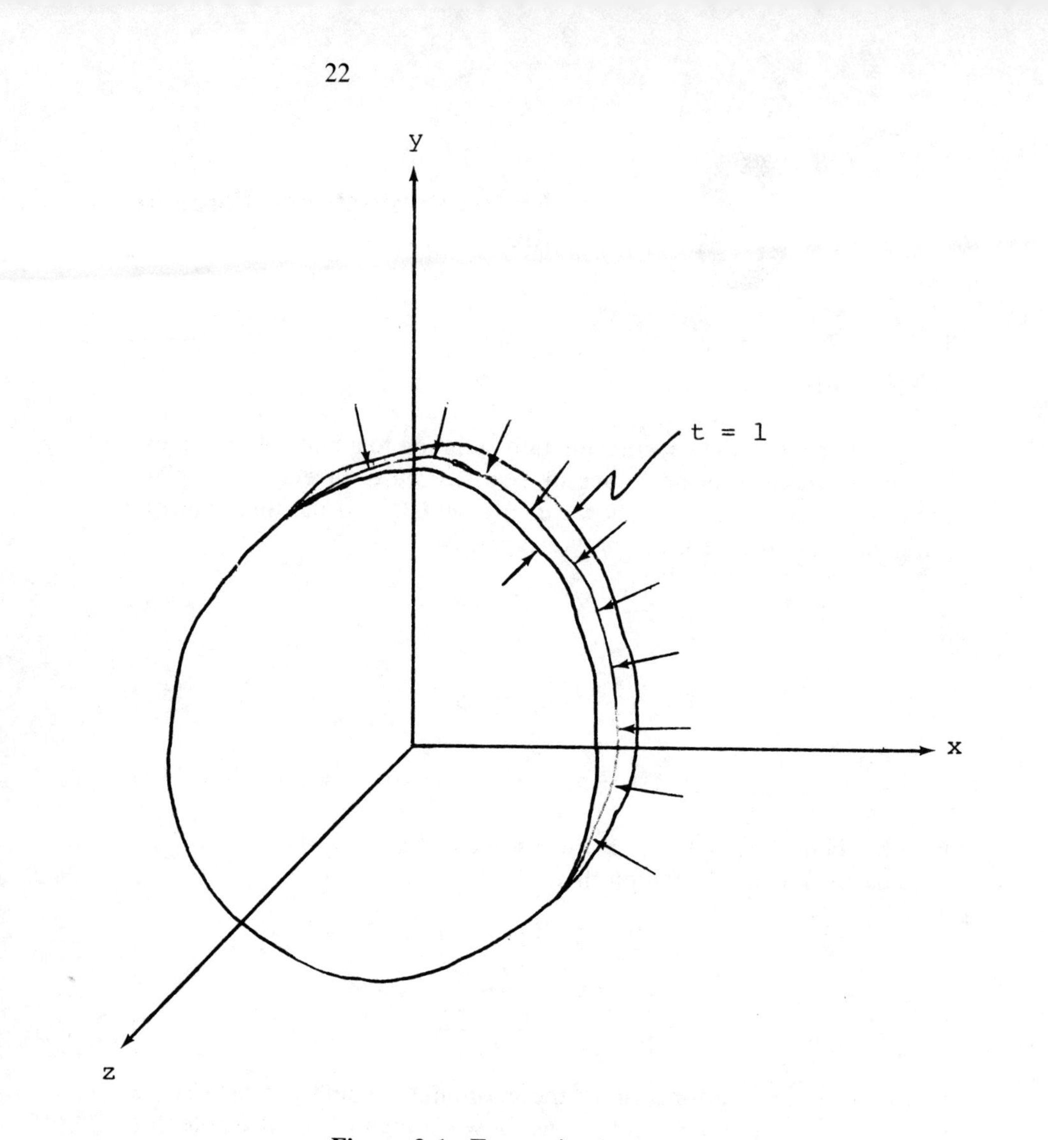

Figure 3-1. Forces in the Plane of a Plate.

Now from equation (2.9), it is known that

$$\tau_{xy} = \tau_{yx}$$

or from equations (3.4) and (3.5),

$$\frac{\partial R(x,\, y)}{\partial x} = \frac{\partial S(x,\, y)}{\partial y} \tag{3.6}$$

The only possible way that equation (3.6) can be met is if:

$$R = \frac{\partial \phi}{\partial y} \qquad \text{and} \qquad S = \frac{\partial \phi}{\partial x} \tag{3.7}$$

thus from equation (3.6)

$$\frac{\partial^2 \phi}{\partial x \, \partial y} \equiv \frac{\partial^2 \phi}{\partial y \, \partial x} \tag{3.8}$$

where $\phi = f(x, y)$. ϕ is known as Airy's stress function [2, 3, 4, 5, 6]. Substituting

$$R = \frac{\partial \phi}{\partial y} \qquad \text{and} \qquad S = \frac{\partial \phi}{\partial x}$$

into equations (3.4) and (3.5) and noting $\tau_{xy} = \tau_{yx}$ gives the stress equations in terms of the one variable $\phi(x, y)$.

$$\sigma_x = \frac{\partial^2 \phi}{\partial y^2} \tag{3.9}$$

$$\sigma_y = \frac{\partial^2 \phi}{\partial x^2} \tag{3.10}$$

$$\tau_{xy} = -\frac{\partial^2 \phi}{\partial x \, \partial y} \tag{3.11}$$

therefore the compatability equation (3.3) can be written similarly by substituting equations (3.9) through (3.11) into equation (3.3), which gives

$$\frac{\partial^2}{\partial y^2}(\sigma_x - \mu \sigma_y) - 2(1 + \mu)\frac{\partial^2 \tau_{xy}}{\partial x \, \partial y} + \frac{\partial^2}{\partial x^2}(\sigma_y - \mu \sigma_x) = 0$$

$$\frac{\partial^2}{\partial y^2}\left(\frac{\partial^2 \phi}{\partial y^2} - \mu \frac{\partial^2 \phi}{\partial x^2}\right) - 2(1 + \mu)\frac{\partial^2}{\partial x \, \partial y}\left(-\frac{\partial^2 \phi}{\partial x \, \partial y}\right)$$

$$+ \frac{\partial^2}{\partial x^2}\left(\frac{\partial^2 \phi}{\partial x^2} - \mu \frac{\partial^2 \phi}{\partial y^2}\right) = 0$$

Expanding this expression and collecting terms gives

$$\frac{\partial^4 \phi}{\partial y^4} + 2\frac{\partial^4 \phi}{\partial x^2 \, \partial y^2} + \frac{\partial^4 \phi}{\partial x^4} = 0 \tag{3.12}$$

or using the notation

$$\nabla^2 = \frac{\partial^2}{\partial x^2} + \frac{\partial^2}{\partial y^2}$$

equation (3.12) can be written as

$$\nabla^4 \phi = 0 \tag{3.13}$$

Equation (3.12) or (3.13) is known as the Biharmonic equation and describes the state of plane stress at a point (x, y). If this equation can be satisfied at all points, then an admissable solution to the plane stress problem has been found.

The boundary equations can also be written in terms of the Airy's function $\phi(x, y)$, by substituting equations (3.9), (3.10), and (3.11) into equations (2.12) and (2.13), which gives:

$$\frac{\partial^2 \phi}{\partial y^2} \cos \alpha - \frac{\partial^2 \phi}{\partial x \, \partial y} \sin \alpha = \bar{x} \tag{3.14}$$

$$\frac{\partial^2 \phi}{\partial x^2} \sin \alpha - \frac{\partial^2 \phi}{\partial x \, \partial y} \cos \alpha = \bar{y} \tag{3.15}$$

It should now be recognized that the problem involving the solution of three unknowns $(\sigma_x, \sigma_y, \tau_{xy})$, has been reduced to a solution of one unknown $\phi(x, y)$. If such a function can be found, and if it satisfies the boundary conditions (3.14) and (3.15), and the biharmonic (3.12), then a proper solution has been found. The task at hand, therefore, is to find the function $\phi(x, y)$.

3.2 Boundary Condition Equations Expanded

The most difficult task, in solving classical plane stress problems, is to evaluate the ϕ function at the boundary. Several methods have been proposed [5, 8], which will permit such an evaluation. These will now be presented.

3.2.1 Integration Method [5]

The equilibrium equations (3.14) and (3.15) represent the state of stress of an element as shown in Figure 3-2. Assuming the slope of this element as shown, with the distance s positive counterclockwise, then equations (3.14) and (3.15), can be written as:

$$\frac{\partial^2 \phi}{\partial y^2}\left(\frac{dy}{ds}\right) - \frac{\partial^2 \phi}{\partial x \, \partial y}\left(-\frac{dx}{ds}\right) = \bar{x} \tag{3.16}$$

$$\frac{\partial^2 \phi}{\partial x^2}\left(-\frac{dx}{ds}\right) - \frac{\partial^2 \phi}{\partial x \, \partial y}\left(\frac{dy}{ds}\right) = \bar{y} \tag{3.17}$$

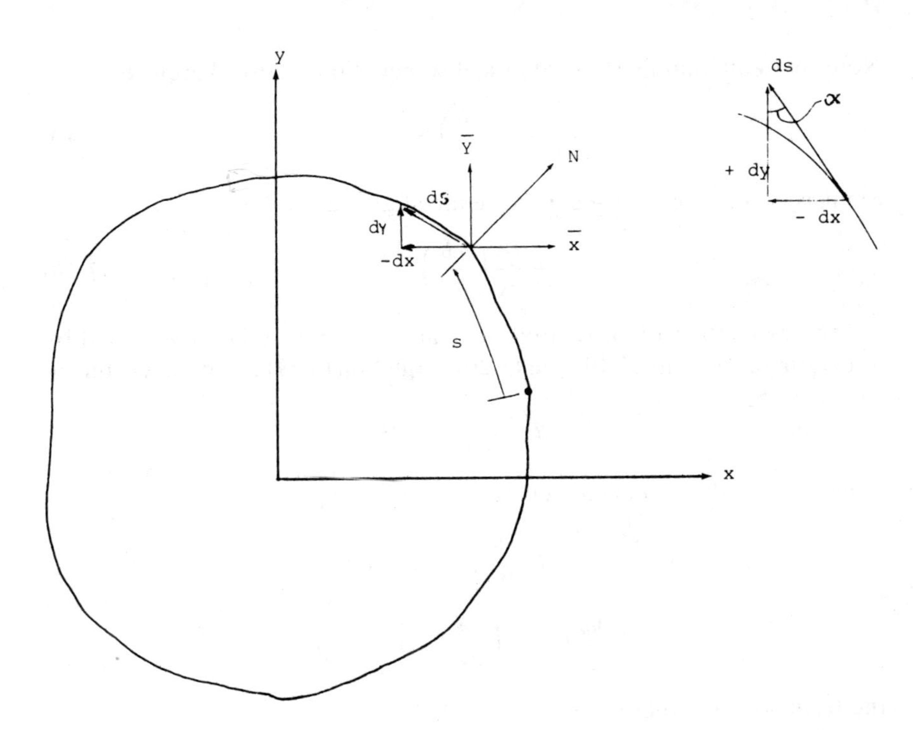

Figure 3-2. External Boundary Forces.

where

$$\sin \alpha = -\frac{dx}{ds}, \qquad \cos \alpha = \frac{dy}{ds}$$

Equation (3.16) can be reduced by using the chain rule expansion [1]:

$$\frac{dF}{ds} = \frac{\partial F}{\partial x}\frac{dx}{ds} + \frac{\partial F}{\partial y}\frac{dy}{ds}$$

where

$$F = \frac{\partial \phi}{\partial y}$$

Therefore:

$$\frac{d(\partial \phi/\partial y)}{ds} = \frac{\partial(\partial \phi/\partial y)}{\partial x}\frac{dx}{ds} + \frac{\partial(\partial \phi/\partial y)}{\partial y}\frac{dy}{ds}$$

$$\frac{d}{ds}\left(\frac{\partial \phi}{\partial y}\right) = \frac{\partial^2 \phi}{\partial x \partial y}\frac{dx}{ds} + \frac{\partial^2 \phi}{\partial y^2}\frac{dy}{ds} \tag{3.18}$$

Note that equation (3.18) is identical to equation (3.16), therefore:

$$\frac{d}{ds}\left(\frac{\partial \phi}{\partial y}\right) = \bar{x} \tag{3.19}$$

Similarly, equation (3.17) can be reduced giving

$$-\frac{d}{ds}\left(\frac{\partial \phi}{\partial x}\right) = \bar{y} \tag{3.20}$$

The evaluation of ϕ, relative to $\bar{x}$ and $\bar{y}$, can now be determined by integrating equations (3.19) and (3.20). Equation (3.19) can be rewritten as

$$d\left(\frac{\partial \phi}{\partial y}\right) = \bar{x}\,ds$$

integrating with respect to s gives

$$\int_0^s d\left(\frac{\partial \phi}{\partial y}\right) = \int_0^s \bar{x}\,ds$$

$$\left(\frac{\partial \phi}{\partial y}\right)_s - \left(\frac{\partial \phi}{\partial y}\right)_0 = \int_0^s \bar{x}\,ds$$

the term $\partial\phi/\partial y|_s$ therefore is

$$\left.\frac{\partial \phi}{\partial y}\right|_s = \int_0^s \bar{x}\,ds + \left(\frac{\partial \phi}{\partial y}\right)_0$$

Let:

$$Q_x = \int_0^s \bar{x}\,ds \tag{3.21}$$

and

$$C_1 = \left(\frac{\partial \phi}{\partial y}\right)_0$$

therefore

$$\left.\frac{\partial \phi}{\partial y}\right|_s = Q_x + C_1 \tag{3.22}$$

Similarly,

$$-\frac{d}{ds}\left(\frac{\partial \phi}{\partial x}\right) = \bar{y}$$

$$\int_0^s d\left(\frac{\partial \phi}{\partial x}\right) = -\int_0^s \bar{y}\,ds$$

$$\left.\frac{\partial \phi}{\partial x}\right|_s = -Q_y + C_2 \tag{3.23}$$

where

$$Q_y = \int_0^s \bar{y}\, ds \tag{3.24}$$

The solution of ϕ, can now be determined by using the chain rule:

$$\frac{d\phi}{ds} = \frac{\partial \phi}{\partial x} \cdot \frac{dx}{ds} + \frac{\partial \phi}{\partial y} \cdot \frac{dy}{ds} \tag{3.25}$$

Substituting equations (3.22) and (3.23) into equation (3.25) gives:

$$\frac{d\phi}{ds} = -Q_y \frac{dx}{ds} + C_2 \frac{dx}{ds} + Q_x \frac{dy}{ds} + C_1 \frac{dy}{ds} \tag{3.26a}$$

Assume $\ell = dy/ds$ and $m = -dx/ds$, then,

$$\frac{d\phi}{ds} = Q_y m + Q_x \ell + C_2 \ell - C_2 m$$

therefore

$$\phi = \int_0^s (Q_y m + Q_x \ell)\, ds + C_1 y - C_2 x + C_3 \tag{3.26b}$$

Since stresses are a function of the second derivative of ϕ, as per equations (3.9), (3.10), and (3.11), therefore set $C_1 = C_2 = C_3 = 0$ and

$$\phi = \int_0^s (Q_y m + Q_x \ell)\, ds \tag{3.26c}$$

which defines the ϕ function on the boundary.

The $\partial\phi/\partial N$, the normal to the element, is found from the chain rule:

$$\frac{\partial \phi}{\partial N} = \frac{\partial \phi}{\partial x} \frac{dx}{dN} + \frac{\partial \phi}{\partial y} \frac{dy}{dN}$$

Substituting in the proper relationships gives

$$\frac{\partial \phi}{\partial N} = -Q_y \ell + Q_x m \tag{3.27}$$

The ϕ at the boundary can now readily be determined as $\bar{x}$, $\bar{y}$ are given, therefore Q_x and Q_y can be found. The following will illustrate the application of these equations.

3.2.2 Problem—ϕ at Boundary—Integration Method

A rectangular plate of dimension $2a \times 2b$, and uniform thickness h is

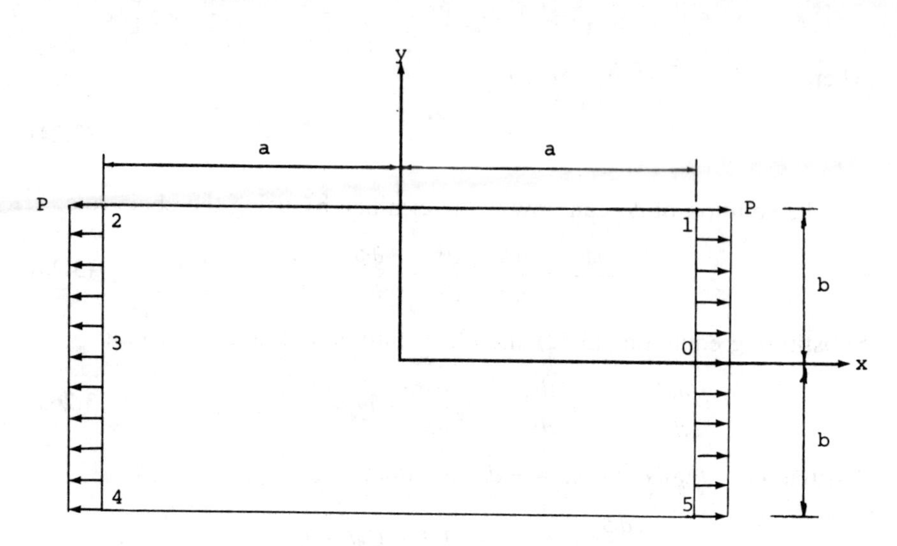

Figure 3-3. Example—In-Plane Stress Distribution.

subjected to uniform end stresses p, as shown in Figure 3-3. It is desired to evaluate the ϕ function along the boundary.

The Q_x and Q_y functions, along the outside edge of the plate will first be obtained:

$$Q_x = \int_0^s \bar{x}\, ds \tag{3.21}$$

$$Q_y = \int_0^s \bar{y}\, ds \tag{3.24}$$

where $\bar{x} = \pm P$ and $\bar{y} = 0$; that is, the applied stresses are only in the x direction. Taking the origin of s arbitrarily at $(a, 0)$ and measuring s counterclockwise

0-1:

$$Q_{x_s} = \int_0^s \bar{x}\, ds = \int_0^y P\, dy = Py \qquad 0 \leq s \leq b$$

$$Q_{x_1} = Pb$$

1-2:

$$Q_{x_s} = Q_1 - \int_a^x \bar{x}\, ds = Pb - \int_a^x (0)\, dx = Pb \qquad b \leq s \leq b + 2a$$

$$Q_{x_2} = Pb$$

2-3-4:

$$Q_{x_s} = Q_2 - \int_b^y \bar{x}\,dy$$

$$= Pb - \int_b^y (-P)\,dy$$

$$= Pb - [-Py + Pb]$$

$$Q_{x_s} = Py \qquad (b + 2a) \leq s \leq (3b + 2a)$$

$$Q_{x_3} = 0$$

$$Q_{x_4} = -Pb$$

4-5:

$$Q_{x_s} = Q_4 + \int_{-a}^{x} (0)\,dx = -Pb \qquad (3b + 2a) \leq x \leq (3b + 4a)$$

$$Q_{x_5} = -Pb$$

5-0:

$$Q_{x_s} = Q_5 + \int_{-b}^{y} P\,dy = -Pb + [Py + Pb] = Py$$

$$(3b + 4a) \leq a \leq (4b + 4a)$$

$$Q_{x_0} = 0$$

A plot of the Q_x terms is given in Figure 3-4. The Q_y diagram would be zero, as $\bar{y} = 0$.

∂φ/∂N: The $\partial\phi/\partial N$ function can now be found by applying equation (3.27)

$$\frac{\partial\phi}{\partial N} = -Q_y\ell + Q_x m \qquad (3.27)$$

where

$$\ell = \frac{dx}{dN} \qquad \text{and} \qquad m = \frac{dy}{dN}$$

For this problem, $Q_y = 0$, $Q_x = \pm P$, and $m = dy/dN = \sin\alpha$, and for the end faces $\alpha = 0, 180°$; therefore $m = 0$. For the top faces $\alpha = 90°, 270°$; therefore $m = \pm 1$, and thus

$$\frac{\partial\phi}{\partial N} = \pm 1(\pm P)$$

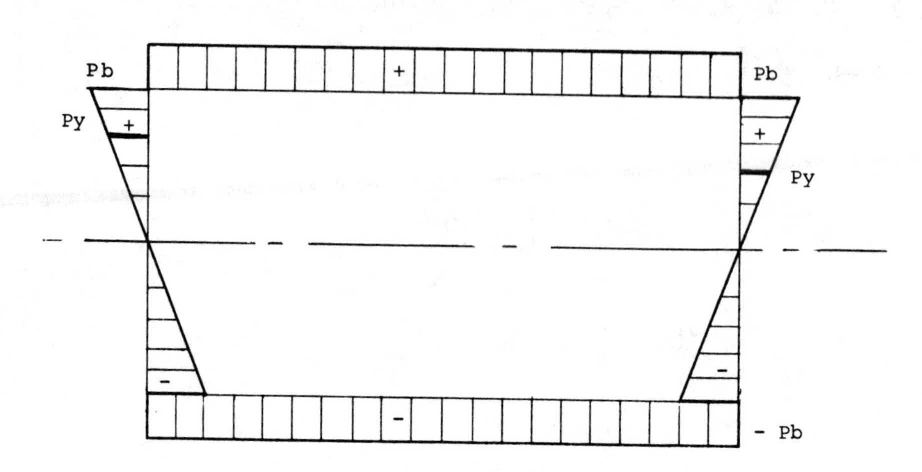

Figure 3-4. Q_x Diagram.

The resulting $\partial\phi/\partial N$ diagram is shown in Figure 3-5.

ϕ: The ϕ function along the plate is now determined as follows. From equation (3.26),

$$\phi = \int_0^s (Q_y m + Q_x \ell)\, ds$$

where

$$m = \frac{dy}{dN} = \sin\alpha \qquad \ell = \frac{dx}{dN} = \cos\alpha$$

and for

$$\alpha = 0,\ 180° \qquad \ell = \pm 1, \qquad \alpha = 90°,\ 270° \qquad \ell = 0$$

0-1:

$$\phi_s = \int_0^y (Py)(1)\, dy = \frac{Py^2}{2} \qquad 0 \le s \le b$$

$$\phi_1 = \frac{Pb^2}{2}$$

1-2:

$$\phi_s = \phi_1 - \int_a Q_x \ell\, dx = \phi_1 - \int_a^x Pb(0)\, dx = \phi \quad b \le s \le (b + 2a)$$

$$\phi_2 = \frac{Pb^2}{2}$$

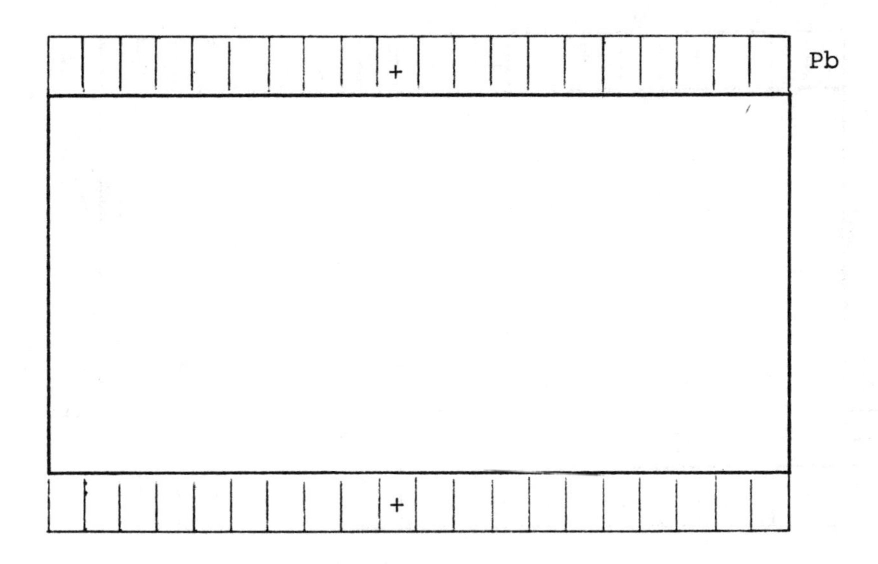

Figure 3-5. $\partial\phi/\partial N$ Diagram.

2-3:

$$\phi_s = \phi_2 - \int_b^y Q_x \ell \, dy = \phi_2 - \int_b^y Py(-1)\, dy$$

$$\phi_s = \frac{Py^2}{2} \qquad (b + 2a) \leq s \leq (3b + 2a)$$

$$\phi_3 = 0$$

3-4:

$$\phi_s = \frac{Py^2}{2}$$

$$\phi_4 = \frac{Pb^2}{2}$$

4-5:

$$\phi_s = \phi_4 + \int_{-a}^{x} -Pb(0)\, dx = \frac{Pb^2}{2} \qquad (3b + 2a) \leq s \leq (3b + 4a)$$

$$\phi_5 = \frac{Pb^2}{2}$$

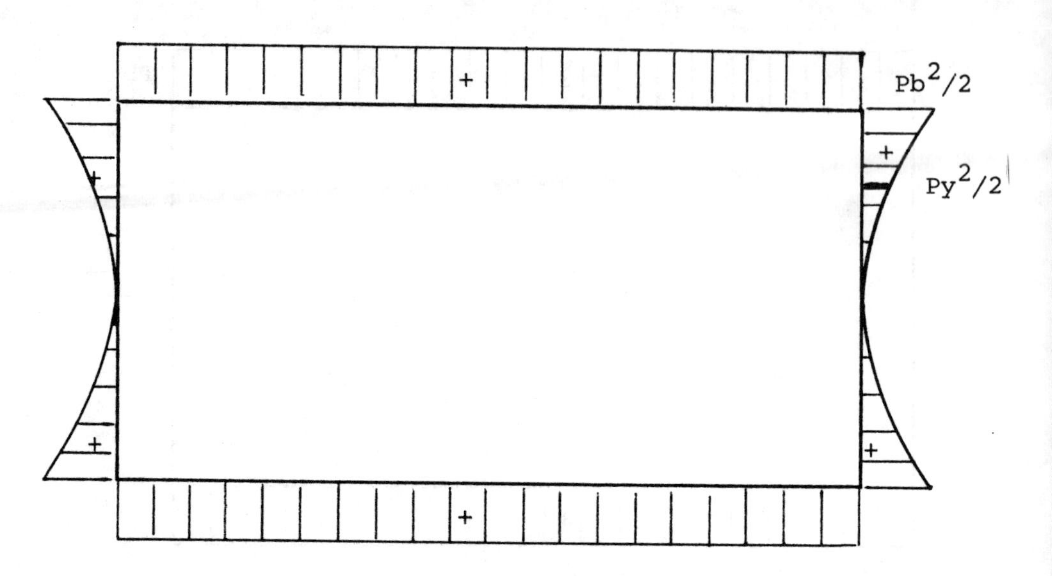

Figure 3-6. ϕ Diagram.

5-0:

$$\phi_s = \phi_5 + \int_{-b}^{y} Py(1)\,dy = \frac{Pb^2}{2} + \frac{Py^2}{2} - \frac{Pb^2}{2} + \frac{Pb^2}{2}$$

$$\phi_s = \frac{Py^2}{2} \qquad (3b + 4a) \leq s \leq (4b + 4a)$$

$$\phi_0 = 0$$

The above listed ϕ values at the various points along the boundary of the plate, are shown in Figure 3-6. These values of ϕ should produce the same magnitude of stresses on the boundary, as originally applied to the plate, i.e., $\sigma_x = P$, at the end surfaces only. The stresses at the end, using the ϕ function are

$$\phi = \frac{Py^2}{2} \qquad \frac{\partial\phi}{\partial y} = Py \qquad \frac{\partial^2\phi}{\partial y^2} = P$$

From equations (3.9), (3.10), and (3.11):

$$\sigma_x = \frac{\partial^2\phi}{\partial y^2} = P$$

$$\sigma_y = \frac{\partial^2\phi}{\partial x^2} = 0$$

$$\tau_{xy} = -\frac{\partial^2\phi}{\partial x\,\partial y} = 0$$

Thus the developed boundary stresses are in agreement with the applied stresses. With the ϕ function known, it is only necessary to evaluate $\nabla^4\phi = 0$, equation (3.13), at the interior of the plate to determine internal stresses. The techniques employed to solve this equation will be described in the following sections.

3.2.3 Frame Analogy [7, 8]

In addition to the direct integration method, as just described, a method known as the "Frame Analogy" [8], will permit evaluation of the ϕ function along the boundary. This method considers the plate as a fictitious frame, subjected to the applied external stresses (forces). The induced moments in this pseudo frame are computed and are equal to the ϕ function. The induced shears and axial forces are equal to derivatives of the ϕ function. The development of this method and application will now follow.

Consider first the general equation (3.26a)

$$\frac{d\phi}{ds} = -Q_y \frac{dx}{ds} + Q_x \frac{dy}{ds} + C_1 \frac{dy}{ds} + C_2 \frac{dx}{ds}$$

or

$$d\phi = (-Q_y + C_2)\, dx + (Q_x + C_1)\, dy$$

where

$$\phi = \int (-Q_y + C_2)\, dx + (Q_x + C_1)\, dy \qquad (3.28)$$

The solution of ϕ at some point $k(s)$ or $k(x, y)$ can be found by intergrating (3.28) by parts [1], i.e., $\phi_k = \int u\, dv = uv - \int v\, du + c$, rewriting (3.28):

$$\phi_k = \int_0^k (Q_x dy - Q_y dx) + \int_0^k (C_1 dy + C_2 dx)$$

Let:

$$u = Q_x \quad dv = dy \qquad \Big| \qquad u = -Q_y \quad dv = dx$$
$$du = dQ_x \quad v = y \qquad \Big| \qquad du = -dQ_y \quad v = x$$

Substituting these variables into the given general uv equation gives

$$\phi_k = (Q_x y - Q_y x)\big|_0^k - \int_0^k (y\, dQ_x - x\, dQ_y) + C_1 k + C_2 k + C_3$$

However, from equations (3.21) and (3.24),

$$Q_x = \int_0^k \bar{x}\, ds \qquad Q_y = \int_0^k \bar{y}\, ds$$

therefore $dQ_x = \bar{x}\,ds$ and $dQ_y = \bar{y}\,ds$, which gives

$$\phi_k = (Q_x y_k - Q_y y_k) - \int_0^k [(y\bar{x}\,ds) - (x\bar{y}\,ds)] + C_1 k + C_2 k + C_3$$

and

$$\phi_k = \left(\int_0^k \bar{x} y_k\,ds - \int_0^k \bar{y} x_k\,ds\right) - \int_0^k (y\bar{x} - x\bar{y})\,ds + C_1 k + C_2 k + C_3$$

Collecting terms gives

$$\phi_k = \int_0^k [\bar{x}(y_k - y) - \bar{y}(x_k - x)]\,ds + C_1 k + C_2 k + C_3$$

or in general,

$$\phi_k = \int_0^k [\bar{x}(y_k - y) - \bar{y}(x_k - x)]\,ds + c \qquad (3.29)$$

where ϕ_k is at any point along the surface at $k(x, y)$, and $\bar{x}$ and $\bar{y}$ are external forces per unit area along the boundary, x and y are integrating variables, and ds is a differential length of the boundary.

With this general equation (3.29) now developed, let us now examine the relationship between the statics of a frame and ϕ_k.

Consider first a plate subjected to in-plane loads parallel to the y-axis, as shown in Figure 3-7. At some distance x, the load has a total magnitude of $\bar{y}\,ds$; at the initial starting point on the surface the plate is cut, thus no internal pseudo forces are assumed. Cutting the plate at a distance (x_k, y_k), there will be assumed forces developed, i.e., M_k, T_k, and V_k. The plate, therefore, is assumed to be a frame whose configuration conforms to the outline of the plate boundary. Arbitrarily the plate is cut at some location, in order to make the pseudo frame determinate. In this instance the location is at the coordinate axes. Examining now a free body, Figure 3-8, summation of moments gives

$$dM_k = -\bar{y}\,ds(x_k - x) \qquad (3.30)$$

Similarly for loads parallel to the x-axis, as shown in Figure 3-9, and for a free body of this system as shown in Figure 3-10, the summation of moments gives

$$dM_k = \bar{x}\,ds(y_k - y) \qquad (3.31)$$

In both equilibrium equations it is assumed that positive moment causes tension on the inside of the pseudo frame. The total moment is thus the combination of equations (3.30) and (3.31), which gives:

$$dM_k = \bar{x}(y_k - y)\,ds - \bar{y}(x_k - x)\,ds$$

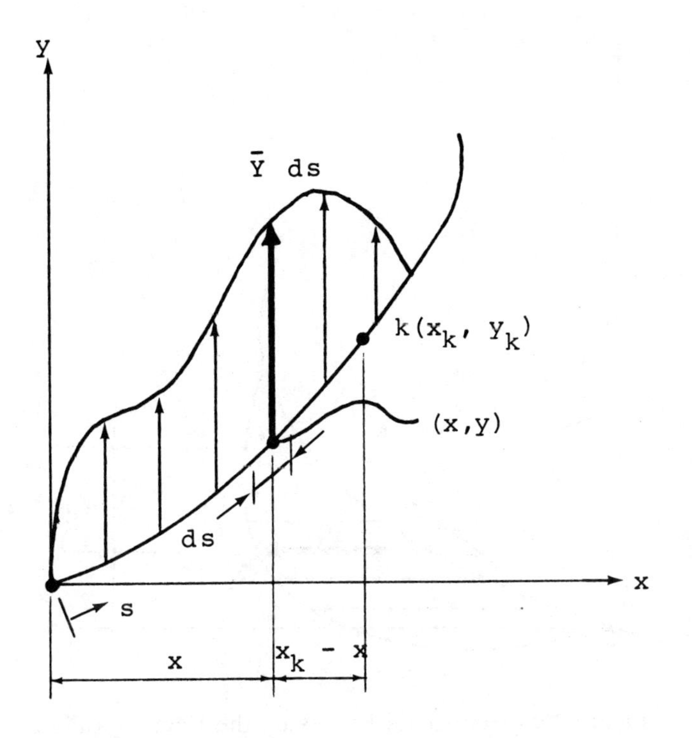

Figure 3-7. Vertical Forces on the Edge of a Plate.

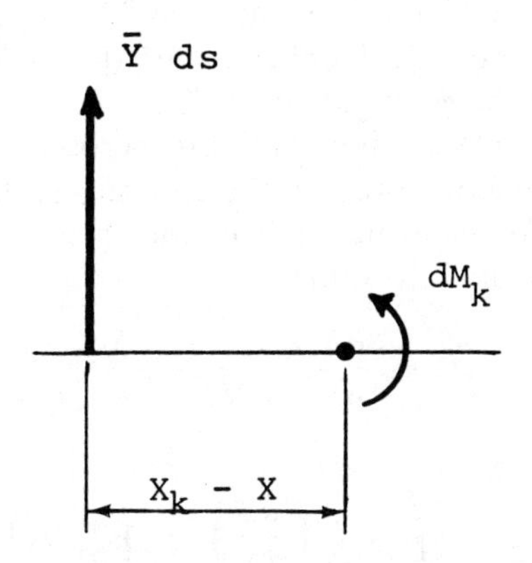

Figure 3-8. Free Body of Plate Element.

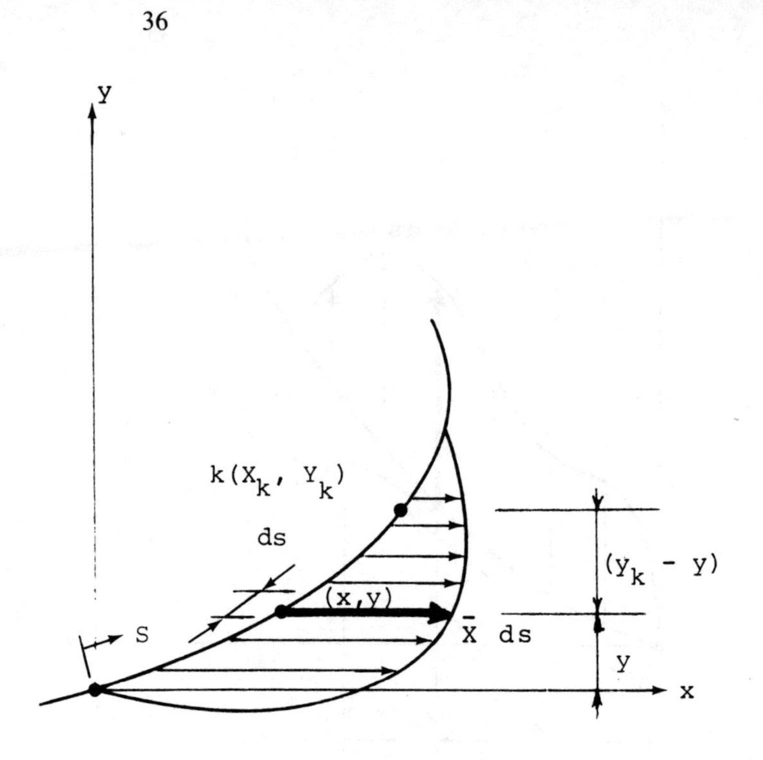

Figure 3-9. Horizontal Forces on the Edge of a Plate.

Integrating dM_k gives:

$$M_k = \int_0^k [\bar{x}(y_k - y) - \bar{y}(x_k - x)]\, ds + c \tag{3.32}$$

A comparison between this equation (3.32) and the previous ϕ_k equation (3.29), shows they are identical, thus the evaluation of M_k on the boundary will give the ϕ_k function.

Two additional relationships can also be obtained by computing the resulting shears and axial forces at the arbitrary section (x_k, y_k) due to the applied boundary forces $\bar{y}$ and $\bar{x}$. Using the chain rule the $\partial\phi/\partial N$ and the $\partial\phi/\partial T$ can be evaluated as follows:

$$\frac{\partial\phi}{\partial N} = \frac{\partial\phi}{\partial x} \cdot \frac{dx}{dN} + \frac{\partial\phi}{\partial y} \cdot \frac{dy}{dN}$$

or

$$\frac{\partial\phi}{\partial N} = -\int_0^s \bar{y}\, ds \left(\frac{dx}{dN}\right) + \int_0^s \bar{x}\, ds \left(\frac{dy}{dN}\right) \tag{3.33}$$

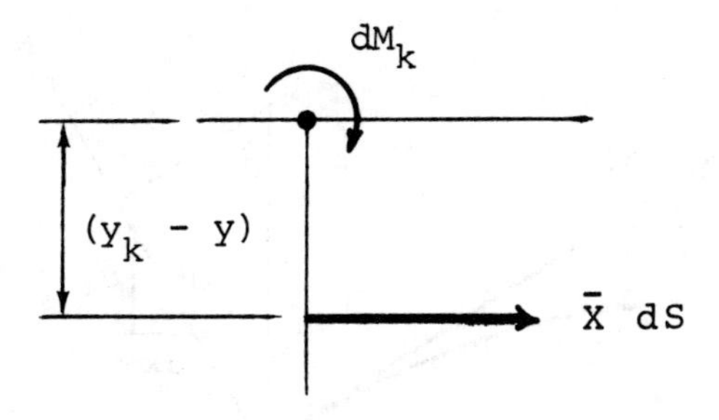

Figure 3-10. Free Body of Plate Element.

Similarly

$$\frac{\partial \phi}{\partial T} = \frac{\partial \phi}{\partial x} \cdot \frac{dx}{dT} + \frac{\partial \phi}{\partial y} \cdot \frac{dy}{dT}$$

or

$$\frac{\partial \phi}{\partial T} = - \int_0^s \bar{y}\, ds \left(\frac{dx}{dT}\right) + \int_0^s \bar{x}\, ds \left(\frac{dy}{dT}\right) \tag{3.34}$$

Return now to the plate, cut at some location, and subjected to force $\bar{x}$ and $\bar{y}$ along its surface. At some location s, the plate is cut and is replaced by pseudo forces V_x and V_y, as shown in Figures 3-11 and 3-12. Each of the total internal forces can be resolved into components, as shown. The summation of these components, relative to the N- and T-axes gives

$$\sum F_n = 0 \qquad V_{yn} + V_{xn} = \sum \bar{x}\, ds,\ \bar{y}\, ds$$

However,

$$V_{yn} = \int_0^s \bar{y}\, ds \left(-\frac{dx}{dT}\right)$$

and

$$V_{xn} = \int_0^s \bar{x}\, ds \left(\frac{dy}{dT}\right)$$

Therefore the total N force is

$$N = \left[- \int_0^s \bar{y}\, ds \left(\frac{dx}{dT}\right) + \int_0^s \bar{x}\, ds \left(\frac{dy}{dT}\right) \right] \tag{3.35}$$

Examination of equations (3.35) and (3.33), shows they are identical,

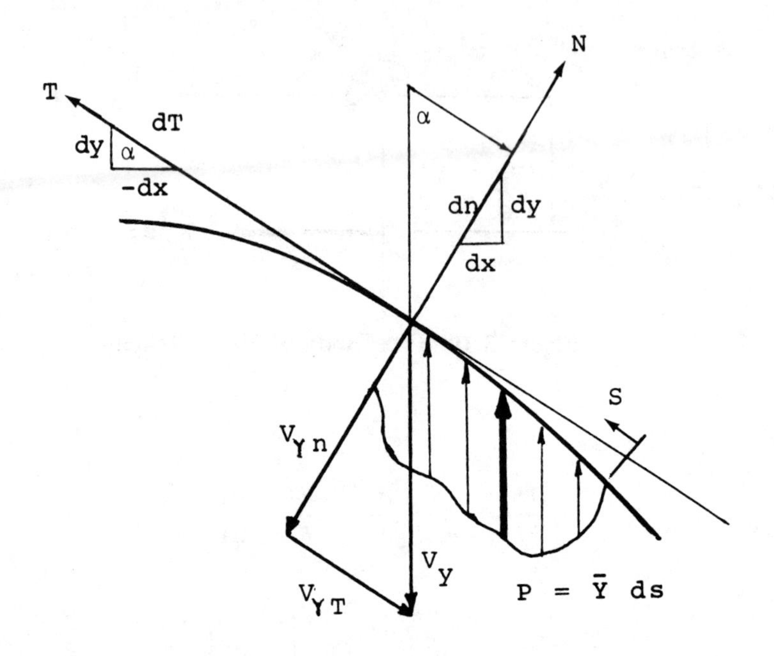

Figure 3-11. Shear Forces on Plate due to Vertical Forces.

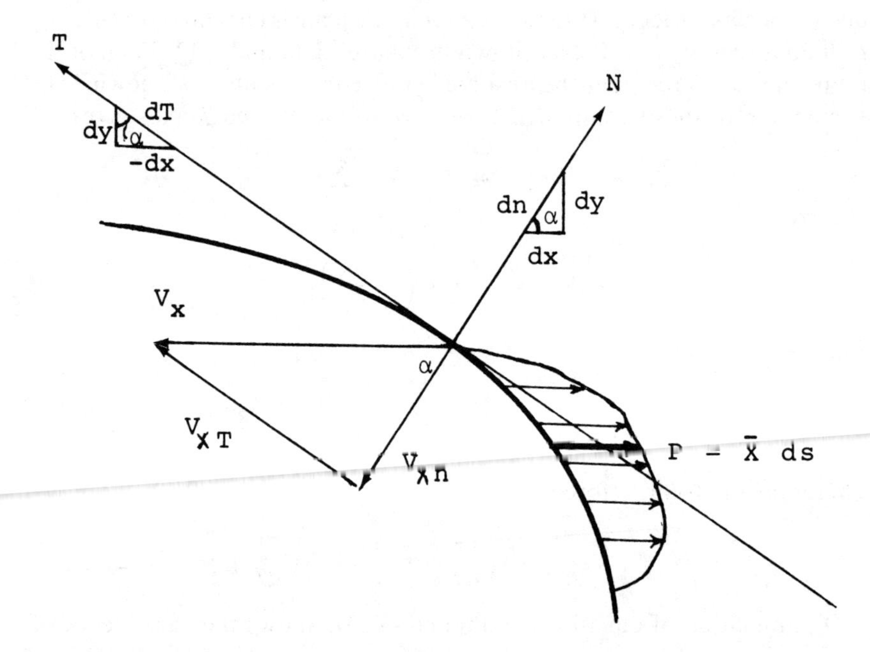

Figure 3-12. Shear Forces on Plate due to Horizontal Forces.

therefore a relationship between the shear N at the cut surface and $\partial\phi/\partial T$ is known:

$$\frac{\partial\phi}{\partial T} = +N \tag{3.36}$$

where positive is in accordance with beam convention: up on the left and down on the right of the cut is positive.

The summation of forces along the T-axis gives

$$\sum F_T = 0; \qquad V_{xT} - V_{yT} = \sum \bar{x}\,ds\,,\ \bar{y}\,ds$$

However,

$$V_{xT} = \int_0^s \bar{x}\,ds\left(\frac{dy}{dN}\right)$$

and

$$V_{yT} = \int_0^s \bar{y}\,ds\left(\frac{dx}{dN}\right)$$

Therefore the total T force is:

$$T = -\int_0^s \bar{y}\,ds\left(\frac{dx}{dN}\right) + \int_0^s \bar{x}\,ds\left(\frac{dy}{dN}\right) \tag{3.37}$$

Examination of equations (3.37) and (3.34) shows these expressions to be identical, therefore the relationship between the axial force at the cut section and $\partial\phi/\partial N$ is known:

$$\frac{\partial\phi}{\partial N} = T \tag{3.38}$$

where an outward axial force is considered positive.

Equations (3.36) and (3.38) will be useful when solving problems requiring additional boundary conditions.

In conclusion, we have developed three expressions which relate the internal forces M, T, N at any cut section on a frame whose outline conforms to the plate configuration to the ϕ function and respective derivatives. Evaluation of these functions at the boundary, will permit solution of the general biharmonic equation, $\nabla^4\phi = 0$ for the interior of the plate. The solution of equation (3.13) and its derivatives at all locations will then permit evaluation of the internal stresses σ_x, σ_y, and τ_{xy} as given by equations (3.9), (3.10), and (3.11).

Application of the frame analogy, in evaluating boundary functions will now be given.

3.2.4 *Problem—ϕ at Boundary—Frame Analogy*

The ϕ along the boundary, $\partial\phi/\partial N$, and $\partial\phi/\partial T$ will now be evaluated for the plate previously analyzed by the integration method, shown in Figure 3-3. The plate will be cut at point 0 and the resulting V, T, and M along the boundary will now be evaluated.

As shown in Figure 3-13(a), a free body of the frame from 0 to y is examined. The sum of moments at some distance y, gives

$$M_y = (Py)\frac{y}{2} = \frac{Py^2}{2}$$

Therefore from the analogy:

$$\phi_s = \frac{Py^2}{2}$$

The sum of forces in the x direction gives

$$V = Py$$

and the sum of forces in the y direction gives

$$T = 0$$

Therefore from the analogy:

$$\frac{\partial\phi}{\partial T} = V = Py \qquad \frac{\partial\phi}{\partial N} = T = 0$$

and at point 1 on the plate, i.e., $y = b$:

$$\phi_1 = \frac{Pb^2}{2} \qquad \frac{\partial\phi}{\partial T} = Pb \qquad \frac{\partial\phi}{\partial N} = 0$$

The values of these functions along the surface 1-2 can now be obtained by examining another free body, as shown in Figure 3-13(b)

$$M = M_1 = \frac{Pb^2}{2}; \qquad \text{therefore,} \qquad \phi_{1-2} = \frac{Pb^2}{2}$$

$$T_{1-2} = V_1 = Pb; \qquad \text{therefore,} \qquad \left.\frac{\partial\phi}{\partial N}\right|_{1-2} = Pb$$

$$V_{1-2} = T_2 = 0; \qquad \text{therefore,} \qquad \left.\frac{\partial\phi}{\partial T}\right|_{1-2} = 0$$

0 - 1

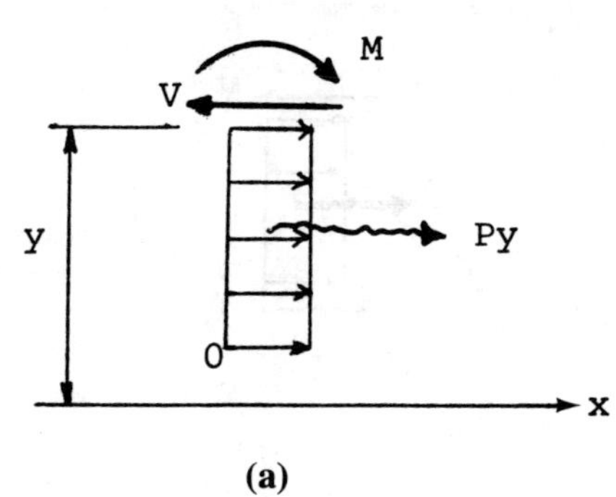

(a)

1 - 2

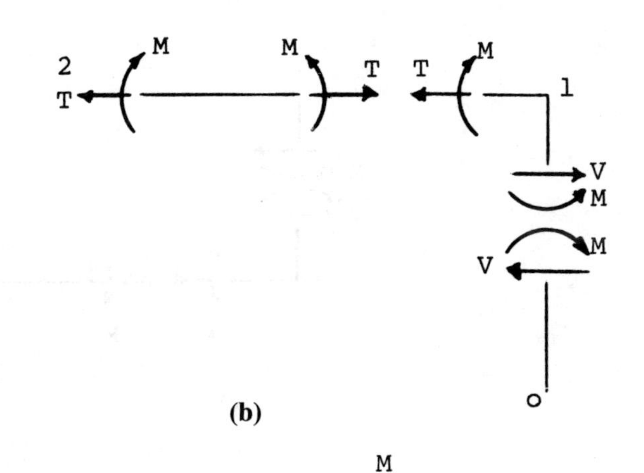

(b)

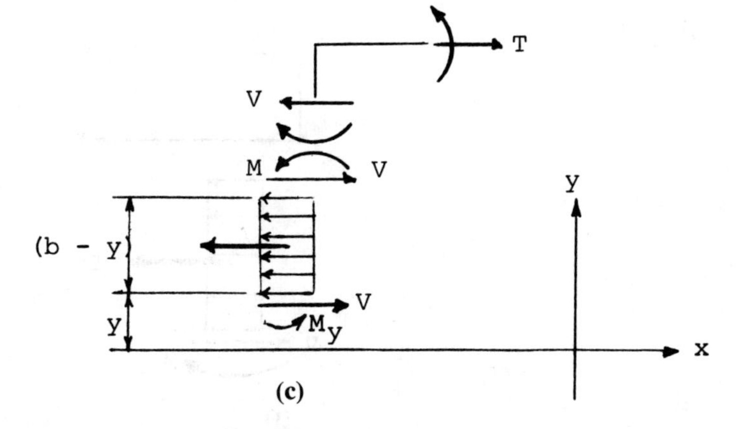

(c)

Figure 3-13. Example—Free Body of Plate Segments.

3 - 4

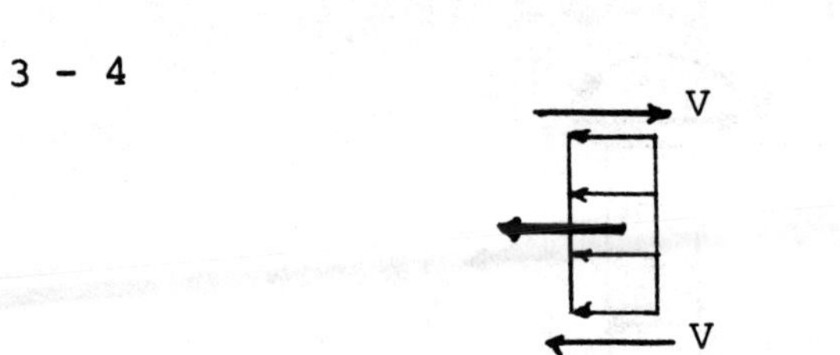

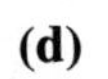

(d)

4 - 5

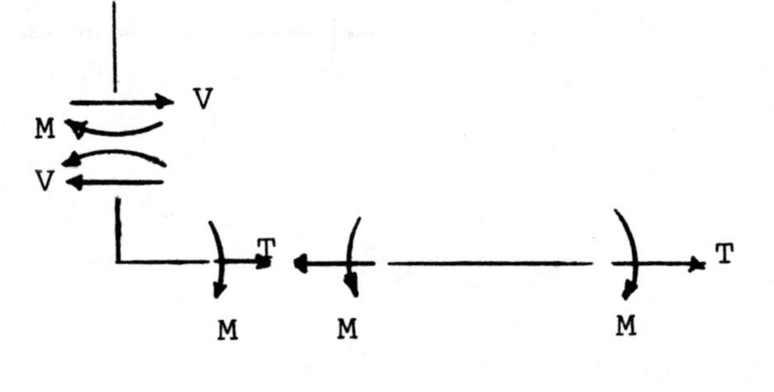

(e)

5 - 0

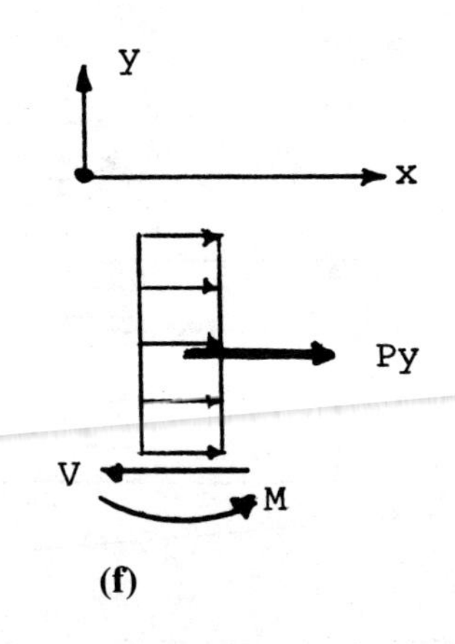

(f)

Figure 3-13. (cont.)

The functions along the boundary 2-3-4 are similarly found by examining a free body shown in Figures 3-13(c) and (d). Using Figure 3-13(c) gives

$$M_y = -\frac{Pb^2}{2} + Pb(b - y) - \frac{P(b - y)^2}{2}$$

$$M_y = -\frac{Py^2}{2}$$

which is opposite to assumed direction shown in Figure 3-13(c), therefore,

$$\phi_s = \frac{Py^2}{2} \quad (+ \text{ if tension on inside face})$$

The sum of horizontal forces gives

$$V = -(Pb - Pb + Py)$$

$$V = -Py$$

therefore V is reversed for the direction assumed and

$$\frac{\partial\phi}{\partial T} = -Py$$

Also

$$T = 0; \qquad \text{therefore,} \qquad \frac{\partial\phi}{\partial N} = 0$$

at point 3, therefore:

$$\phi_3 = 0 \qquad \left.\frac{\partial\phi}{\partial T}\right|_3 = 0 \qquad \left.\frac{\partial\phi}{\partial N}\right|_3 = 0$$

A free body of the frame from point 3 to a distance y, where the forces at point 3 are all zero from the above, gives

$$M_y = (Py)\left(\frac{y}{2}\right) = \frac{Py^2}{2}; \qquad \text{therefore,} \qquad \phi_s = \frac{Py^2}{2}$$

$$V = Py; \qquad \text{therefore,} \qquad \frac{\partial\phi}{\partial T} = Py$$

$$T = 0; \qquad \text{therefore,} \qquad \frac{\partial\phi}{\partial N} = 0$$

At point 4, therefore:

$$\phi_4 = \frac{Pb^2}{2} \qquad \left.\frac{\partial\phi}{\partial T}\right|_4 = Pb \qquad \left.\frac{\partial\phi}{\partial N}\right|_4 = 0$$

The functions along surface 4-5 are obtained from the free body shown in Figure 3-13(e).

$$M_y = \phi_4 = \frac{Pb^2}{2}$$

$$V = T_4 = 0$$

$$T = V_4 = Pb$$

Therefore,

$$\phi_{4\text{-}5} = \frac{Pb^2}{2} \qquad \frac{\partial \phi}{\partial N} = Pb \qquad \frac{\partial \phi}{\partial T} = 0$$

The variables from points 5-0 are computed as per Figure 3-13(f).

$$M_y = (Py)\left(\frac{y}{2}\right) = \frac{Py^2}{2}$$

$$V = -Py$$

$$T = 0$$

Therefore,

$$\phi_s = \frac{Py^2}{2} \qquad \frac{\partial \phi}{\partial T} = -Py \qquad \frac{\partial \phi}{\partial N} = 0$$

At point 0,

$$\phi_0 = 0 \qquad \frac{\partial \phi}{\partial T} = 0 \qquad \frac{\partial \phi}{\partial N} = 0$$

The resulting ϕ, $\partial\phi/\partial N$, and $\partial\phi/\partial T$ along the boundary are plotted in Figures 3-14, 3-15, and 3-16.

These values are in agreement with the previous values computed by the integration method shown in Figures 3-5 and 3-6.

3.3 Boundary Conditions Represented by Fourier Series

In the previous section, two methods were presented by which the ϕ function could be evaluated along the boundary of the plate. These methods will prove most useful when solving the in-plane stress problem by a numerical method, e.g., by finite differences. As an alternative to evaluating the ϕ function along the boundary and then solving the biharmonic $\nabla^4\phi = 0$ in the interior, it is possible to select the boundary stress in such a manner to agree with the solution of the biharmonic and thus solve the

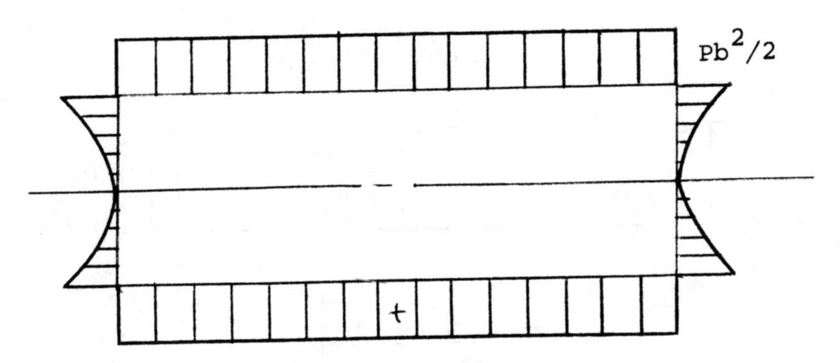

Figure 3-14. ϕ Diagram.

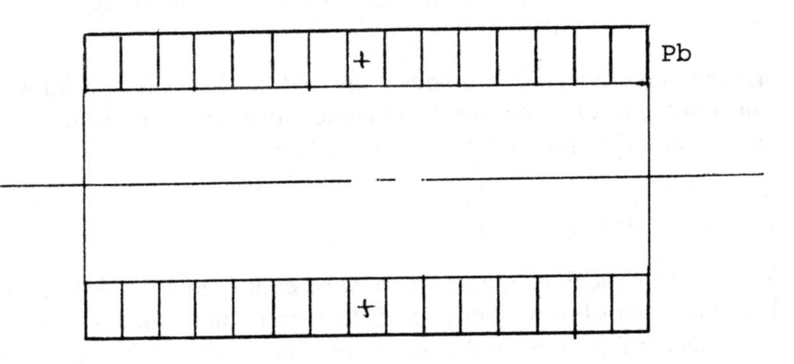

Figure 3-15. $\partial\phi/\partial N$ Diagram.

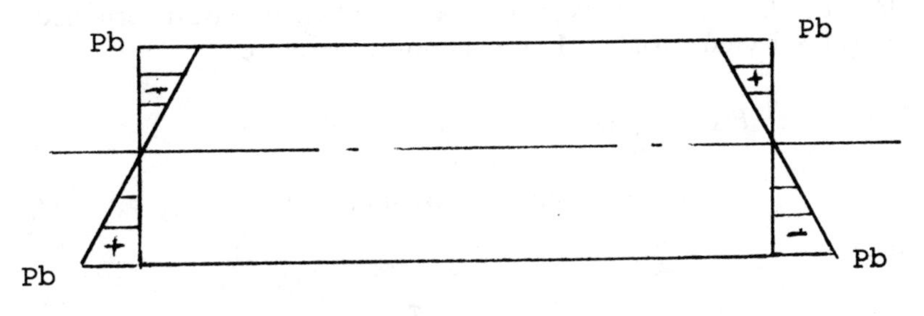

Figure 3-16. $\partial\phi/\partial T$ Diagram.

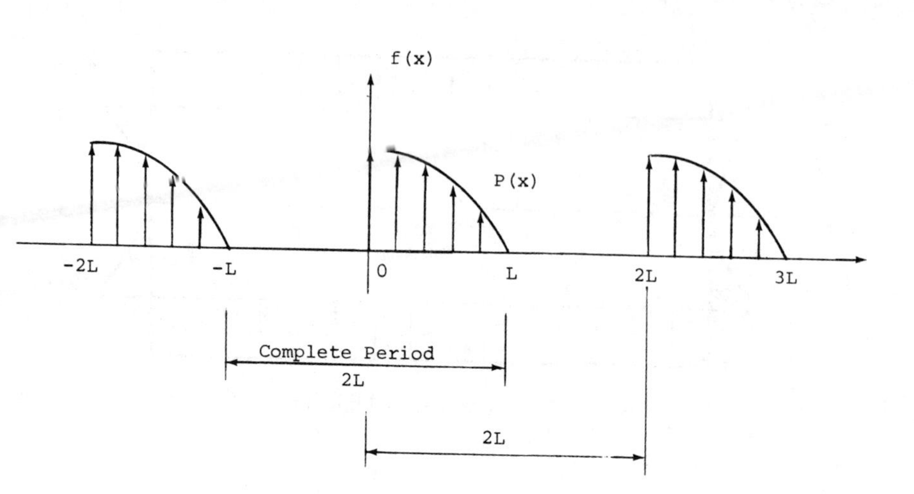

Figure 3-17. General Loading Configuration Series.

problem directly. This technique is limited, in that only special problems can be solved. A more general technique, utilizing the frame analogy and a numerical technique, will be explained later.

3.3.1 General Fourier Series

This mathematical technique will now be explained in order to familiarize the reader with Fourier series and their basic fundamental concepts.

Consider first a general loading function as shown in Figure 3-17. As seen by this figure, the loading is repetitive or periodic in the x direction. That is, the loading will be the same at every distance $2L$ for an infinite length along the x-axis. It is now desirable to express this periodic loading in some mathematical form, in order to evaluate the load at some arbitrary distance x. Assume the loading

$$P(x) = \sum A_n f_n(x) \tag{3.39}$$

where $n = 0, 1, 2, \ldots, A_n$ is an arbitrary coefficient to be determined, and $f_n(x)$ is a selected variable. Let us choose, therefore

$$P(x) = \frac{1}{2} A_0 + A_1 \cos \lambda x + A_2 \cos 2\lambda x + \ldots$$

$$+ B_1 \sin \lambda x + B_2 \sin 2\lambda x + \ldots \tag{3.40a}$$

where

$$\lambda = \frac{\pi}{L}$$

In more general terms, equation (3.40a) can be written as

$$P(x) = \frac{1}{2}A_0 + A_1 \cos \frac{\pi x}{L} + \ldots A_n \cos \frac{n\pi x}{L}$$

$$+ B_1 \sin \frac{\pi x}{L} + \ldots + B_n \sin \frac{n\pi x}{L} \tag{3.40b}$$

If the terms A_0, A_1, ... A_n, B_1 ..., B_n can be determined, then the $P(x)$ coefficient can be evaluated. The following describes the evaluation of these terms.

A_0 **Term:** Multiply both sides of equation (3.40b) by dx and integrate, which will eliminate atl terms except the first term, or

$$\int_0^{2L} P(x)\, dx = \frac{1}{2}A_0 \int_0^{2L} dx + \ldots$$

Solving for A_0 gives

$$A_0 = \frac{1}{L}\int_0^{2L} P(x)\, dx \tag{3.41}$$

A_n **Term:** The A_n term is found by multiplying both sides of equation (3.40b) by $(\cos m\pi/L)\, dx$ and integrating, where m is an integer. The results will give

$$\int_0^{2L} P(x) \cos \frac{m\pi x}{L}\, dx = A_n \int_0^{2L} \cos \frac{n\pi x}{L} \cos \frac{m\pi x}{L}\, dx$$

All terms of the general equation reduce to zero when $n \neq m$, and when $n = m$:

$$\int_0^{2L} P(x) \cos \frac{n\pi x}{L}\, dx = A_n L$$

or

$$A_n = \frac{1}{L}\int_0^{2L} P(x) \cos \frac{n\pi x}{L}\, dx \tag{3.42}$$

B_n **Term:** The evaluation of the B_n term is performed by multiplying both sides of equation (3.40b) by $(\sin m\pi x/L)\, dx$ and integrating. This gives

$$\int_0^{2L} P(x) \sin \frac{n\pi x}{L}\, dx = B_n L$$

or

$$B_n = \frac{1}{L}\int_0^{2L} P(x) \sin \frac{n\pi x}{L} dx \tag{3.43}$$

The final function $P(x)$, is then computed as:

$$P(x) = \frac{1}{2}A_0 + \sum A_n \cos \frac{n\pi x}{L} + \sum B_n \sin \frac{n\pi x}{L} \tag{3.44}$$

where A_0, A_n and B_n are as defined by equations (3.41), (3.42), and (3.43).

3.3.2 Even and Odd Functions

The general equation (3.44) can express any periodic function in terms of $\sin \lambda x$ and $\cos \lambda x$, providing the terms A_0, A_n, and B_n are evaluated. However, if the periodic functions are of a particular form, the need for using the entire general equation and terms A_0, A_n and B_n may not be required.

Consider now the function to be even, as shown in Figure 3-18, then the function $f(x) = f(-x)$ or $\cos(x) = \cos(-x)$ is such a function. However, $\sin(-x) = -\sin(x)$ does not agree with the function shown, therefore the B_n term can be eliminated, and due to symmetry:

$$A_n = \frac{1}{L}\int_0^{2L} f(x) \cos \frac{n\pi}{L} x\, dx$$

or

$$A_n = \frac{1}{L}\int_{-L}^{0} f(x) \cos \frac{n\pi}{L} x\, dx + \frac{1}{L}\int_0^{L} f(x) \cos \frac{n\pi}{L} x\, dx$$

$$A_n = \frac{2}{L}\int_0^{L} f(x) \cos \frac{n\pi}{L} x\, dx \tag{3.45}$$

Similarly,

$$A_0 = \frac{2}{L}\int_0^{L} f(x)\, dx \tag{3.46}$$

If the function is an odd variable, as shown in Figure 3-19, then $f(-x) = -f(x)$. This type of function corresponds to the sine function, i.e., $\sin(-x) = -\sin(x)$; however the cosine function is not in agreement; therefore, $A_n = A_0 = 0$, and

$$B_n = \frac{2}{L}\int_0^{L} f(x) \sin \frac{n\pi x}{L} dx \tag{3.47}$$

Equations (3.45), (3.46), and (3.47), are used in conjunction with the general equation (3.44) for $P(x)$.

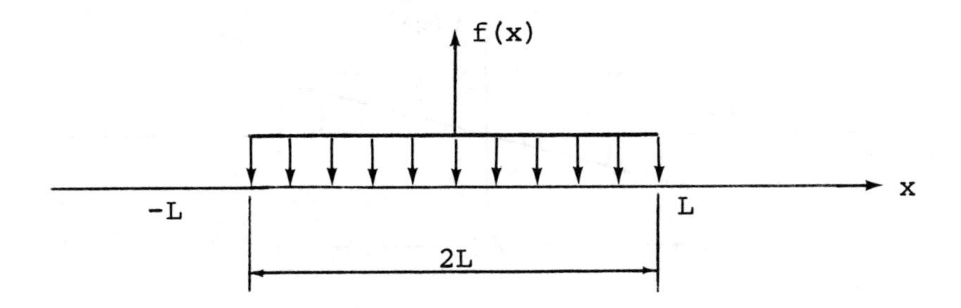

Figure 3-18. Even Loading Function. (cos)

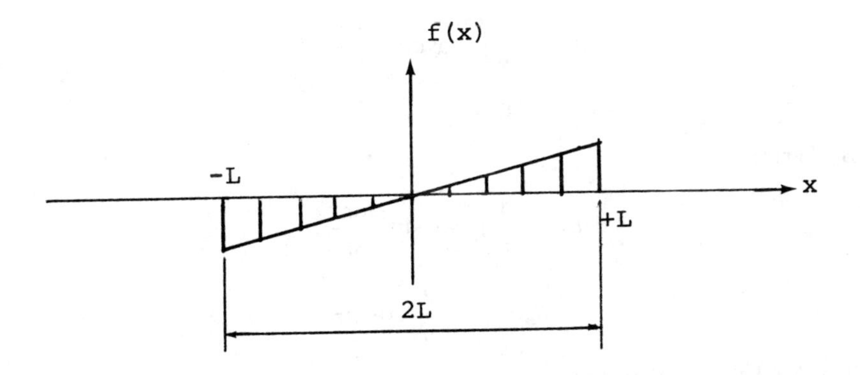

Figure 3-19. Odd Loading Function. (sin)

3.3.3 Example 1—Linear Function

In order to illustrate the Fourier series technique, let us examine the function shown in Figure 3-20. It is desirable to express this function in Fourier series form. The function is $f(x) = x$ over the period $-\pi$ to π. The integration length will be changed from (0 to 2π) to ($-\pi$ to π).

A_0 **Term:**

$$A_0 = \frac{1}{L}\int_0^{2L} f(x)\,dx$$

Since $2L = 2\pi$, $L = \pi$,

$$A_0 = \frac{1}{\pi}\int_{-\pi}^{\pi} x\,dx = 0$$

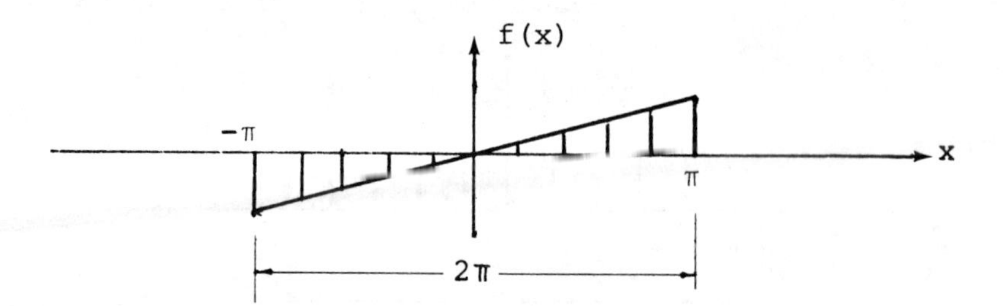

Figure 3-20. Example—Function Distribution.

A_n **Term:**

$$A_n = \frac{1}{\pi}\int_{-\pi}^{\pi} x \cos \frac{n\pi x}{\pi}\, dx$$

$$A_n = 0$$

B_n **Term:**

$$B_n = \frac{1}{\pi}\int_{-\pi}^{\pi} x \sin \frac{n\pi x}{\pi}\, dx$$

$$B_n = -\frac{2}{n} \cos n\pi$$

The final $f(x)$ is therefore,

$$f(x) = \sum -\frac{2}{n} \cos n\pi \left(\sin \frac{n\pi x}{\pi} \right)$$

3.3.4 Example 2—Plate Loading

The general Fourier series equation (3.44), can always be applied for any periodic function. If the function has some unique characteristics in that it is symmetrical (even) or antisymmetrical (odd), then a special case of the general equation can be utilized, i.e., (3.45), (3.46), or (3.47). As will be illustrated in this example, it is possible to create even or odd periodic functions [7] and thus reduce the number of terms required for a solution.

Consider a plate or boundary of length a, as shown in Figure 3-21, subjected to a partial load P over a length $2c$. If the load is periodic, and in accordance with the general loading function, Case I (Figure 3-22) will be the load arrangement. The solution of this case is as follows.

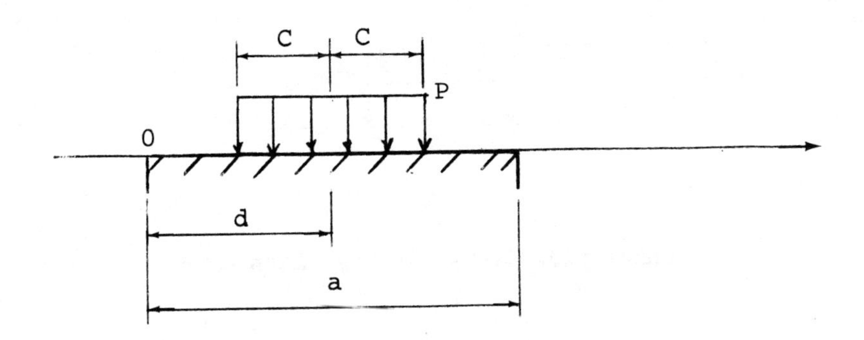

Figure 3-21. General Loading

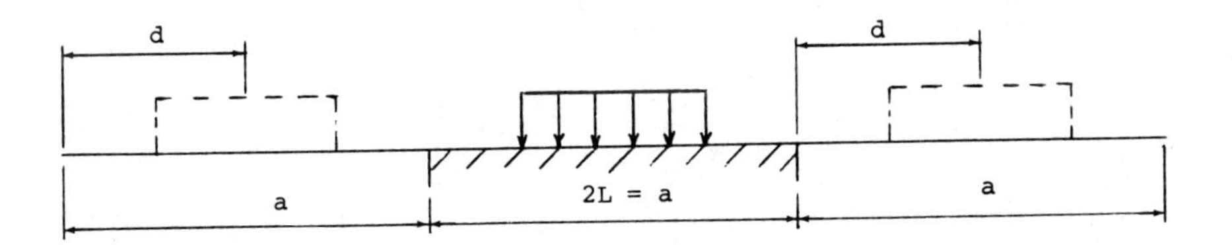

Figure 3-22. Case 1 Loading—All Terms.

Case I: There is no symmetry in this loading case, therefore both A_0, A_n, and B_n terms will be required.

$$\frac{1}{2}A_0 = \frac{1}{2L}\int_{d-c}^{d+c} P(x)\,dx \tag{3.41}$$

$$\frac{1}{2}A_0 = \frac{1}{a}\int_{d-c}^{d+c} P\,dx$$

$$\frac{1}{2}A_0 = \frac{P}{a}[d + c - d + c] = \frac{2Pc}{a}$$

$$A_n = \frac{1}{L}\int_{d-c}^{d+c} P(x)\cos\frac{n\pi x}{L}\,dx \tag{3.42}$$

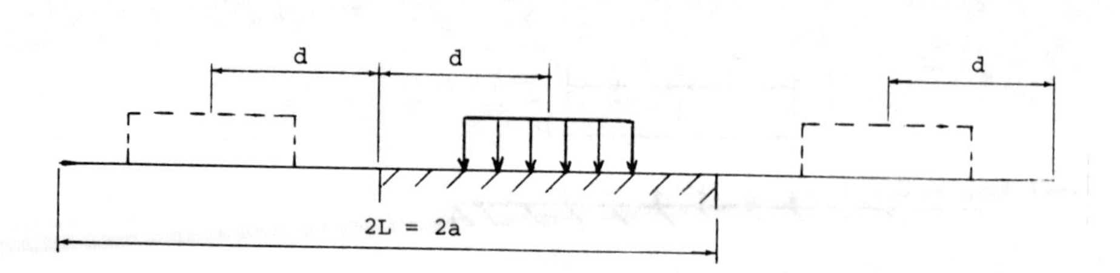

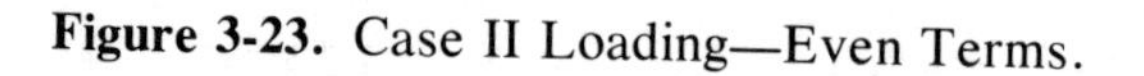

Figure 3-23. Case II Loading—Even Terms.

where $\quad 2L = a \quad$ so $\quad L = \dfrac{a}{2}$

$$A_n = \frac{1}{a/2}\int_{d-c}^{d+c} P \cos \frac{n\pi x}{a/2}\, dx$$

$$A_n = \frac{2P}{a}\int_{d-c}^{d+c} \cos \frac{2n\pi}{a} x\, dx$$

$$A_n = \frac{2P}{n\pi} \sin \frac{2n\pi c}{a} \cos \frac{2n\pi d}{a}$$

(3.43)

$$B_n = \frac{1}{L}\int_{d-c}^{d+c} P(x) \sin \frac{n\pi x}{L}\, dx$$

$$B_n = \frac{1}{a/2}\int_{d-c}^{d+c} P \sin \frac{2n\pi}{a} x\, dx$$

$$B_n = \frac{2P}{n\pi} \sin \frac{2n\pi c}{a} \sin \frac{2n\pi d}{a}$$

Substituting $A_0/2$, A_n, and B_n into the general equation (3.44) for $P(x)$ gives

$$P(x) = \frac{2Pc}{a} + \frac{2P}{\pi}\sum \frac{1}{n} \sin \frac{2n\pi c}{a} \cos \frac{2n\pi d}{a} \cos \frac{2n\pi x}{a}$$

$$+ \frac{2P}{\pi}\sum \frac{1}{n} \sin \frac{2n\pi c}{a} \sin \frac{2n\pi d}{a} \sin \frac{2n\pi x}{a}$$

If the periodic loading is now arranged such that an even series is created, the solution can be reduced to evaluating only the A_0 and A_n terms. Examining Figure 3-23, we have:

Case II: The period length is now $2L = 2a$, therefore $L = a$, and $B_n = 0$, so

$$\frac{1}{2}A_0 = \left[\frac{2}{L}\int P(x)\,dx\right]\frac{1}{2} \tag{3.46}$$

$$\frac{1}{2}A_0 = \frac{1}{a}\int_{d-c}^{d+c} P\,dx$$

$$\frac{1}{2}A_0 = \frac{2Pc}{a}$$

$$A_n = \frac{2}{L}\int P(x)\cos\frac{n\pi x}{L}\,dx \tag{3.45}$$

$$A_n = \frac{2}{a}\int_{d-c}^{d+c} P\cos\frac{n\pi x}{a}\,dx$$

$$A_n = \frac{4P}{n\pi}\sin\frac{n\pi c}{a}\cos\frac{n\pi d}{a}$$

Substituting A_0, A_n, and B_n into the general equation (3.44) gives the Fourier series representation for the loading or:

$$P(x) = \frac{2Pc}{a} + \frac{4P}{\pi}\sum\frac{1}{n}\sin\frac{n\pi c}{a}\cos\frac{n\pi d}{a}\cos\frac{n\pi x}{a}$$

The last alternative in the development of a Fourier series function, is to create an odd function, as shown in Case III (Figure 3-24). With such a function $A_0 = A_n = 0$ and B_n is as defined by equation (3.47).

Case III:

$$B_n = \frac{2}{L}\sum P(x)\sin\frac{n\pi x}{L}\,dx$$

$$B_n = \frac{2}{a}\int_{d-c}^{d+c} P\sin\frac{n\pi x}{a}\,dx$$

$$B_n = \frac{4P}{n\pi}\sin\frac{n\pi c}{a}\sin\frac{n\pi d}{a}$$

Substituting B_n into the general $P(x)$ equation gives

$$P(x) = \frac{4P}{\pi}\sum\frac{1}{n}\sin\frac{n\pi c}{a}\sin\frac{n\pi d}{a}\sin\frac{n\pi x}{a}$$

What has just been illustrated in this example is the variation that can be created in the solution by selecting the proper period length. In this in-

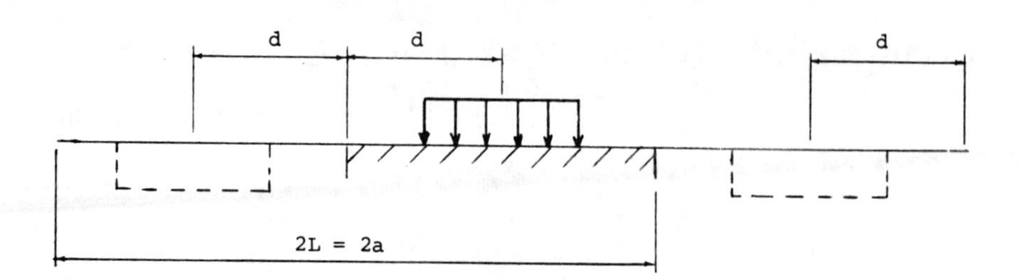

Figure 3-24. Case III Loading—Odd Terms.

stance the entire function was used or even/odd functions created, thus reducing the complexity of the general function.

3.4 Semi-Infinite Plate with Periodic Loading

Utilizing the Fourier series formulation of the boundary loading, it is now possible to solve directly the biharmonic equation (3.13) in series form. Consider a plate that has infinite length and infinite height subjected to a repetitive edge boundary load, as shown in Figure 3-25. This type of loading may represent a wall supported on columns spaced at a distance of $2a$, with uniform wall thickness h.

The boundary conditions, for this plate are

$$\text{at } y = 0 \begin{cases} \sigma_y = \dfrac{P(x)}{h} \\ \tau_{xy} = 0 \end{cases}$$

The function $P(x)$ can be expressed as a Fourier series, utilizing the even function relationships equations (3.45) and (3.46). Applying these equations gives the following:

$$A_n = \frac{2}{L}\int_0^L f(x) \cos \frac{n\pi x}{L}\, dx$$

$$A_n = \frac{2}{a}\int_0^{a-c} P \cos \frac{n\pi x}{a}\, dx + \frac{2}{a}\int_{a-c}^{a} - P\left(\frac{a-c}{a}\right) \cos \frac{n\pi x}{a}\, dx$$

$$A_n = \frac{2P}{a}\left[\frac{\sin(n\pi x/a)}{n\pi/a}\right]_0^{a-c} - \frac{2}{a}P\left(\frac{a-c}{a}\right)\left[\sin \frac{n\pi x/a}{n\pi/a}\right]_{a-c}^{a}$$

$$A_n = \frac{2Pa}{n\pi c} \sin \frac{n\pi}{a}(a-c)$$

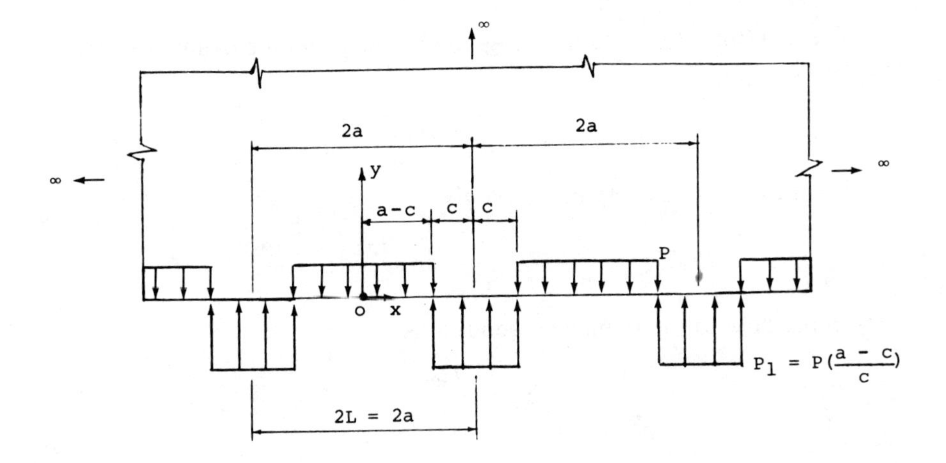

Figure 3-25. Semi-Infinite Plate with Periodic Loading.

A_n can be further reduced by using the relationship, $\sin(\alpha - \beta) = \sin\alpha \cos\beta - \cos\alpha \sin\beta$, which gives

$$A_n = \frac{2Pa}{n\pi c}\left[\sin\frac{n\pi}{a}a\cos\frac{n\pi}{a}(-c) - \cos\frac{n\pi}{a}a\sin\frac{n\pi}{a}(-c)\right]$$

$$A_n = \frac{-2Pa}{n\pi c}\left[\cos n\pi \sin\frac{n\pi c}{a}\right]$$

$$A_n = -\frac{2Pa}{n\pi c}(-1)^n \sin\frac{n\pi}{a}c$$

The evaluation of the A_0 term is determined similarly, giving

$$A_0 = \frac{2}{L}\int_0^L f(x)\,dx$$

$$A_0 = \frac{2}{a}\int_0^{a-c} P\,dx + \frac{2}{a}\int_{a-c}^{a} -P\left(\frac{a-c}{c}\right)dx$$

$$A_0 = \frac{2P}{a}(a-c) + \frac{2}{a}\left(-P\frac{a-c}{c}\right)[a-(a-c)]$$

$$A_0 = \frac{+2P}{a}(a-c) - \frac{2P}{a}(a-c)$$

$$A_0 = 0$$

The resulting $P(x)$ function, as given by the general equation (3.44), is therefore:

$$P(x) = \frac{1}{2}A_0 + \sum A_n \cos \frac{n\pi x}{L}$$

Substituting A_0 and A_n into $P(x)$ gives

$$P(x) = -\frac{2Pa}{\pi c}\sum\frac{(-1)^n}{n} \sin \frac{n\pi c}{a} \cos \frac{n\pi x}{a} \tag{3.48}$$

Returning now to the boundary conditions:

$$\sigma_y = \frac{P(x)}{h}$$

or

$$\sigma_y = \frac{1}{h}\sum \bar{A}_n \cos \lambda_n x \tag{3.49}$$

where $\lambda_n = \dfrac{n\pi}{a}$, $\bar{A}_n = -\dfrac{2Pa}{\pi c}\dfrac{(-1)^n}{n} \sin \dfrac{n\pi c}{a}$

The general stress equation, relative to the Airy's stress function is $\sigma_y = \partial^2\phi/\partial x^2$, equation (3.10). Examination of the stress equation (3.49) suggests that the ϕ function be of the form:

$$\phi = \sum Y_n \cos \lambda_n x \tag{3.50}$$

The reason for this selection is seen by taking the $\partial^2\phi/\partial x^2$ of equation (3.50) which gives

$$\frac{\partial^2\phi}{\partial x^2} = -Y_n \lambda_n^2 \cos \lambda_n x = \sigma_y$$

This relationship is of the form given by the boundary stress equation (3.49), and thus is the reason for the selection of the cos λx variable. The variable Y_n is included since the stress must vary as a function of the coordinate y as well as x. What is now required is the solution of the Y_n variable.

As will be recalled, it is required in a plane stress problem to evaluate the $\nabla^4\phi = 0$ equation at all points, where

$$\nabla^4\phi = \frac{\partial^4\phi}{\partial x^4} + 2\frac{\partial^4\phi}{\partial x^2\,\partial y^2} + \frac{\partial^4\phi}{\partial y^4} = 0$$

Substituting the ϕ equation (3.50) into $\bar{\nabla}^4\phi$ gives

$$Y_n\lambda_n^4 \cos \lambda_n x - 2\,\lambda_n^2\frac{\partial^2 Y_n}{\partial y^2} \cos \lambda_n x + \frac{\partial^4 Y_n}{\partial y^4} \cos \lambda_n x = 0$$

or for any value of x,

$$\lambda_n^4 Y_n - 2\lambda_n^2 Y_n'' + Y_n'''' = 0 \tag{3.51}$$

where $\quad Y_n' = \dfrac{\partial Y_n}{\partial y}, \qquad Y_n'' = \dfrac{\partial^2 Y_n}{\partial y^2}, \qquad$ etc.

The original partial differential biharmonic equation has now been reduced to the linear differential equation (3.51) with constant coefficients, which is readily solvable.

The solution of linear differential equations with constant coefficients can be obtained by assuming the solution of the form [1]:

$$Y_n = e^{ky} \tag{3.52}$$

The derivative of (3.52) gives

$$Y_n'' = k^2 e^{ky}, \qquad Y_n'''' = k^4 e^{ky}$$

now substituting these derivatives into equation (3.51), gives

$$\lambda_n^4 - 2\lambda_n^2 k^2 + k^4 = 0 \tag{3.53}$$

which represents the characteristic equation, whose roots are the solution to the linear differential equation (3.52). Factoring equation (3.53) gives

$$(\lambda_n^2 - k^2)(\lambda_n^2 - k^2) = 0$$

or

$$k = \pm\lambda_n$$

The solution to (3.52) is therefore a combination of the individual solutions multiplied by arbitrary constants, or:

$$Y_n = Ae^{\lambda_n y} + B\,\lambda_n y e^{\lambda_n y} + Ce^{-\lambda_n y} + D\,\lambda_n y e^{-\lambda_n y} \tag{3.54}$$

Note that the solution Y_n contains coefficients λ_n for those terms that have similar roots. This is required if four roots are to be retained for the fourth order differential equation. Equation (3.54) can be written in trigometric form, using the relationships [1]:

$$e^{\lambda_n y} = \cosh \lambda_n y + \sinh \lambda_n y$$

$$e^{-\lambda_n y} = \cosh \lambda_n y - \sinh \lambda_n y$$

Substituting these relationships into equation (3.54) will yield the following:

$$\begin{aligned} Y_n = {} & A(\cosh \lambda_n y + \sinh \lambda_n y) + B\lambda_n y(\cosh \lambda_n y + \sinh \lambda_n y) \\ & + C(\cosh \lambda_n y - \sinh \lambda_n y) + D\lambda_n y(\cosh \lambda_n y \end{aligned}$$

Collecting terms,

$$Y_n = (\sinh \lambda_n y)(A - C) + (\cosh \lambda_n y)(A + C) + (\lambda_n y \sinh \lambda_n y)(B - D)$$
$$+ (\lambda_n y \sinh \lambda_n y)(B - D) + (\lambda_n y \cosh \lambda_n y)(B + D)$$

The combined constant terms A, B, C, and D, are modified constants, therefore

$$Y_n = E \sinh \lambda_n y + F \cosh \lambda_n y + Gy \sinh \lambda_n y + Hy \cosh \lambda_n y \qquad (3.55)$$

The two solutions for Y_n are used when the boundaries are infinite—equation (3.54)—or finite—equation (3.55).

The problem we are presently trying to solve has boundaries that are infinite; therefore equation (3.54) will be used, i.e.,

$$\phi = \sum Y_n \cos \lambda_n X$$

or

$$\phi = \sum(Ae^{\lambda_n y} + B\lambda_n y^{\lambda_n y} + Ce^{-\lambda_n y} + D\lambda_n Ye^{-\lambda_n y}) \cos \lambda_n x \qquad (3.56)$$

It is now required to evaluate the four constants A, B, C, and D, utilizing the boundary conditions. Examination of the loads indicates that as y approaches infinity, the stresses should reduce to zero. The variables that influence the stresses are Y_n and $\cos \lambda_n X$, with Y_n being the prime variable to reduce the stresses to zero for a given value of X. A plot of the function $Y_n = e^y$, Figure 3-26, shows how the function varies with the conditions:

$$\text{as } y \to -\infty \qquad Y_n \to 0$$

$$\text{as } y \to +\infty \qquad Y_n \to \infty$$

The requirement of letting Y_n approach zero as y approaches infinity can only be satisfied if $e^{+\lambda y}$ terms are set to zero; therefore the constants A and B are zero, and only constants C and D need to be determined using the boundary conditions at $y = 0$ ($\sigma_y = [P(x)]/h$, $\tau = 0$). The ϕ equation can now be written as

$$\phi = \sum(C + \lambda_n yD)\, e^{-\lambda_n y} \cos \lambda_n x \qquad (3.57)$$

The stresses, using equations (3.9), (3.10), and (3.11), are therefore:

$$\sigma_y = \frac{\partial^2 \phi}{\partial x^2} = \sum -\lambda_n^2(C + \lambda_n yD)\, e^{-\lambda_n y} \cos \lambda_n X \qquad (3.58)$$

$$\sigma_x = \frac{\partial^2 \phi}{\partial y^2} = \sum \lambda_n^2 \left[C + D\left(y - \frac{2}{\lambda_n}\right)\right] e^{-\lambda_n y} \cos \lambda_n x \qquad (3.59)$$

$$\tau_{xy} = -\frac{\partial^2 \phi}{\partial x\, \partial y} = -\sum \lambda_n^2 \left[C + D\lambda_n\left(y - \frac{1}{\lambda_n}\right)\right] e^{-\lambda_n y} \sin \lambda_n x \qquad (3.60)$$

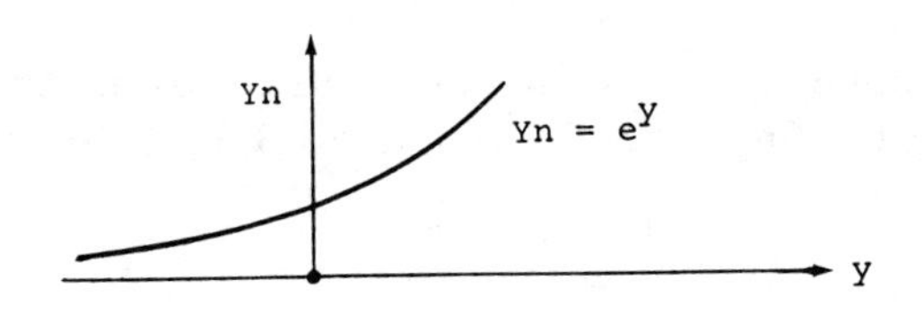

Figure 3-26. Variation of the Function $Y_n = e^y$.

Apply now the boundary conditions at $y = 0$:

$$\sigma_y = \frac{P(x)}{h}$$

$$\sigma_y = \sum \bar{A}_n \cos \frac{\lambda_n x}{h}$$

However, σ_y is given by equation (3.58), therefore considering an nth term:

$$-\lambda_n^2(C_n + \lambda_n y D_n)\, e^{-\lambda_n y} \cos \lambda_n x = \frac{\bar{A}_n}{h} \cos \lambda_n x$$

For $y = 0$ and any value of x, C_n is computed as

$$C_n = -\frac{\bar{A}_n}{h\lambda_n^2}$$

The other boundary condition at $y = 0$ is

$$\tau = 0$$

and from equation (3.60) this gives

$$\lambda_n^2\left[C_n + D_n\lambda_n\left(y - \frac{1}{\lambda_n}\right)\right] e^{-\lambda_n y} \sin \lambda_n x = 0$$

and when $y = 0$, for any value of x:

$$\lambda_n^2(C_n - D_n) = 0$$

$$C_n = D_n$$

The final ϕ equation, substituting in D_n and C_n, gives

$$\phi = \sum - \frac{\bar{A}_n}{h\lambda_n^2}(1 + \lambda_n y)e^{-\lambda_n y} \cos \lambda_n x \tag{3.57}$$

It should be noted that when the coefficients were evaluated the variable in x cancelled, which was part of the intuitive reason for assuming the ϕ variable to be similar to the loading series term.

The resulting equations can now be listed, using the ϕ equation (3.57) and the relationships given by equations (3.9), (3.10), and (3.11):

$$\sigma_y = \frac{1}{h}\sum \bar{A}_n (1 + \lambda_n y) e^{-\lambda_n y} \cos \lambda_n x \tag{3.58}$$

$$\sigma_x = \frac{1}{h}\sum \bar{A}_n(1 - \lambda_n y)e^{-\lambda_n y} \cos \lambda_n x \tag{3.59}$$

$$\tau_{xy} = -\frac{1}{h}\sum \bar{A}_n y e^{-\lambda_n y} \sin \lambda_n x \tag{3.60}$$

where $\lambda_n = \dfrac{n\pi}{a}$

$$\bar{A}_n = -\frac{2Pa}{\pi c}\frac{(-1)^n}{n} \sin \frac{n\pi c}{a}$$

The solution of this problem and subsequent tabulations have been given by Dischinger [9] and in Portland Cement Association publication [10]. Table 3-1 lists the resulting σ_x values at various distances y/a for ratios of c/a at the support and midspan of the plate. It should be noted that for a depth of plate $y = 1.5a$, the stresses are essentially zero; thus a plate of this depth would give the same results as an infinite plate solution. Also listed and shown in Figure 3-27 are the resultant forces T under the area of the stress block and the distance d between these resultants and locations relative to the edge of the plate d_0.

3.5 Infinitely Long Strip with Periodic Loading

Consider a plate with similar edge loading as given in the previous section, but with a finite depth of $2b$, as shown in Figure 3-28. This case involves finite edges, therefore equation (3.55), will be employed; thus

$$\phi = \sum Y_n \cos \lambda X$$

where $Y_n = E \sinh \lambda_n y + F \cosh \lambda_n y + Gy \sinh \lambda_n y + Hy \cosh \lambda_n y$. The boundary conditions required are:

$$\text{at } y = +b: \qquad \sigma_y = 0 \qquad \tau_{xy} = 0$$

$$\text{at } y = -b: \qquad \sigma_y = \frac{P(x)}{h} \qquad \tau_{xy} = 0$$

With these four boundary conditions, the four constants E, F, G, and H

Table 3-1
Semi-Infinite Plate Data

		Midspan ($x = 0$)			*Over Support* ($x = a$)			*Multip.*
	$\frac{y}{a}$	$\frac{c}{a}=\frac{1}{2}$	$\frac{c}{a}=\frac{1}{5}$	$\frac{c}{a}=\frac{1}{20}$	$\frac{c}{a}=\frac{1}{2}$	$\frac{c}{a}=\frac{1}{5}$	$\frac{c}{a}=\frac{1}{20}$	
	0	1.00	1.00	1.00	−1.00	−4.00	−19.00	
	0.25	0.17	0.27	0.29	−0.17	+0.15	+0.68	
$\frac{\sigma_x}{P/h}$	0.50	−0.14	−0.11	−0.11	0.14	0.40	0.51	$\frac{P}{h}$
	0.75	−0.16	−0.19	−0.20	0.16	0.30	0.33	
	1.00	−0.12	−0.16	−0.17	0.12	0.19	0.20	
	2.00	−0.01	−0.02	−0.02	0.01	0.02	0.02	
T		0.143	0.171	0.177	0.143	0.322	0.495	Pa
d_0		0.108	0.121	0.122	0.108	0.059	0.024	a
d		0.874	0.930	0.938	0.874	0.746	0.612	a

can be determined [9]. The results of this solution and subsequent determination of the primary bending stress σ_x have been plotted [10] for various aspect ratios of $\beta = b/a$ and $\varepsilon = c/a$ as given in Figures 3-29 and 3-30.

Figure 3-29 shows the distribution of stress σ_x at midspan for a constant $\varepsilon = 1/10$. However, the curves for $\varepsilon = 1/5$ and $\varepsilon = 1/20$ are nearly alike, thus $\varepsilon = 1/10$ may be used as a good average. The results of these plots show that for $\beta \leq 1/2$, linear theory is quite applicable for shallow beams as shown by the dotted linear line and the actual stress line for $\beta = 1/2$. When β increases from 1/2 to 1, however, then stress changes significantly, and linear theory cannot be applied.

Figure 3-30 shows the complete distribution of stress σ_x over the support for $\varepsilon = 1/5$ while for $\varepsilon = 1/10$ and $\varepsilon = 1/20$ only the tensile stress is plotted.

3.6 Deep Beam

The final problem to be considered is a deep beam, supported by two supports as shown in Figure 3-31. This plate does not have a periodic loading, as the length of the plate is finite; however, in order to apply the Fourier series solution, a simulated periodic condition will be created. Consider the plate to be periodic over period $2L = 4a$, as shown. The loading function will therefore be of the form

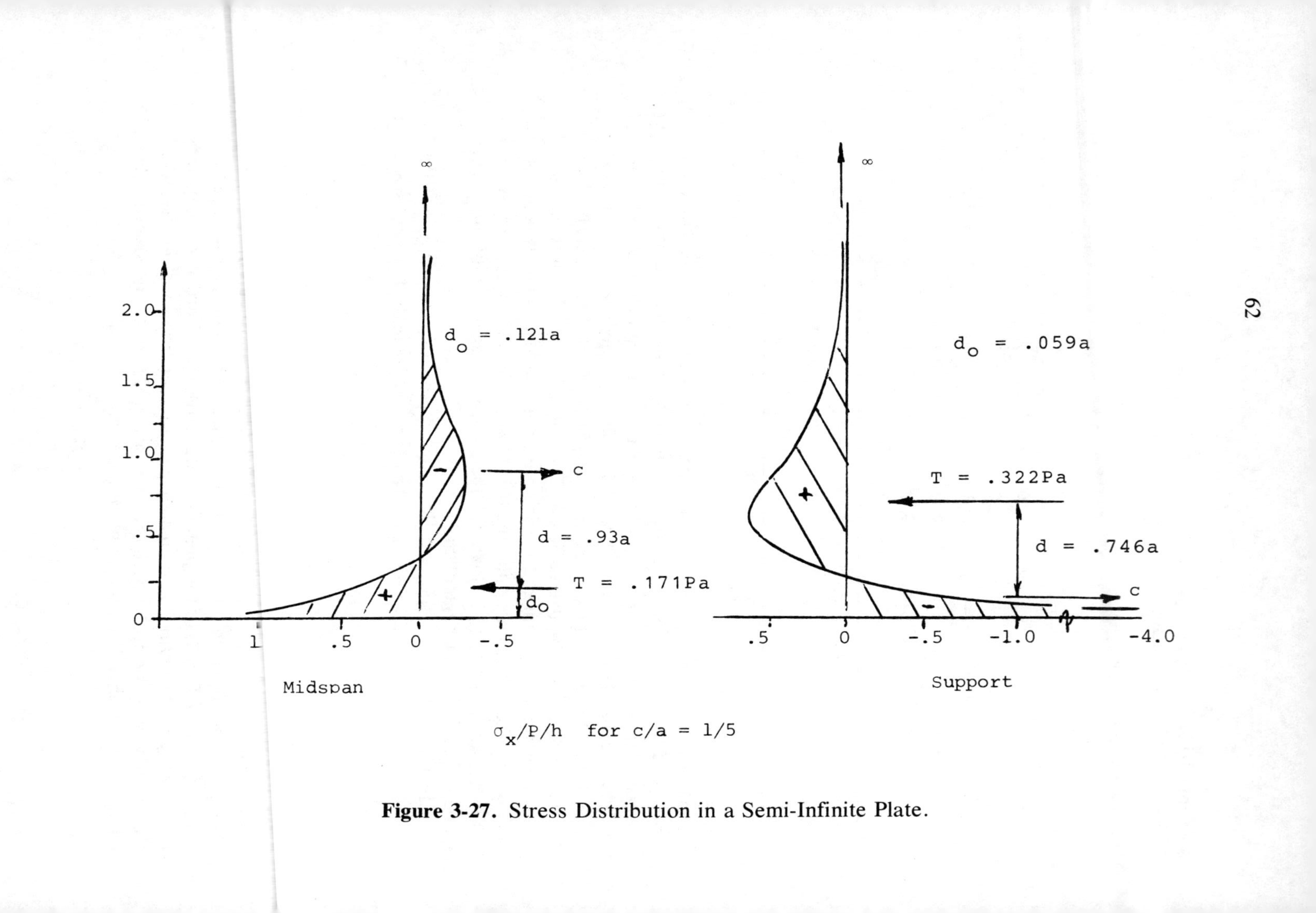

Figure 3-27. Stress Distribution in a Semi-Infinite Plate.

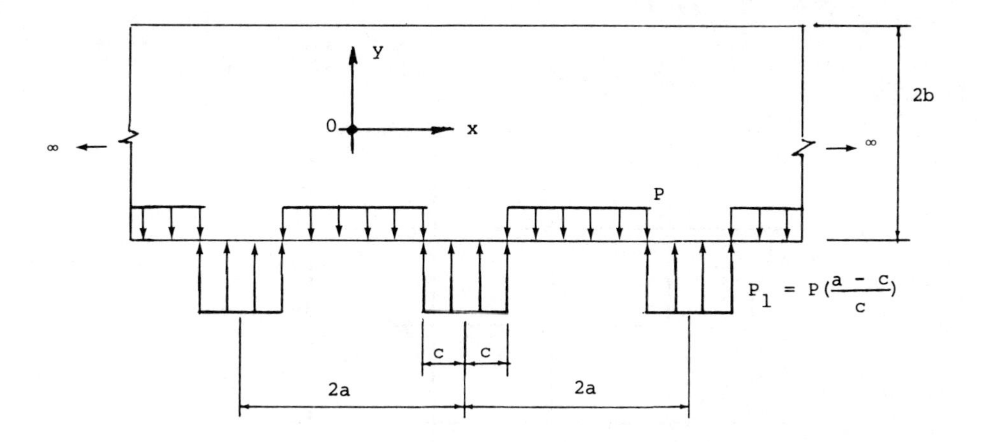

Figure 3-28. Infinitely Long Strip with Periodic Loading.

$$P(x) = K \cdot \cos \frac{n\pi x}{2a}$$

The boundary conditions for this plate are

$$\text{at } y = b: \qquad \sigma_y = \tau_{xy} = 0$$

$$\text{at } y = -b: \qquad \sigma_y = \frac{P(x)}{h}, \qquad \tau = 0$$

$$\text{at } x = \pm a: \qquad \sigma_x = \tau_{xy} = 0$$

and if the ϕ function is to agree with the load function, then

$$\phi = \sum Y_n \cos \frac{n\pi x}{2a}$$

The stresses, using this ϕ function are

$$\sigma_x = \frac{\partial^2 \phi}{\partial y^2} = \sum \frac{\partial^2 Y_n}{\partial y^2} \cos \frac{n\pi x}{2a}$$

$$\sigma_y = \frac{\partial^2 \phi}{\partial x^2} = -\sum Y_n \left(\frac{n\pi}{2a}\right)^2 \cos \frac{n\pi x}{2a}$$

$$\tau_{xy} = -\frac{\partial^2 \phi}{\partial x \partial y} = \sum \frac{\partial Y_n}{\partial y} \left(\frac{n\pi}{2a}\right) \sin \frac{n\pi x}{2a}$$

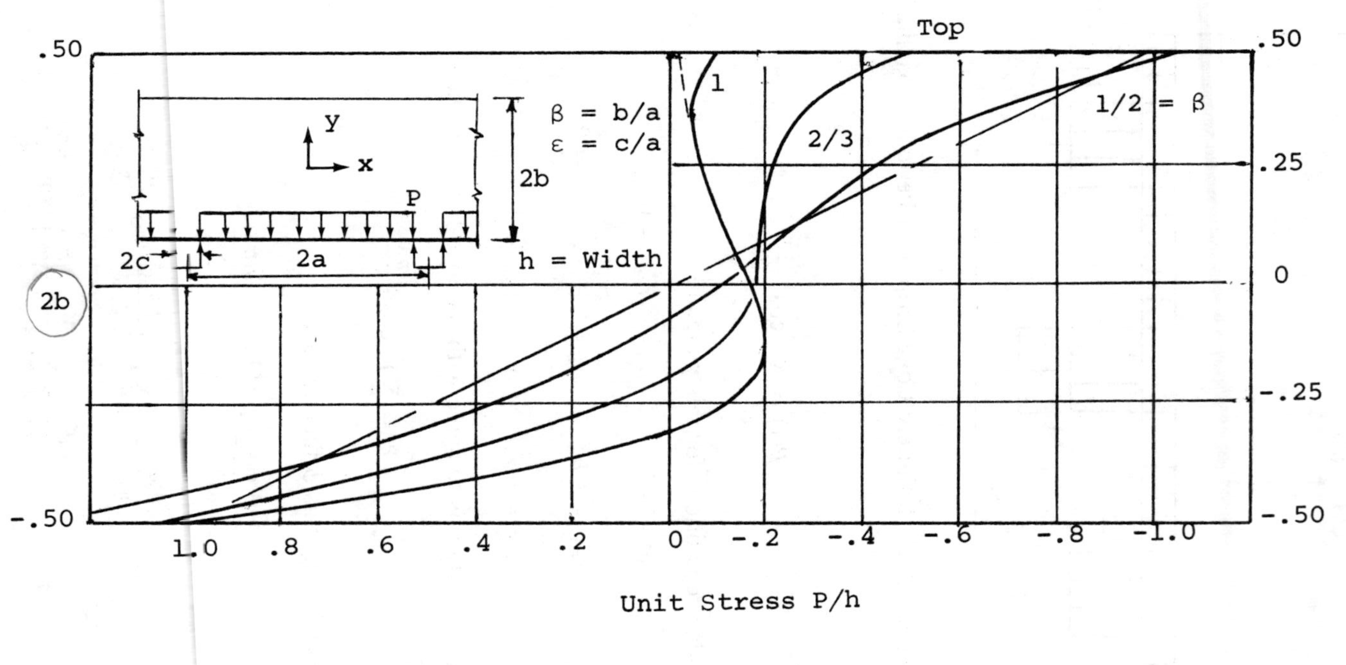

Figure 3-29. Variation of Stress σ_x at Midspan ($x = 0$).

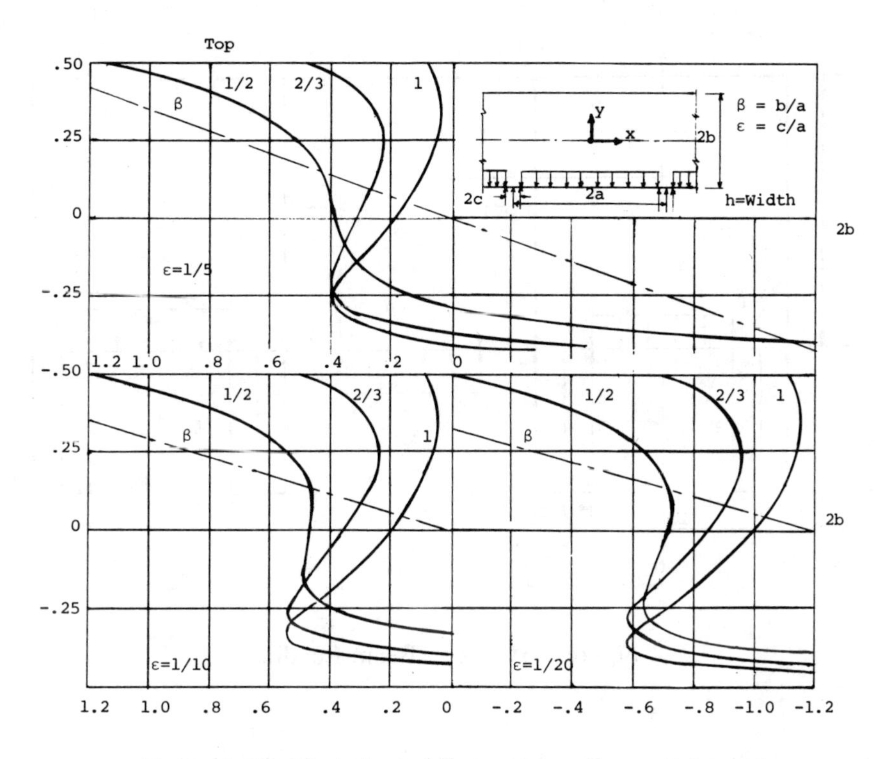

Figure 3-30. Variation of Stress σ_x at Support ($x = a$).

Checking the boundary conditions gives

$$\text{at } x = \pm\, a: \qquad \sigma_x = 0$$

$$\sigma_x = \sum \frac{\partial^2 Y_n}{\partial y^2} \cos \frac{n\pi a}{2a}$$

or

$$\sigma_x = \sum Y_n'' \cos \frac{n\pi}{2} = 0 \qquad n = 1, 3, 5, \ldots$$

therefore this boundary condition is satisfied. The other boundary condition is

$$\text{at } x = \pm\, a: \qquad \tau_{xy} = 0$$

However,

$$\text{at } x = \pm a: \tau_{xy} = \sum \frac{\partial Y_n}{\partial y} \left(\frac{n\pi}{2a} \right) \sin \frac{n\pi}{2} \neq 0$$

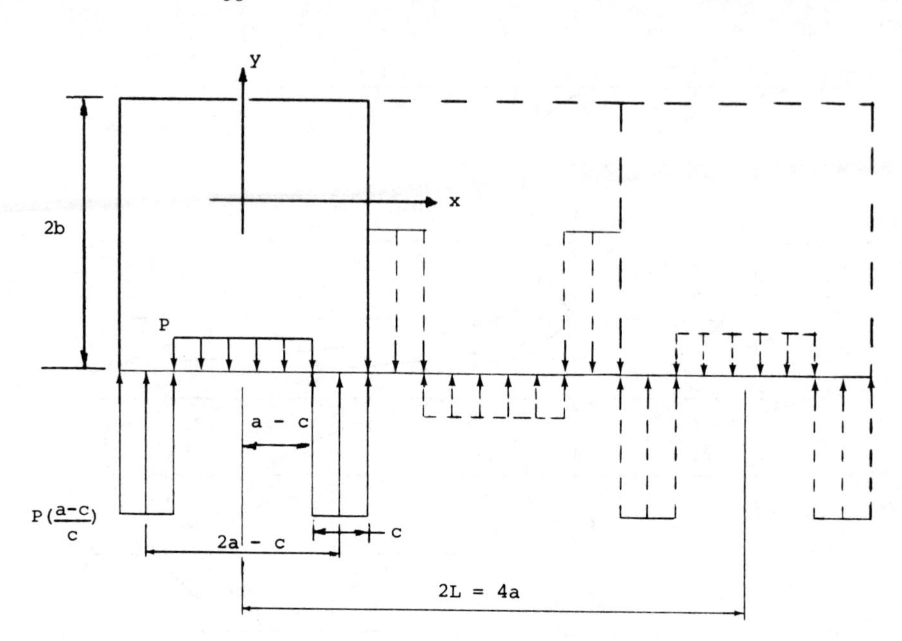

Figure 3-31. Deep Beam Loading.

therefore the shear along the boundary is not met, although

$$\int_{-b}^{+b} \tau_{xy}\, dy = 0$$

along the boundary is satisfied. Thus St. Venant's principle applies and the results will be good at a distance away from the boundary.

The other four boundary conditions to be considered are at $y = \pm b$. These conditions are met by solving for the coefficients (E, F, G, H) in the Y_n equation (3.55).

The evaluation of the load $P(x)$ function is as follows, referring to Figure 3-31:

$$A_n = \frac{2}{L}\int_0^L P(x) \cos \frac{n\pi}{L} x\, dx$$

$$A_n = \frac{2}{2a} P\left[\int_0^{a-c} \cos \lambda x\, dx - \left(\frac{a-c}{c}\right)\int_{a-c}^{a} \cos \lambda x\, dx + \left(\frac{a-c}{c}\right)\int_a^{a+c} \cos \lambda x\, dx - \int_{a+c}^{2a} \cos \lambda x\, dx\right]$$

Integrating and collecting terms gives

$$A_n = \frac{P}{\lambda a}\left[-\frac{a}{c}(\sin \lambda a - \sin \lambda(a - c)) + 2 \sin \lambda a \right.$$

$$\left. + \frac{a}{c}(\sin \lambda(a + c) - \sin \lambda a) - \sin \lambda 2a \right]$$

using the trigometric identities,

$$\sin(\alpha + \beta) = \sin \alpha \cos \beta + \cos \alpha \sin \beta$$

$$\sin(\alpha - \beta) = \sin \alpha \cos \beta - \cos \alpha \sin \beta$$

The A_n term can be reduced to the following:

$$A_n = \frac{-4Pa}{\pi c}\left(\frac{1}{n}\right)(-1)^{m+1/2}\left[\cos \frac{n\pi c}{2a} - \left(\frac{a-c}{a}\right)\right] \qquad n = 1, 3, 5, \ldots$$

$$A_0 = 0$$

The resulting $P(x)$ load is therefore

$$P(x) = -\frac{4Pa}{\pi c}\sum \frac{1}{n}(-1)^{(n+1)/2}\left[\cos \frac{n\pi c}{2a} - \left(\frac{a-c}{a}\right)\right] \cos \frac{n\pi x}{2a}$$

$$n = 1, 3, 5, \ldots$$

This function is used in conjunction with the boundary condition at $y = -b$, $\sigma_y = P(x)/h$.

The total solution to this problem is given in references [11, 12]. A modification to the problem, in which the support distance $2c = 0$, has also been evaluated [10] giving the data shown in Figure 3-32.

3.7 Multi-Material Beam

Many engineering problems deal with structural elements that are constructed of various materials, such as composite concrete-steel girders. The evaluation of the stresses throughout such a beam can be determined by using plain stress theory providing the beam has negligable lateral shear distortion.

Consider now a three material multi-material beam, shown in Figure 3-33. The solution to this problem [13] will demonstrate the procedure required, but by no means is the technique limited to three materials. The

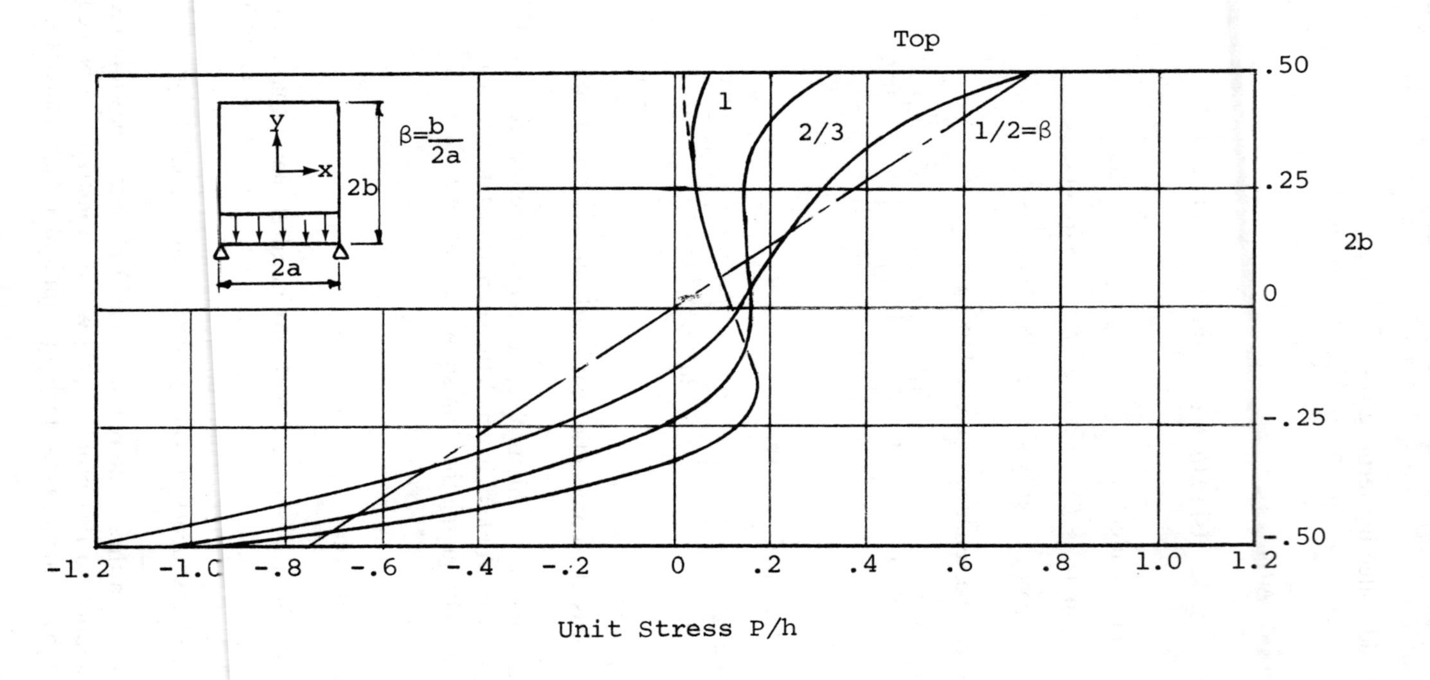

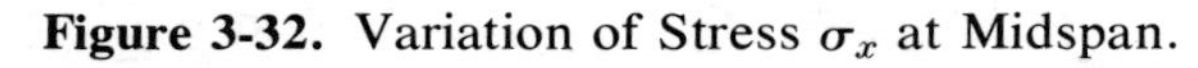
Figure 3-32. Variation of Stress σ_x at Midspan.

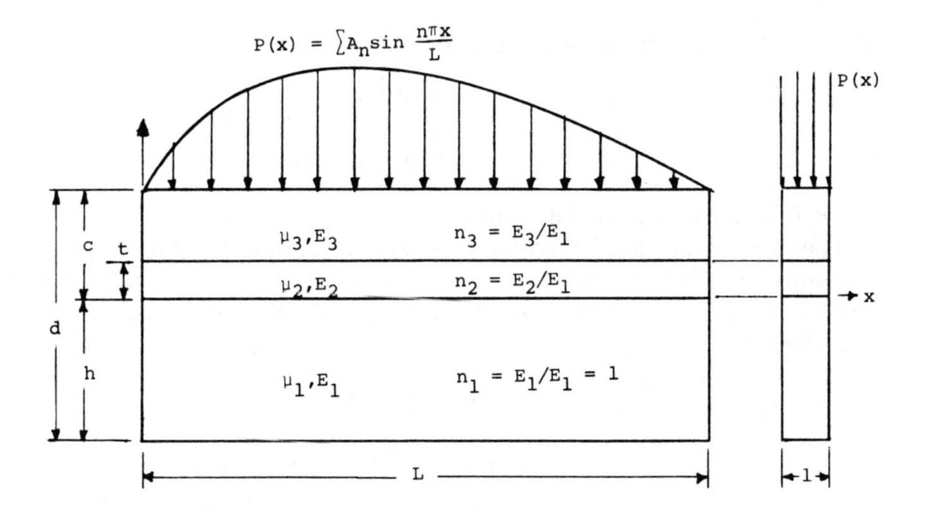

Figure 3-33. Multi-Material Beam Geometry and Loading.

beam shown has referenced properties μ and E, for each of the three materials, and the ϕ function will be assumed to be of the form:

$$\phi = \sum Y_n \sin \frac{n\pi x}{L}$$

where Y_n is given by equation (3.55). Therefore:

$$\phi = \sum(E \sinh \lambda_n y + F \cosh \lambda_n y + Gy \sinh \lambda_n y + Hy \cosh \lambda_n y) \sin \lambda_n x \tag{3.61}$$

where $\lambda_n = n\pi/L$.

The assumption of the ϕ function is similar to that of the previous plate problems, with the exception that $\sin \lambda x$ is used. This function was selected because the function satisfies identically the boundary condition of zero moment and axial forces at the end support of a simple beam. In general these conditions can be stated as follows:

$$\int_A \sigma_x (0, y)\, dy = 0$$

$$\int_A \sigma_x(L, y)\, dy = 0$$

$$\int_A \sigma_x(0, y) y\, dy = 0$$

$$\int_A \sigma_x(L, y) y\, dy = 0$$

Also stipulated are the support reactions, or

$$\int_A \tau_{xy}(0, y)\,dy = \text{Reaction}$$

$$\int_A \tau_{xy}(L, y)\,dy = \text{Reaction}$$

where $\int_A$ = cross sectional area of beams.

The stresses can now be written in general form, using the Airy's stress equations (3.9), (3.10), and (3.11), and the ϕ equation (3.61) with the coefficients E, F, G, and H substituted for by C_2, C_1, C_4 and C_3 respectively, as follows:

$$\sigma_x(x, y) = \frac{\partial^2\phi}{\partial y^2} = \sin\lambda x[C_1\lambda^2\cosh\lambda y + C_2\lambda^2\sinh\lambda y + C_3\lambda(2\sinh\lambda y + \lambda y\cosh\lambda y) + C_4\lambda(2\cosh\lambda y + \lambda y\sinh\lambda y)] \quad (3.62a)$$

$$\sigma_y(x, y) = \frac{\partial^2\phi}{\partial x^2} = -\lambda^2\sin\lambda x[C_1\cosh\lambda y + C_2\sinh\lambda y + C_3 y\cosh\lambda y + C_4 y\sinh\lambda y] \quad (3.62b)$$

$$\tau_{xy}(x, y) = -\frac{\partial^2\phi}{\partial x\,\partial y} = -\lambda\cos\lambda x[C_1\lambda\sinh\lambda y + C_2\lambda\cosh\lambda y + C_3(\cosh\lambda y + \lambda y\sinh\lambda y) + C_4(\sinh\lambda y + \lambda y\cosh\lambda y)] \quad (3.62c)$$

Because the composite beam shown in Figure 3-33 has three different materials, there are therefore three different sets of stress functions having the same form as equation (3.61) but having different constants C_1, C_2, C_3, and C_4. To determine these constants the bending (σ_x, σ_y) and shearing (τ_{xy}) stresses and displacements (u, v) must be equated at the interfaces of the adjoining material layers. The horizontal displacement u and the vertical displacement v, therefore, must be considered in the boundary conditions, and thus have to be determined.

The relationships between strains, displacements and stresses as given by equations (2.29) through (3.21) are:

$$\varepsilon_x = \frac{\partial u}{\partial x} = \frac{1}{E}(\sigma_x - \mu\sigma_y) \quad (3.63a)$$

$$\varepsilon_y = \frac{\partial v}{\partial y} = \frac{1}{E}(\sigma_y - \mu\sigma_x) \quad (3.63b)$$

$$\tau_{xy} = \frac{\partial u}{\partial y} + \frac{\partial v}{\partial x} = \frac{2(1+\mu)}{E}\tau_{xy} \quad (3.63c)$$

where E is the modulus of elasticity and μ is Poisson's ratio. Now, substituting equations (3.62) into the strain equations (3.63) and integrating (3.63a) and (3.63b), the displacements u and v are determined. However, the integrations of equations (4.63a) and (4.63b) yield two arbitrary constants which are found by substituting the displacement expressions u and v into equation (4.63c). As a result the following expressions for u and v are obtained:

$$\begin{aligned} u = & -\frac{\cos \lambda x}{E}\{C_1\lambda(1 + \mu) \cosh \lambda y + C_2\lambda(1 + \mu) \sinh \lambda y \\ & + C_3[2 \sinh \lambda y + \lambda y(1 + \mu) \cosh \lambda y] \\ & + C_4[2 \cosh \lambda y + \lambda y(1 + \mu) \sinh \lambda y]\} \end{aligned} \tag{3.64}$$

$$\begin{aligned} v = & -\frac{\sin \lambda x}{E}\{C_1\lambda(1 + \mu) \sinh \lambda y + C_2\lambda(1 + \mu) \cosh \lambda y \\ & + C_3[\lambda y(1 + \mu) \sinh \lambda y + (-1 + \mu) \cosh \lambda y] \\ & + C_4[\lambda y(1 + \mu) \cosh \lambda y + (-1 + \mu) \sinh \lambda y]\} \end{aligned} \tag{3.65}$$

where the constants of integration associated with rigid body displacements are set equal to zero.

The remaining portion of the problem involves evaluation of the three sets of constants C_1, C_2, C_3, and C_4 from the boundary conditions. As mentioned previously there is a stress function for each material and, from equation (3.61), four constants for each stress function; therefore, there are twelve independent constants for which there must be twelve independent boundary conditions in order that the solution be obtained. The three stress functions will be designated as ϕ_1, ϕ_2, and ϕ_3, where the subscripts correspond to the material layers with the same number as shown in Figure 3-33. The constants associated with each layer will be designated as C_1^1, C_2^1, etc., corresponding to ϕ_1; C_1^2, C_2^2, etc., corresponding to ϕ_2 etc.

The boundary conditions required are listed as follows, where the superscripts denote the corresponding material layers as indicated in Figure 3-33.

$$\tau_{xy}^1(x, -h) = 0 \tag{3.66a}$$

$$\sigma_y^1(x, -h) = 0 \tag{3.66b}$$

$$\tau_{xy}^1(x, 0) = \tau_{xy}^2(x, 0) \tag{3.66c}$$

$$\sigma_y^1(x, 0) = \sigma_y^2(x, 0) \tag{3.66d}$$

$$u^1(x, 0) = u^2(x, 0) \tag{3.66e}$$

$$v^1(x, 0) = v^2(x, 0) \tag{3.66f}$$

$$\tau_{xy}^2(x, t) = \tau_{xy}^3(x, t) \tag{3.66g}$$

$$\sigma_y^2(x, t) = \sigma_y^3(x, t) \tag{3.66h}$$

$$u^2(x, t) = u^3(x, t) \tag{3.66i}$$

$$v^2(x, t) = v^3(x, t) \tag{3.66j}$$

$$\tau_{xy}^3(x, c) = 0 \tag{3.66k}$$

$$\sigma_y^3(x, c) = \sum A_n \sin \frac{n\pi x}{L} \tag{3.66l}$$

Since $C_1^2 = C_1$
boundary conditions reduce to eleven equations in the eleven coefficients C_i^j which are summarized as follows [13]:

Boundary Condition Equations

Equation Number	*Equation*	*Derived From*
1	$-(\sinh \lambda h + \lambda h \cosh \lambda h)C_4^1 + (\cosh \lambda h + \lambda h \sinh \lambda h)C_3^1 + \lambda \cosh \lambda h C_2^1 - \lambda \sinh \lambda h C_1^1 = 0$	(3.62c)(3.66a)
2	$h \sinh \lambda h C_4^1 - h\cosh \lambda h C_3^1 - \sinh \lambda h C_2^1 + \cosh \lambda h C_1^1 = 0$	(3.62b)(3.66b)
3	$2C_4^1 - \frac{2n_1}{n_2} C_4^2 + \lambda(1 + \mu_1)C_1^1 - \frac{n_1}{n_2}\lambda(1 + \mu_2)C_1^2 = 0$	(3.64)(3.66e)
4	$C_3^1 + \lambda C_2^1 - C_3^2 - \lambda C_2^2 = 0$	(3.62c)(3.66c)
5	$(-1 + \mu_1)C_3^1 + \lambda(1 + \mu_1)C_2^1 - \frac{n_1}{n_2}(-1 + \mu_2) C_3^2 - \frac{n_1}{n_2}\lambda(1 + \mu_2)C_2^2 = 0$	(3.65)(3.66f)
6	$(\sinh \lambda t + \lambda t \cosh \lambda t)C_4^2 + (\cosh \lambda t + \lambda t \sinh \lambda t)C_3^2 + \lambda \cosh \lambda t C_2^2 - \lambda \sinh \lambda t C_1^3 - \lambda \cosh \lambda t C_2^3 - (\cosh \lambda t + \lambda t \sinh \lambda t)C_3^3 - (\sinh \lambda t + \lambda t \cosh \lambda t)C_4^3 + \lambda \sinh \lambda t C_1^2 = 0$	(3.62c)(3.66g)
7	$t \sinh \lambda t C_4^2 + t \cosh \lambda t C_3^2 + \sinh \lambda t C_2^2 - \cosh \lambda t C_1^3 - \sinh \lambda t C_2^3 - t \cosh \lambda t C_3^3 - t \sinh \lambda t C_4^3 + \cosh \lambda t C_1^2 = 0$	(3.62b)(3.66h)

Equation Number	*Equation*	*Derived From*
8	$\frac{n_1}{n_3}[2 \cosh \lambda t + \lambda t(1 + \mu_2) \sinh \lambda t]C_4^2 + \frac{n_1}{n_3}[2 \sinh \lambda t + (1 + \mu_2) \cosh \lambda t]C_3^2 + \frac{n_1}{n_3}\lambda(1 + \mu_2) \sinh \lambda t C_2^2 - \lambda(1 + \mu_3) \cosh \lambda t C_1^3 - \lambda(1 + \mu_3) \sinh \lambda t C_2^3 - [2 \sinh \lambda t + \lambda t(1 + \mu_3) \cosh \lambda t]C_3^3 - [2 \cosh \lambda t + \lambda t(1 + \mu_3) \sinh \lambda t)C_4^3 + \frac{n_1}{n_3}\lambda(1 + \mu_2) \cosh \lambda t C_1^2 = 0$	(3.64)(3.66i)
9	$-\frac{n_1}{n_3}[\lambda t(1 + \mu_2) \cosh \lambda t + (-1 + \mu_2) \sinh \lambda t]C_4^2 - \frac{n_1}{n_3}[\lambda t(1 + \mu_2) \sinh \lambda t + (-1 + \mu_2) \cosh \lambda t]C_3^2 - \frac{n_1}{n_3}\lambda(1 + \mu_2) \cosh \lambda t C_2^2 - \frac{n_1}{n_3}\lambda(1 + \mu_2) \sinh \lambda t C_1^2 + \lambda(1 + \mu_3) \cosh \lambda t C_2^3 + [\lambda t(1 + \mu_3) \sinh \lambda t + (-1 + \mu_3) \cosh \lambda t]C_3^3 + [\lambda t(1 + \mu_3) \cosh \lambda t + (-1 + \mu_3) \sinh \lambda t]C_4^3 + \lambda(1 + \mu_2) \sinh \lambda t C_1^3 = 0$	(3.65)(3.66j)
10	$\lambda \sinh \lambda c C_1^3 + \lambda \cosh \lambda c C_2^3 + (\cosh \lambda c + \lambda c \sinh \lambda c)C_3^3 + (\sinh \lambda c + \lambda c \cosh \lambda c)C_4^3 = 0$	(3.62c)(3.66k)
11	$\cosh \lambda c C_1^3 + \sinh \lambda c C_2^3 + c \cosh \lambda c C_3^3 + c \sinh \lambda c C_4^3 = A_n$	(3.62b)(3.66l)

These eleven equations can now readily be solved for the eleven constants noting $C_1^1 = C_1^2$ using a computer program. Upon solution of these eleven constants, the stresses, strains, and displacements can be determined from equations (3.62) and equations (3.63). Results for epoxy bonded composite T-beams using these equations have been presented by Keiegh and Richard [13].

3.8 Finite Difference Solution of Plane Stress Problems

3.8.1 Equations

In the past several sections, the solution to plane stress problems were obtained by using the Fourier series technique. As an alternate method to solving such problems a numerical procedure called "finite differences" [14, 15] can be employed. This procedure generally results in the generation of a series of numerous simultaneous linear equations, and thus in the need for a digital computer.

As will be recalled, in all plane stress problems two major types of equations need to be solved:

1. Boundary equations: (3.14), (3.15)
2. The biharmonic equation: (3.13)

An expedient technique for determining the boundary equations, i.e., the ϕ function along the boundary will be the frame analogy. The solution of the biharmonic $\nabla^4\phi = 0$, at all points on the plate will be obtained by first assuming designated points on the plate with a corresponding ϕ surface, as shown in Figure 3-34. If the ϕs are known at these points, then the slopes at point 0 can be written as

$$\tan \Theta_0 = \frac{\phi_r - \phi_l}{2\Delta x}$$

$$\tan \Theta_0 = \frac{\phi_a - \phi_b}{2\Delta y}$$

where the spacing between these points is arbitrarily assumed equal to Δx and Δy referenced to the x- and y-axis respectively. These slope equations represent differentials at the central reference point 0 or

$$\left.\frac{\Delta\phi}{\Delta x}\right|_0 = \frac{\phi_r - \phi_l}{2\Delta x} \qquad \text{(a)}$$

$$\left.\frac{\Delta\phi}{\Delta y}\right|_0 = \frac{\phi_a - \phi_b}{2\Delta y} \qquad \text{(b)}$$

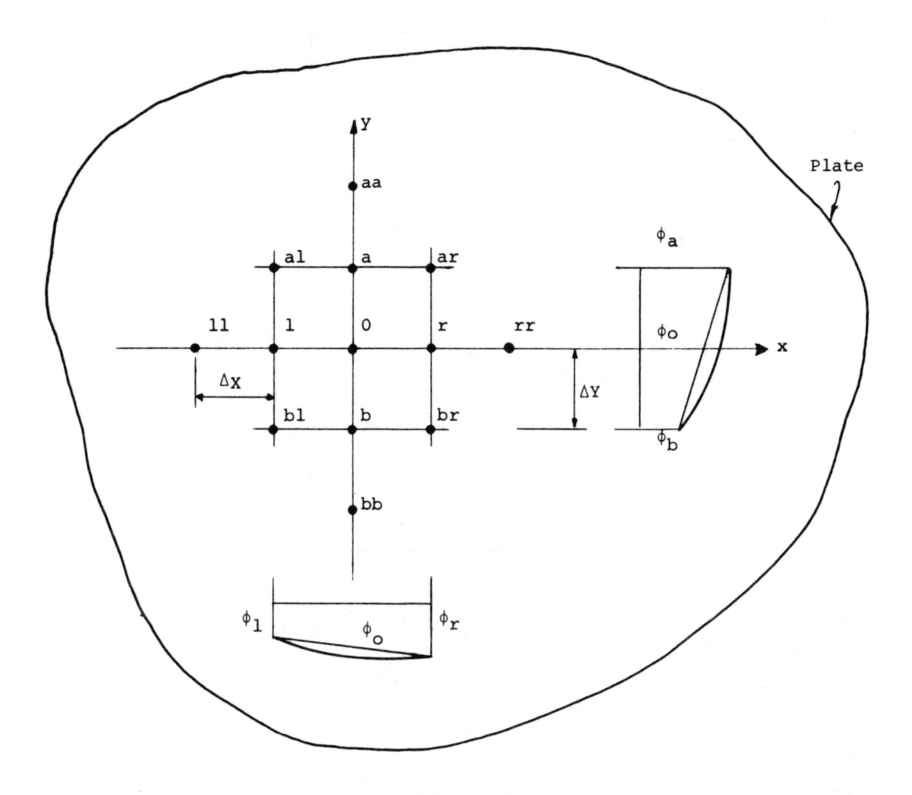

Figure 3-34. General Plate Mesh Pattern.

Equations (a) and (b) are finite difference equations representing differentials at a prescribed point 0. Higher order differentials can be found by using the relationship

$$\frac{\Delta}{\Delta x}\left(\frac{\Delta\phi}{\Delta x}\right)\Bigg|_0 = \text{rate of change in slope}$$

where the change in slope is now required. This slope can be determined by considering the general ϕ surface shown in Figure 3-35(a), which has prescribed points at spacing Δx to the left and right of ϕ_0. Assume now that the ϕs at $\Delta x/2$ are prescribed, as shown in Figure 3-55(b). The change in slope is therefore

$$\Delta\Theta = \Theta_l^- = \frac{\phi_0 - \phi_l}{\Delta x} \tag{c}$$

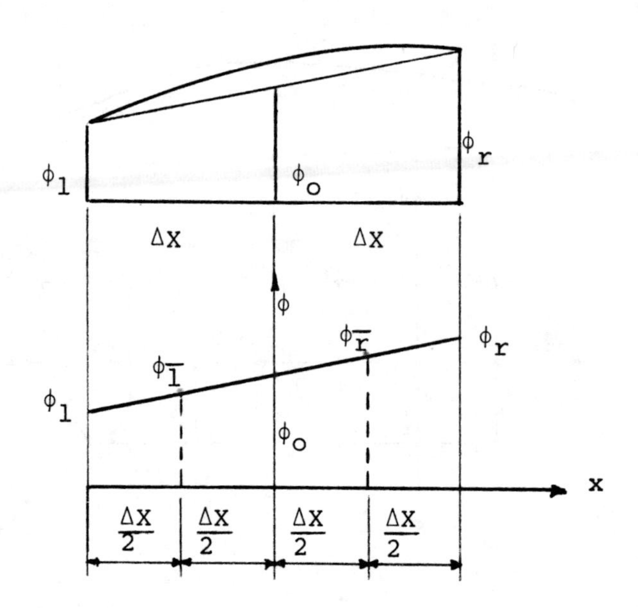

Figure 3-35. ϕ Surface Variation.

or

$$\Delta\Theta = \Theta_{\bar{r}} = \frac{\phi_r - \phi_0}{\Delta x} \tag{d}$$

However, the general slope equation is

$$\Theta_0 = \left.\frac{\Delta\phi}{\Delta x}\right|_0 = \frac{\phi_r - \phi_l}{2\Delta x} \tag{a}$$

therefore;

$$\left.\frac{\Delta}{\Delta x}(\Delta\Theta)\right|_0 = \frac{\Theta_{\bar{r}} - \Theta_{\bar{l}}}{2\left(\frac{\Delta x}{2}\right)}$$

Substituting in equations (c) and (d), gives

$$\left.\frac{\Delta}{\Delta x}\left(\frac{\Delta\phi}{\Delta x}\right)\right|_0 = \frac{\phi_r - \phi_0 - \phi_0 + \phi_l}{\Delta x^2}$$

or

$$\left.\frac{\Delta^2\phi}{\Delta x^2}\right|_0 = \frac{\phi_r - 2\phi_0 + \phi_l}{\Delta x^2} \tag{e}$$

Similarly,

$$\left.\frac{\Delta^2\phi}{\Delta_y{}^2}\right|_0 = \frac{\phi_a - 2\phi_0 + \phi_b}{\Delta y^2} \tag{f}$$

Following a similar procedure, the fourth order differential is found as

$$\left.\frac{\Delta^4\phi}{\Delta x^4}\right|_0 = \frac{\Delta^2}{\Delta x^2}\left(\frac{\Delta^2\phi}{\Delta x^2}\right)$$

where $\Delta^2\phi/\Delta x^2$ is given by equation (c), therefore,

$$\left.\frac{\Delta^4\phi}{\Delta x^4}\right|_0 = \frac{\Delta^2}{\Delta x^2}\left[\frac{\phi_r - \phi_0 + \phi_l}{\Delta x^2}\right]$$

Expanding this equation gives

$$\left.\frac{\Delta^4\phi}{\Delta x^4}\right|_0 = \left[\left.\frac{\Delta^2\phi}{\Delta x^2}\right|_r - 2\left.\frac{\Delta^2\phi}{\Delta x^2}\right|_0 + \left.\frac{\Delta^2\phi}{\Delta x^2}\right|_l\right]\frac{1}{\Delta x^2}$$

The $\Delta^2\phi/\Delta x^2$ terms with respect to each of the central terms r, 0, and l can now be written in difference terms by using the general equation (e), expanding about these referenced points relative to the surrounding points, as given in Figure 3-34; therefore,

$$\left.\frac{\Delta^4\phi}{\Delta x^4}\right|_0 =$$

$$\left[\frac{\phi_{rr} - 2\phi_r + \phi_0}{\Delta x^2} - 2\frac{\phi_r - 2\phi_0 + \phi_l}{\Delta x^2} + \frac{\phi_0 - 2\phi_l + \phi_{ll}}{\Delta x^2}\right]\frac{1}{\Delta x^2}$$

Expanding and collecting terms gives

$$\left.\frac{\Delta^4\phi}{\Delta x^4}\right|_0 = \frac{[\phi_{rr} - 4\phi_r + 6\phi_0 - 4\phi_l + \phi_{ll}]}{\Delta x^4} \tag{g}$$

Similarly the differential $\Delta\phi^4/\Delta y^4$, referring the points on the plate shown in Figure 3-34 gives

$$\left.\frac{\Delta^4\phi}{\Delta y^4}\right|_0 = \frac{[\phi_{aa} - 4\phi_a + 6\phi_0 - 4\phi_b + \phi_{bb}]}{\Delta y^4} \tag{h}$$

The determination of the fourth order differentials $\Delta^4\phi/\Delta x^4$ and $\Delta^4\phi/\Delta y^4$ are required as they are two terms in the biharmonic equation. The other term that is required is the mixed differential, i.e., $\Delta^4\phi/\Delta x^2\Delta y^2$. This term is found by using the relationship:

$$\left.\frac{\Delta^4\phi}{\Delta x^2\Delta y^2}\right|_0 = \frac{\Delta^2}{\Delta y^2}\left|\frac{\Delta^2\phi}{\Delta x^2}\right|_0$$

The term $\Delta^2\phi/\Delta x^2$ is given by equation (e), therefore,

$$\left.\frac{\Delta^4\phi}{\Delta x^2\,\Delta y^2}\right|_0 = \frac{\Delta^2}{\Delta y^2}\left[\frac{\phi_r - 2\phi_0 + \phi_l}{\Delta x^2}\right]$$

Expanding this equation about points r, 0, and l gives

$$\left.\frac{\Delta^4\phi}{\Delta x^2\,\Delta y^2}\right|_0 = \frac{\left.\dfrac{\Delta^2\phi}{\Delta y^2}\right|_r - 2\left.\dfrac{\Delta^2\phi}{\Delta y^2}\right|_0 + \left.\dfrac{\Delta^2\phi}{\Delta y^2}\right|_l}{\Delta x^2}$$

The terms $\Delta^2\phi/\Delta y^2$ are given in general by equation (f), referenced about point 0. Therefore expanding back $\Delta^2\phi/\Delta y^2$ about the reference points r, 0, and ℓ including the surrounding points as shown in Figure 3-34 gives

$$\left.\frac{\Delta^4\phi}{\Delta x^2\,\Delta y^2}\right|_0 = \frac{\dfrac{\phi_{ar} - 2\phi_r + \phi_{br}}{\Delta y^2} - 2\dfrac{\phi_a - 2\phi_0 + \phi_b}{\Delta y^2} + \dfrac{\phi_{al} - 2\phi_l + \phi_{bl}}{\Delta y^2}}{\Delta x^2}$$

Expanding this expression and combining terms gives

$$\left.\frac{\Delta^4\phi}{\Delta x^2\,\Delta y^2}\right|_0 = \frac{4\phi_0 - 2(\phi_l + \phi_r + \phi_a + \phi_b) + (\phi_{al} + \phi_{bl} + \phi_{ar} + \phi_{br})}{\Delta x^2\,\Delta y^2} \tag{i}$$

With equations (g), (h), and (i), the biharmonic can now be expressed in difference form:

$$\nabla^4\phi = \frac{\partial^4\phi}{\partial x^4} + 2\frac{\partial^4\phi}{\partial x^2\,\partial y^2} + \frac{\partial^4\phi}{\partial y^4} = 0$$

Let $\Delta x/\Delta y = n$ or $n\,\Delta y = \Delta x$, substituting in equations (g), (h), and (i) gives:

biharmonic

$$\begin{aligned}\phi_0[6 + 8n^2 + 6n^4] &+ (\phi_l + \phi_r)[-4(1 + n^2)] \\ &+ (\phi_{al} + \phi_{bl} + \phi_{ar} + \phi_{br})[2n^2] + (\phi_{aa} + \phi_{bb})[n^4] \\ &+ (\phi_{ll} + \phi_{rr})[1] + (\phi_a + \phi_b)[-4n^2(1 + n^2)] = 0\end{aligned} \tag{3.67}$$

This equation can also be written in a more convenient form, designated as a mesh pattern, as shown in Figure 3-36.

The stresses can similarly be written in difference form, using equations (3.9), (3.10), and (3.11) and the appropriate difference expressions as·

$$\sqrt{}\ \sigma_x = \left.\frac{\partial^2\phi}{\partial y^2}\right|_0 = \frac{\phi_a - 2\phi_0 + \phi_b}{\Delta y^2} \tag{3.68a}$$

$$\sqrt{}\ \sigma_y = \left.\frac{\partial^2\phi}{\partial x^2}\right|_0 = \frac{\phi_r - 2\phi_0 + \phi_\ell}{n^2\,\Delta y^2} \tag{3.68b}$$

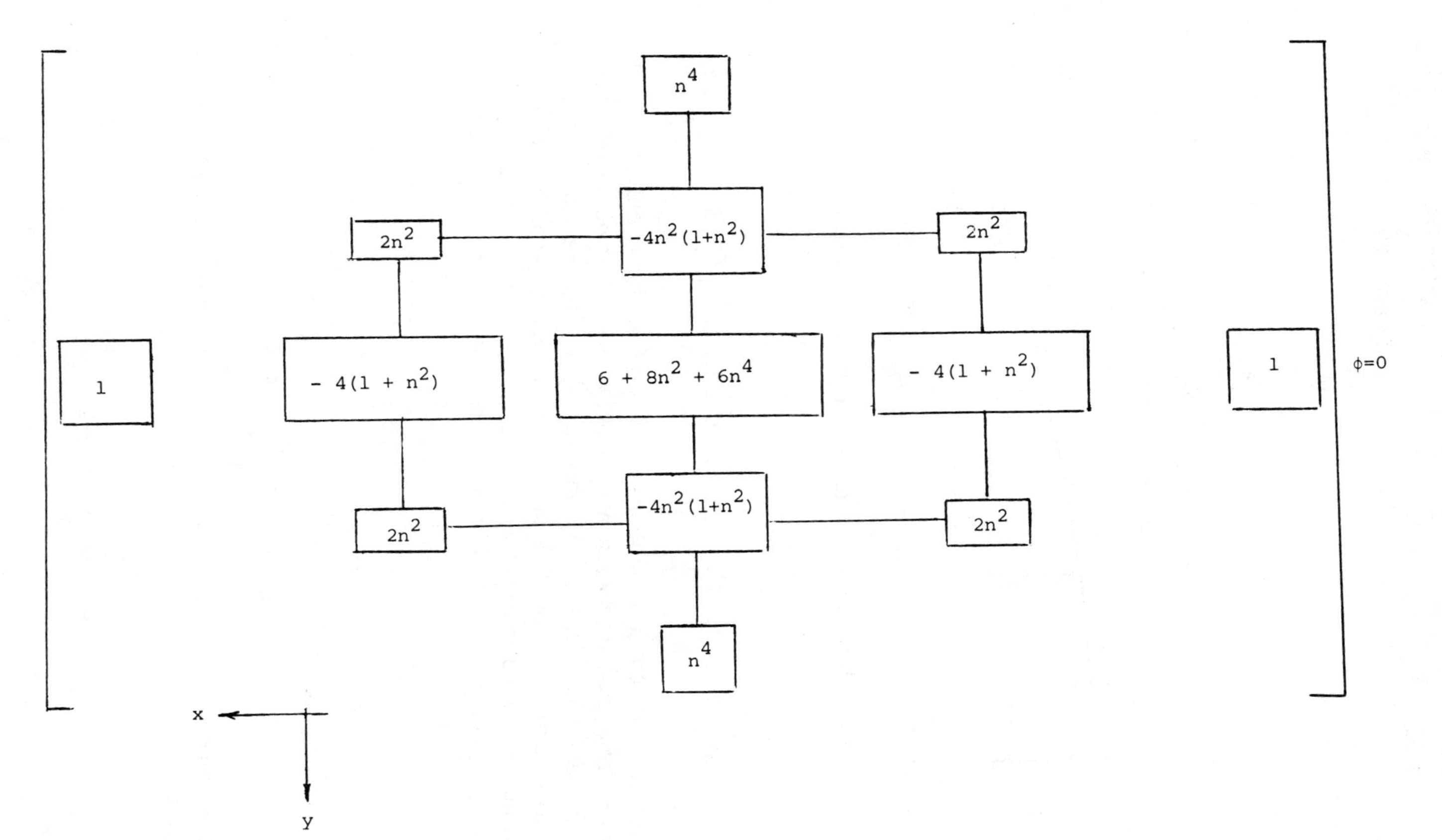

Figure 3-36. Biharmonic Mesh Pattern.

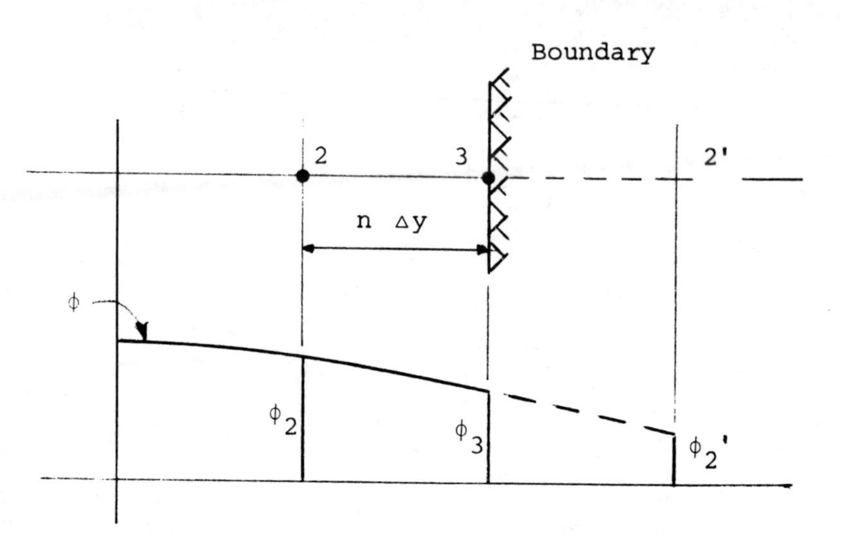

Figure 3-37. Finite Difference Boundary Condition.

$$\tau_{xy} = -\left.\frac{\partial^2\phi}{\partial x\,\partial y}\right|_0 = -\frac{\phi_{ar} - \phi_{br} - \phi_{a\ell} + \phi_{b\ell}}{4\,\Delta x\,\Delta y} \tag{3.68c}$$

In many problems the solution of $\nabla^4\phi = 0$ equation in difference form requires a relationship between interior and exterior mesh points relative to the boundary. Consider several points on a plate, adjacent to a boundary as shown in Figure 3-37. Point 2′ is beyond the boundary, points 3 and 2 are on or within the boundary. The ϕ surface is also shown in Figure 3-37. The slope ϕ at point 3, using central difference notation can be written as

$$\left.\frac{\partial\phi}{\partial x}\right|_3 = \frac{\phi_2' - \phi_2}{2n\,\Delta y}$$

or

$$\phi_2' = \phi_2 + \left.\frac{\partial\phi}{\partial x}\right|_3 \cdot (2n\,\Delta y) \tag{3.69a}$$

The slope at point 3 can also be written as follows, using backward differences [14, 15]:

$$\left.\frac{\partial\phi}{\partial x}\right|_3 = \frac{\phi_2' - \phi_3}{n\,\Delta y}$$

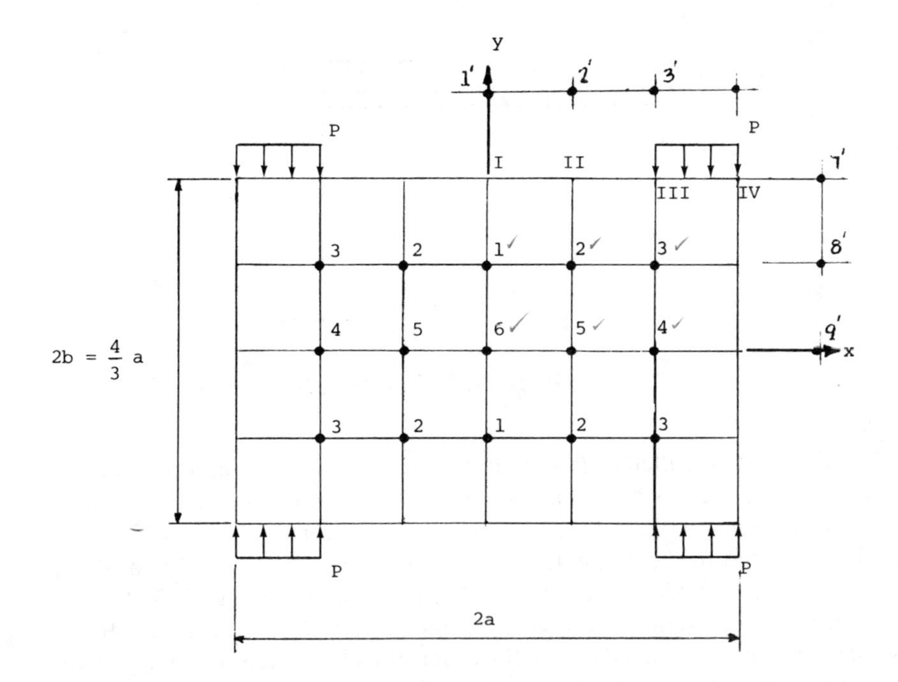

Figure 3-38. Example—Plate Loading and Mesh Pattern.

or

$$\phi_2' = \phi_3 + \left.\frac{\partial \phi}{\partial x}\right|_3 (n\,\Delta y) \tag{3.69b}$$

3.8.2 Example Problem

In order to illustrate the method of finite differences as applied to plane stress problems, a plate subjected to end loads with intensity p, plate depth $2b$, and plate width $2a$, as shown in Figure 3-38, will be studied [7]. The plate will be divided into equal spacings of Δy or λ, where $\lambda = a/3$, $n = 1$, and the applied load p will be in units of force per length. As noted in Figure 3-38, the loading is symmetrical; therefore, only one fourth of the plate need be considered. The biharmonic equation will therefore be written at mesh points 1 through 6, with the ϕs along the boundary (points I through VI in Figure 3-39) to be evaluated by the frame analogy, i.e., by M. Also to be computed will be the slopes along the boundary, i.e., V and T, which will be used to modify those points which are outside the boundary surface.

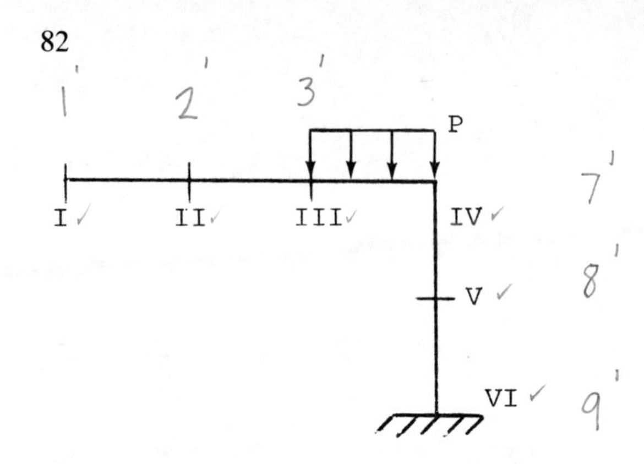

Figure 3-39. Frame Analogy.

As will be recalled from section 3.2.3, a plate subjected to in-plane loads can be idealized as a determinate plane frame. Such a frame can then be analyzed for moments M, shears V, and axial forces T. These forces are then equated to the ϕ function and respective slopes $\partial\phi/\partial T$ and $\partial\phi/\partial N$. The frame considered for this problem is shown in Figure 3-39, which represents one quarter of the actual plate. The plate is arbitrarly fixed at one end and considered free at the other; thus it is a determinate structure. The evaluation of moments, shears, and axial forces along the frame yields the following:

$$M_{\mathrm{I}} = M_{\mathrm{II}} = M_{\mathrm{III}} = 0$$

$$M_{\mathrm{IV}} = -\frac{P}{2}\lambda^2 = -P_0\frac{a^2}{18}$$

$$M_{\mathrm{V}} = M_{\mathrm{VI}} = M_{\mathrm{IV}}$$

$$V_{\mathrm{I}} = V_{\mathrm{II}} = V_{\mathrm{III}} = 0 \qquad V_{\mathrm{IV}} = -P\lambda \qquad \text{member I-IV}$$

$$V_{\mathrm{IV}} = V_{\mathrm{V}} = V_{\mathrm{VI}} = 0 \qquad \text{member IV-VI}$$

$$T_{\mathrm{I}} = T_{\mathrm{II}} = T_{\mathrm{III}} = T_{\mathrm{IV}} = 0 \qquad \text{member I-IV}$$

$$T_{\mathrm{IV}} = T_{\mathrm{V}} = T_{\mathrm{VI}} = -P\lambda \qquad \text{member IV-VI}$$

Using the frame analogy relationships, $M = \phi$, $\partial\phi/\partial N - T$, and $\partial\phi/\partial T = V$ from the free body shown in Figure 3-40, gives

$$\phi_{\mathrm{I}} = \phi_{\mathrm{II}} = \phi_{\mathrm{III}} = 0$$

$$\phi_{\mathrm{IV}} = \phi_{\mathrm{V}} = \phi_{\mathrm{VI}} = -\frac{Pa^2}{18}$$

$$\frac{\partial\phi}{\partial N} = \frac{\partial\phi}{\partial y} = T = 0 \qquad \text{for point I through IV of the top member}$$

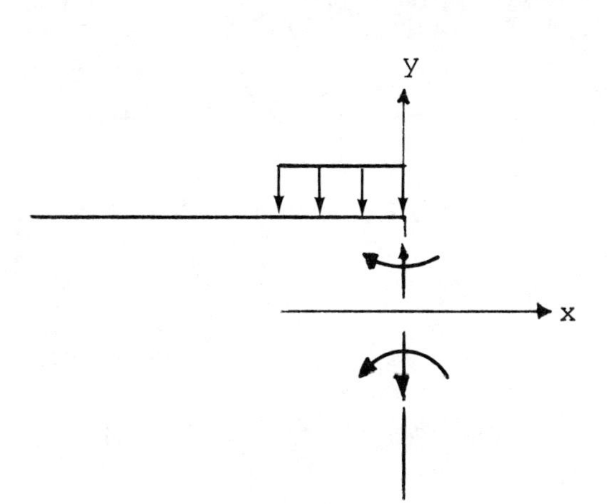

Figure 3-40. Frame Free Body.

$\dfrac{\partial \phi}{\partial N} = \dfrac{\partial \phi}{\partial x} = T = -P\lambda$ for points IV through VI of the vertical member

$\dfrac{\partial \phi}{\partial T} = \dfrac{\partial \phi}{\partial x} = V = 0$ ✓ for points I, II, and III

$\dfrac{\partial \phi}{\partial x} = V = -P\lambda$ ✓ for point IV (top member)

$\dfrac{\partial \phi}{\partial T} = \dfrac{\partial \phi}{\partial y} = V = 0$ ✓ for points IV through VI of the vertical member

Relationships between mesh points 1′, 2′, 3′, 7′, 8′, and 9′ relative to the interior and boundary points can now be established by applying equation (3.69a).

Mesh Point 1′, 2′, 3′:

$$\phi_{1'} = \phi_1 + \left.\frac{\partial \phi}{\partial y}\right|_{\mathrm{I}} (2n\lambda)$$

However, from the frame analogy $\phi_{\mathrm{I}} = 0$ and the $(\partial \phi / \partial y)|_{\mathrm{I}} = 0$, therefore:

$$\phi_{1'} = \phi_1 \tag{a}$$

Similarly $\phi_{\mathrm{II}} = \phi_{\mathrm{III}} = 0$ and $(\partial \phi / \partial y)|_{\mathrm{II}} = \partial \phi / \partial y|_{\mathrm{III}} = 0$, therefore:

$$\phi_{2'} = \phi_2 \tag{b}$$

and

$$\phi_{3'} = \phi_3 \tag{c}$$

Mesh Point $7'$, $8'$, $9'$: Applying equation (3.69a) gives

$$\phi_{7'} = \phi_{\text{III}} + \left.\frac{\partial \phi}{\partial x}\right|_{\text{IV}} (2n\lambda)$$

However, $\lambda = a/3$, $n = 1$, $\partial\phi/\partial x|_{\text{IV}} = -P\lambda$, and therefore:

$$\phi_{7'} = \phi_{\text{III}} - P\left(\frac{a}{3}\right)\left(2 \cdot 1 \cdot \frac{a}{3}\right)$$

$$\phi_{7'} = \phi_{\text{III}} - P\frac{2a^2}{9} \qquad \text{where } \phi_{\text{III}} = 0 \tag{d}$$

Similarly,

$$\phi_{8'} = \phi_3 - P\frac{2a^2}{9} \tag{e}$$

where ϕ_3 is to be determined. Likewise,

$$\phi_{9'} = \phi_4 - P\frac{2a^2}{9} \tag{f}$$

where ϕ_4 is to be determined.

Applying the general finite biharmonic difference equation (3.67) at each of the interior mesh points 1 through 6, noting the relationship between the ϕs along or outside the boundary gives:

Mesh Point 1:

$$20\phi_1 - 8\phi_{\text{I}} + 2\phi_{\text{II}} + 2\phi_{\text{II}} + \phi_{\text{I}} + \phi_{1'} + \phi_3 - 8\phi_2$$
$$- 8\phi_2 + \phi_3 + 2\phi_5 - 8\phi_6 + 2\phi_5 + \phi_1 = 0 \tag{g}$$

Mesh Point 2:

$$20\phi_2 - 8(\phi_1 + \phi_3 + \phi_5 + \phi_{\text{II}}) + 2(\phi_4 + \phi_6 + \phi_{\text{I}} + \phi_{\text{III}})$$
$$+ (2\phi_2 + \phi_{\text{V}} + \phi_{2'}) = 0 \tag{h}$$

Mesh Point 3:

$$20\phi_3 - 8(\phi_2 + \phi_4 + \phi_{\text{III}} + \phi_{\text{V}}) + 2(\phi_5 + \phi_{\text{II}} + \phi_{\text{IV}} + \phi_{\text{VI}})$$
$$+ (\phi_1 + \phi_3 + \phi_{3'} + \phi_{8'}) = 0 \tag{i}$$

Mesh Point 4:

$$20\phi_4 - 8(2\phi_3 + \phi_5 + \phi_{VI}) + 2(2\phi_2 + 2\phi_V) + (\phi_6 + \phi_{9'} + 2\phi_{III}) = 0 \qquad \text{(j)}$$

Mesh Point 5:

$$20\phi_5 - 8(2\phi_2 + \phi_4 + \phi_6) + 2(2\phi_1 + 2\phi_3) + (\phi_5 + 2\phi_{II} + \phi_{VI}) = 0 \qquad \text{(k)}$$

Mesh Point 6:

$$20\phi_6 - 8(2\phi_1 + 2\phi_5) + 2(4\phi_2) + (2\phi_4 + 2\phi_I) = 0 \qquad \text{(l)}$$

Substituting the boundary values into equations (g) through (l) gives the final ϕ equations:

$$\text{B.C.} \begin{cases} \phi_I = \phi_{II} = \phi_{III} = 0 \\ \phi_{IV} = \phi_V = \phi_{VI} = -P_0\dfrac{a^2}{18} \\ \phi_{8'} = \phi_3 - P_0\dfrac{2a^2}{9} \\ \phi_{9'} = \phi_4 - P_0\dfrac{2a^2}{9} \\ \phi_{7'} = \phi_{III} - P_0\dfrac{2a^2}{9} \\ \phi_{1'} = \phi_1,\ \phi_{2'} = \phi_2,\ \phi_{3'} = \phi_3 \end{cases}$$

The final equations are therefore:

$$11\phi_1 - 8\phi_2 + \phi_3 + 0 + 2\phi_5 - 4\phi_6 = 0 \quad ?$$

$$-8\phi_1 + 23\phi_2 - 8\phi_3 + 2\phi_4 - 8\phi_5 + 2\phi_6 = P_0\frac{a^2}{18}$$

$$\phi_1 - 8\phi_2 + 23\phi_3 - 8\phi_4 + 2\phi_5 + 0 = 0$$

$$0 + 4\phi_2 - 16\phi_3 + 21\phi_4 - 8\phi_5 + \phi_6 = 0$$

$$4\phi_1 - 16\phi_2 + 4\phi_3 - 8\phi_4 + 21\phi_5 - 8\phi_6 = P_0\frac{a^2}{18}$$

$$-8\phi_2 + 4\phi_2 + 0 + \phi_4 - 8\phi_5 + 10\phi_6 = 0$$

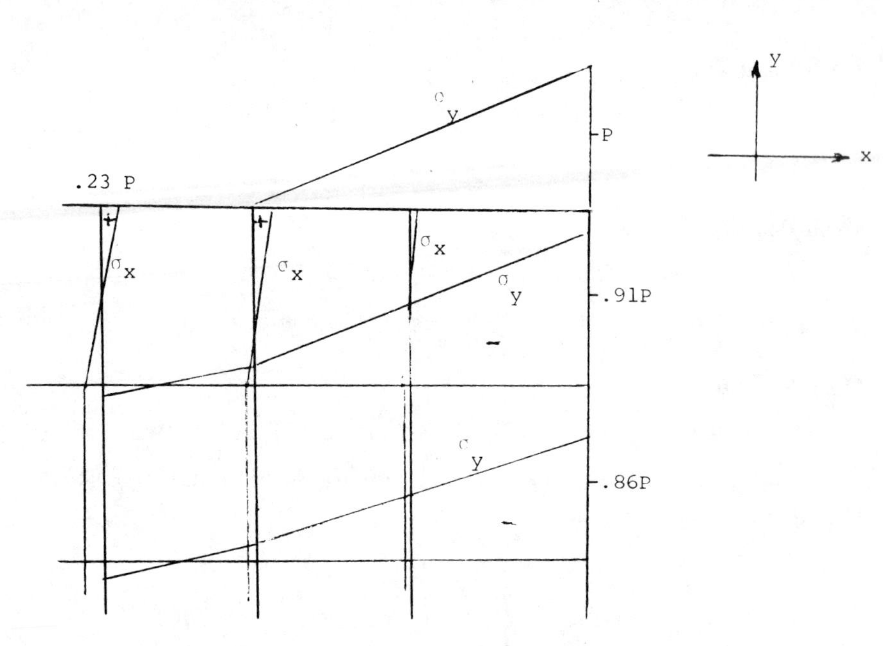

Figure 3-41. Stress Distribution throughout Plate.

Solving these equations,

$$\phi_1 = 0.2299\ M$$
$$\phi_2 = 0.2359\ M$$
$$\phi_3 = 0.0909\ M$$
$$\phi_4 = 0.1439\ M$$
$$\phi_5 = 0.3591\ M$$
$$\phi_6 = 0.3645\ M$$
$$\phi_7 = -4\ M$$
$$\phi_8 = -3.9091\ M$$
$$\phi_9 = -3.8561\ M$$

where $M = P_0 \dfrac{a^2}{18}$

Substituting these stress functions into the stress difference equations (3.68) will give σ_x, σ_y, and τ_{xy} at the various mesh points. A plot of the normal stresses are shown in Figure 3-41.

3.9 Effective Width of T-Beams

3.9.1 Theory

In a section composed of a thin flange and rib or stem such as a T-beam, as shown in Figure 3-42, the question arises as to how effective is the flange in resisting bending. Because the solution and evaluation of the actual normal in-plane bending stresses, and thus effectiveness, is complex, the designer will generally assume a portion of the flange to be effective and then isolate the resulting T-beam, thus forming a determinate structure, then the designer can use the conventional strength of material equation ($\sigma = My/I$) to evaluate stresses. The resulting effective flange width that should be used for various plate-stem boundary conditions will thus be the objective of this section. The solution of such will incorporate the Airy's stress function and Fourier series [16, 17, 18] as described previously.

In order to apply the in-plane stress equations the following assumptions will be made:

1. Ordinary bending theory of beams holds for the ribs.
2. Flange thickness h is small so that the bending stiffness of the flange alone is negligible compared with rib stiffness.
3. Stresses in the flange do not vary across thickness h of the flange (an average flange stress can be used).
4. Loads act only in the plane of the web (transverse deflections of the flange and singularities due to concentrated loads are not considered).
5. No buckling of the flange occurs (this may not be true for very thin wide steel flanges).

In addition to the indeterminancy of the plate-beam problem there is the problem that the normal stresses are not constant across the width of the flange but decrease when moving away from the rib or beam. This is due to the condition that the stresses are transmitted by shear as shown in Figure 3-43. The shearning deformations tend to reduce the normal stresses as they progress from the rib. This phenomenon of transmitting reduced normal stresses due to shearing strains is known as shear lag. This phenomenon can be explained further by examining the differential elements shown in Figure 3-43, of the plan of the compressed flange. When compression is exerted in the flange through positive bending of the ribs, the distance between a and b is decreased more than the distance between c and d; shearing strains thus deform the elements as shown, and therefore compression is less along fiber cd. The resulting flexural normal stress distribution is as shown in Figure 3-44 for the case in which all ribs are bent in the same direction.

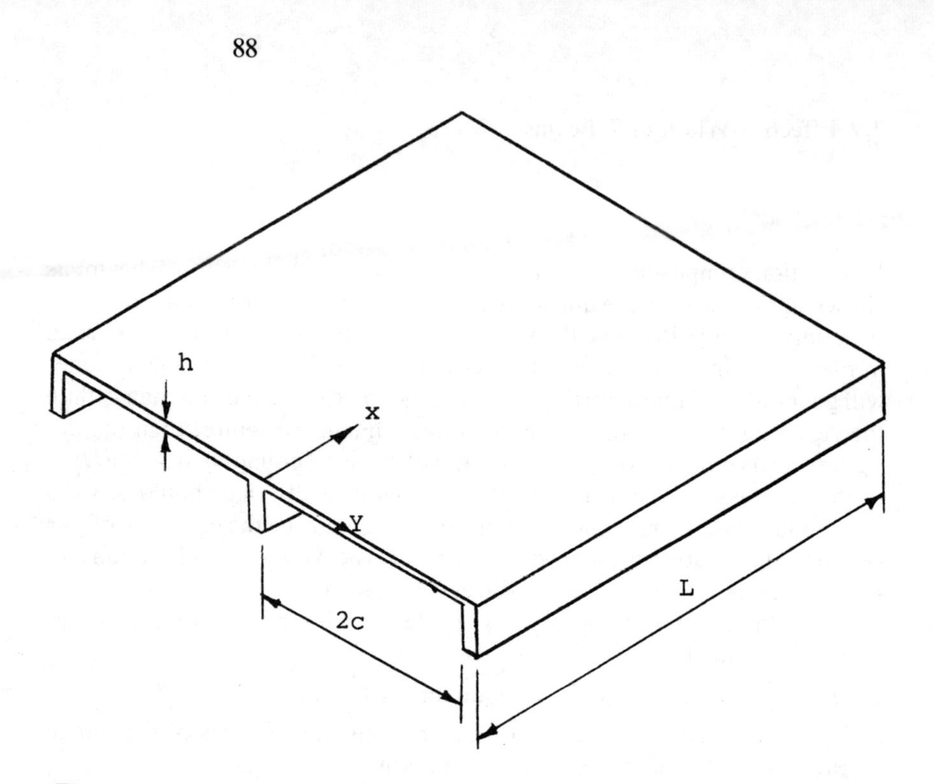

Figure 3-42. Cross Sectional Coordinates and Dimensions of T-Beams.

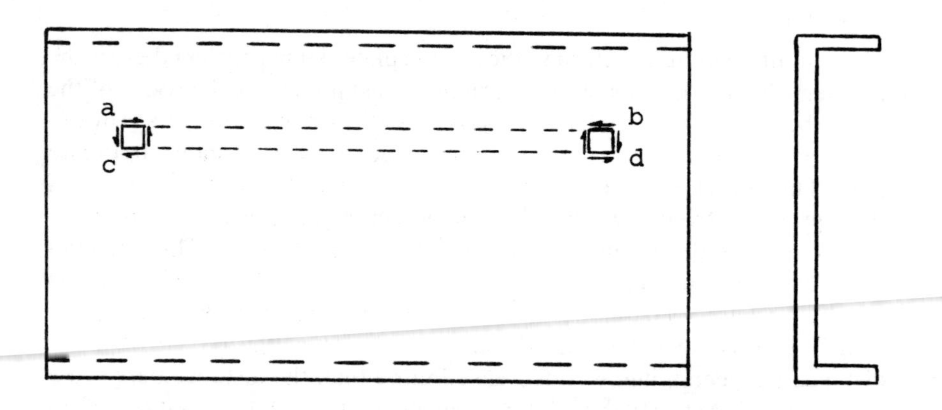

Figure 3-43. Plan View of Top Flange.

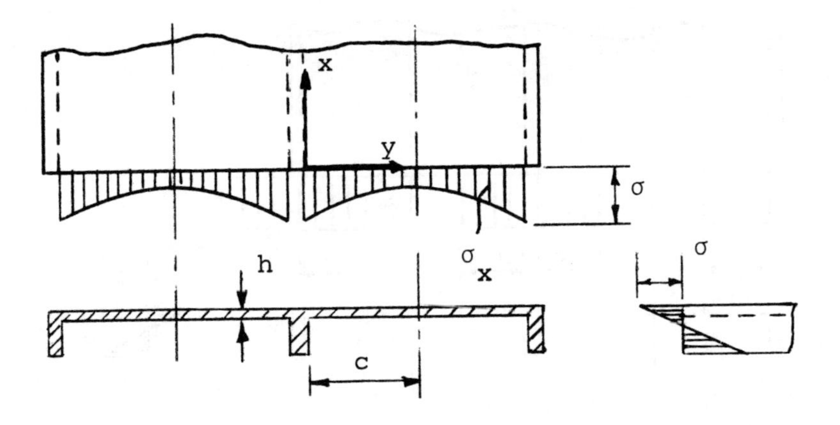

Figure 3-44. Flexural Stress Distribution.

Now, instead of evaluating the variable stress σ_x along the flange-beam surface, it would be simpler to create an equivalent stress distribution, as shown in Figure 3-45. In order to maintain the same resistance in the original systems (Figure 3-43) as in the equivalent system (Figure 3-44), the resulting flange forces in each system will be equated, or:

$$[\text{Flange Force}]_{\text{equivalent}} = [\text{Flange Force}]_{\text{actual}}$$

or

$$(\sigma_m)(2b_e)h = h \int_{\text{Flange}} \sigma_x \, dy \tag{3.70a}$$

Solving for b_e gives

$$b_e = h \int_0^{2c} \frac{\sigma_x \, dy}{2\sigma_m h} \tag{3.70b}$$

The problem now is to determine the expression for σ_x so that b_e can be evaluated. The actual expression for bending moment M_x along the length of the beam can be expanded into an odd term Fourier series of the type:

$$M_x = \sum M_n \sin \frac{n\pi x}{L}$$

M_n = Fourier coefficient

$2L$ = period

The choice of sine terms was made for the case of a simply supported beam

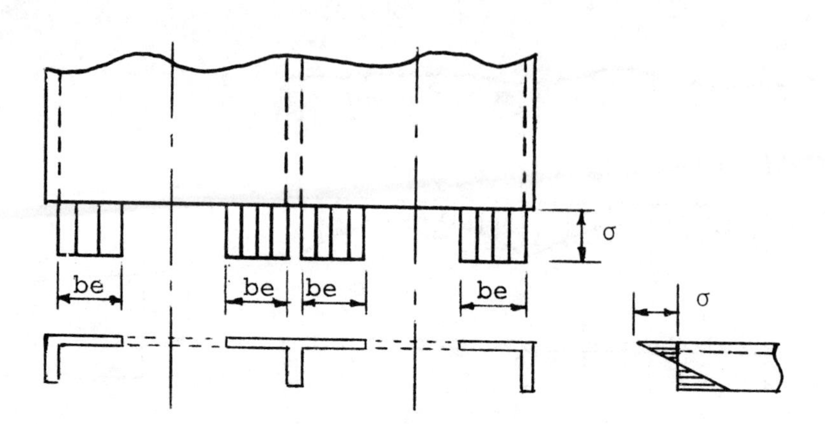

Figure 3-45. Statically Equivalent Stress Distribution.

since the moments at the ends will be zero. Thus the sine terms satisfy the boundary conditions at the supports, as shown in Figure 3-46.

Considering the flange as a plate loaded at the junction of the rib and the flange, the stresses in the flange must then satisfy the biharmonic plate equation (3.13)

$$\frac{\partial^4 \phi}{\partial x^4} + 2\frac{\partial^2 \phi}{\partial x^2 \partial y^2} + \frac{\partial^4 \phi}{\partial y^4} = 0 \tag{3.13}$$

where ϕ = Airy's stress function, and the stresses as given previously are:

$$\sigma_x = \frac{\partial^2 \phi}{\partial y^2} \tag{3.9}$$

$$\sigma_y = \frac{\partial^2 \phi}{\partial x^2} \tag{3.10}$$

$$\tau_{xy} = -\frac{\partial^2 \phi}{\partial x \, \partial y} \tag{3.11}$$

Analyzing the stresses resulting from the nth term of the Fourier expansion of moment,

$$M_x = M_n \sin \frac{n\pi x}{L}$$

According to bending theory the flexural stresses along any longitudinal fiber vary directly as the moment and hence as $\sin n\pi x/L$. Widthwise the stresses are assumed to vary according to some function Y_n, which is

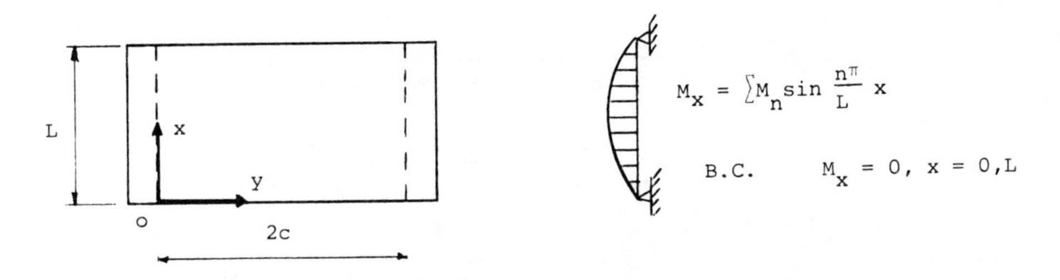

Figure 3-46. Moment Distribution.

independent of coordinate x. Therefore a stress function ϕ in the product form is used for the solution of the plate equation:

$$\phi_n = Y_n \sin \frac{n\pi x}{L}$$

which is the same equation as used for the multi-material beam solution in section 3.7. The general solution for the stress function is then:

$$\phi_n = \left(E \sinh \frac{n\pi y}{L} + F \cosh \frac{n\pi y}{L} + Gy \sinh \frac{n\pi y}{L} + Hy \cosh \frac{n\pi y}{L}\right) \sin \frac{n\pi x}{L} \quad (3.61)$$

where E, F, G, and H are constants to be determined.

The boundary conditions that the derivatives of ϕ must satisfy according to equations (3.9) through (3.11) are for the case of a channel, simply supported along the ends, free along the lateral edges, and symmetrically loaded (both ribs bent in same direction).

at $y = 0$ and $y = 2c$ (junction of flange and rib):

$$\sigma_x = \frac{\partial^2 \phi}{\partial y^2} = \sigma_m \sin \frac{n\pi x}{L}$$

where σ_m is the maximum stress along flange segment above rib.

3.9.2 Channel Section

The first case that will be examined will be a simply supported channel section, as shown in Figures 3-46 and 3-47. If the stem or rib and plate are

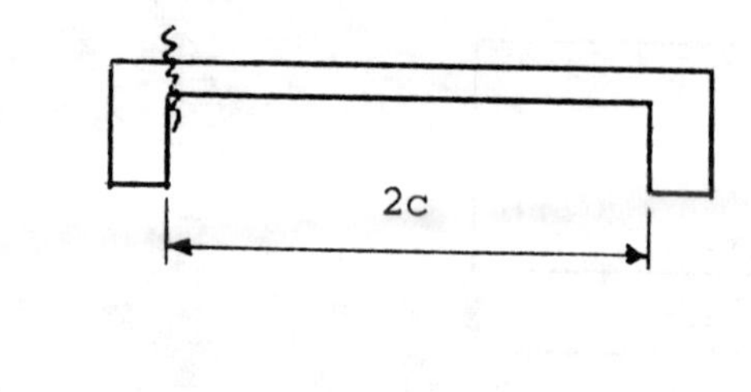

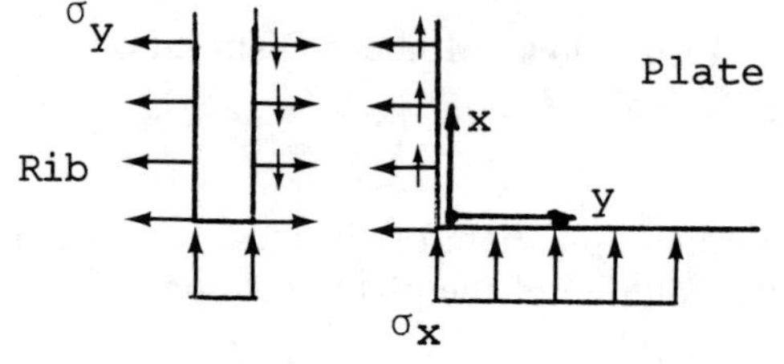

Figure 3-47. Joint Forces.

now separated, as shown in Figure 3-47, compatibility of stresses at the junction gives

$$\sigma_{x\ \mathrm{rib}} = \sigma_{x\ \mathrm{flange}} = \sigma_m \sin \frac{n\pi x}{L} \tag{1}$$

$$\sigma_y = 0 \tag{2}$$

$$\tau_{\mathrm{rib}} = \tau_{\mathrm{flange}} \tag{3}$$

Taking the proper derivatives of the stress equation (3.61) in accordance with equations (3.9) through (3.11) (see equation 3.62) and applying the boundary conditions (1) and (2) at $y = 0$ and $y = 2c$ gives:

$y = 0$: $\quad \sigma_y = 0 = -\lambda^2 (F); \quad F = 0$

$y = 2c$: $\quad \sigma_y = 0 \quad = -\lambda^2(E \sinh \lambda 2c + G2c \sinh \lambda 2c + H2c \cosh \lambda 2c)$

$y = 0$: $\quad \sigma_x = \sigma_m \sin \lambda x = [G\lambda(2)] \sin \lambda x$

$y = 2c$: $\quad \sigma_x = \sigma_m \sin \lambda x = [E\lambda^2 \sinh \lambda 2c + G\lambda(2 \cosh \lambda 2c + 2c\lambda \sinh \lambda 2c) + H\lambda(2 \sinh \lambda 2c + 2c\lambda \cosh \lambda 2c)] \sin \lambda x$

The solution of these equations gives the following for the coefficients E, F, G, and H:

$$E = \frac{\sigma_m c}{\lambda} \frac{(1 - \cosh 2\lambda c)}{\sinh^2 2\lambda c}$$

$$F = 0$$

$$G = \frac{\sigma_m}{2\lambda}$$

$$H = \frac{\sigma_m}{2\lambda}\left(\frac{1 - \cosh 2\lambda c}{\sinh 2\lambda c}\right)$$

where $\lambda = \dfrac{n\pi}{L}$

Substituting now these coefficients into the stress equation for σ_x,

$$\sigma_x = \frac{\partial^2\phi}{\partial y^2} = [E\lambda^2 \sinh \lambda y + F\lambda^2 \cosh \lambda y$$
$$+ G\lambda(2 \cosh \lambda y + y\lambda \sinh \lambda y)$$
$$+ H\lambda(2 \sinh \lambda y + y\lambda \cosh \lambda y)] \sin \lambda x$$

and integrating across the flange width gives

$$R = h\int_0^{2c} \sigma_x \, dy = \frac{\sigma_m}{\lambda}\left[\frac{2\lambda c + \sinh 2\lambda c}{\cosh 2\lambda c + 1}\right] \sin \lambda x \tag{3.71}$$

This resultant represents the numerator of equation (3.70b); therefore the effective width is

$$b_e = \frac{h\displaystyle\int_0^{2c} \sigma_x \, dy}{2\sigma_m h} \qquad \text{where } \sigma_m = \sigma_m \sin \lambda x$$

$$b_e = \frac{\sigma_m \sin \lambda x}{\lambda}\left[\frac{2\lambda c + \sinh 2\lambda c}{\cosh 2\lambda c + 1}\right]\frac{1}{2\sigma_m \sin \lambda x}$$

$$b_e = \sum \frac{L}{2n\pi}\left[\frac{\sinh \dfrac{2n\pi c}{L} + \dfrac{2n\pi c}{L}}{\cosh \dfrac{2n\pi c}{L} + 1}\right] \tag{3.72}$$

For the extreme case when c/L approaches infinity and $n = 1$,

$$b_e = \frac{L}{2n\pi} = 0.159\ L$$

Also the effective flange width is not dependent on flange thickness and is constant for each Fourier series term.

The shear stresses, which are compatible at the junction of the flange and stem are computed as follows:

$$\tau = -\frac{\partial^2\phi}{\partial x\,\partial y} = -\lambda[E\cosh\lambda y + F\lambda\sinh\lambda y$$
$$+ G(\sinh\lambda y + y\lambda\cosh\lambda y)$$
$$+ H(\cosh\lambda y + y\lambda\sinh\lambda y)]\cos\lambda x$$

as given by equation (3.62c). Setting $y = 0$ and substituting in E, F, G, H as defined previously gives

$$\tau\Big|_{y=0} = \frac{\sigma_m}{2}\left[\frac{\sinh 2\lambda c + 2\lambda c}{\cosh 2\lambda c + 1}\right]\cos\lambda x$$

The term

$$\left[\frac{\sinh 2\lambda c + 2\lambda c}{\cosh 2\lambda c + 1}\right]$$

is identical to the term in equation (3.72) or $2b_e\lambda$; therefore:

$$\tau\Big|_{y=0} = \frac{\sigma_m\cos\lambda x}{2}[2b_e\lambda]$$

$$\tau\Big|_{y=0} = \sum b_e\lambda\sigma_m\cos\lambda x \qquad (3.73)$$

For a value c/L approaching infinity and $x = 0$:

$$\tau = 0.159L\left(\frac{\pi}{L}\right)\sigma_m$$

$$\tau = \frac{\sigma_m}{2}$$

Thus τ at the ends is not zero, and therefore a stiffener is required; otherwise the effective width would be reduced.

A final plot of the stresses in the channel section are shown in Figure 3-48.

3.9.3 Multi-Beams

The effective width problem just described is only one of several that may be encountered in practice. Consider now the case where a series of

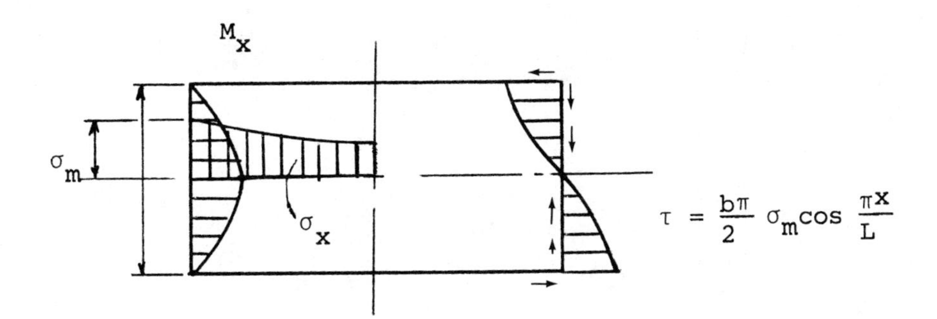

Figure 3-48. Moment and Stress Distributions.

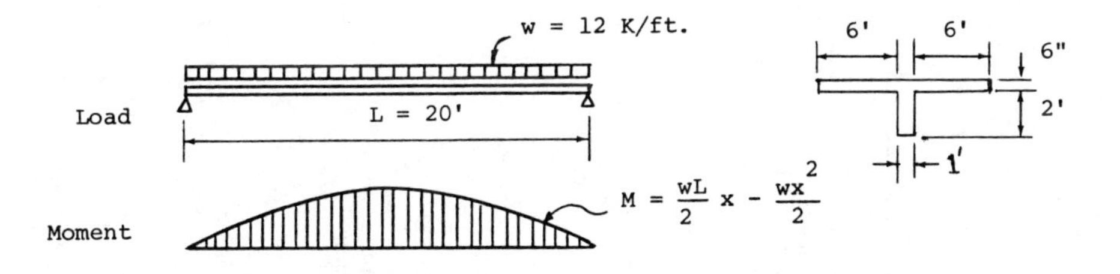

Figure 3-49. Example T-Beam Problem.

multi-beams are in the system, as shown in Figure 3-49. The boundary condition in this case involves displacements, i.e.,

$$\text{at } y = 0,\ y = 2c: \qquad v = 0$$

or the lateral displacement is zero, in addition to the stress condition of

$$\text{at } y = 0,\ y = 2c: \qquad \sigma_x = \sigma_m \sin \lambda x$$

The equations for stress σ_x have previously been defined; the displacement equation, however, is now required. This equation is found as described in section 3.7 and reference [18] resulting in

$$\left(\frac{\partial^3 \phi}{\partial y^3} + (2 + \mu) \frac{\partial^3 \phi}{\partial x^2 \, \partial y} \right) = - \frac{\partial^2 v}{\partial x^2} \tag{3.74}$$

Table 3-2
Effective Width Solutions

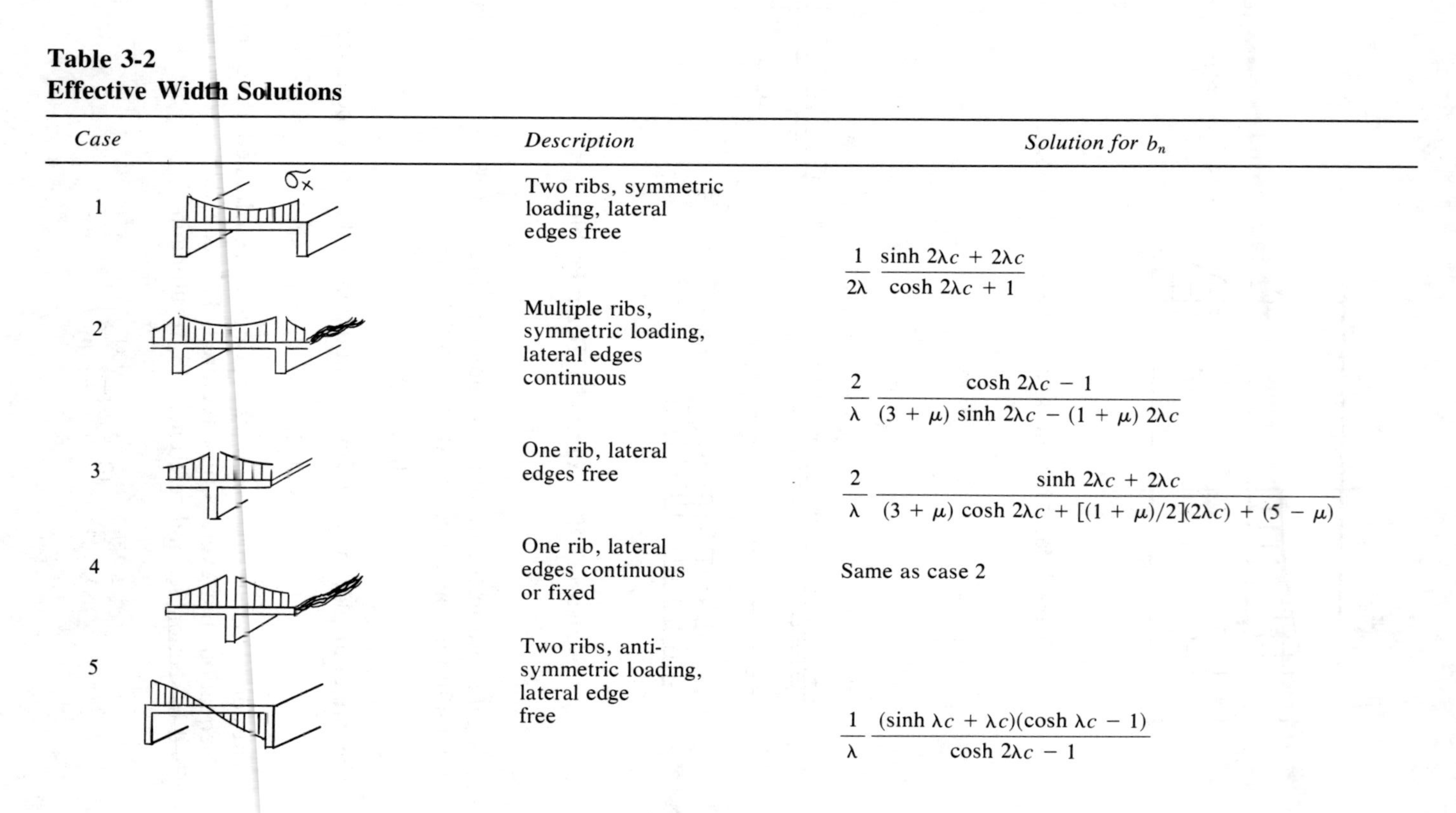

Case	*Description*	*Solution for b_n*
1	Two ribs, symmetric loading, lateral edges free	$\frac{1}{2\lambda} \frac{\sinh 2\lambda c + 2\lambda c}{\cosh 2\lambda c + 1}$
2	Multiple ribs, symmetric loading, lateral edges continuous	$\frac{2}{\lambda} \frac{\cosh 2\lambda c - 1}{(3 + \mu) \sinh 2\lambda c - (1 + \mu)\, 2\lambda c}$
3	One rib, lateral edges free	$\frac{2}{\lambda} \frac{\sinh 2\lambda c + 2\lambda c}{(3 + \mu) \cosh 2\lambda c + [(1 + \mu)/2](2\lambda c) + (5 - \mu)}$
4	One rib, lateral edges continuous or fixed	Same as case 2
5	Two ribs, anti-symmetric loading, lateral edge free	$\frac{1}{\lambda} \frac{(\sinh \lambda c + \lambda c)(\cosh \lambda c - 1)}{\cosh 2\lambda c - 1}$

6		Multiple ribs, antisymmetric loading, lateral edges continuous	$\frac{2}{\lambda} \frac{2 \cosh \lambda c(\cosh \lambda c - 1) + (1 + \mu)\lambda c \sinh \lambda c}{(3 + \mu) \sinh 2\lambda c + (1 + \mu)2\lambda c}$

Notes: μ is Poisson's ratio, $\lambda = n\pi/L$.

All beams simply supported at ends.

Symmetric loading means all ribs bent in same direction.

Antisymmetric loading means alternate ribs bent in same direction.

Lateral edges free means no lateral stresses σ_y.

Lateral edges continuous means no lateral displacements of ribs.

Poisson's ratio does not enter into cases 1 and 5 since the ribs are considered free to rotate axially as well as to deflect laterally, hence there are no stresses resulting from suppression of Poisson's ratio.

Table 3-3
Effective Plate Width Coefficients

	Values of β_n										
c/L_n											
Case No.	0.05	0.10	0.15	0.20	0.25	0.30	0.35	0.40	0.45	0.50	∞
1	0.049	0.094	0.130	0.158	0.176	0.186	0.190	0.191	0.189	0.186	0.159
2 & 4	0.049	0.094	0.130	0.158	0.178	0.191	0.199	0.204	0.207	0.208	0.205
3	0.049	0.093	0.129	0.157	0.175	0.187	0.195	0.199	0.201	0.201	0.205
5	0.025	0.049	0.072	0.094	0.113	0.130	0.145	0.158	0.168	0.176	0.159
6	0.025	0.049	0.073	0.095	0.116	0.134	0.151	0.165	0.177	0.187	0.205

and for all values of x the displacement v and thus $\partial^2 v/\partial x^2$ equal zero; therefore it is required to solve

$$\left(\frac{\partial^3 \phi}{\partial y^3} + (2 + \mu)\frac{\partial^3 \phi}{\partial x^2 \partial y}\right) = 0$$

at $y = 0$ and $y = 2c$. The solution to this problem has been obtained and the resulting effective width evaluated [16], giving

$$b_e = \sum \frac{2}{\lambda}\left[\frac{(\cosh 2\lambda c - 1)}{(3 + \mu)\sinh 2\lambda c - (1 + \mu)\, 2\lambda c}\right] \tag{3.75}$$

The evaluation of plate effective widths, for other geometry and beam arrangements have also been computed [16], and are listed in Table 3-2. The numerical evaluations of these equations, for each series term n, have also been defined, as listed in Table 3-3. The application of these tables will be given in the following example.

3.9.4 Example—Effective Width Evaluation and Resulting Stressts

Consider now the problem of a T-beam section as shown in Figure 3-49 supporting a uniform load of 12 k/ft. The moment will first have to be written in terms of a Fourier series, as the effective width equation is a term by term solution.

The moment expressed as a Fourier series is found using the general equation (3.47), for an odd function:

$$M = \sum M_n \sin \frac{n\pi x}{L}$$

where

$$M_n = \frac{2}{L}\int_0^L M(x) \sin \frac{n\pi x}{L}\, dx \tag{3.47}$$

However,

$$M(x) = \frac{wL}{2}x - \frac{wx^2}{2}$$

Therefore,

$$M_n = \frac{2}{L}\int_0^L \left(\frac{wLx}{2} - \frac{wx^2}{2}\right) \sin \frac{n\pi x}{L}\, dx$$

$$M_n = \frac{4wL^2}{\pi^3 n^3} \qquad n = 1, 3, 5$$

Therefore,

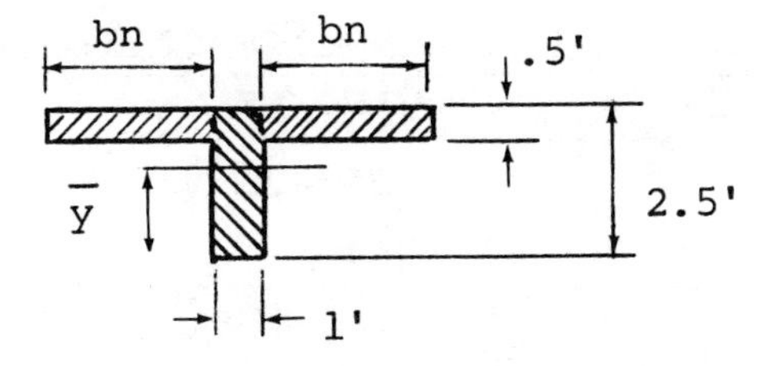

Figure 3-50. T-Beam Geometry.

$$M = \sum_{n=1,3,5} \frac{4wL^2}{\pi^3 n^3} \sin \frac{n\pi x}{L}$$

The magnitude of the moment at $x = L/2$ for $n = 1, 3, 5$ computed as:

$$M_1 = \frac{4(1)(240)^2}{\pi^3 \times 1} = 7430 \text{ k-in}$$

$$M_3 = \frac{7430}{27} = 270 \text{ k-in}$$

$$M_5 = \frac{7430}{125} = 60 \text{ k-in}$$

The effective width corresponding to each integer can now be computed from Table 3-3, Case III. The procedure for using this table is to evaluate $b_n = \beta_n L_n$, where β_n is taken from the table using appropriate $c/L_n = cn/L$ where $L_n = L/n$ values, i.e.,

For $n = 1$

$$L_n = \frac{L}{n} = \frac{240}{1} = 240$$

and

$$\frac{c}{L_n} = \frac{cn}{L} = \frac{72 \times 1}{240} = 0.30$$

Now examining Table 3-3, Case III, $\beta_n = 0.187$ for $c/L_n = 0.30$, and the effective width is

$$b_n = \beta_n L_n$$

$$b_n = 0.187(240)$$

$$b_n = 44.9 \text{ in}$$

With this value of the effective width, the inertia I can now be evaluated, examining Figure 3-50:

$$\bar{y} = \frac{\Sigma A_i y_i}{\Sigma A_i} = \frac{2.5 \times 1.25 + 3.74 \times 2.25}{3.74 + 2.5} = \frac{3.13 + 8.41}{6.24}$$

$$= 1.85 = 22.2 \text{ in}$$

$$I = 2\left(\frac{3.74 \times 0.125}{12}\right) + \frac{15.625}{12} + 3.74 \times 0.16 + 2.5 \times 0.36$$

$$= \frac{0.936 + 15.625}{12} + 0.598 + 0.9 = 1.38 + 1.498$$

$$= 2.88 \text{ ft}^4 = 59{,}600 \text{ in}^4$$

The stress in the flange therefore, for the first term $n = 1$, is computed as

$$\sigma_n = \frac{MC}{I} = \frac{7430 \times 7.8}{59{,}600} = -0.973 \text{ ksi}$$

Following a similar procedure, the effective widths, inertia, and finally stresses have been computed, as given in Table 3-4. The final resulting stress, is $\sigma = -0.933$ kis.

Compare the computed effective width to the AASHTO Code: Effective flange width ($2b_n$ + stem) is limited to

(a) $\frac{L}{4}$

(b) $12 \times t_f$ + stem

whichever is less, therefore, since

$$\frac{L}{4} = \frac{240}{4} = 60 \text{ in} = 5 \text{ ft}$$

$$12 \times t_f + \text{stem} = 12 \times 6 + 12 = 7 \text{ ft}$$

the effective width = 5 ft or b = 2 ft compared to 3.74 ft (44.9 in) for theoretical first term, therefore AASHTO stresses will be much greater than actual stresses.

3.9.5 Elastic Effective Slab Width—Flange Stiffness Included

The previous section on effective slab widths utilized in-plane stress theory and assumed the flange thickness h was small, so that the bending of the flange was negligible compared to the rib stiffness. If the flange stiffness relative to the stem stiffness is included, then the resulting effective widths will be modified. The results of these effects have been reported by Levi in

Table 3-4
Effective Plate Width Solution

n	$L_n = \frac{L}{n}$	$\frac{c}{L_n} = \frac{cn}{L}$	β_n	$b_n = \beta_n L_n$	I	C_{top}	$\sigma_n = \frac{Mc}{I}$
	(in)			*(in)*	*(in⁴)*	*(in)*	*(ksi)*
1	240	0.30	0.187	44.9	59,600	7.8	−0.973
3	80	0.90	0.201	16.1	57,600	10.8	+0.052
5	48	1.50	0.201	9.6	54,000	12.1	−0.012

$\sigma = -0.933$ ksi
maximum flange compression

reference [19], which shows the effective widths to be functions of the following parameters, as shown in Figures 3-51 and 3-52.

$\left(\frac{h}{b'}\right)$ h = flange thickness
b' = stem width

$\left(\frac{L}{b'}\right)$ L = span length

$\left(\frac{L}{c}\right)$ c = half the clear distance between stems

Figures 3-51 and 3-52, also show the effects these parameters have when flange stiffnesses are included. As noted, when the flexural rigidity of the flange is considered, a greater effective flange width b_e can be assumed. Thus it is conservative to use the previously derived effective width equations listed in Table 3-2 which neglect the flange stiffness.

Also given by Levi [19], are tabular values of the effective flange widths for the two cases given in Figures 3-51 and 3-52, as listed in Tables 3-5 and 3-6. In using these data the following conditions are suggested [19]:

1. For freely supported T-beams having either a single rib or a number of parallel ribs attached to the same slab, the effective width b_e of the compression flange may be determined by using the numerical values given in Tables 3-5 and 3-6.

   The values indicated are valid for uniformly distributed (or practically uniformly distributed) applied loads. They are also valid for the cases of triangular, parabolic, or sinusoidal distribution of the loading, and also for the case of a constant bending moment.

   On the other hand, if the beam under consideration carries a local

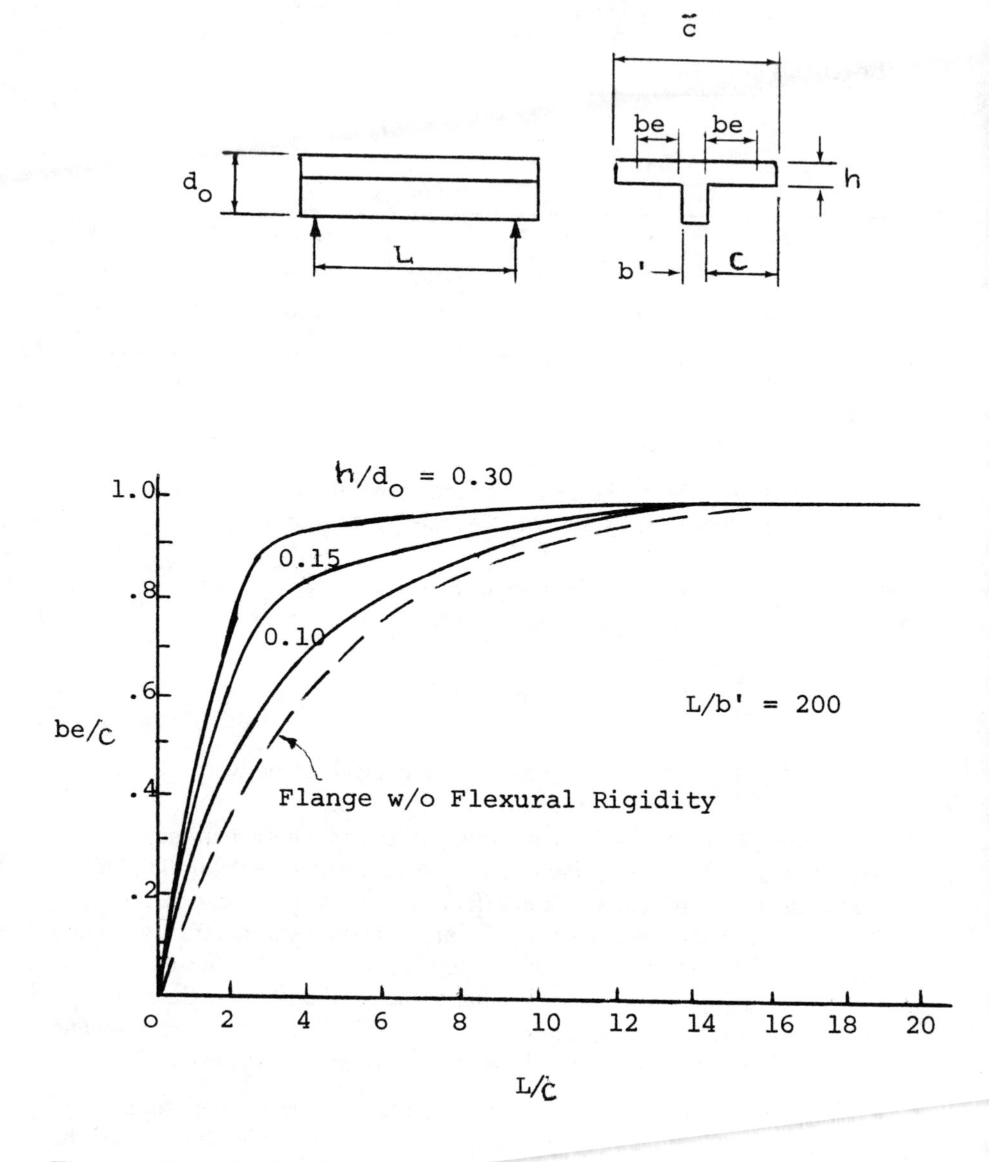

Figure 3-51. Effective Width of Flange at Midspan for Uniformly Loaded Isolated T-Beam.

load, the value of the effective width b_e of the compression flange should, at the section where the load is located, be reduced in the following proportions with respect to the foregoing values:

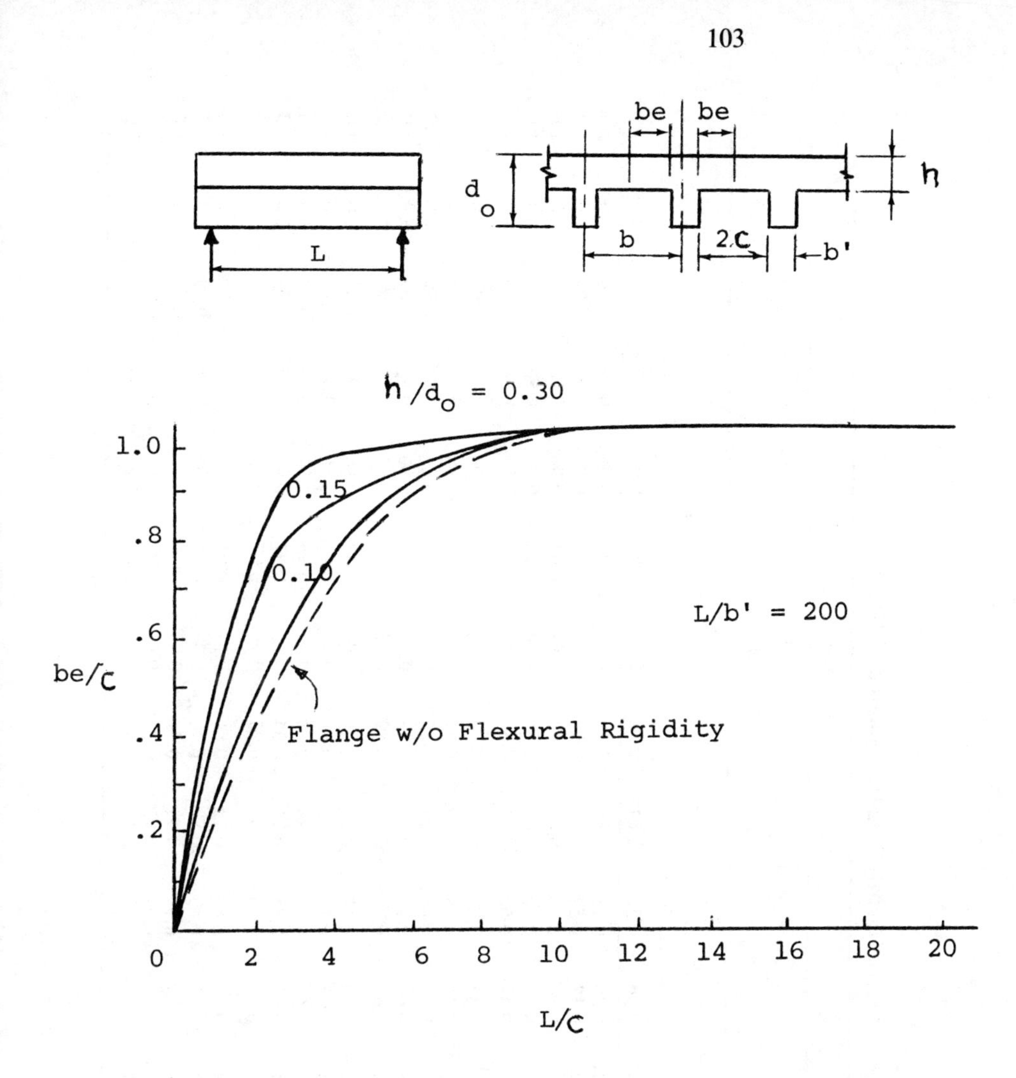

Figure 3-52. Effective Width of Flange at Midspan for Uniformly Loaded Multiple T-Beam.

	Values of L/c		20	10	0
Values of the effective width	for:	$a \geq 1/10$	$1.0b_e$	$1.0b_e$	$1.0b_e$
	for:	$0 < a < 1/10$	calculate by linear interpolation		
	for:	$a = 0$	$0.9b_e$	$0.7b_e$	$0.5b_e$

Table 3-5
Effective Width of Flange at Midspan for Uniformly Loaded Isolated T-Beam

h/d_0	L/b'	Values of b_e/c												
		L/c												
		0	*1*	*2*	*3*	*4*	*6*	*8*	*10*	*12*	*14*	*16*	*18*	*20*
Flange without Flexural Stiffness		0	0.18	0.36	0.52	0.64	0.78	0.86	0.92	0.95	0.97	0.98	0.99	1.00
	10	0	0.18	0.36	0.53	0.65	0.78	0.87	0.92	0.95	0.98	0.99	1.00	1.00
	50	0	0.19	0.37	0.54	0.66	0.79	0.87	0.92	0.95	0.98	0.99	1.0	1.0
0.10	100	0	0.21	0.40	0.56	0.67	0.80	0.87	0.92	0.96	0.98	0.99	1.0	1.0
	150	0	0.23	0.43	0.59	0.69	0.81	0.88	0.92	0.96	0.98	0.99	1.0	1.0
	200	0	0.27	0.47	0.62	0.71	0.81	0.88	0.93	0.96	0.98	0.99	1.0	1.0
	10	0	0.19	0.37	0.53	0.66	0.79	0.87	0.92	0.95	0.98	0.99	1.0	1.0
	50	0	0.22	0.42	0.58	0.69	0.81	0.88	0.92	0.96	0.98	0.99	1.0	1.0
0.15	100	0	0.30	0.51	0.66	0.74	0.83	0.89	0.93	0.96	0.98	0.99	1.0	1.0
	150	0	0.36	0.60	0.73	0.80	0.86	0.91	0.94	0.96	0.98	0.99	1.0	1.0
	200	0	0.40	0.65	0.79	0.85	0.89	0.92	0.95	0.97	0.98	0.99	1.0	1.0
	10	0	0.21	0.40	0.57	0.68	0.81	0.87	0.92	0.96	0.98	0.99	1.0	1.0
	50	0	0.30	0.52	0.69	0.78	0.86	0.90	0.94	0.96	0.98	0.99	1.0	1.0
0.20	100	0	0.40	0.65	0.79	0.86	0.89	0.92	0.95	0.97	0.98	0.99	1.0	1.0
	150	0	0.44	0.70	0.85	0.91	0.94	0.95	0.97	0.97	0.98	0.99	1.0	1.0
	200	0	0.45	0.73	0.89	0.93	0.95	0.96	0.97	0.98	0.99	1.00	1.0	1.0
	10	0	0.28	0.48	0.63	0.72	0.81	0.87	0.92	0.96	0.98	0.99	1.0	1.0
	50	0	0.42	0.65	0.83	0.87	0.90	0.92	0.94	0.96	0.98	0.99	1.0	1.0
0.30	100	0	0.45	0.73	0.90	0.92	0.94	0.95	0.96	0.97	0.98	0.99	1.0	1.0
	150	0	0.46	0.75	0.91	0.93	0.95	0.97	0.97	0.98	0.99	0.99	1.0	1.0
	200	0	0.46	0.77	0.92	0.94	0.96	0.97	0.98	0.99	0.99	0.99	1.0	1.0

Table 3-6
Effective Width of Flange at Midspan for Multiple T-Beam All Uniformly Loaded

h/d_0	L/b'	Values of b_e/c							
		L/c							
		0	*1*	*2*	*3*	*4*	*6*	*8*	*10*
Flange without Flexural Stiffness		0	0.19	0.38	0.57	0.71	0.88	0.96	0.99
	10	0	0.19	0.38	0.57	0.72	0.88	0.96	0.99
	50	0	0.19	0.39	0.58	0.73	0.89	0.96	1.0
0.10	100	0	0.21	0.42	0.60	0.75	0.89	1.0	1.0
	150	0	0.24	0.45	0.62	0.75	0.90	1.0	1.0
	200	0	0.27	0.48	0.64	0.77	0.90	1.0	1.0
	10	0	0.19	0.39	0.58	0.72	0.89	0.97	1.0
	50	0	0.23	0.44	0.62	0.74	0.90	0.97	1.0
0.15	100	0	0.31	0.53	0.68	0.78	0.91	0.97	1.0
	150	0	0.37	0.61	0.74	0.83	0.92	0.97	1.0
	200	0	0.41	0.66	0.80	0.87	0.93	0.98	1.0
	10	0	0.21	0.42	0.61	0.74	0.90	0.97	1.0
	50	0	0.30	0.54	0.71	0.82	0.92	0.97	1.0
0.20	100	0	0.41	0.66	0.80	0.87	0.94	0.98	1.0
	150	0	0.44	0.71	0.86	0.91	0.96	0.98	1.0
	200	0	0.45	0.74	0.89	0.93	0.97	0.99	1.0
	10	0	0.28	0.50	0.65	0.77	0.91	0.97	1.0
	50	0	0.42	0.69	0.83	0.88	0.93	0.97	1.0
0.30	100	0	0.45	0.74	0.90	0.94	0.96	0.98	1.0
	150	0	0.46	0.76	0.92	0.95	0.97	0.99	1.0
	200	0	0.47	0.77	0.92	0.96	0.98	0.99	1.0

where L = span of beam between simple supports

b_e = effective width of overhanging portion of compression flange

a = length of zone of application of local load along span of beam

2. For T-beams in which under a given set of loads changes of sign occur in the bending moment (e.g., for continuous T-beams), the recommendations contained in item 1 will be applicable, provided that the distance between the points of zero bending moment is adopted for the value L of the span.
3. In the vicinity of a simple support the effective width b_e of the compression flange of a T-beam must not exceed the distance between the support and the section under consideration.

4. In the case where the web is connected to the compression flange by haunches, the actual width b' of the rib should be replaced by an imaginary width b_1, which should be taken to be equal to:

$$b_1 = b' + 2b_s \quad \text{if } b_s < d_s$$

$$b_1 = b' + 2d_s \quad \text{if } b_s > d_s$$

where b_s = the width of the haunch

d_s = the depth of the haunch

5. For the case where different sets of loads exist simultaneously, which, if they acted separately, would determine the moments M_1, M_2, . . . and the effective widths b_{e1}, b_{e2}, . . ., then the effective width b_e of the T-beam should be calculated by the formula:

$$b_e = \frac{M_1 + M_2 + \ldots}{\dfrac{M_1}{b_{e1}} + \dfrac{M_2}{b_{e2}} + \ldots}$$

6. For an approximate calculation of the effective width b_e of a T-beam, it can be assumed as a first approximation that b_e/c is proportional to L/c, i.e., it can be assumed that

$$\frac{b_e}{c} = \frac{1}{10}\frac{L}{c}$$

$$b_e = \frac{L}{10} \quad \text{but not more than } c$$

Applying the tabular data given in Table 3-5, to the problem discussed inthe previous section, thus considering this flange stiffness, gives the following:

$$\frac{L}{c} = \frac{20}{6} = 3.3$$

$$\frac{L}{b'} = \frac{20 \text{ ft}}{1 \text{ ft}} = 20$$

$$\frac{h}{d_0} = \frac{6 \text{ in}}{30 \text{ in}} = 0.20$$

For these parameters, referencing Table 3-5 and interpolating linearly gives

$$\frac{b_e}{c} = 0.63$$

or

$$b_e = 0.63 \times 6 \text{ ft} = 3.78 \text{ ft}$$

or

$$b_e = 45.36 \text{ in}$$

which is greater than the first term $b_e = 44.9$ in, obtained by using the simplified theory.

3.9.6 Effective Slab Width at Ultimate Load

In sections 3.9.1 through 3.9.5, the effective slab width interacting with a beam were examined and evaluated at elastic conditions. If full plastification of the sections were considered, the theories utilized in this chapter would be supplemented with plate bending equations and an incremental load method of analysis [20]. Such an approach has been developed [20] and has resulted in a computer program to analyze a series of composite girders interacting during elastic-plastic action. The results were applied to the behavior of composite bridges resulting in the following design equations.

Interior Girders:

$$\frac{2b_e}{C} = 0.617\frac{C}{L} + 0.702 \tag{3.76}$$

where $2b_e > \dfrac{\text{Bridge width}}{\text{No. of Girders}}$

Exterior Girders:

$$\frac{2b_e}{C} = -1.98\frac{C}{L} + 0.873 \tag{3.77}$$

where $b_e < \dfrac{\text{Bridge width}}{\text{No. of Girders}}$

A comparison between these equations and the elastic effective width equations indicates that at ultimate load or complete plastification of the multi-girder system the effective width increases for the interior girders and decreases for the exterior girders. These trends were similarly noted by Levi [19], in the discussion of his work.

3.10 Finite Element Analysis

The concept of the finite element method [21, 22, 23, 24] is the idealization of a continuous media by an articulated structural system to which the matrix methods of structural analysis are directly applicable. There are two matrix method approaches associated with the finite element method.

1. The displacement finite element method (stiffness method), where the displacements of the nodes are chosen as the unknowns. In this approach the compatibility conditions in and among elements are initially satisfied. Then the governing equations in terms of the node displacements are written for each nodal point using the equilibrium conditions.
2. The force finite element method (flexibility method), where the internal forces are chosen as the unknowns. In this approach the equilibrium conditions are used first to generate the governing equations. The compatibility conditions are then introduced to develop additional equations that might be necessary to obtain a solution.

The experience of engineers has shown that the displacement finite element method (stiffness method) is more desirable since its formulation is simpler for the majority of structural analysis problems. Therefore, this book will use the stiffness method for the treatment of the topic and for the analysis of the plates.

This displacement finite element method requires the performance of five major steps to obtain a solution for the given system. They are:

(1) Idealization of the structural system (plate)
(2) Definition of the geometric and elastic properties of the system
(3) Description of the boundary conditions
(4) Definition of the loading conditions
(5) Operations on the generated matrix quantities to obtain the displacements and stresses

The development of the displacement finite element method is given in the following sections 3.10.1 and 3.10.2.

3.10.1 Introduction to the Stiffness Method

The technique is based on the evaluation of displacements or rotations that are required to create equilibrium in a system, when an assumed equilibrium condition is initiated. This concept is familiar in the context of using the slope deflection and moment distribution techniques.

To expand further, examine the following problem as shown in Figure 3-53. The redundant in this problem θ_B will be assumed to be with the following equilibrium condition, as given in Figure (3-54). The equilibrium

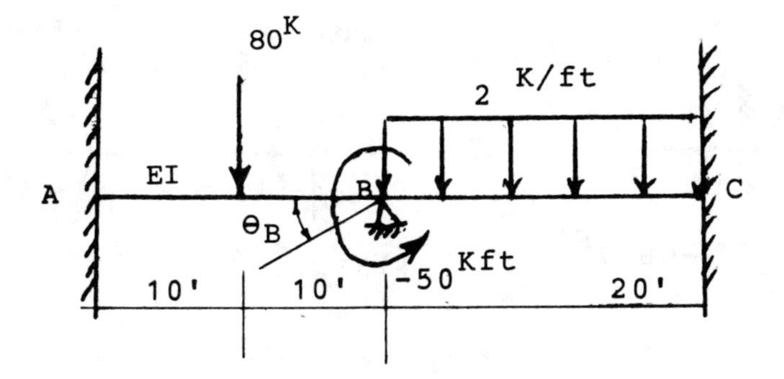

Figure 3-53. General Beam Loads.

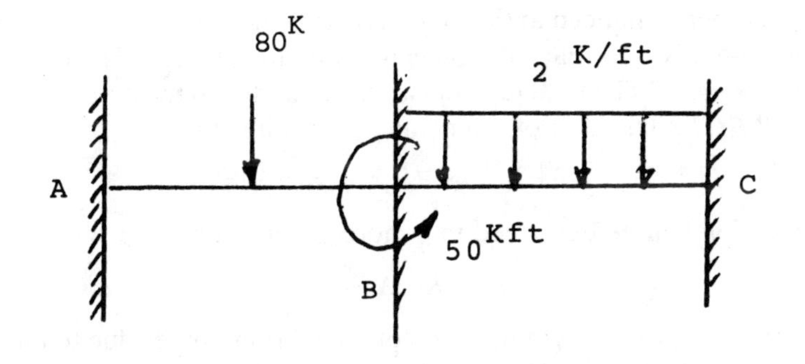

Figure 3-54. Restrained End Supports.

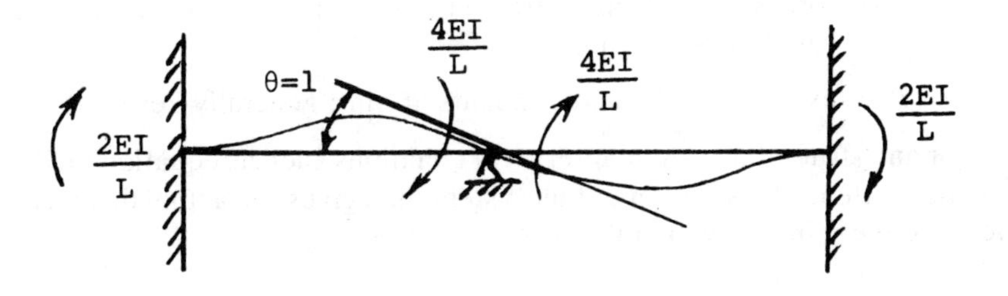

Figure 3-55. Unit Rotation Effects.

end moments at joint B are +200 k-ft, −66.7 k-ft (internal), and −50 k-ft (external). However, the structure is not fully restrained at B, but has a possible slope θ_B. The relative value of the induced moment due to a rotation of unity at B is shown in Figure 3-55.

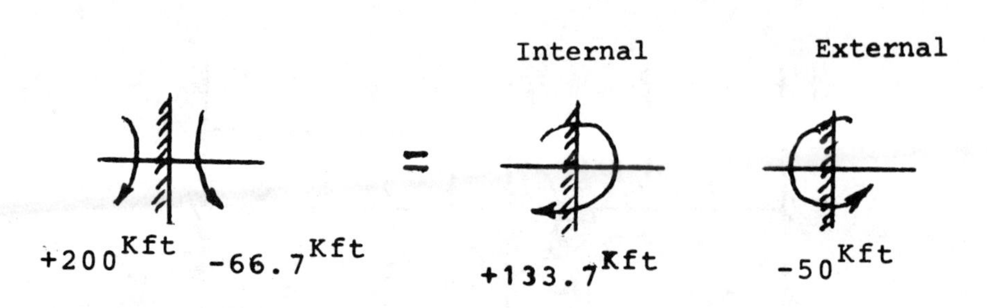

Figure 3-56. Internal Equilibrium.

The moments induced at the joint, assuming no rotation, are described in Figure 3-56. Now, what rotation is required to balance this internal and external moment? The total moment at the joint, due to the unit θ ($\theta = 1$), is $M_T = +8\ EI/L$. The sum of moments at the joint B is

$$133.7 + (8EI/L)\ \theta_B = +50 \tag{3.78}$$

This equation can be described in a more general form; i.e.,

$$P = K \cdot \Delta + \bar{P} \tag{3.79}$$

where K = represents contribution of equilibrium forces due to unit displacements (θ, Δ) (stiffness matrix)

Δ = displacements required to create equilibrium, or actual displacements in system

$\bar{P}$ = forces induced at restrained joint due to external loads, located between joints

P = external loads or redundants at oint; generally zero

For any structural system, a series of equations such as equation (3.79) may be written. The solution of this system then gives the actual displacements of the joints in which fixity was imposed.

3.10.2 Virtual Work Concept

In developing an expression for an element of a plate, which might be similar to equation (3.79), the virtual work concept and strain energy will be utilized [24]. Examining Figure (3-57), assume that the above system is subjected to loads $P_1, P_2, \ldots, P_m$. Now apply a virtual strain at load P_1 of magnitude $\delta\Delta_1$, such that equilibrium is not disturbed. This virtual dis-

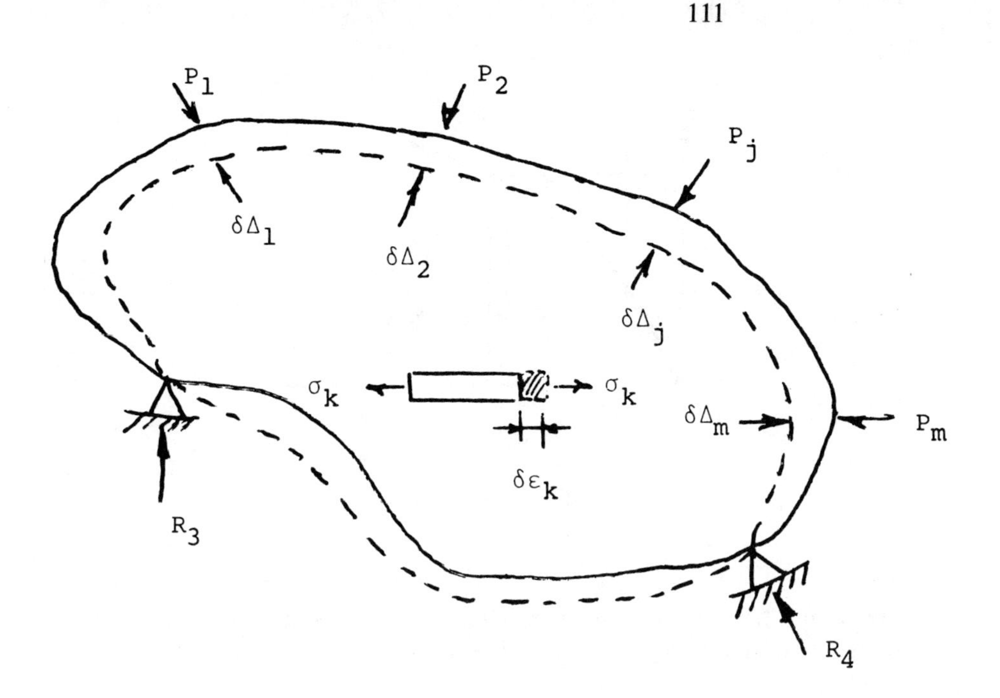

Figure 3-57. General Distorted Beam.

placement will cause a virtual strain in some stressed element σ_k (caused by original loads $P_1, \ldots, P_m$) of magnitude $\delta\varepsilon_k$. Examine now, the $\sigma - \varepsilon$ curve of the material of the system as shown in Figure (3-58).

The area under the stress-elastic strain curve, shown shaded horizontally in Figure 3-58 represents the density of strain energy U, which may be measured in pound-inches per cubic inch. The elastic strain energy U stored in the plate can be obtained by integrating the strain-energy density over the whole volume of the plate. Hence the strain energy can be expressed as

$$U = \int_v (\sigma - \varepsilon)\, dV \tag{3.80}$$

If now the displacements are increased from u to $u + \delta u$, there will be an accompanying increase in strains from ε to $\varepsilon + \delta\varepsilon$, and the corresponding increment in the strain energy density will be given by

$$\delta u = \sum_{k=1}^{n} \Delta \bar{u}_k \cdot AL \tag{3.81}$$

where $\Delta \bar{u}_k = \sigma_k \delta\varepsilon_k + \dfrac{1}{2}\, \delta\varepsilon_k \sigma_k$ (3.82)

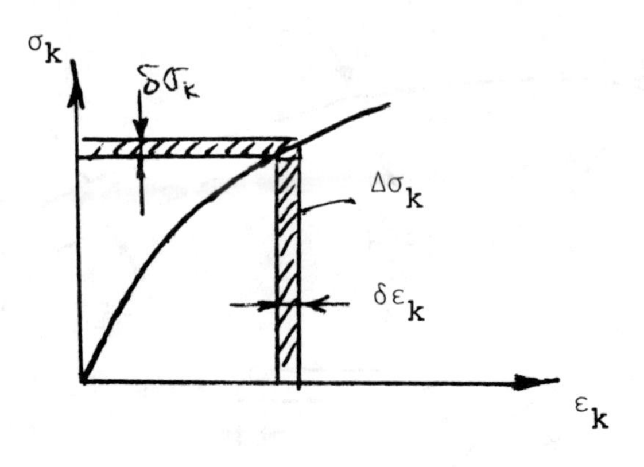

Figure 3-58. Stress-Strain Diagram.

Neglecting the terms of higher order in equation (3.82), the incremental change in strain energy density will be given from equation (3.81) by

$$\delta u = \sum_{k=1}^{n} \sigma_k \, \delta\varepsilon_k AL \tag{3.83}$$

or in terms of integrals

$$\delta u = \int_v \sigma \, \delta\varepsilon \, dV \tag{3.84}$$

These expressions state that the total incremental change in internal strain energy is equal to the sum of all stresses (due to loads P_i) times the incremental virtual strain $\delta\varepsilon$, induced by the virtual displacement $\delta\Delta$.

The external work is

$$\delta w = \sum_{j=1}^{m} P_j \, \delta\Delta_j \tag{3.85}$$

as caused by the virtual displacements. The relationship between work and energy gives

$$\sum P_j \, \delta\Delta_j = \sum \sigma_k \, \delta\varepsilon_k AL \tag{3.86}$$

Now if the virtual displacement $\delta\Delta_j = 1$, the virtual strains $\delta\varepsilon$ can be set equal to ε_k as caused by $\delta\Delta_j = 1$; σ_k are stresses caused by real loads P_m.

$$\sum P_j = \sum \sigma_k \varepsilon_k AL \tag{3.87}$$

In terms of the volume integral

$$\sum P = \int_v \sigma\varepsilon \, dV \tag{3.88}$$

3.10.3 Plane Stress-Strain Equations

There are two types of two-dimensional stress distributions, plane-stress and plane strain. The first type is used for thin flat plates loaded in the plane of the plate, as given in chapter 3. The second type is used for elongated bodies of constant cross sections subjected to uniform loading. This book will deal only with plane stress elements for plates.

Plane Stress: The plane stress distribution is based on the assumptions that

$$\sigma_z = \sigma_{zx} = \sigma_{zy} = 0 \tag{3.89}$$

where the z direction represents the direction perpendicular to the plane of the plate, and that no stress components vary through the plate thickness. These assumptions are sufficiently accurate for practical applications if the plate is thin.

The basic equation for plane stress for a plate, as given by equations (2.29) through (2.31), using matrix notation is

$$\begin{bmatrix} \sigma_x \\ \sigma_y \\ \tau_{xy} \end{bmatrix} = \frac{E}{1-\mu^2} \begin{bmatrix} 1 & \mu & 0 \\ \mu & 1 & 0 \\ 0 & 0 & \dfrac{1-\mu}{2} \end{bmatrix} \begin{bmatrix} \varepsilon_x \\ \varepsilon_y \\ \gamma_{xy} \end{bmatrix} \tag{3.90}$$

which in matrix notation can be presented as

$$\sigma = \chi e \tag{3.91}$$

where:

$$\sigma = \{\sigma_x \sigma_y \tau_{xy}\} \tag{3.92}$$

$$e = \{\varepsilon_x \varepsilon_y \gamma_{xy}\} \tag{3.93}$$

$$\chi = D \begin{bmatrix} 1 & \mu & 0 \\ \mu & 1 & 0 \\ 0 & 0 & \dfrac{1-\mu}{2} \end{bmatrix} \tag{3.94}$$

$$D = \frac{E}{1-\mu^2} \tag{3.95}$$

The braces used in equations (3.92) and (3.93) represent column matrices written horizontally to save space.

Plane Strain: For the plane strain, the distribution is based on the assumption that $\partial/\partial Z = U_z = 0$ where z represents the thickness of an elastic elongated body of constant cross section subjected to uniform loading. With the above assumption, it follows then that

$$\varepsilon_z = \varepsilon_{zx} = \varepsilon_{zy} = 0 \tag{3.96}$$

For plane strain, as given by equations (2.29) through (2.31):

$$\begin{bmatrix} \varepsilon_x \\ \varepsilon_y \\ \gamma_{xy} \end{bmatrix} = \frac{1+\mu}{E} \begin{bmatrix} 1-\mu & -\mu & 0 \\ -\mu & 1-\mu & 0 \\ 0 & 0 & 2 \end{bmatrix} \begin{bmatrix} \sigma_x \\ \sigma_y \\ \tau_{xy} \end{bmatrix} \tag{3.97}$$

$$e = \phi\sigma \tag{3.98}$$

where

$$e = \{\varepsilon_x \varepsilon_y \gamma_{xy}\} \tag{3.99}$$

$$\sigma = \{\sigma_x \sigma_y \tau_{xy}\} \tag{3.100}$$

$$\phi = \frac{1+\mu}{E} \begin{bmatrix} 1-\mu & -\mu & 0 \\ -\mu & 1-\mu & 0 \\ 0 & 0 & 2 \end{bmatrix} \tag{3.101}$$

3.10.4 Stiffness Equation for Plane Stress Problems

The general equation, of the form of equation (3.79), can be evaluated for an element of a plate by applying the work energy equations (3.87) or (3.88). Writing equation (3.88) in matrix notation we have

$$P = \int_v \bar{\varepsilon}^T \sigma \, dV \tag{3.102}$$

where $P = \{P_1, P_2, \ldots, P_m\}$ = external loads or reactions

$\sigma = \{\sigma_x, \sigma_y, \tau_{xy}\}$ = true stresses due to P

$\bar{\varepsilon}^T = [\bar{\varepsilon}_1, \bar{\varepsilon}_2, \ldots, \bar{\varepsilon}_m]$ = strains due to unit displacements at each load position $P_1, P_2, \ldots, P_m$ in direction of load

Equation (3.102) will result in an expression of the form of equation (3.79), i.e., $P = K \cdot \Delta + \bar{P}$, by use of the constant-strain triangular element.

3.10.5 Constant-Strain Triangular Element

This is the first element extensively employed in engineering analysis projects. It is used for thc analysis of plane stress and plane strain plate problems. This triangular element will have prescribed deflections at each node point 1, 2, and 3 as shown in Figure 3-59.

The element nodes are numbered in a clockwise direction starting at any node.

The assumed displacement variation will be taken as [24]:

$$
\begin{aligned}
u &= C_1 x + C_2 y + C_3 \\
v &= C_4 x + C_5 y + C_6
\end{aligned}
\tag{3.103}
$$

where the six arbitrary coefficients $C_1, \ldots, C_6$ can be found from the displacement of the three vertices of the triangle shown in Figure 3-59. Using the boundary conditions

$$
\begin{aligned}
&\text{at } (x_1, y_1): \quad && u = \Delta_1 \quad && v = \Delta_2 \\
&\text{at } (x_2, y_2): \quad && u = \Delta_3 \quad && v = \Delta_4 \\
&\text{at } (x_3, y_3): \quad && u = \Delta_5 \quad && v = \Delta_6
\end{aligned}
$$

and substituting these into equations (3.103) to evaluate the unknown coefficients gives

$$
\begin{aligned}
u = \frac{1}{2A_{123}} \{ & [y_{32}(x - x_2) - x_{32}(y - y_2)]\Delta_1 \\
& + [-y_{31}(x - x_3) + x_{31}(y - y_3)]\Delta_3 \\
& + [y_{21}(x - x_1) - x_{21}(y - y_1)\Delta_5]\}
\end{aligned}
\tag{3.104}
$$

$$
\begin{aligned}
v = \frac{1}{2A_{123}} \{ & [y_{32}(x - x_2) - x_{32}(y - y_2)]\Delta_2 \\
& + [-y_{31}(x - x_3) + x_{31}(y - y_3)]\Delta_4 \\
& + [y_{21}(x - x_1) - x_{21}(y - y_1)]\Delta_6\}
\end{aligned}
\tag{3.105}
$$

where:

$$
\begin{aligned}
2A_{123} &= x_{32}y_{21} - x_{21}y_{32} \\
&= 2(\text{area of triangle 123})
\end{aligned}
\tag{3.106}
$$

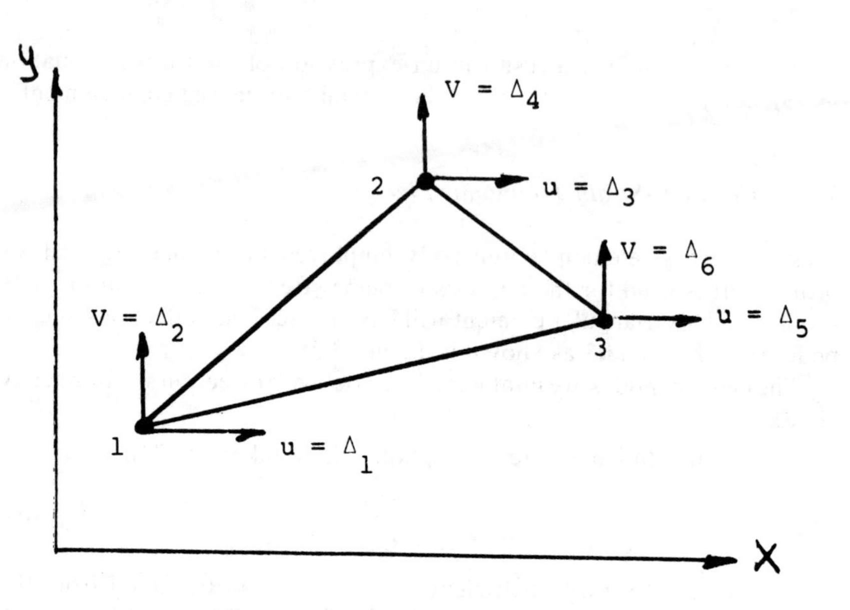

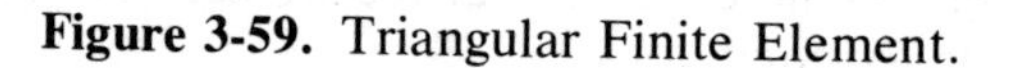

Figure 3-59. Triangular Finite Element.

and

$$x_{ij} = x_i - x_j \qquad y_{ij} = y_i - y_j \tag{3.107}$$

From equations (3.104) and (3.105) it follows that the assumed displacements along any edge vary linearly, and they depend only on the displacements of the two vertices of the particular edge; this ensures the satisfaction of the compatibility of displacements on two adjacent triangular elements with a common boundary.

Using the equations (3.104) and (3.105) and the relationship between strain and displacement,

$$\begin{bmatrix} \varepsilon_x \\ \\ \varepsilon_y \\ \\ \tau_{xy} \end{bmatrix} = \begin{bmatrix} \dfrac{\partial u}{\partial x} \\ \\ \dfrac{\partial x}{\partial y} \\ \\ \dfrac{\partial u}{\partial y} + \dfrac{\partial v}{\partial x} \end{bmatrix} \tag{3.108}$$

a strain displacement equation is obtained for ε_x, ε_y, and τ_{xy}, which in matrix notation gives

$$\begin{bmatrix} \varepsilon_x \\ \varepsilon_y \\ \tau_{xy} \end{bmatrix} = \frac{1}{2A_{123}} \begin{bmatrix} y_{32} & 0 & -y_{31} & 0 & y_{21} & 0 \\ 0 & -x_{32} & 0 & x_{31} & 0 & -x_{21} \\ -x_{32} & y_{32} & x_{31} & -y_{31} & -x_{21} & y_{21} \end{bmatrix} \begin{bmatrix} \Delta_1 \\ \Delta_2 \\ \Delta_3 \\ \Delta_4 \\ \Delta_5 \\ \Delta_6 \end{bmatrix} \tag{3.109}$$

This can be expressed as

$$e = B\Delta \tag{3.110}$$

where:

$$e = \{\varepsilon_x \varepsilon_y \tau_{xy}\} \tag{3.111}$$

$$\Delta = \{\Delta_1 \Delta_2 \Delta_3 \Delta_4 \Delta_5 \Delta_6\} \tag{3.112}$$

$$B = \frac{1}{2A_{123}} \begin{bmatrix} y_{32} & 0 & -y_{31} & 0 & y_{21} & 0 \\ 0 & -x_{32} & 0 & x_{31} & 0 & -x_{21} \\ -x_{32} & y_{32} & x_{31} & -y_{31} & -x_{21} & y_{21} \end{bmatrix} \tag{3.113}$$

However, from equation (3.91)

$$\sigma = \chi e$$

The total strains e can now be substituted into equation (3.91) to give the stress-displacement relationship

$$\sigma = \chi B \Delta \tag{3.114}$$

or equation (3.115),

$$\begin{bmatrix} \sigma_x \\ \sigma_y \\ \tau_{xy} \end{bmatrix} = \frac{E}{2A_{123}(1 - \mu^2)} \begin{bmatrix} \quad \end{bmatrix} \begin{bmatrix} \Delta_1 \\ \Delta_2 \\ \Delta_3 \\ \Delta_4 \\ \Delta_5 \\ \Delta_6 \end{bmatrix} \tag{3.115}$$

where [ ] matrix equals:

$$\begin{bmatrix} y_{32} & -\mu x_{32} & -y_{31} & \mu x_{31} & y_{21} & -\mu x_{21} \\ \mu y_{32} & -x_{32} & -\mu y_{31} & x_{31} & \mu y_{21} & -x_{21} \\ -Tx_{32} & Ty_{32} & Tx_{31} & -Ty_{31} & -Tx_{21} & Ty_{21} \end{bmatrix}$$

and $T = (1 - \mu)/2$.

The general equation relating actions to internal stress and virtual strain is

$$P = \int_v \bar{\varepsilon}^T \sigma \, dV \tag{3.116}$$

Substituting $\sigma = \chi B \Delta$ into equation (3.116) gives

$$P = \int_v \bar{\varepsilon}^T \chi B \Delta \, dV \tag{3.117}$$

However, $\bar{\varepsilon}^T$ are internal strains due to virtual displacements equal to one, i.e., $\Delta_1 = \Delta_2 = \ldots = \Delta_6 = 1$. Therefore,

$$e = B\Delta$$

$$e^T = \Delta B^T = B^T$$

Therefore,

$$e^T = B^T = \bar{\varepsilon}^T$$

Applying equation (3.117), and noting $\bar{\varepsilon}^T = B^T$ gives

$$P = \int_v B^T \chi B \, dV \cdot \Delta \tag{3.118}$$

where the general matrix is of the form:

$$P = K \cdot \Delta$$

and

$$K = \int_v B^T \chi B \, dV \tag{3.119}$$

Equation (3.118) is of the form of equation (3.79) and the stiffness matrix is given by equation (3.119).

The stiffness matrix can now be evaluated for the triangular element. The stiffness matrix is defined by equation (3.119) where

$$B^T = \frac{1}{2A_{123}} \begin{bmatrix} y_{32} & 0 & -x_{32} \\ 0 & -x_{32} & y_{32} \\ -y_{31} & 0 & x_{31} \\ 0 & x_{31} & -y_{31} \\ y_{21} & 0 & -x_{21} \\ 0 & -x_{21} & y_{21} \end{bmatrix} \tag{3.120}$$

For convenience of presentation the stiffness matrix K can be separated into two parts, stiffness due to normal stress and stiffness due to shearing stress,

$$K = K_n + K_s \tag{3.121}$$

Therefore, applying equation (3.119) in part, the term χB is first determined using equations (3.94) and (3.113), which gives

$$\chi B_n = \frac{E}{2A_{123}(1 - \mu^2)} \begin{bmatrix} y_{32} & -\mu x_{32} & -y_{31} & \mu x_{31} & y_{21} & -\mu x_{21} \\ \mu y_{32} & -x_{32} & -\mu y_{31} & x_{31} & \mu y_{21} & -x_{21} \\ 0 & 0 & 0 & 0 & 0 & 0 \end{bmatrix} \tag{3.122}$$

and

$$\chi B_s = \frac{E}{2A_{123}(1 - \mu^2)} \begin{bmatrix} 0 & 0 & 0 & 0 & 0 & 0 \\ 0 & 0 & 0 & 0 & 0 & 0 \\ -\psi x_{32} & \psi y_{32} & \psi x_{31} & -\psi y_{31} & -\psi x_{21} & \psi y_{21} \end{bmatrix} \tag{3.123}$$

where

$$\psi = \frac{1 - \mu}{2} \tag{3.124}$$

Now, evaluating

$$K = \int_v B^T \chi B \, dV$$

using equations (3.120), (3.122), and (3.123), where

$$\int_v dV = A_{123} \cdot t$$

gives the following [24]:

$$K_n = \frac{Et}{4A_{123}(1 - \mu^2)} \begin{bmatrix} \quad \end{bmatrix} \tag{3.125}$$

where [ ] matrix equals:

$$\begin{bmatrix} y_{32}^2 & & & & & \\ -\mu y_{32}x_{32} & x_{32}^2 & & \text{symmetric} & & \\ -y_{32}y_{31} & \mu x_{32}y_{31} & y_{31}^2 & & & \\ \mu y_{32}x_{31} & -x_{32}x_{31} & -\mu y_{31}x_{31} & x_{31}^2 & & \\ y_{32}y_{21} & -\mu x_{32}y_{21} & -y_{31}y_{21} & \mu x_{31}y_{21} & y_{21}^2 & \\ -\mu y_{32}x_{21} & x_{32}x_{21} & \mu x_{21}y_{31} & -x_{31}x_{21} & -\mu y_{21}x_{21} & x_{21}^2 \end{bmatrix}$$

$$K_s = \frac{Et}{8A_{123}(1 + \mu)} \begin{bmatrix} \quad \end{bmatrix} \tag{3.126}$$

where [ ] matrix equals:

$$\begin{bmatrix} x_{32}^2 & & & & & \\ -x_{32}y_{32} & y_{32}^2 & & \text{symmetric} & & \\ -x_{32}x_{31} & y_{32}x_{31} & x_{31}^2 & & & \\ x_{32}y_{31} & -y_{32}y_{31} & -x_{31}y_{31} & y_{31}^2 & & \\ x_{32}x_{21} & -y_{32}x_{21} & -x_{31}x_{21} & y_{31}x_{21} & x_{21}^2 & \\ -x_{32}y_{21} & y_{32}y_{21} & y_{21}x_{31} & -y_{31}y_{21} & -x_{21}y_{21} & y_{21}^2 \end{bmatrix}$$

Example: Using the finite element method, the stresses in the cantilevered plate shown in Figure 3-60 will now be determined. Subdividing the plate into two elements as shown in Figure 3-61, the forces acting at the nodes are 5λ. The individual displacements or actions for each of these elements are shown in Figure 3-62. The development of the stiffness matrix K, for each of the elements 1 and 2 will now be determined by using equations (3.125) and (3.126).

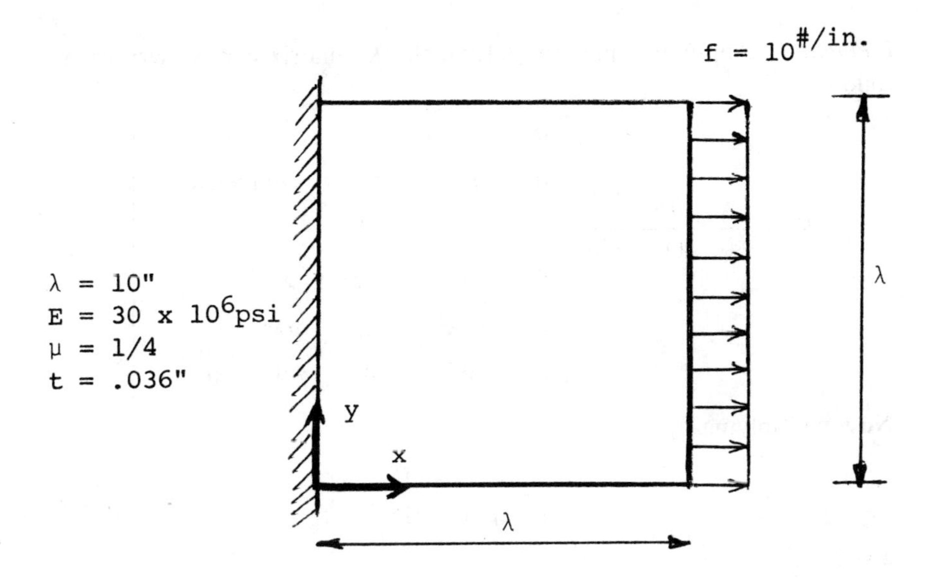

Figure 3-60. Example—End Loaded Plate.

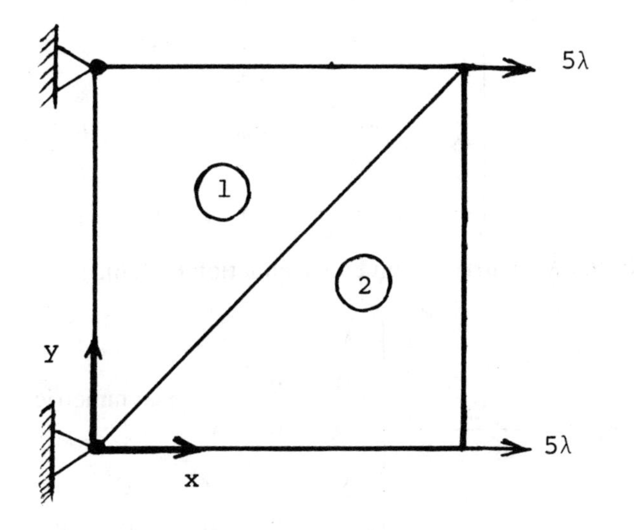

Figure 3-61. Finite Elements.

Element 1: Applying equation (3.125), the K_n matrix can be written as follows:

$$K_n = \frac{Et}{4A_{123}(1-\mu^2)}\begin{bmatrix} 0 & & & & & \\ 0 & \lambda^2 & & \text{symmetric} & & \\ 0 & \mu\lambda^2 & \lambda^2 & & & \\ 0 & -\lambda^2 & -\mu\lambda^2 & \lambda^2 & & \\ 0 & -\mu\lambda^2 & -\lambda^2 & \mu\lambda^2 & \lambda^2 & \\ 0 & 0 & 0 & 0 & 0 & 0 \end{bmatrix}$$

Now multiplying by

$$\frac{1}{(1-\mu^2)} = \frac{16}{15}$$

gives

$$K_n = \frac{Et}{4A}\begin{bmatrix} 0 & & & & & \\ 0 & \frac{16\lambda^2}{15} & & \text{symmetric} & & \\ 0 & \frac{4\lambda^2}{15} & \frac{16\lambda^2}{15} & & & \\ 0 & -\frac{16\lambda^2}{15} & -\frac{4\lambda^2}{15} & \frac{16\lambda^2}{15} & & \\ 0 & -\frac{4\lambda^2}{15} & -\frac{16\lambda^2}{15} & \frac{4\lambda^2}{15} & \frac{16\lambda^2}{15} & \\ 0 & 0 & 0 & 0 & 0 & 0 \end{bmatrix}$$

Similarly the K_s matrix, as given by equation (3.126), is written as

$$K_s = \frac{Et}{8A_{123}(1+\mu)}\begin{bmatrix} \lambda^2 & & & & & \\ 0 & 0 & & \text{symmetric} & & \\ -\lambda^2 & 0 & \lambda^2 & & & \\ \lambda^2 & 0 & -\lambda^2 & \lambda^2 & & \\ 0 & 0 & 0 & 0 & 0 & \\ -\lambda^2 & 0 & \lambda^2 & -\lambda^2 & 0 & \lambda^2 \end{bmatrix}$$

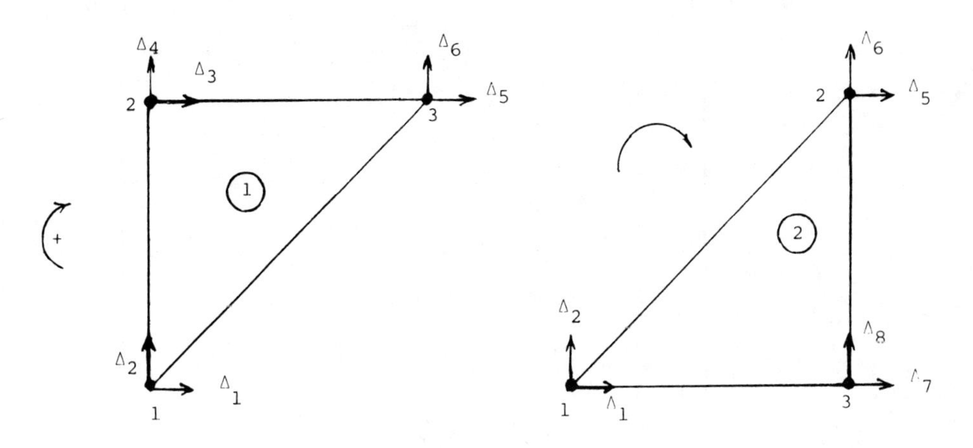

Figure 3-62. Distortion of Elements.

Multiplying through by

$$\frac{1}{2(1 + \mu)} = \frac{2}{5}$$

gives

$$K_s = \frac{Et}{4A}\begin{bmatrix} \frac{2\lambda^2}{5} & & & & & \\ 0 & 0 & & \text{symmetric} & & \\ -\frac{2\lambda^2}{5} & 0 & \frac{2\lambda^2}{5} & & & \\ \frac{2\lambda^2}{5} & 0 & -\frac{2\lambda^2}{5} & \frac{2\lambda^2}{5} & & \\ 0 & 0 & 0 & 0 & 0 & \\ -\frac{2\lambda^2}{5} & 0 & \frac{2\lambda^2}{5} & -\frac{2\lambda^2}{5} & 0 & \frac{2\lambda^2}{5} \end{bmatrix}$$

Now adding the K_s and K_n matrix gives the final K matrix of element 1:

$$K = \frac{Et}{4A}\begin{bmatrix} \frac{2\lambda^2}{5} & & & & & \\ 0 & \frac{16\lambda^2}{15} & & & \text{symmetric} & \\ -\frac{2\lambda^2}{5} & \frac{4\lambda^2}{15} & \frac{22\lambda^2}{15} & & & \\ \frac{2\lambda^2}{5} & -\frac{16\lambda^2}{15} & -\frac{10\lambda^2}{15} & \frac{22\lambda^2}{15} & & \\ 0 & -\frac{4\lambda^2}{15} & -\frac{16\lambda^2}{15} & \frac{4\lambda^2}{15} & \frac{15\lambda^2}{15} & \\ -\frac{2\lambda^2}{5} & 0 & \frac{2\lambda^2}{5} & -\frac{2\lambda^2}{5} & 0 & \frac{2\lambda^2}{5} \end{bmatrix} \tag{3.127}$$

Element 2: Again applying equations (3.125) and (3.126), relative to element 2 location and nodes gives the following matrices K_n and K_s:

$$K_n = \frac{Et}{4A}\begin{bmatrix} \frac{16\lambda^2}{15} & & & & & \\ 0 & 0 & & \text{symmetric} & & \\ 0 & 0 & 0 & & & \\ -\frac{4\lambda^2}{15} & 0 & 0 & \frac{16\lambda^2}{15} & & \\ -\frac{16\lambda^2}{15} & 0 & 0 & \frac{4\lambda^2}{15} & \frac{16\lambda^2}{15} & \\ \frac{4\lambda^2}{15} & 0 & 0 & -\frac{16\lambda^2}{15} & -\frac{4\lambda^2}{15} & \frac{16\lambda^2}{15} \end{bmatrix}$$

$$K_s = \frac{Et}{4A}\begin{bmatrix} 0 & & & & & \\ 0 & \frac{2\lambda^2}{5} & & \text{symmetric} & & \\ 0 & -\frac{2\lambda^2}{5} & \frac{2\lambda^2}{5} & & & \\ 0 & 0 & 0 & 0 & & \\ 0 & \frac{2\lambda^2}{5} & -\frac{2\lambda^2}{5} & 0 & \frac{2\lambda^2}{5} & \\ 0 & -\frac{2\lambda^2}{5} & \frac{2\lambda^2}{5} & 0 & -\frac{2\lambda^2}{5} & \frac{2\lambda^2}{5} \end{bmatrix}$$

Adding equations for K_s and K_n, the final K matrix of element 2 is:

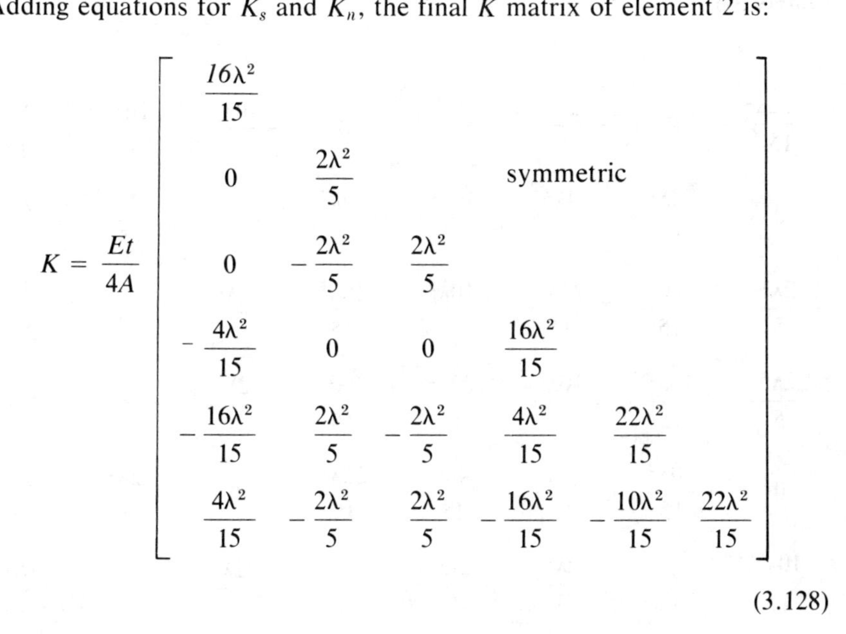

$$K = \frac{Et}{4A}\begin{bmatrix} \frac{16\lambda^2}{15} & & & & & \\ 0 & \frac{2\lambda^2}{5} & & \text{symmetric} & & \\ 0 & -\frac{2\lambda^2}{5} & \frac{2\lambda^2}{5} & & & \\ -\frac{4\lambda^2}{15} & 0 & 0 & \frac{16\lambda^2}{15} & & \\ -\frac{16\lambda^2}{15} & \frac{2\lambda^2}{5} & -\frac{2\lambda^2}{5} & \frac{4\lambda^2}{15} & \frac{22\lambda^2}{15} & \\ \frac{4\lambda^2}{15} & -\frac{2\lambda^2}{5} & \frac{2\lambda^2}{5} & -\frac{16\lambda^2}{15} & -\frac{10\lambda^2}{15} & \frac{22\lambda^2}{15} \end{bmatrix} \tag{3.128}$$

The entire stiffness matrix of the assembled two elements can now be obtained by adding equations (3.127) and (3.128) and noting the displacements at nodes 1 and 3 in element 1 are equal to the displacements at nodes 1 and 2 in element 2, and that $\Delta_1 = \Delta_2 = \Delta_3 = \Delta_4 = 0$. This gives the following general equation (3.129):

$$\frac{Et}{4A}\left[\begin{array}{c} \\ \\ \\ \text{matrix} \\ \text{3-1} \\ \\ \\ \\ \end{array}\right]\begin{bmatrix} \Delta_1 \\ \Delta_2 \\ \Delta_3 \\ \Delta_4 \\ \Delta_5 \\ \Delta_6 \\ \Delta_7 \\ \Delta_8 \end{bmatrix} = \begin{bmatrix} P_1 \\ P_2 \\ P_3 \\ P_4 \\ 5\lambda \\ 0 \\ 5\lambda \\ 0 \end{bmatrix} \tag{3.129}$$

which is of the form [K][Δ] = [P]. The [ ] matrix in equation (3.129) is equal to Matrix 3-1.

Matrix 3-1

$$\begin{bmatrix}
\frac{22\lambda^2}{15} & 0 & -\frac{2\lambda^2}{5} & \frac{2\lambda^2}{5} & 0 & -\frac{10\lambda^2}{15} & -\frac{16\lambda^2}{15} & \frac{4\lambda^2}{15} \\
0 & \frac{22\lambda^2}{15} & \frac{4\lambda^2}{15} & -\frac{16\lambda^2}{15} & -\frac{10\lambda^2}{15} & 0 & \frac{2\lambda^2}{5} & -\frac{2\lambda^2}{5} \\
-\frac{2\lambda^2}{5} & \frac{4\lambda^2}{15} & \frac{22\lambda^2}{15} & -\frac{10\lambda^2}{15} & -\frac{16\lambda^2}{15} & \frac{2\lambda^2}{5} & 0 & 0 \\
\frac{2\lambda^2}{5} & -\frac{16\lambda^2}{15} & -\frac{10\lambda^2}{15} & \frac{22\lambda^2}{15} & \frac{4\lambda^2}{15} & -\frac{2\lambda^2}{5} & 0 & 0 \\
0 & -\frac{10\lambda^2}{15} & -\frac{16\lambda^2}{15} & \frac{4\lambda^2}{15} & \frac{22\lambda^2}{15} & 0 & -\frac{2\lambda^2}{5} & \frac{2\lambda^2}{5} \\
-\frac{10\lambda^2}{15} & 0 & \frac{2\lambda^2}{5} & -\frac{2\lambda^2}{5} & 0 & \frac{22\lambda^2}{15} & \frac{4\lambda^2}{15} & -\frac{16\lambda^2}{15} \\
\frac{16\lambda^2}{15} & \frac{2\lambda^2}{5} & 0 & 0 & -\frac{2\lambda^2}{5} & \frac{4\lambda^2}{15} & \frac{22\lambda^2}{15} & -\frac{10\lambda^2}{15} \\
\frac{4\lambda^2}{15} & -\frac{2\lambda^2}{5} & 0 & 0 & \frac{2\lambda^2}{5} & -\frac{16\lambda^2}{15} & -\frac{10\lambda^2}{15} & \frac{22\lambda^2}{15}
\end{bmatrix}$$

Applying equation (3.129) gives the following eight equations:

$$P_1 = \left[-\frac{10}{15}\lambda^2\Delta_6 - \frac{16}{15}\lambda^2\Delta_7 + \frac{4}{15}\lambda^2\Delta_8 \right]\frac{Et}{4A} \tag{a}$$

$$P_2 = \left[-\frac{10}{15}\lambda^2\Delta_5 + \frac{2\lambda^2}{5}\Delta_7 - \frac{2\lambda^2}{5}\Delta_8 \right]\frac{Et}{4A} \tag{b}$$

$$P_3 = \left[-\frac{16\lambda^2}{15}\Delta_5 + \frac{2\lambda^2}{5}\Delta_6 \right]\frac{Et}{4A} \tag{c}$$

$$P_4 = \left[\frac{4\lambda^2}{15}\Delta_5 - \frac{2\lambda^2}{5}\Delta_6 \right]\frac{Et}{4A} \tag{d}$$

$$\frac{4A}{Et}(5\lambda) = \frac{22\lambda^2}{15}\Delta_5 - \frac{2\lambda^2}{5}\Delta_7 + \frac{2\lambda^2}{5}\Delta_8 \tag{e}$$

$$0 = \frac{22\lambda^2}{15}\Delta_6 + \frac{4\lambda^2}{15}\Delta_7 - \frac{16\lambda^2}{15}\Delta_8 \tag{f}$$

$$\frac{4\text{A}}{Et}(5\lambda) = -\frac{2}{5}\lambda^2\Delta_5 + \frac{4\lambda^2}{15}\Delta_6 + \frac{22\lambda^2}{15}\Delta_7 - \frac{10}{15}\lambda^2\Delta_8 \tag{g}$$

$$0 = \frac{2}{5}\lambda^2\Delta_5 - \frac{16\lambda^2}{15}\Delta_6 - \frac{10\lambda^2}{15}\Delta_7 + \frac{22\lambda^2}{15}\Delta_8 \tag{h}$$

The deflections Δ_5, Δ_6, Δ_7, and Δ_8 can be found by solving equations (e) through (h); or

$$\left(\frac{4A}{Et}\right)75 = 22\lambda\Delta_5 - 6\lambda\Delta_7 + 6\lambda\Delta_8$$

$$0 = 22\lambda\Delta_6 + 4\lambda\Delta_7 - 16\lambda\Delta_8$$

$$\left(\frac{4A}{Et}\right)75 = -6\lambda\Delta_5 + 4\lambda\Delta_6 + 22\lambda\Delta_7 - 10\lambda\Delta_8$$

$$0 = 6\lambda\Delta_5 - 16\lambda\Delta_6 - 10\lambda\Delta_7 + 22\lambda\Delta_8$$

Solving these equations gives,

$$\frac{Et\lambda\Delta_5}{4A} = 4.56932; \qquad \text{or} \qquad \Delta_5 = 8.461704\,\lambda \cdot 10^{-6} \rightarrow$$

$$\frac{Et\lambda\Delta_6}{4A} = -3.15125 \times 10^{-1}; \qquad \text{or} \qquad \Delta_6 = -0.583565\,\lambda \cdot 10^{-6}\downarrow$$

$$\frac{Et\lambda\Delta_7}{4A} = 5.09453; \qquad \text{or} \qquad \Delta_7 = 9.434316\ \lambda \cdot 10^{-6} \rightarrow$$

$$\frac{Et\lambda\Delta_8}{4A} = 8.40336 \times 10^{-1}; \qquad \text{or} \qquad \Delta_8 = 1.556178\ \lambda \cdot 10^{-6} \uparrow$$

Using these deflection results, the supports forces can now be evaluated, using the first four force-displacement equations (a), (b), (c), (d):

$$\begin{aligned}
P_1 &= +0.210083\lambda - 5.434165\lambda + 0.224090\lambda &&= -5.0000\lambda \\
P_2 &= -3.046213\lambda + 2.037812\lambda - 0.336134\lambda &&= -1.3445\lambda \\
P_3 &= -4.873941\lambda - 0.126050\lambda &&= -5.0000\lambda \\
P_4 &= 1.218485\lambda = 0.126050\lambda &&= 1.3445\lambda
\end{aligned}$$

With the displacements and actions known, the strains at the centroid of the elements can be determined. From equation (3.110) the strain matrix is:

$$e = B \cdot \Delta$$

Applying equation (3.110), gives

Element 1:

$$\begin{bmatrix} \varepsilon_x \\ \varepsilon_y \\ \gamma_{xy} \end{bmatrix} = \frac{1}{2(\lambda^2/2)} \begin{bmatrix} \quad \end{bmatrix} \begin{bmatrix} 0 \\ 0 \\ 0 \\ 0 \\ 8.462\lambda \times 10^{-6} \\ -0.5836\lambda \times 10^{-6} \end{bmatrix} \begin{matrix} \Delta_1 \\ \Delta_2 \\ \Delta_3 \\ \Delta_4 \\ \Delta_5 \\ \Delta_6 \end{matrix}$$

where [ ] matrix equals:

$$\begin{bmatrix} 0 & 0 & -\lambda & 0 & \lambda & 0 \\ 0 & -\lambda & 0 & \lambda & 0 & 0 \\ -\lambda & 0 & \lambda & -\lambda & 0 & \lambda \end{bmatrix}$$

which gives

$$\begin{bmatrix} \varepsilon_x \\ \varepsilon_y \\ \gamma_{xy} \end{bmatrix} = \frac{1}{\lambda^2}\begin{bmatrix} 8.462\lambda^2 \times 10^{-6} \\ 0 \\ -0.5836\lambda^2 \times 10^{-6} \end{bmatrix} = \begin{bmatrix} 8.462 \times 10^{-6} \\ 0 \\ -0.5836 \times 10^{-6} \end{bmatrix}$$

Element 2:

$$\begin{bmatrix} \varepsilon_x \\ \varepsilon_y \\ \gamma_{xy} \end{bmatrix} = \frac{1}{2(\lambda^2/2)}\begin{bmatrix} \quad \end{bmatrix}\begin{bmatrix} 0 \\ 0 \\ 8.4620\lambda \times 10^{-6} \\ -0.5836\lambda \times 10^{-6} \\ 9.4343\lambda \times 10^{-6} \\ 1.5562\lambda \times 10^{-6} \end{bmatrix}\begin{matrix} \Delta_1 \\ \Delta_2 \\ \Delta_5 \\ \Delta_6 \\ \Delta_7 \\ \Delta_8 \end{matrix}$$

where [ ] matrix equals:

$$\begin{bmatrix} -\lambda & 0 & 0 & 0 & \lambda & 0 \\ 0 & 0 & 0 & \lambda & 0 & -\lambda \\ 0 & -\lambda & \lambda & 0 & -\lambda & \lambda \end{bmatrix}$$

which gives

$$\begin{bmatrix} \varepsilon_x \\ \varepsilon_y \\ \gamma_{xy} \end{bmatrix} = \frac{1}{\lambda^2}\begin{bmatrix} 9.4343\lambda^2 \times 10^{-6} \\ -2.1398\lambda^2 \times 10^{-6} \\ -0.5836\lambda^2 \times 10^{-6} \end{bmatrix} = \begin{bmatrix} +9.4343 \times 10^{-6} \\ -2.1398 \times 10^{-6} \\ -0.5836 \times 10^{-6} \end{bmatrix}$$

The stress can now be determined, using these strains or, from the stress matrix, σ is

$$\sigma = \chi \cdot e$$

Applying equation (3.115) gives

Element 1:

$$\begin{bmatrix} \sigma_x \\ \sigma_y \\ \tau_{xy} \end{bmatrix} = \frac{30 \times 10^6}{2\left(\frac{\lambda^2}{2}\right)\left[1 - \left(\frac{1}{4}\right)^2\right]} \begin{bmatrix} \quad \end{bmatrix} \begin{bmatrix} 0 \\ 0 \\ 0 \\ 0 \\ 8.462\lambda \times 10^{-6} \\ -0.5836\lambda \times 10^{-6} \end{bmatrix}$$

where [ ] matrix equals:

$$\begin{bmatrix} 0 & \frac{-\lambda}{4} & -\lambda & \frac{\lambda}{4} & \lambda & 0 \\ 0 & -\lambda & \frac{-\lambda}{4} & \lambda & \frac{\lambda}{4} & 0 \\ \frac{-3\lambda}{8} & 0 & \frac{3\lambda}{8} & \frac{-3\lambda}{8} & 0 & \frac{3\lambda}{8} \end{bmatrix}$$

which gives

$$\begin{bmatrix} \sigma_x \\ \sigma_y \\ \tau_{xy} \end{bmatrix} = \frac{32 \times 10^6}{\lambda^2} \begin{bmatrix} \quad \end{bmatrix} = \begin{bmatrix} 270.8 \\ 67.70 \\ 7.002 \end{bmatrix} \text{psi}$$

where [ ] matrix equals:

$$\begin{bmatrix} 8.4620\lambda^2 \times 10^{-6} \\ 2.1155\lambda^2 \times 10^{-6} \\ -0.2118\lambda^2 \times 10^{-6} \end{bmatrix}$$

Element 2:

$$\begin{bmatrix} \sigma_x \\ \sigma_y \\ \tau_{xy} \end{bmatrix} = \frac{30 \times 10^6}{2(\lambda^2/2)[1 - (1/4)^2]} \begin{bmatrix} \quad \end{bmatrix} \begin{bmatrix} 0 \\ 0 \\ 8.4260\lambda \times 10^{-6} \\ -0.5836\lambda \times 10^{-6} \\ 9.4343\lambda \times 10^{-6} \\ 1.5562\lambda \times 10^{-6} \end{bmatrix} \begin{matrix} \Delta_1 \\ \Delta_2 \\ \Delta_5 \\ \Delta_6 \\ \Delta_7 \\ \Delta_8 \end{matrix}$$

where [ ] matrix equals:

$$\begin{bmatrix} -\lambda & 0 & 0 & \frac{\lambda}{4} & \lambda & \frac{-\lambda}{4} \\ \frac{-\lambda}{4} & 0 & 0 & \lambda & \frac{\lambda}{4} & -\lambda \\ 0 & \frac{-3\lambda}{8} & \frac{3\lambda}{8} & 0 & \frac{-3\lambda}{8} & \frac{3\lambda}{8} \end{bmatrix}$$

which gives

$$\begin{bmatrix} \sigma_x \\ \sigma_y \\ \tau_{xy} \end{bmatrix} = \frac{32 \times 10^6}{\lambda^2} \begin{bmatrix} \quad \end{bmatrix} = \begin{bmatrix} 284.8 \\ 7.002 \\ 7.002 \end{bmatrix} \text{psi}$$

where [ ] matrix equals:

$$\begin{bmatrix} 8.8994\lambda^2 \times 10^{-6} \\ 0.2188\lambda^2 \times 10^{-6} \\ 0.2189\lambda^2 \times 10^{-6} \end{bmatrix}$$

4 Plane Bending of Plates

4.1 General

In the previous Chapter, the behavior of plates was considered when such plates were subjected to in-plane loads. Such plate action might occur in practice, as illustrated by several example problems. In general, however, the loads that are applied to plates are normal to the surface and thus induce bending moments, torsional moments, and shears, as shown in Figure 4-1. It is this action and the evaluation of these induced stresses that will be discussed in this chapter [25, 26, 27, 28]. The application of such solutions will be given in the next chapter as associated with orthotropic bridges. The development of the basic equations will be given in two parts, i.e., for isotropic plates and for orthotropic plates. The definition of each class of plate will be described prior to each development The application of these equations, as mentioned, will be given in chapter 5.

4.2 Isotropic Plate Equations

Consider a small differential element dx by dy, subjected to an external loading q. Such a loading will institute stresses on the element as shown in Figure 4-1. If the resulting deformations are small, i.e., $w < 0.3t$ where t = plate thickness, then there will be an unstrained neutral surface in the plate at mid-depth $t/2$. If the deformations are large, then this neutral surface will develop a strain and thus additional in-plane stresses. Such a condition requires incorporation of the in-plane force equations, which will be examined later.

These plate stresses, shown in Figure 4-1, can be resolved into stress resultants or forces as shown in Figures 4-2 and 4-3. Each of these forces are given in units of force per unit of length; therefore the resultant force can be expressed as a total force by multiplying by the plate width. The changes in the forces are also given, as the result of examining one face of the plate relative to the other. The applied vertical load q, is shown on Figure 4-3 only as are the vertical shears. In actuality the plate forces are the combined effects of those shown in Figure 4-2 and 4-3.

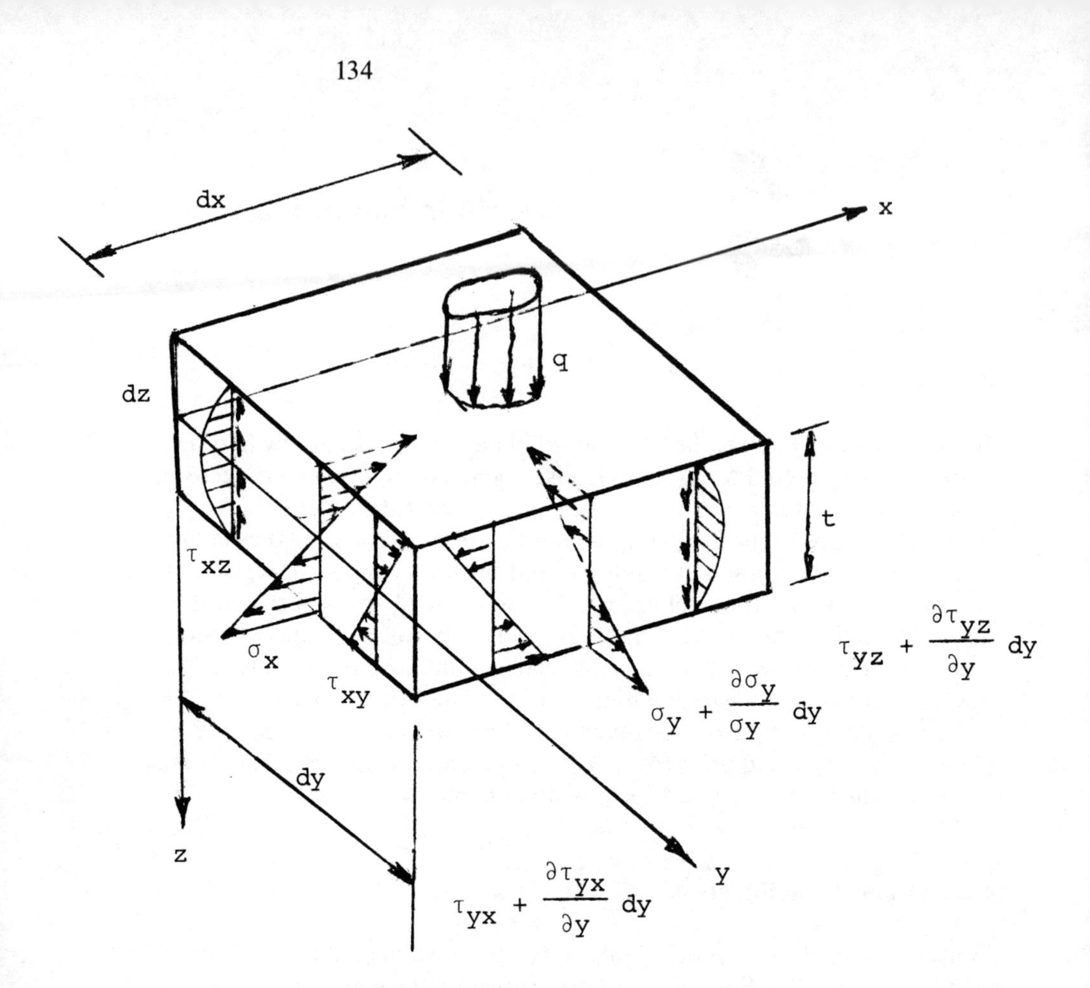

Figure 4-1. Plate Bending Internal Stresses.

4.2.1 Force-Stress Relationship

The forces shown in Figure 4-2 and 4-3 are related to the stresses given in Figure 4-1, by the following:

$$M_x = \int_{-t/2}^{t/2} \sigma_x z \, dz \tag{4.1}$$

$$M_y = \int_{-t/2}^{t/2} \sigma_y z \, dz \tag{4.2}$$

$$M_{xy} = \int_{-t/2}^{t/2} \tau_{xy} z \, dz \tag{4.3}$$

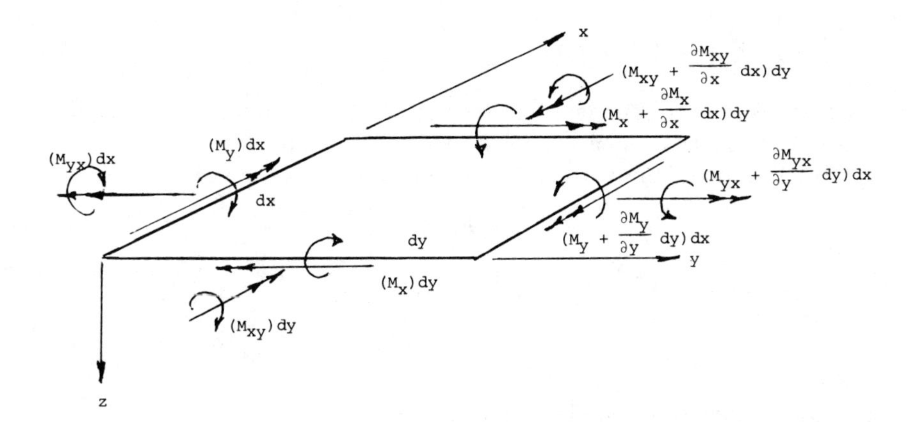

Figure 4-2. Plate Bending Moments and Torques.

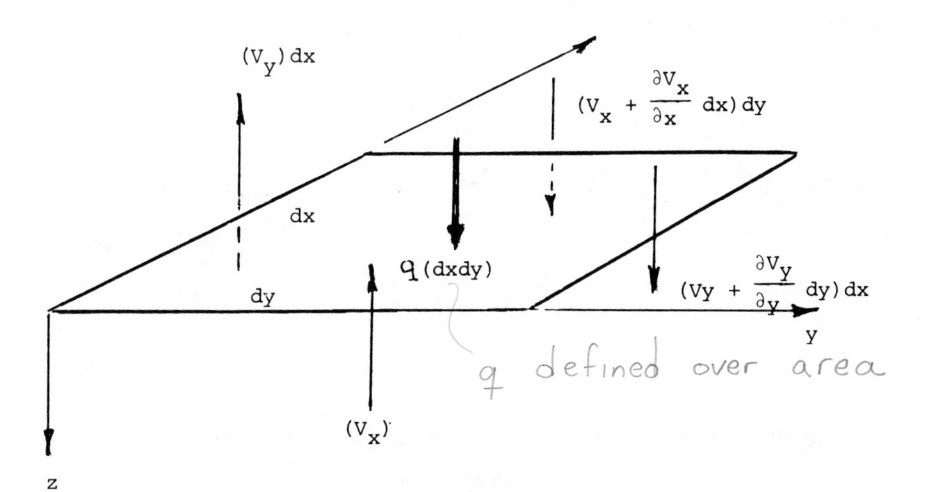

Figure 4-3. Plate Bending Shears.

$$V_x = \int_{-t/2}^{t/2} \tau_{xz}\, dz \qquad (4.4)$$

$$V_y = \int_{-t/2}^{t/2} \tau_{yz}\, dz \qquad (4.5)$$

where the stresses $\tau_{xy} = \tau_{yx}$ as shown in chapter 2, therefore $M_{xy} = M_{yx}$.

4.2.2 Equilibrium Equation

Relationships between these forces can be obtained by considering equilibrium of the plate elements, shown in Figure 4-2 and 4-3.

Summation of the vertical forces, per Figure 4-3, gives

$$\sum F_z = 0$$

$$-V_x\,dy + \left(V_x + \frac{\partial V_x}{\partial x}\,dx\right)dy - V_y\,dx + \left(V_y + \frac{\partial V_y}{\partial y}\,dy\right)dx + q\,dx\,dy = 0$$

multiplying terms and dividing through by $dx\,dy$ gives

$$\frac{\partial V_x}{\partial x} + \frac{\partial V_y}{\partial y} + q = 0 \tag{4.6}$$

The sum of moments through the center of the plate parallel to the y-axis, neglecting higher order terms and referring to Figure 4-2 and 4-3, gives

$$\sum M'_y = 0$$

$$M_x\,dy - \left(M_x + \frac{\partial M_x}{\partial x}\,dx\right)dy + M_{yx}\,dx - \left(M_{yx} + \frac{\partial M_{yx}}{\partial y}\,dy\right)dx + V_x\,dy\,dx = 0$$

or

$$\frac{\partial M_x}{\partial x} + \frac{\partial M_{xy}}{\partial y} - V_x = 0 \tag{4.7}$$

Similarly summing moments parallel to the x-axis gives

$$\sum M'_x = 0$$

$$M_y\,dx - \left(M_y + \frac{\partial M_y}{\partial y}\,dy\right)dx + M_{xy}\,dy - \left(M_{xy} + \frac{\partial M_{xy}}{\partial x}\,dx\right)dy + V_y\,dx\,dy = 0$$

and neglecting higher order terms, gives

$$\frac{\partial M_y}{\partial y} + \frac{\partial M_{xy}}{\partial x} - V_y = 0 \tag{4.8}$$

Equations (4.6), (4.7), and (4.8) can be combined into one equation by taking $\partial/\partial x$ [equation (4.7)] and $\partial/\partial y$ [equation (4.8)] and substituting into equation (4.6), which gives

$$\frac{\partial^2 M_x}{\partial x^2} + 2\frac{\partial^2 M_{xy}}{\partial x \partial y} + \frac{\partial^2 M_y}{\partial y^2} = -q(x, y) \tag{4.9}$$

It should be noted that each of these four equilibrium equations are similar to those developed in strength of materials or static courses, when examining a plane beam under bending. The general equation for a differential beam element under uniform load is

$$\frac{dM}{dx} = V \qquad dM = V\,dx$$

That is, the area under the shear diagram gives the change in moment. This equation assumes only plane bending moment occurs, therefore $M_{xy} = M_{yx} = 0$ (no torsion). Examining equations (4.7) and (4.8), when $M_{xy} = 0$ gives

$$\frac{\partial M_x}{\partial x} = V_x \qquad \text{and} \qquad \frac{\partial M_y}{\partial y} = V_y$$

which is identical to the strength equation given above, as the partial is now an exact derivative because there is only one variable x or y. Continuing, the shear in a plane beam is related to load q by

$$dV = q\,dx \qquad \text{but} \qquad V = \frac{dM}{dx} \qquad \text{or} \qquad \frac{dV}{dx} = \frac{d^2M}{dx^2}$$

therefore, $d^2M/dx^2 = q$. Examination of equation (4.7) shows a similar relationship of $M_{xy} = 0$ and plane bending along one axis is considered.

The solution of equation (4.9) is now required. If the moments were known and they satisfied equation (4.9) then a valid solution could be considered. However, an infinite number of combinations are possible, while only one will uniquely satisfy the imposed boundary conditions. The imposing of the boundary conditions can readily be done by using deformations, rather than forces. Therefore, the relationship between forces and deformation (geometry) is required. The following will describe these relationships and the necessary equations.

4.2.3 Geometry

Considering the plate subjected to the load q, this load will cause the original plate to deform, as shown in Figures 4-4 and 4-5. This deformation will involve both vertical movement w and horizontal movement u. The horizontal movement occurs due to the rotation ϕ_x of the plate, as shown in

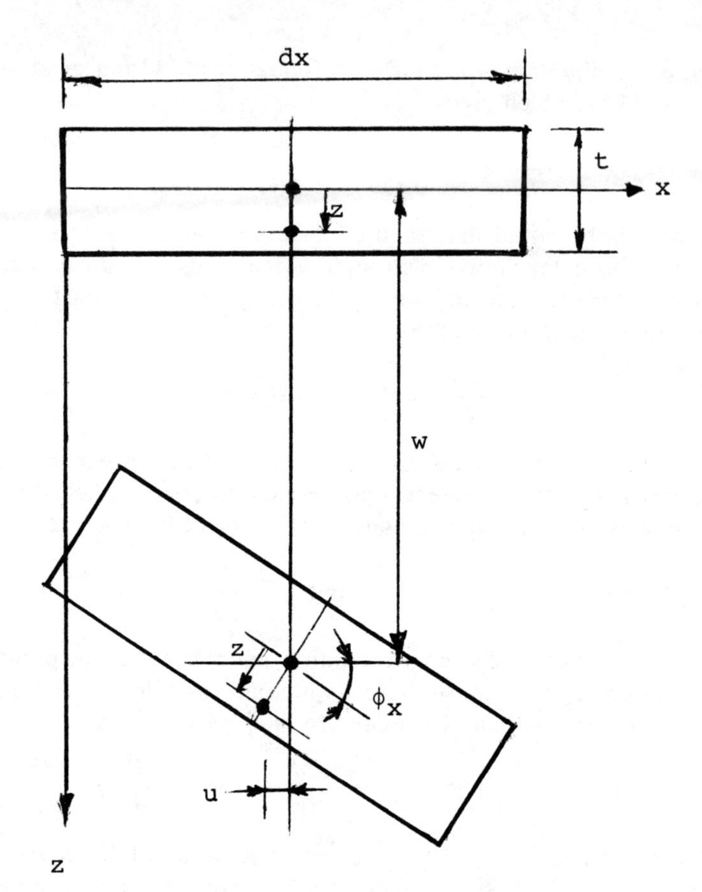

Figure 4-4. Geometry of Plate.

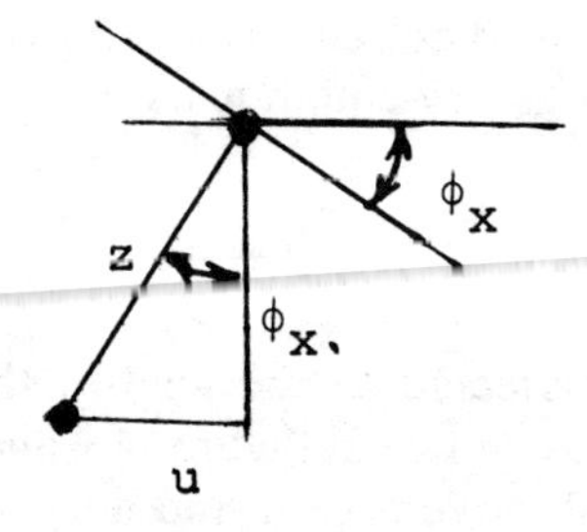

Figure 4-5. Distorted Plate Element.

Figure 4-5. The rotation in this instance is for a plate section along the x-axis. A similar rotation ϕ_y, will occur for a plate section along the y-axis, due to displacement w and v.

Referencing again Figure 4-5, the notation or slope is given by

$$\phi_x = \frac{\partial x}{\partial w}$$

The slope ϕ_x is related to the horizontal displacement u, by the geometry of the section as shown in Figure 4-5,

$$\sin \phi_x = -\frac{u}{z}$$

and for small displacements,

$$u = -z\phi_x$$

or

$$u = -z\frac{\partial w}{\partial x} \tag{4.10a}$$

In the $y-z$ plane the geometry similarly gives

$$v = -z\frac{\partial w}{\partial y} \tag{4.10b}$$

The displacements u and v, were related to strains by equations (2.23) through (2.25) in chapter 2 or

$$\varepsilon_x = \frac{\partial u}{\partial x}$$

$$\varepsilon_y = \frac{\partial v}{\partial y}$$

$$\gamma_{xy} = \frac{\partial u}{\partial y} + \frac{\partial v}{\partial x}$$

Taking the derivatives of (4.10a) and (4.10b) and substituting in equations (2.23) through (2.25) gives

$$\varepsilon_x = -z\frac{\partial^2 w}{\partial x^2} \tag{4.11}$$

$$\varepsilon_y = -z\frac{\partial^2 w}{\partial y^2} \tag{4.12}$$

$$\gamma_{xy} = -2z\frac{\partial^2 w}{\partial x\, \partial y} \tag{4.13}$$

4.2.4 *Stress Equations*

The relationship between stress and strain in a material is given by material constants. If the material is linearly elastic and isotropic then the constants are related by the following equations:

$$\sigma_x = \frac{E}{(1 - \mu^2)}(\varepsilon_x + \mu\ \varepsilon_y)$$

$$\sigma_y = \frac{E}{(1 - \mu^2)}(\varepsilon_y + \mu\ \varepsilon_x) \tag{4.14}$$

$$\tau_{xy} = \frac{E}{2(1 + \mu)}\gamma_{xy} = G\ \gamma_{xy}$$

where E = Modulus of Elasticity

G = Modulus of Rigidity

μ = Poisson's Ratio

Substituting now equations (4.11) through (4.13) into equation (4.14) gives

$$\sigma_x = -\frac{Ez}{(1 - \mu^2)}\left[\frac{\partial^2 w}{\partial x^2} + \mu\frac{\partial^2 w}{\partial y^2}\right]$$

$$\sigma_y = -\frac{Ez}{(1 - \mu^2)}\left[\frac{\partial^2 w}{\partial y^2} + \mu\frac{\partial^2 w}{\partial x^2}\right] \tag{4.15}$$

$$\tau_{xy} = -2Gz\frac{\partial^2 w}{\partial x \partial y} = -\frac{E}{(1 + \mu)}z\frac{\partial^2 w}{\partial x \partial y}$$

4.2.5 *Force Equations*

The relationship between the deformation w and forces M_x, M_y, M_{xy} can now be determined by substituting equations (4.15) into the stress resultant equations (4.1) through (4.3). This gives

$$M_x = \int_{-t/2}^{t/2} \sigma_x z\, dz = -\frac{E}{(1 - \mu^2)}\left[\frac{\partial^2 w}{\partial x^2} + \mu\frac{\partial^2 w}{\partial y^2}\right]\int_{-t/2}^{t/2} z^2\, dz$$

or

$$M_x = -\frac{Et^3}{12(1 - \mu^2)}\left[\frac{\partial^2 w}{\partial x^2} + \mu\frac{\partial^2 w}{\partial y^2}\right] \tag{4.16}$$

$$M_y = \int_{-t/2}^{t/2} \sigma_y z\, dz = -\frac{E}{(1 - \mu^2)}\left[\frac{\partial^2 w}{\partial y^2} + \mu\frac{\partial^2 w}{\partial x^2}\right]\int_{-t/2}^{t/2} z^2\, dz$$

or

$$M_y = -\frac{Et^3}{12(1 - \mu^2)}\left[\frac{\partial^2 w}{\partial y^2} + \mu\frac{\partial^2 w}{\partial x^2}\right] \tag{4.17}$$

$$M_{xy} = \int_{-t/2}^{t/2} \tau_{xy} z\, dz = -G\frac{\partial^2 w}{\partial x\,\partial y}\int_{-t/2}^{t/2} z^2\, dz$$

or

$$M_{xy} = -\frac{Gt^3}{6} \cdot \frac{\partial^2 w}{\partial x\,\partial y} = -(1 - \mu) \cdot \frac{Et^3}{12(1 - \mu^2)} \cdot \frac{\partial^2 w}{\partial x\,\partial y} \tag{4.18}$$

The term $Et^3/[12(1 - \mu^2)]$ will be defined as D, plate stiffness.

Taking the derivatives of equations (4.16) through (4.18), according to the moment-load equation (4.9), gives

$$\frac{\partial^2 M_x}{\partial x^2} = -D\left[\frac{\partial^4 w}{\partial x^4} + \mu\frac{\partial^4 w}{\partial x^2\,\partial y^2}\right]$$

$$\frac{\partial^2 M_{xy}}{\partial x\,\partial y} = -D(1 - \mu) \cdot \left[\frac{\partial^4 w}{\partial x^2\,\partial y^2}\right]$$

$$\frac{\partial^2 M_y}{\partial y^2} = -D\left[\frac{\partial^4 w}{\partial y^4} + \mu\frac{\partial^4 w}{\partial y^2\,\partial x^2}\right]$$

Substituting these derivatives into equation (4.9) gives

$$-D\left[\frac{\partial^4 w}{\partial x^4} + \mu\frac{\partial^4 w}{\partial x^2\,\partial y^2} + 2\frac{\partial^4 w}{\partial x^2\,\partial y^2} - 2\,\mu\frac{\partial^4 w}{\partial x^2\,\partial y^2} + \mu\frac{\partial^4 w}{\partial x^2\,\partial y^2} + \frac{\partial^4 w}{\partial y^4}\right] = -q(x, y)$$

Collecting terms gives the general biharmonic isotropic plate equation (4.19).

$$\frac{\partial^4 w}{\partial x^4} + 2\frac{\partial^4 w}{\partial x^2\,\partial y^2} + \frac{\partial^4 w}{\partial y^4} = \frac{q(x, y)}{D} \tag{4.19}$$

or

$$\nabla^4 w = \frac{q}{D}$$

The form of this plate bending equation (4.19) is similar to the plane stress equation (3.12). Therefore, the homogeneous solution of this equation, i.e., $q/D = 0$, if Fourier series are used, should yield the same solution. This will be explained in the following section.

One interesting aspect of equation (4.19) is the relationship of this expression to strength of materials if $M_{xy} = 0$. The basic strength of material equations for load-deformation of a beam is

$$\frac{d^4y}{dx^4} = \frac{q}{EI}$$

where y = Deformation along the x-Axis

q = Uniform Load

EI = Bending Stiffness

In equation (4.19), considering only action along one axis, $M_y = 0$, then

$$\frac{d^4w}{dx^4} = \frac{q}{D}$$

which is similar to the above expression where D respresents the beam stiffness EI.

The objective of any plate problem is the solution of equation (4.19). The solution of such, as applied to various problems, will be explained in the next sections.

Continuing with the other force equations, V_x, and V_y, then relationships in terms of deflections are found by taking derivatives of equations (4.16) through (4.18) and substituting into equations (4.7) and (4.8) which gives

$$V_x = -D\left[\frac{\partial^3 w}{\partial x^3} + \frac{\partial^3 w}{\partial x\,\partial y^2}\right] \tag{4.20}$$

$$V_y = -D\left[\frac{\partial^3 w}{\partial y^3} + \frac{\partial^3 w}{\partial y\,\partial x^2}\right] \tag{4.21}$$

4.2.6 Boundary Conditions

The solution of the general plate equation (4.19) requires specification of the plate boundaries. Therefore, depending on the plate restraint and deformation, certain conditions can be stipulated. Three basic conditions will be considered. (1) Fixed or built-in support, (2) Simple support, (3) Free edge. The boundary conditions for these support cases are as follows.

Fixed Support: As shown in Figure 4-6, the edge conditions are

$$\left.\begin{aligned} w &= 0 \quad \text{at} \quad x = 0 \\ \frac{\partial w}{\partial x} &= 0 \quad \text{at} \quad x = 0 \end{aligned}\right\} \tag{4.22}$$

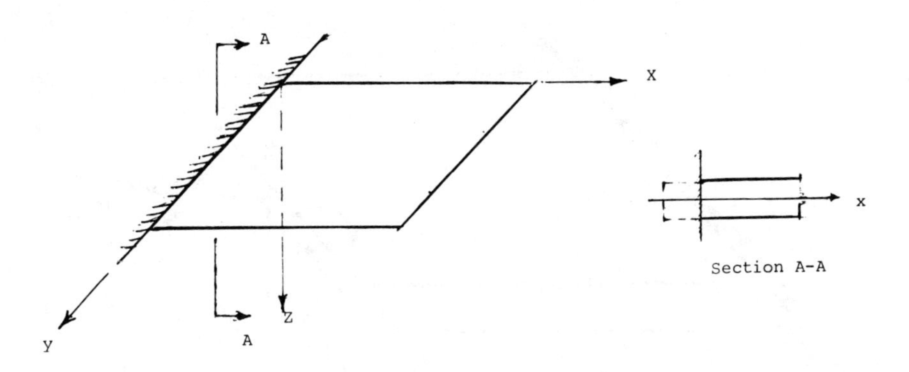

Figure 4-6. Fixed Ended Plate.

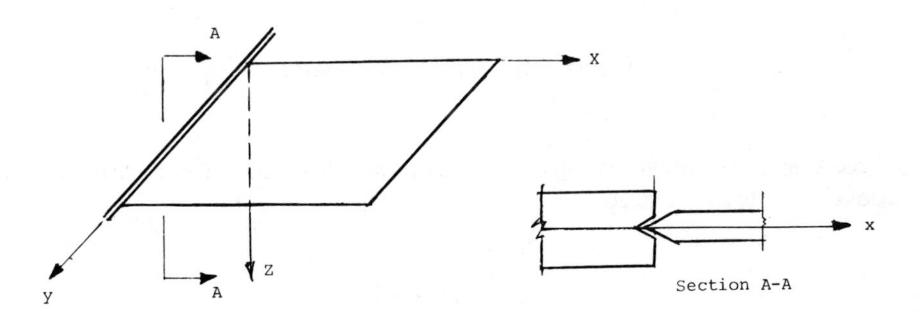

Figure 4-7. Pinned Ended Plate.

Simple Support: As shown in Figure 4-7, the edge conditions are

$$w = 0 \qquad \text{at} \qquad x = 0$$

$$M_x \Big|_{x=0} = \frac{\partial^2 w}{\partial x^2} + \mu \frac{\partial^2 w}{\partial y^2} = 0$$

however,

$$\frac{\partial w}{\partial y} = \frac{\partial^2 w}{\partial y^2} = 0$$

therefore

$$\frac{\partial^2 w}{\partial x^2} = 0 \tag{4.23}$$

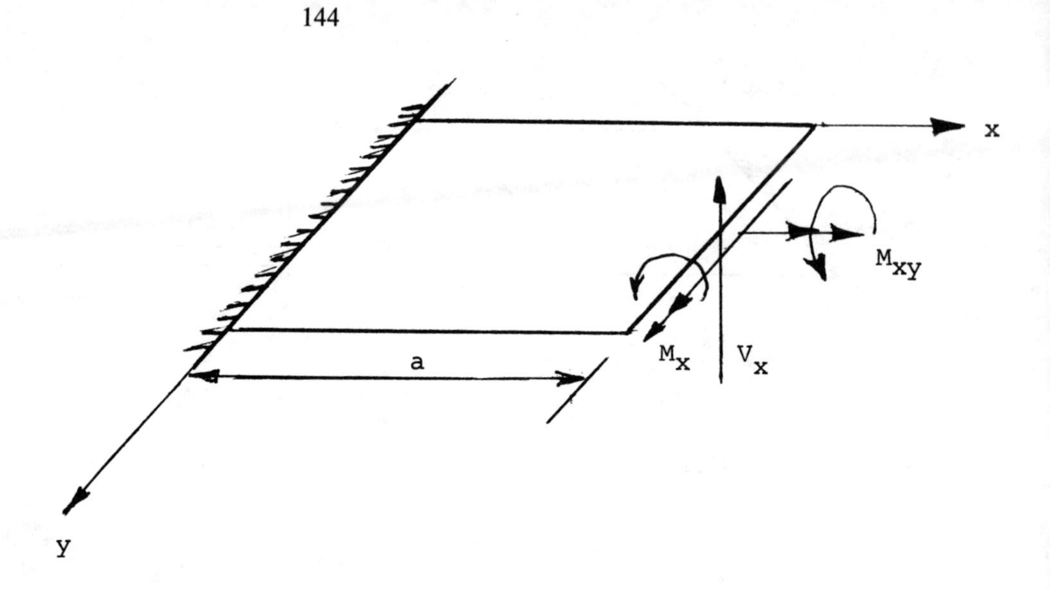

Figure 4-8. Free Ended Plate.

Free Edge: As shown in Figure 4-8, there are three edge forces that can develop, which are zero.

$$\left.\begin{aligned} M_x &= 0 \\ V_x &= 0 \\ M_{xy} &= 0 \end{aligned}\right\} \quad \text{at} \quad x = a$$

Two of the edge forces, V_x and M_{xy}, can be combined into one equation, which define a reaction R_x or R_y term [4, 6]. As an example, the R_x equation will be developed by examination of a free body of the y face of the plate, as shown in Figure 4-9(a). Consider the shear force applied $V_x dy$, and to either side the torsional moment and the increase in the torsional moment on a differential length dy. These moments are made equal to couples by dividing by the length dy, which gives the forces shown in Figure 4-9(b). The total resultant force R at the section is, thereforc,

$$R = M_{xy} + \frac{\partial M_{xy}}{\partial y} dy - M_{xy} + V_x dy$$

$$R = \frac{\partial M_y}{\partial y} dy + V_x dy$$

The force per unit length R_x is found by dividing R by dy,

$$R_x = \frac{R}{dy} = V_x + \frac{\partial M_{xy}}{\partial y} \tag{4.24a}$$

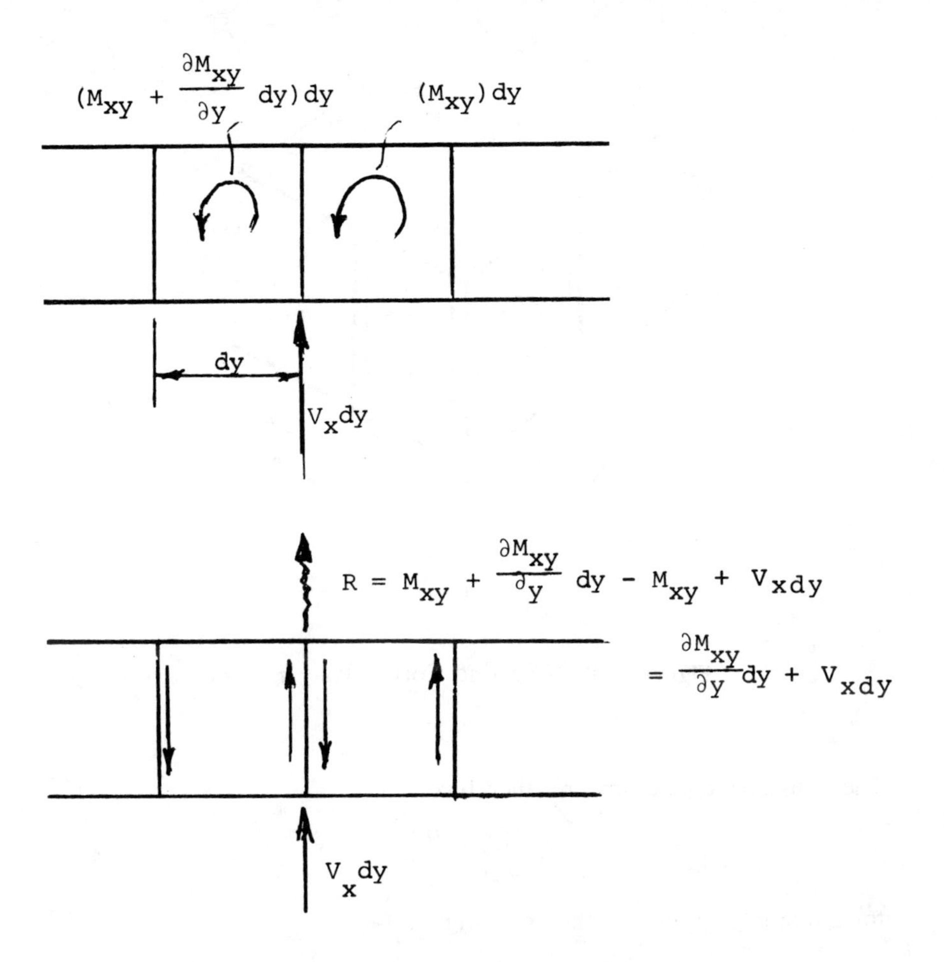

Figure 4-9. Free End Plate Forces.

Substituting in equation (4.20) and the $\partial/\partial y$ of equation (4.18) gives

$$R_x = -D\left[\frac{\partial^3 w}{\partial x^3} + (2 - \mu)\frac{\partial^3 w}{\partial x\,\partial y^2}\right] \qquad (4.24b)$$

Similarly for the x face:

$$R_y = V_y + \frac{\partial M_{xy}}{\partial x} \qquad (4.25a)$$

or

$$R_y = -D\left[\frac{\partial^3 w}{\partial y^3} + (2 - \mu)\frac{\partial^3 w}{\partial y\,\partial x^2}\right] \qquad (4.25b)$$

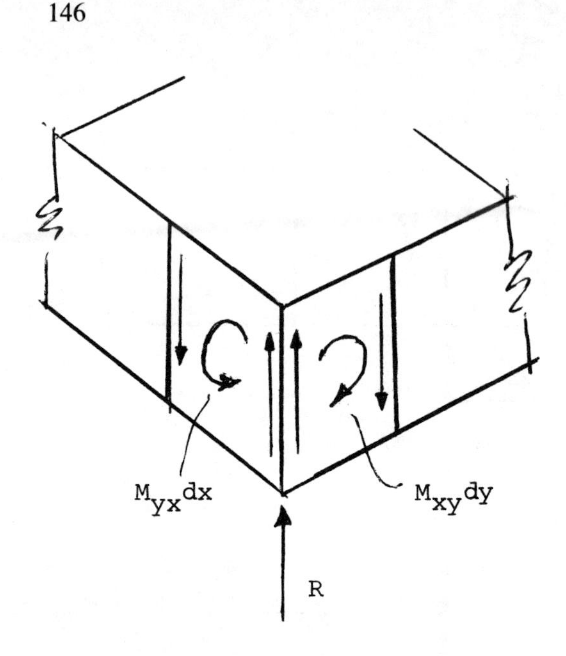

Figure 4-10. Free End Corner Plate Forces.

The boundary conditions are, therefore,

$$\left.\begin{aligned} M_x &= 0 \\ R_x &= 0 \end{aligned}\right\} \tag{4.26}$$

for a free edged plate as shown in Figure 4-8.

Corner: As in the case of a free edge, the torsional moment developed at a corner can be equated to a couple, as shown in Figure 4-10. The resultant force at the corner is, therefore,

$$\left.\begin{aligned} R_{xy} &= M_{xy} + M_{yx} \\ R_{xy} &= 2M_{xy} \end{aligned}\right\} \tag{4.27}$$

where M_{xy} is as defined by equation (4.18).

4.3 Orthotropic Plate Equations

The development of the general plate equation (4.19) considered the equilibrium and deformation of a differential plate element that had isotropic material properties, i.e., E, G, and μ. If we consider a more general

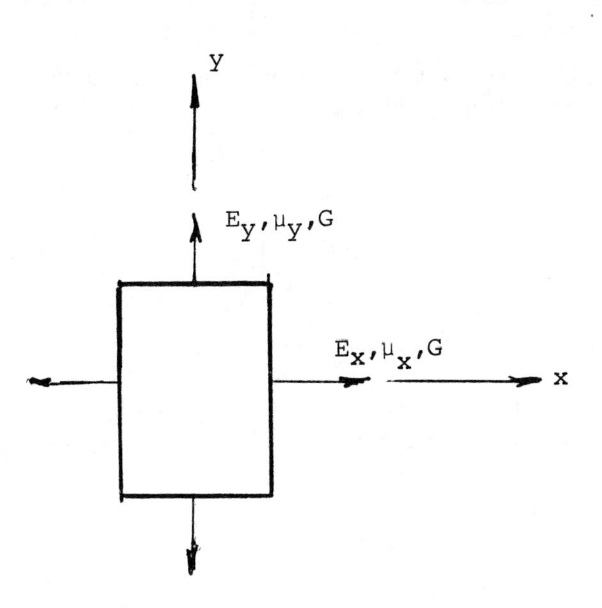

Figure 4-11. Orthotropic Plate Material Properties.

state of the material in which various properties may exist at various angles referenced to the $x-y$ coordinates then the plate material is considered anisotropic [4, 6, 25]. If the material has a variation at right angles, then the material is considered orthogonal-anisotropic, or as abbreviated, "orthotropic." This special case of anisotrophy will be used in the modification of the isotropic plate equations. Such a material and its properties is shown in Figure 4-11.

The general strain equations for this material can be written as

$$\begin{aligned}\varepsilon_x &= \sigma_x/E_x - \mu_y\sigma_y/E_y \\ \varepsilon_y &= \sigma_y/E_y - \mu_x\sigma_x/E_x \\ \gamma_{xy} &= \tau_{xy}/G\end{aligned} \tag{4.28}$$

These strain equations can be solved for stresses, giving

$$\begin{aligned}\sigma_x &= \frac{E_x}{(1 - \mu_x\mu_y)}(\varepsilon_x + \mu_y\varepsilon_y) \\ \sigma_y &= \frac{E_y}{(1 - \mu_x\mu_y)}(\varepsilon_y + \mu_x\varepsilon_x) \\ \tau_{xy} &= G\gamma_{xy}\end{aligned} \tag{4.29}$$

Substituting the strain-displacement relationship, equations (4.11) through (4.13), into equation (4.29) gives

$$\sigma_x = -\frac{E_x z}{(1-\mu_x\mu_y)}\left[\frac{\partial^2 w}{\partial x^2} + \mu_y \frac{\partial^2 w}{\partial x\,\partial y^2}\right]$$

$$\sigma_y = -\frac{E_y z}{(1-\mu_x\mu_y)}\left[\frac{\partial^2 w}{\partial y^2} + \mu_x \frac{\partial^2 w}{\partial x^2\,\partial y}\right] \tag{4.30}$$

$$\tau_{xy} = -2\,Gz\frac{\partial^2 w}{\partial x\,\partial y^2}$$

Integrating these stress equations (4.30) through the plate depth, according to equations (4.1) through (4.3), gives the following moment expressions:

$$M_x = -\frac{E_x t^3}{12(1-\mu_x\mu_y)}\left[\frac{\partial^2 w}{\partial x^2} + \mu_y \frac{\partial^2 w}{\partial y^2}\right] \tag{4.31}$$

$$M_y = -\frac{E_y t^3}{12(1-\mu_x\mu_y)}\left[\frac{\partial^2 w}{\partial y^2} + \mu_x \frac{\partial^2 w}{\partial x^2}\right] \tag{4.32}$$

$$M_{xy} = -\frac{Gt^3}{6}\cdot\frac{\partial^2 w}{\partial x\,\partial y} \tag{4.33}$$

Substituting equations (4.31) through (4.33) into the moment equation (4.9)

$$\frac{\partial^2 M_x}{\partial x^2} + 2\frac{\partial^2 M_{xy}}{\partial x\,\partial y} + \frac{\partial^2 M_y}{\partial y^2} = -q(x, y) \tag{4.9}$$

gives

$$-\frac{E_x t^3}{12(1-\mu_x\mu_y)}\left[\frac{\partial^4 w}{\partial x^4} + \mu_y \frac{\partial^4 w}{\partial x^2\,\partial y^2}\right] - \frac{E_y t^3}{12(1-\mu_x\mu_y)}\left[\frac{\partial^4 w}{\partial y^4} + \mu_x \frac{\partial^4 w}{\partial y^2\,\partial x^2}\right] - 4\frac{Gt^3}{12}\frac{\partial^4 w}{\partial x^2\,\partial y^2} = -q(x, y)$$

Let

$$D_x = \frac{E_x t^3}{12(1-\mu_x\mu_y)} \tag{4.34}$$

$$D_y = \frac{E_y t^3}{12(1-\mu_x\mu_y)} \tag{4.35}$$

$$D_{xy} = \frac{Gt^3}{12} \tag{4.36}$$

therefore,

$$- D_x\left(\frac{\partial^4 w}{\partial x^4}\right) + (-D_x\mu_y - D_y\mu_x - 4D_{xy})\frac{\partial^4 w}{\partial x^2\,\partial y^4} - D_y\left(\frac{\partial^4 w}{\partial y^4}\right) = -q(x,\ y)$$

and letting

$$2H = 4D_{xy} + \mu_y D_x + \mu_x D_y \tag{4.37}$$

$$D_x\frac{\partial^4 w}{\partial x^4} + 2H\frac{\partial^4 w}{\partial x^2\,\partial y^2} + D_y\frac{\partial^4 w}{\partial y^4} = q(x,\ y) \tag{4.38}$$

Equation (4.38) should agree with the isotropic equation (4.19), if $\mu_x = \mu_y = \mu$ and $E_x = E_y = E$, therefore,

$$D = D_x = D_y = \frac{Et^3}{12(1 - \mu^2)} \qquad \text{(4.34) and (4.35)}$$

and

$$D_{xy} = \frac{Gt^3}{12} \tag{4.36}$$

Therefore,

$$2H = 4D_{xy} + \mu_y D_x + \mu_y D_y \tag{4.37}$$

gives

$$2H = \frac{2Gt^3}{6} + \mu\left[\frac{Et^3}{12(1 - \mu^2)} + \frac{Et^3}{12(1 - \mu^2)}\right]$$

and

$$G = \frac{E}{2(1 + \mu)}$$

Therefore,

$$2H = \frac{2Et^3}{12(1 + \mu)} + \frac{2\mu Et^3}{12(1 - \mu^2)}$$

or

$$2H = (2(1 - \mu) + 2\mu)\frac{Et^3}{12(1 - \mu^2)}$$

and

$$2H = \frac{2Et^3}{12(1 - \mu^2)}$$

$$2H = 2\frac{Et^3}{12(1 - \mu^2)} = 2D$$

Therefore $H = D$.

Substituting $D_x = D$, $D_y = D$, and $H = D$ into equation (4.38) gives

$$D\frac{\partial^4 w}{\partial x^4} + 2D\frac{\partial^4 w}{\partial x^2\,\partial y^2} + D\frac{\partial^4 w}{\partial y^4} = q(x, y)$$

$$\nabla^4 w = q/D$$

which is identical to equation (4.19).

Expressions for V_x, V_y, R_x, and R_y for the orthotropic material can also be developed [25]. A summary of all of the equations are given as follows.

$$M_x = -D_x \cdot \left(\frac{\partial^2 w}{\partial x^2} + \mu_y\frac{\partial^2 w}{\partial y^2}\right) \tag{4.31}$$

$$M_y = -D_y \cdot \left(\frac{\partial^2 w}{\partial y^2} + \mu_x\frac{\partial^2 w}{\partial x^2}\right) \tag{4.32}$$

$$M_{xy} = M_{yx} = -2D_{xy}\frac{\partial^2 w}{\partial x\,\partial y} \tag{4.33}$$

$$V_x = -D_x \cdot \left[\frac{\partial^3 w}{\partial x^3} + \left(\mu_y + 2\frac{D_{xy}}{D_x}\right) \cdot \frac{\partial^3 w}{\partial x\,\partial y^2}\right] \tag{4.39}$$

$$V_y = -D_y \cdot \left[\frac{\partial^3 w}{\partial y^3} + \left(\mu_x + 2\frac{D_{xy}}{D_y}\right) \cdot \frac{\partial^3 w}{\partial y\,\partial x^2}\right] \tag{4.40}$$

$$R_x = -D_x \cdot \left[\frac{\partial^3 w}{\partial x^3} + \left(\mu_y + 4\frac{D_{xy}}{D_x}\right) \cdot \frac{\partial^3 w}{\partial x\,\partial y^2}\right] \tag{4.41}$$

$$R_y = -D_y \cdot \left[\frac{\partial^3 w}{\partial y^3} + \left(\mu_x + 4\frac{D_{xy}}{D_y}\right) \cdot \frac{\partial^3 w}{\partial x^2\,\partial y}\right] \tag{4.42}$$

4.4 Infinitely Long Plate with Parallel Simple Supports and Constant Load

The solution of the isotropic plate equation (4.19) or the orthotropic plate equation (4.38) can be solved by direct integration or by Fourier series

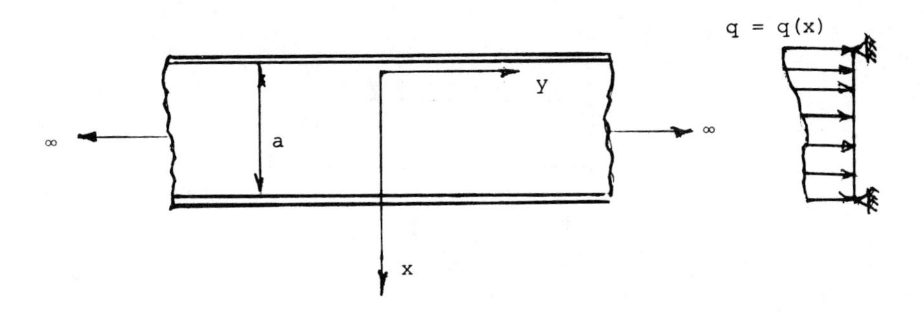

Figure 4-12. Simply Supported Infinite Plate.

when the plate is infinite in length, has simple supports, and has a load that is constant with respect to the y-axis, as shown in Figure 4-12. These two methods of solution will now be presented.

4.4.1 Plate Analysis Using Direct Integration

Examining again Figure 4-12, $q = q(x)$ thus $w = w(x)$ only, and therefore $\partial w/\partial y = \partial^2 w/\partial y^2 = \partial^3 w/\partial y^3 = \partial^4 w/\partial y^4 = 0$. The plate equation $\nabla^4 w = q/D$, considering an isotropic plate, becomes:

$$\frac{\partial^4 w}{\partial x^4} = \frac{q(x)}{D}$$

and

$$M_x = -D\frac{\partial^2 w}{\partial x^2}$$

$$M_y = -D\,\mu\frac{\partial^2 w}{\partial x^2} \qquad (4.43a)$$

$$M_{xy} = -M_{yx} = 0$$

$$V_x = -D\frac{\partial^3 w}{\partial x^3}$$

$$V_y = 0$$

If $q(x)$ is a continuously integrable function, then the solution to $d^4w/dx^4 = q(x)$ is found by repeated integration with application of the proper boundary conditions.

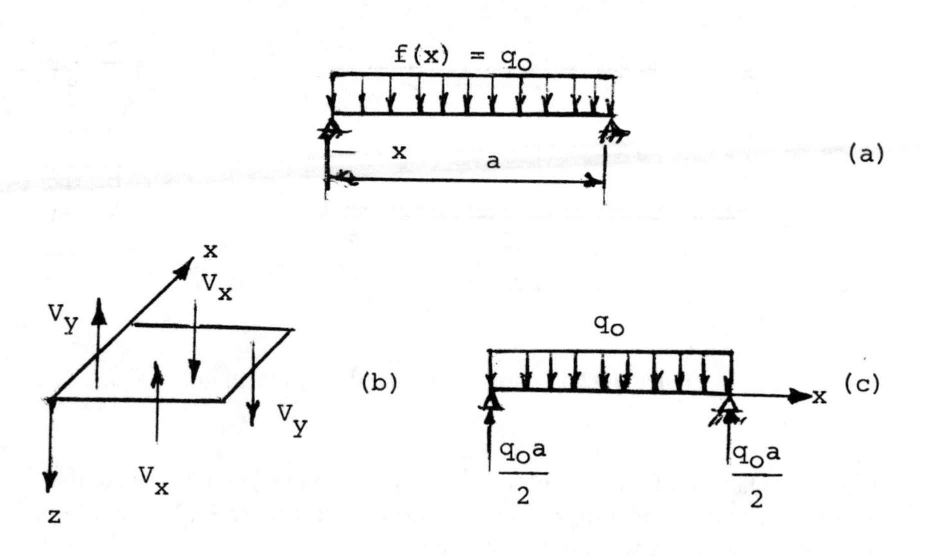

Figure 4-13. Uniformly Loaded Plate.

Example: Consider the problem of a uniformly loaded plate, as shown in Figure 4-13. The load $q(x) = q_0$; the boundary conditions are:

$$\text{at} \quad x = 0, \quad w = 0: \quad \partial^2 w/\partial x^2 = 0$$

$$\text{at} \quad x = a, \quad w = 0: \quad \partial^2 w/\partial x^2 = 0$$

which refer to simple supports, as given by equation (4.23). The plate equation (4.19) is, therefore,

$$\frac{\partial^4 w}{\partial x^4} = \frac{q_0}{D}$$

Integrating continuously gives

$$w''' = \frac{q_0}{D}x + C_1$$

$$w'' = \frac{q_0}{D}x^2/2 + C_1 x + C_2$$

$$w' = \frac{q_0}{D}x^3/6 + C_1 x^2/2 + C_2 x + C_3$$

$$w = \frac{q_0}{D}x^4/24 + C_1 x^3/6 + C_2 x^2/2 + C_3 x + C_4 \qquad (4.43b)$$

Applying the boundary conditions gives

$$\text{at } x = 0: \qquad w = 0$$

$$C_4 = 0$$

$$\text{at } x = 0: \qquad \partial^2 w / \partial x^2 = 0$$

$$C_2 = 0$$

$$\text{at } x = a: \qquad w = 0$$

$$0 = \frac{q_0}{D}\frac{a^4}{24} + \frac{C_1 a^3}{6} + C_3 a$$

$$\text{at } x = a: \qquad w'' = 0$$

$$0 = \frac{q_0}{D}\frac{a^2}{2} + C_1 a; \qquad C_1 = -\frac{q_0}{D}\frac{a}{2}$$

$$C_3 = -\frac{q_0}{D}\frac{a^3}{24} + \frac{q_0}{D}\frac{a^3}{12}; \qquad C_3 = \frac{q_0}{D}\frac{a^3}{24}$$

Substituting the constants C_1, C_2, C_3, and C_4 into equation (4.43b) gives

$$w = \frac{q_0}{D}\left[\frac{x^4}{24} - \frac{a}{2}\frac{x^3}{6} + \frac{a^3}{24}x\right]$$

or

$$w = \frac{q_0}{24D}[x^4 - 2ax^3 + a^3 x]$$

With this general deflection known, the shear or reactions can be found.

$$V_x = -D\left[\frac{\partial^3 w}{\partial x^3} + \frac{\partial^3 w}{\partial x \partial y^2}\right]$$

however

$$\frac{\partial^3 w}{\partial x \partial y^2} = 0$$

therefore,

$$V_x = -D\left[\frac{\partial^3 w}{\partial x^3}\right] = R_x,$$

therefore,

$$V_x = -\frac{q_0}{24}[24x - 12a]$$

or

$$V_x = \frac{q_0}{2}[a - 2x]$$

$$V_y = -D\left[\frac{\partial^3 w}{\partial y^3} + \frac{\partial^3 w}{\partial y\, \partial x^2}\right]$$

or

$$V_y = 0$$

The reactions or shears acting on the plate are shown in Figure 4-13(a):

$$\text{at } x = 0, \qquad R_x = V_x = q_0 a/2$$

and

$$\text{at } x = a, \qquad R_x = V_x = -q_0 a/2$$

The directions of the forces are as shown in Figure 4-13(c). The moments are computed as follows:

$$M_x = -D\frac{\partial^2 w}{\partial x^2} = -\frac{q_0}{24}(12x^2 - 12ax]$$

or

$$M_x = \frac{q_0}{2}[ax - x^2]$$

$$M_y = -D\,\mu\frac{\partial^2 w}{\partial x^2} = \mu\frac{q_0}{2}[ax - x^2]$$

4.4.2 Plate Analysis Using Fourier Series

Generally, the load is such that it can't be continuously integrated; therefore, it is convenient to represent the loading approximately in the form of the easily integrated Fourier series, as given in section 3.3.1:

$$q(x) = \frac{1}{2}a_0 + \sum a_n \cos \lambda_n x + \sum b_n \sin \lambda_n x$$

Considering the plate under study, as shown in Figure 4-12, the loading $q(x)$ can be represented as an odd function, as shown in Figure 4-14. Using an odd function of period length $2L = 2a$, the loading and deformation will be assumed to have the following form

$$q(x) = \sum q_n \sin \lambda_n x$$

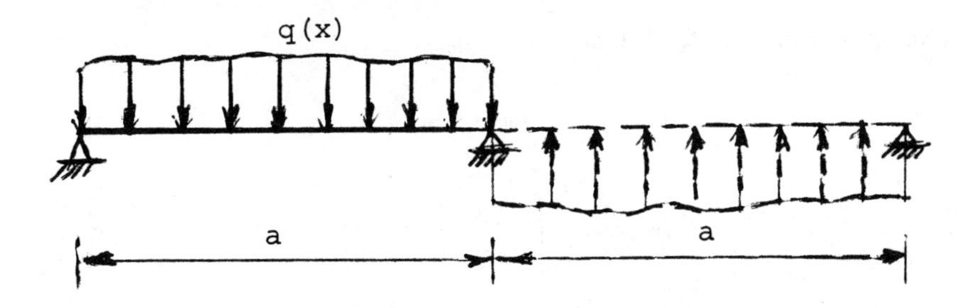

Figure 4-14. Fourier Series Loading.

where

$$\lambda_n = \frac{n\pi}{a}$$

$$w(x) = \sum w_n \sin \lambda_n x$$

which automatically satisfies the boundary conditions. The plate equation becomes

$$\frac{\partial^4 w}{\partial x^4} = \frac{q}{D}$$

or

$$\lambda^4 w_n \sin \lambda_n x = \frac{q_n}{D} \sin \lambda_n x$$

Solving for w_n gives

$$w_n = \frac{q_n}{D} \Big/ \lambda_n^4 = \frac{q_n}{D} \cdot \frac{a^4}{n^4 \pi^4}$$

$$w(x) = \frac{a^4}{\pi^4 D} \sum \frac{q_n}{n^4} \sin \lambda_n x \qquad (4.44)$$

This equation (4.44) is now general for any loading $q(x)$. All that has to be evaluated is the Fourier series term q_n for the respective loading $q(x)$. The term q_n is found from the odd series equation (3.47), or:

$$q_n = \frac{2}{L} \int_0^L f(x) \sin \lambda_n x \, dx$$

where $L = a$, $\lambda_n = n\pi/a$, and the $f(x)$ equal $q(x)$.

Example: For a uniform load $q(x) = q_0$,

$$q_n = \frac{2}{a}\int_0^a q_0 \sin \lambda_n x\, dx$$

$$q_n = \frac{2q_0}{a}\left[\frac{1}{\lambda_n}(-\cos \lambda_n x)\right]_0^a$$

$$q_n = \frac{4q_0}{n\pi} \qquad n = 1, 3, 5, \ldots$$

$$q_n = 0 \qquad n = 2, 4, 6, \ldots$$

$$q(x) = \sum q_n \sin \lambda_n x$$

$$q(x) = \frac{4q_0}{\pi}\sum \frac{1}{n} \sin \lambda_n x$$

Substituting $q(x)$ into equation (4.44) gives

$$w(x) = \frac{4q_0 a^4}{\pi^5 D} \sum_{n=1,3,5,\ldots} \left(\frac{1}{n^5} \sin \lambda_n x\right) \tag{4.45}$$

The moments and shears, using equations (4.43a), are therefore computed as

$$M_x = -D\frac{\partial^2 w}{\partial x^2}$$

$$M_x = +\frac{4q_0 a^4}{\pi^5}\frac{1}{n^5}\lambda_n^2 \sin \lambda_n x$$

$$M_x = \frac{4q_0 a^4}{\pi^5}\frac{1}{n^5}\frac{n^2\pi^2}{a^2} \sin \lambda_n x$$

$$M_x = \frac{4q_0 a^2}{\pi^3}\sum \frac{1}{n^3} \sin \lambda_n x$$

$$M_y = \mu\frac{4q_0 a^2}{\pi^3}\sum \frac{1}{n^3} \sin \lambda_n x$$

$$V_x = -D\frac{\partial^3 w}{\partial x^3}$$

$$V_x = \frac{4q_0 a^4}{\pi^5}\frac{1}{n^5}\lambda_n^3 \cos \lambda_n x$$

$$V_x = \frac{4q_0 a^4}{\pi^5}\frac{1}{n^5}\frac{n^3\pi^3}{a^3} \cos \lambda_n x$$

$$V_x = \frac{4q_0a}{\pi^2}\sum\frac{1}{n^2}\cos\lambda_n x$$

4.4.3 Comparison of Results

In the previous sections two methods were employed to solve the plate equation. These results will now be compared for the resulting deflections and forces at midspan, i.e., at $x = a/2$ for deflection and moment and at $x = 0$ for shear.

Direct Integration:

(1) *Deflection* $x = a/2$

$$w = \frac{q_0}{24D}[x^4 - 2ax^3 + a^3x]$$

$$w = \frac{q_0}{24D}\left[\frac{a^4}{16} - 2a\frac{a^3}{8} + a^3\frac{a}{2}\right]$$

$$w = \frac{q_0}{24D}\left[a^4 - 4a^4 + 8a^4\right]\frac{1}{16}$$

$$w = \frac{q_0}{384D}5a^4$$

(2) *Moment* $x = a/2$

$$M_x = \frac{q_0}{2}\left[\frac{a^2}{2} - \frac{a^2}{4}\right]$$

$$M_x = \frac{q_0a^2}{8}$$

(3) *Shears* $x = 0$

$$V_x = \frac{q_0a}{2}$$

Fourier Series:

(1) *Deflection*

$$w = \frac{4q_0a^4}{\pi^5 D}\sum\frac{1}{n^5}\sin\lambda_n x$$

$$x = \frac{a}{2}, \qquad n = 1$$

$$w = \frac{4q_0a^4}{\pi^5 D} \sin\frac{\pi}{a}\frac{a}{2}$$

$$w = \frac{4q_0a^4}{\pi^5 D} = \frac{4q_0a^4}{304D}$$

$$w = \frac{5.87q_0a^4}{384D} \quad \text{versus} \quad \frac{5.0q_0a^4}{384D}$$

(2) *Moment*

$$M_x = \frac{4q_0a^2}{\pi^3}\sum\frac{1}{n^3} \sin \lambda_n x$$

$$M_x = \frac{4q_0a^2}{\pi^3} = \frac{4q_0a^2}{31}$$

$$M_x = \frac{1.03q_0a^2}{8} \quad \text{versus} \quad \frac{q_0a^2}{8}$$

(3) *Shears*

$$V_x = \frac{4q_0a}{\pi^2}\sum\frac{1}{n^2} \cos \lambda_n x \qquad \text{at } x = 0$$

$$V_x = \frac{4q_0a}{\pi^2}$$

$$V_x = \frac{4}{9.85} q_0a$$

$$V_x = 0.814\frac{q_0a}{2} \quad \text{versus} \quad \frac{q_0a}{2}$$

As seen by these comparisons, the Fourier series technique yields good results considering that only one term is used.

4.5 Simply Supported Plates on All Sides

4.5.1 Navier's Solution

Consider a plate that has simple supports on all sides, as shown in Figure 4-15. The boundary conditions for this plate are

$$\text{at } x = 0, \qquad x = L:$$

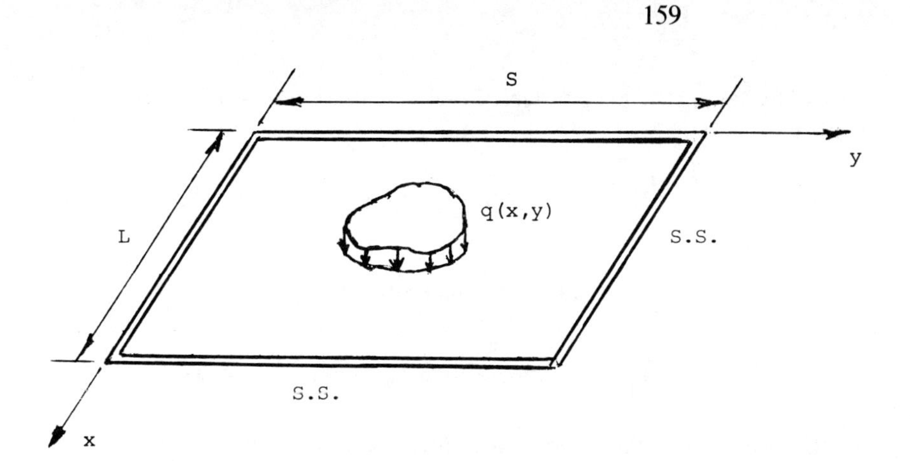

Figure 4-15. Navier Plate.

$$w = 0, \qquad \partial^2 w/\partial x^2 = 0$$

$$\text{at } y = 0, \qquad y = S:$$

$$w = 0, \qquad \partial^2 w/\partial y^2 = 0$$

with an arbitrary load $q(x, y)$ applied. The solution to this problem is called "Navier's Solution" [4, 6, 22, 23, 24] and utilizes double Fourier series.

Assume first that the loading has a double series form or:

$$q(x, y) = \sum_m \sum_n a_{mn} \sin \lambda_m x \sin \alpha_n y \tag{4.46}$$

where $\quad m = 1, 2, 3, 4, \ldots; \qquad \lambda_m = \dfrac{m\pi}{L}$

$$n = 1, 2, 3, 4, \ldots; \qquad \alpha_n = \frac{n\pi}{S}$$

Assume also the deflected surface is of the form

$$w = \sum_m \sum_n b_{mn} \sin \lambda_m x \sin \alpha_n y \tag{4.47}$$

which automatically satisfies the boundary conditions, i.e.,

$$w = 0, \qquad \partial^2 w/\partial x^2 = 0 \qquad x = 0, L$$

$$w = 0, \qquad \partial^2 w/\partial y^2 = 0 \qquad y = 0, S$$

Now the solution of the isotropic plate equation (4.19)

$$\frac{\partial^4 w}{\partial x^4} + 2\frac{\partial^4 w}{\partial x^2 \partial y^2} + \frac{\partial^4 w}{\partial y^4} = \frac{q}{D} \tag{4.19}$$

is found by taking the proper derivatives of equation (4.47) for a given m, n term, or

$$\frac{\partial^4 w}{\partial x^4} = b_{mn}\lambda_n^4 \sin \lambda_m x \sin \alpha_n y$$

$$\frac{\partial^4 w}{\partial y^4} = b_{mn}\alpha_m^4 \sin \lambda_m x \sin \alpha_n y$$

$$\frac{\partial^4 w}{\partial x^2\, \partial y^2} = b_{mn}(2\alpha_n^2\lambda_m^2) \sin \lambda_m x \sin \alpha_n y$$

Substituting these relationships into (4.19) gives

$$\nabla^4 w = b_{mn}(\lambda_m^4 + 2\alpha_n^2\lambda_m^2 + \alpha_n^4) \sin \lambda_m x \sin \alpha_n y = \frac{q}{D}$$

$$= \frac{1}{D} a_{mn} \sin \lambda_m x \sin \alpha_n y$$

Solving for the coefficient b_{mn} gives

$$b_{mn} = \frac{a_{mn}}{D[\lambda_m^4 + 2\alpha_n^2\lambda_m^2 + \alpha_n^4]}\,; \qquad \lambda_m = \frac{m\pi}{L}\,; \qquad \alpha_n = \frac{n\pi}{S}$$

and substituting b_{mn} into the deflection equation (4.47) gives

$$w = \sum_m \sum_n \frac{a_{mn}}{D\left[\dfrac{m^4\pi^4}{L^4} + 2\dfrac{m^2n^2\pi^4}{L^2S^2} + \dfrac{n^4\pi^4}{S^4}\right]} \sin m\lambda_n \sin n\alpha_n$$

or

$$w = \frac{S^4}{\pi^4 D} \sum_m \sum_n \frac{a_{mn} \sin \lambda_m x \sin \alpha_n y}{[m^4(S/L)^4 + 2m^2n^2(S/L)^2 + n^4]} \tag{4.48a}$$

where a_{mn} is a coefficient to be defined by a Fourier series solution. Equation (4.48a) can also be written as

$$w = \frac{S^4}{\pi^4 D} \sum_m \sum_n \frac{a_{mn} \sin m\lambda_m \sin n\alpha_n}{[m^2(S/L)^2 + n^2)]^2} \tag{4.48b}$$

The solution may also be obtained by using the orthotropic plate equation:

$$D_x \frac{\partial^4 w}{\partial x^4} + 2H \frac{\partial^4 w}{\partial x^2\, \partial y^2} + D_y \frac{\partial^4 w}{\partial y^4} = q \tag{4.38}$$

Substituting the derivatives of w into equation (4.38) gives

$$b_{mn}[D_x\lambda_m^4 + 2H\lambda_m^2\alpha_n^2 + D_y\alpha_n^4] \sin \lambda_m x \sin \alpha_n y = a_{mn} \sin \lambda_m x \sin \alpha_n y$$

Solving for b_{mn}:

$$b_{mn} = \frac{a_{mn}}{[D_x\lambda_m^4 + 2H\lambda_m^2\alpha_n^2 + D_y\alpha_n^4]}$$

where

$$\lambda_m = \frac{m\pi}{L}; \qquad \alpha_n = \frac{n\pi}{S}$$

Assuming

$$\alpha = H/\sqrt{D_x D_y}; \qquad \beta = H/D_y$$

Therefore,

$$(\beta/\alpha)^2 = \frac{H^2}{D_y^2}\frac{D_x D_y}{H^2} = \frac{D_x}{D_y}$$

and b_{mn} becomes

$$b_{mn} = \frac{a_{mn}}{D_y\left[\frac{D_x}{D_y}\frac{m^4\pi^4}{L^4} + 2\frac{H}{D_y}\frac{m^2n^2\pi^4}{L^2S^2} + \frac{n^4\pi^4}{S^4}\right]}$$

or

$$b_{mn} = \frac{S^4}{\pi^4 D_y}\frac{a_{mn}}{[(\beta/\alpha)^2(S/L)^4m^4 + m^2n^2(S/L)^2 2\beta + n^4]}$$

Substituting b_{mn} into the deflection equation (4.47), we have

$$w = \frac{S^4}{\pi^4 D_y}\sum_m\sum_n\frac{a_{mn}\sin \lambda_m x \sin \alpha_n y}{[(\beta/\alpha)^2(S/L)^4m^4 + m^2n^2(S/L)^2 2\beta + n^4]} \tag{4.49}$$

the general deflection response of a simply supported plate on all sides composed of an orthotropic material.

In order to solve equations (4.48) or (4.49), the Fourier series term a_{mn} must be evaluated. As in the case of a single variable, given in section 3.3, the term a_{mn} is found by expressing the general load term as a continuous series:

$$q(x, y) = \sum\sum a_{mn} \sin \lambda_m x \sin \alpha_n y$$

where

$$\lambda_m = \frac{m\pi}{L}; \qquad \alpha_n = \frac{n\pi}{S}$$

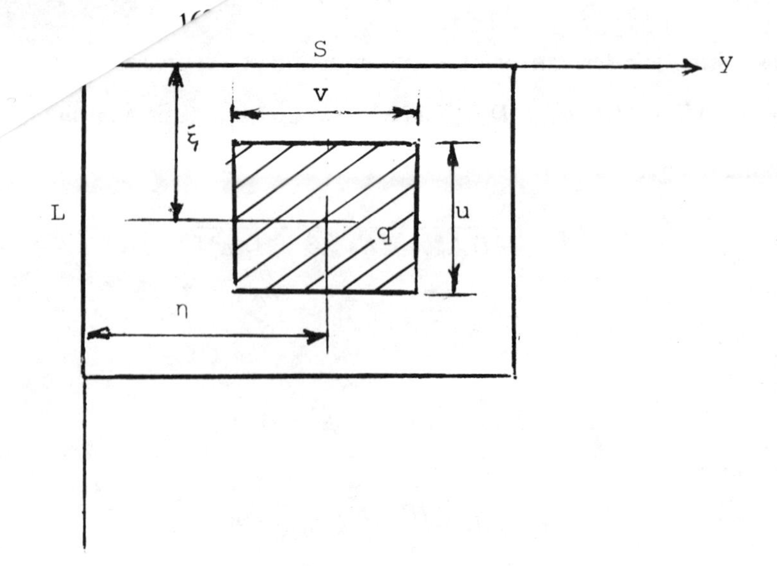

Figure 4-16. Patch Loading.

Multiplying both sides of this expression by $\sin \lambda x \sin \alpha y$ and integrating for $x|0$ to $L|$ and $y|0$ to $S|$ gives the solution for a_{mn} as [1]:

$$a_{mn} = \frac{4}{SL}\int_0^L \int_0^S q(x, y) \sin \lambda_m x \sin \alpha_n y \, dx \, dy \tag{4.50}$$

general form

Example:

(1) *Patch Load:* In order to illustrate the solution of equation (4.50), consider a simply supported plate subjected to a patch loading q, as shown in Figure 4-16. The load $q = P/uv$, where P = total load, a_{mn} is found by applying equation (4.50):

$$a_{mn} = \frac{4}{SL}\frac{P}{uv}\int_{\xi-u/2}^{\xi+u/2} \int_{\eta-v/2}^{\eta+v/2} \sin \lambda_m x \sin \alpha_n y \, dx \, dy$$

Integrating this equation gives

$$a_{mn} = \frac{16P}{\pi^2 mnuv} \sin \frac{m\pi\xi}{L} \sin \frac{n\pi\eta}{S} \sin \frac{m\pi u}{2L} \sin \frac{n\pi v}{2S} \tag{4.51}$$

(2) *Uniform Loading:* If the loading is uniform across the entire plate, then $P/uv = q_0$ and

$$\xi = L/2, \qquad \eta = S/2, \qquad u = L, \qquad v = S$$

therefore from equation (4.51),

$$a_{mn} = \frac{16q_0}{\pi^2 mn} \sin \frac{m\pi}{2} \sin \frac{n\pi}{2} \sin \frac{m\pi}{2} \sin \frac{n\pi}{2}$$

or

$$a_{mn} = \frac{16q_0}{\pi^2 mn} \tag{4.52}$$

(3) *Concentrated Load:* If the loading P is concentrated and is located at $x = \xi, y = \eta$, then applying equation (4.51) where u and v approach zero gives

$$a_{mn} = \frac{4P}{SL} \sin \frac{m\pi\xi}{L} \sin \frac{n\pi\eta}{S} \tag{4.53}$$

Equations (4.51), (4.52), and (4.53) give the Fourier series coefficients for three of the most common loadings. These terms can then be used in conjunction with deflection equations (4.48) or (4.49).

4.5.2 Example: Simply Supported Rectangular Plate under Sinusoidal Load

In order to illustrate these general Fourier series equations and the resulting forces, consider a single (one term) sinusoidal load q_0 applied to a simply supported plate, as shown in Figure 4-17(a). The load is given by

$$q = q_0 \sin \frac{\pi x}{a} \sin \frac{\pi y}{b} \tag{a}$$

and the deflection

$$w = C \sin \frac{\pi x}{a} \sin \frac{\pi y}{b} \tag{b}$$

The assumed deflection w must satisfy the plate equation:

$$\nabla^4 w = \frac{\partial^4 w}{\partial x^4} + 2\frac{\partial^4 w}{\partial x^2\, \partial y^2} + \frac{\partial^4 w}{\partial y^4} = \frac{q}{D}$$

Taking the derivative of equation (b) and substituting these derivatives and equation (a) into the $\nabla^4 w = q/D$ equation gives

$$C\left[\left(\frac{\pi}{b}\right)^4 + 2\left(\frac{\pi^2}{ab}\right)^2 + \left(\frac{\pi}{a}\right)^4\right] \sin \frac{\pi x}{a} \sin \frac{\pi y}{b} = \frac{q_0}{D} \sin \frac{\pi x}{a} \sin \frac{\pi y}{b}$$

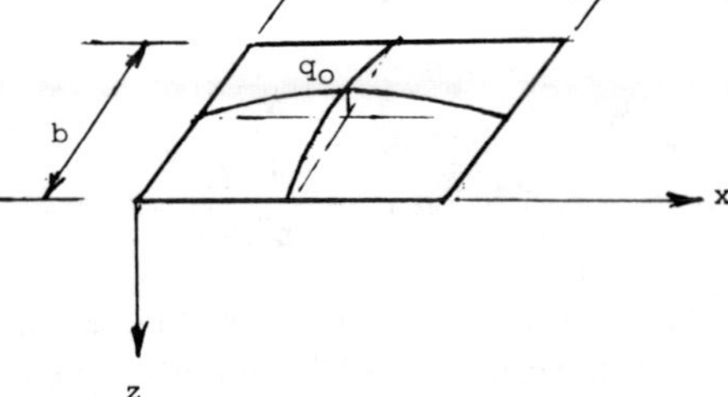

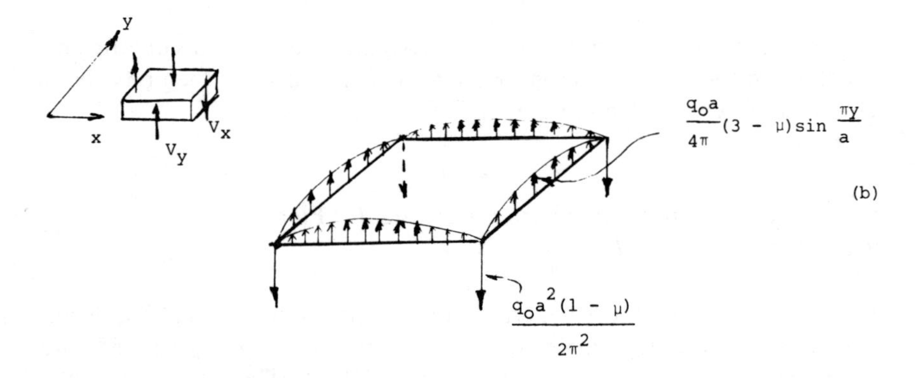

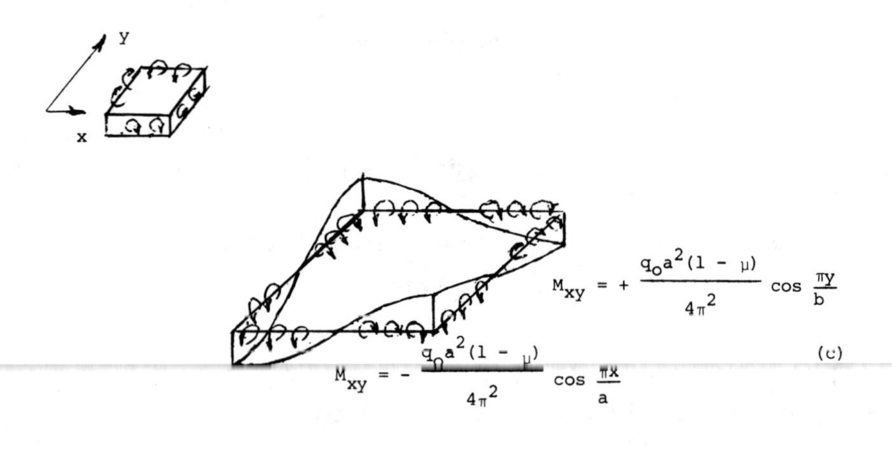

Figure 4-17. Sinusoidal Loaded Plate Loading and Forces.

Solving for C gives

$$C = \frac{q_0}{D} \frac{1}{\pi^4[1/a^2 + 1/b^2]^2}$$

therefore,

$$w = C \sin \frac{\pi x}{a} \sin \frac{\pi y}{b} = \frac{q_0 \sin \pi x/a \sin \pi y/b}{D\pi^4[1/a^2 + 1/b^2]^2} \tag{c}$$

Let

$$K = \frac{q_0}{D\pi^4} \frac{1}{[1/a^2 + 1/b^2]^2}$$

therefore,

$$w = K \sin \frac{\pi x}{a} \sin \frac{\pi y}{b} \tag{d}$$

The moments can now be written from

$$M_x = -D\left[\frac{\partial^2 w}{\partial x^2} + \mu \frac{\partial^2 w}{\partial y^2}\right]$$

or

$$M_x = KD\pi^2\left[\frac{1}{a^2} + \frac{\mu}{b^2}\right] \sin \frac{\pi x}{a} \sin \frac{\pi y}{b} \tag{e}$$

$$M_y = -D\left[\frac{\partial^2 w}{\partial y^2} + \mu \frac{\partial^2 w}{\partial x^2}\right]$$

or

$$M_y = KD\pi^2\left[\frac{1}{b^2} + \frac{\mu}{a^2}\right] \sin \frac{\pi x}{a} \sin \frac{\pi y}{b} \tag{f}$$

$$M_{xy} = -D(1 - \mu) \frac{\partial^2 w}{\partial x \partial y}$$

or

$$M_{xy} = - DK(1 - \mu)\left(\frac{\pi^2}{ab}\right) \cos \frac{\pi x}{a} \cos \frac{\pi y}{b} \tag{g}$$

$$V_x = -D\frac{\partial}{\partial x}(\nabla w)$$

or

$$V_x = \frac{q_0}{\pi a(1/a^2 + 1/b^2)} \cos \frac{\pi x}{a} \sin \frac{\pi y}{b} \tag{h}$$

$$V_y = -D\frac{\partial}{\partial y}(\nabla w)$$

or

$$V_y = \frac{q_0}{\pi b(1/a^2 + 1/b^2)} \sin \frac{\pi x}{a} \cos \frac{\pi y}{b} \tag{i}$$

$$R_x = \left(V_x + \frac{\partial M_{xy}}{\partial y}\right)$$

or

$$R_x = \frac{q_0}{\pi a(1/a^2 + 1/b^2)^2}\left[\frac{1}{a^2} + \frac{2-\mu}{b^2}\right] \cos \frac{\pi x}{a} \sin \frac{\pi y}{b} \tag{j}$$

$$R_y = \left(V_y + \frac{\partial M_{xy}}{\partial x}\right)$$

or

$$R_y = \frac{q_0}{\pi b(1/a^2 + 1/b^2)^2}\left[\frac{1}{b^2} + \frac{2-\mu}{a^2}\right] \cos \frac{\pi y}{b} \sin \frac{\pi x}{a} \tag{k}$$

The corner force $R_{xy} = 2M_{xy}$ is

$$R_{xy} = \frac{2q_0(1-\mu)}{\pi^2 ab(1/a^2 + 1/b^2)^2}$$

The resulting distribution of the reactions and twisting moments along the plate are shown in Figures 4-17(b) and 4-17(c), when $b = a$.

4.6 Infinitely Long Plate with Parallel Supports and Varying Line Load

4.6.1 Isotropic Plate Solution

In section 4.4, the behavior of a simply supported infinite length plate subjected to a nonvarying load along the length was examined. Section 4.5 showed the solution of a plate simply supported on all sides. This section

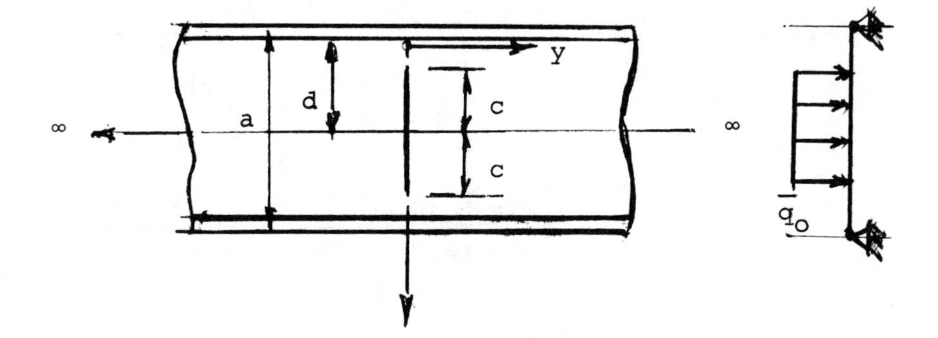

Figure 4-18. Line Loaded Plate.

will describe the classic solution of the same plate presented in section 4.4, but with a line loading, as shown in Figure 4-18.

It will be required to solve the plate equation (4.19):

$$\nabla^4 w = \frac{q}{D}$$

where q is defined over an area. The load in this problem is a line load q_0(k/in) applied on a strip of length $2c$ and zero width. The solution will require the Fourier series technique, using odd series, or

Zero area

$$\bar{q}(x) = \sum \bar{q}_n \sin \frac{n\pi}{a} x$$

where, as defined previously:

$$\bar{q}_n = \frac{2}{L}\int_0^L f(x) \sin \lambda_n x \, dx$$

or

$$\bar{q}_n = \frac{2}{a}\int_{d-c}^{d} \bar{q}_0 \sin \frac{n\pi}{a} x \, dx + \int_{d}^{d+c} \sin \lambda_n x \, dx$$

therefore,

$$\bar{q}_n = \frac{2}{a}\bar{q}_0 \frac{a}{n\pi}\left\{\left[-\cos\frac{n\pi}{a}x\right]_{d-c}^{d} + \left[-\cos\frac{n\pi}{a}x\right]_{d}^{d+c}\right\}$$

and

$$\bar{q}_n = \frac{2q_0}{n\pi}\left[-\cos\frac{n\pi}{a}d + \cos\frac{n\pi}{a}(d-c) - \cos\frac{n\pi}{a}(d+c) + \cos\frac{n\pi}{a}d\right]$$

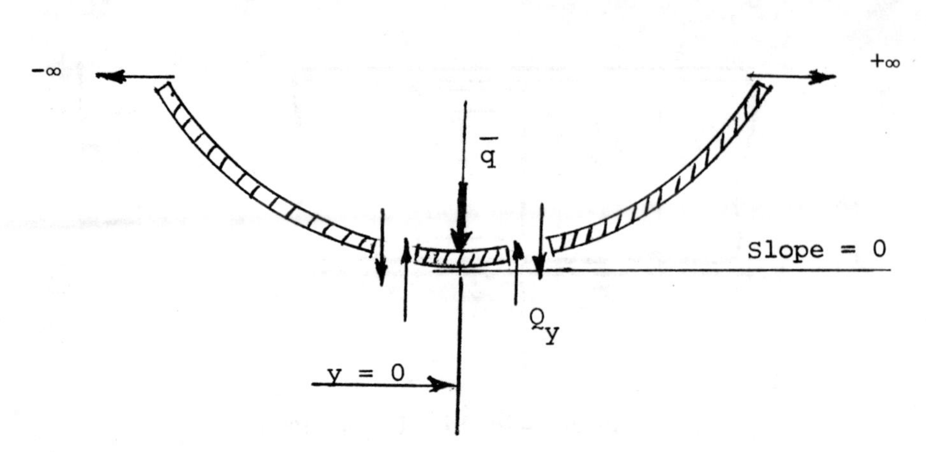

Figure 4-19. Line Loaded Plate Free Body.

Using the proper trigometric identity gives

$$\bar{q}_n = \frac{2\bar{q}_0}{n\pi}\left\{ \cos\frac{n\pi}{a}d \cos\frac{n\pi}{a}c + \sin\frac{n\pi}{a}d \sin\frac{n\pi}{a}c \right.$$

$$\left. - \cos\frac{n\pi}{a}d \cos\frac{n\pi}{a}c + \sin\frac{n\pi}{a}d \sin\frac{n\pi}{a}c \right\}$$

Therefore,

$$\bar{q}_n = \frac{4\bar{q}_0}{n\pi}\left\{ \sin\frac{n\pi d}{a} \sin\frac{n\pi c}{a} \right\} \tag{4.54a}$$

$$\bar{q}(x) = \sum\left(\frac{4\bar{q}_0}{n\pi} \sin\frac{n\pi d}{a} \sin\frac{n\pi c}{a}\right) \sin\frac{n\pi}{a}x \tag{4.54b}$$

Now operating with the nth term of the series, $\bar{q}(x)$ will be introduced as part of the boundary conditions. Figure 4-19 shows a cross section through the plate. Due to symmetry of the loading, the following are noted

Boundary Conditions

at $y = 0$:

slope

$$\frac{\partial w}{\partial y} = 0$$

$$Q_y = -\frac{q}{2} = -D\left(\frac{\partial^3 w}{\partial y^3} + \frac{\partial^3 w}{\partial y\,\partial x^2}\right)$$

at $x = 0, a$:

$$w = 0$$

$$\frac{\partial^2 w}{\partial x^2} = 0$$

at $y = \infty$:

$$w = \frac{\partial w}{\partial y} = \frac{\partial^2 w}{\partial y^2} = \frac{\partial^2 w}{\partial x^2} = 0$$

Returning now to the basic plate equation (4.19), it is necessary to evaluate w, with the incorporation of the above boundary conditions. Consider a solution for w of the form

$$w = \sum_n w_n = \sum_n Y_n X_n = \sum_n Y_n \sin \lambda_n x \tag{4.55}$$

where Y_n is a function of y only and $X_n = \sin \lambda_n x$ is a function of x only and $\lambda_n = n\pi/a$. Equation (4.55) was assumed because it meets, in part, the boundary conditions at $x = 0$ and $x = a$.

The conditions at $y = 0$ and $y = \infty$ can be found by solution of

$$\nabla^4 w = 0$$

Substituting the derivatives of equation (4.55), gives

$$\nabla^4 w = (\lambda_n^4 Y_n - 2\lambda_n^2 Y_n'' + Y_n'''') = 0 \tag{4.56}$$

The solution to this linear differential equation, as given in section 3.4 is found by assuming

$$Y_n = k_n e^{ky}$$

Substituting the derivatives of Y_n into equation (4.56) gives the characteristic function:

$$k^4 - 2\lambda_n^2 k^2 + \lambda_n^4 = 0$$

and solving for the roots k gives

$$k_{1,2} = +\lambda_n$$

$$k_{3,4} = -\lambda_n$$

Therefore,

$$Y_n = (a_n + y b_n) e^{k_{1,2} y} + (c_n + y d_n) e^{k_{3,4} y}$$

Letting $b_n = B_n \lambda_n$ and $d_n = D_n \lambda_n$ gives

$$Y_n = A_n e^{\lambda_n y} + B_n \lambda_n y e^{\lambda_n y} + C_n e^{-\lambda_n y} + D_n \lambda_n y e^{-\lambda_n y} \tag{4.57a}$$

and using trigometric identities, transforms (4.57a) into

$$Y_n = E_n \sinh \lambda_n y + F_n \cosh \lambda_n y + G_n \lambda_n y \sinh \lambda_n y + H_n \alpha_n \lambda_n \cosh \lambda_n y \tag{4.57b}$$

If the problem involves finite boundaries at y, then equation (4.57b) would be used; for this problem we will use (4.57a). Note that since w approaches zero as y approaches infinity, the positive exponentials must be zero, therefore, $A_n = B_n = 0$.

The deflection equation is, therefore,

$$w = \sum (C_n e^{-\lambda_n y} + D_n \lambda_n y e^{-\lambda_n y}) \sin \lambda_n x$$

Apply now the boundary conditions.

$$\text{at } y = 0, \quad \frac{\partial w}{\partial y} = 0:$$

$$\left.\frac{\partial w}{\partial y}\right|_{y=0} = 0 = -\lambda_n C_n e^{-\lambda_n y} - D_n \lambda_n^2 y e^{-\lambda_n y} + D_n \lambda_n e^{-\lambda_n y} \Big|_{y=0}$$

Therefore

$$C_n = D_n \quad \text{and} \quad w_n = C_n(1 + \lambda_n y)e^{-\lambda_n y} \sin \lambda_n x$$

$$\text{at } y = 0, \quad x = 0$$

$$-\frac{\bar{q}}{2} = -D\frac{\partial}{\partial y}\nabla^2 w$$

$$-D\frac{\partial}{\partial y}\nabla^2 w = -\frac{1}{2}\sum \bar{q}_n \sin \lambda_n x$$

Now examine the nth term.

$$\frac{\partial w_n}{\partial y} = C_n(-\lambda_n + \lambda_n - \lambda_n^2 y)e^{-\lambda_n y} \sin \lambda_n x$$

$$\frac{\partial^2 w_n}{\partial y^2} = C_n(+\lambda_n^3 y - \lambda_n^2)e^{\ \lambda_n y} \sin \lambda_n x$$

$$\frac{\partial^3 w_n'''}{\partial y^3} = C_n(\lambda_n^3 - \lambda_n^4 y + \lambda_n^3)e^{-\lambda_n y} \sin \lambda_n x$$

$$\frac{\partial^3 w_n}{\partial x^2 \partial y} = C_n(-\lambda_n^2 y e^{-\lambda_n y})(-\lambda_n^2 \sin \lambda_n x)$$

$$-D\frac{\partial}{\partial y}\nabla^2 w = -D(2\lambda_n^3 - \lambda_n^4 y + \lambda_n^4 y)e^{-\lambda_n y}(\sin\lambda_n X)C_n$$

$$= -D(2\lambda_n^3)e^{-\lambda_n y}(\sin\lambda_n X)C_n$$

$$= -2D\lambda_n^3(\sin\lambda_n X)C_n\big|_{y=0}$$

However,

$$-D\frac{\partial}{\partial y}\nabla^2 w = -\frac{1}{2}\bar{q}_n \sin\lambda_n x$$

Substituting in $-D(\partial/\partial y)\nabla^2 w$ and $\bar{q}_n$, as given by equation (4.54a), gives

$$[-2D\lambda_n^3]C_n \sin\lambda_n x = -\frac{1}{2}\frac{4\bar{q}_0}{n\pi}\sin\lambda_n d \sin\lambda_n c \sin\lambda_n x$$

or

$$C_n = \frac{\bar{q}_0}{n\pi}\frac{1}{D}\frac{1}{\lambda_n^3}\sin\lambda_n d \sin\lambda_n c$$

therefore,

$$C_n = \frac{\bar{q}_0 a^3}{n^4\pi^4 D}\sin\lambda_n d \sin\lambda_n c$$

The solution for w is, therefore,

$$w = \frac{\bar{q}_0 a^3}{\pi^4 D}\sum_n \frac{\sin\lambda_n d \sin\lambda_n c}{n^4}(1 + \lambda_n y)e^{-\lambda_n y}\sin\lambda_n x$$

$$\lambda_n = \frac{n\pi}{a}; \qquad n = 1, 2, 3, \ldots \tag{4.58}$$

If the line load $2c$ increases to the full width of the plate or decreases to a point, the following special solutions are obtained.

Uniform Line Load across Plate: Examining Figure 4-18, let $d = c = a/2$, therefore,

$$w = \frac{\bar{q}_0 a^3}{\pi^4 D}\sum\frac{1}{n^4}\sin\frac{n\pi}{a}d \sin\frac{n\pi c}{a}(1 + \lambda_n y)e^{-\lambda_n y}\sin\lambda_n x$$

$$w = \frac{\bar{q}_0 a^3}{\pi^4 D}\sum\frac{1}{n^4}\sin\frac{n\pi}{2}\sin\frac{n\pi}{2}(1 + \lambda_n y)e^{-\lambda_n y}\sin\lambda_n x$$

$$w = \frac{\bar{q}_0 a^3}{\pi^4 D}\sum_{n=1,3,5}\frac{1}{n^4}(1 + \lambda_n y)e^{-\lambda_n y}\sin\lambda_n x \tag{4.59}$$

Concentrated Load at Any Distance d**:** Examining Figure 4-18, when $c = 0$, the load $2c\bar{q}_0 \rightarrow P$. The load is given by equation (4.54) or

$$\bar{q}_n = \frac{4\bar{q}_0}{n\pi} \sin \frac{n\pi d}{a} \sin \frac{n\pi}{a} c$$

therefore,

$$\bar{q}_n \Big|_{c \to 0} = \frac{P}{2c} \frac{4}{n\pi} \sin \frac{n\pi d}{a} \sin \frac{n\pi}{a} c$$

or

$$\bar{q}_n = \frac{2P}{a} \sin \frac{n\pi d}{a} \cdot (\sin \lambda_n c/\lambda_n c) \Big|_{c \to 0}$$

Applying L'Hospital's rule [1] gives

$$\frac{d \sin \lambda_n c}{d\lambda_n c} \Big|_{c \to 0} = \frac{\lambda_n \cos \lambda_n c}{\lambda_n} \Big|_{c \to 0} = 1$$

therefore,

$$\bar{q}_n = \frac{2P}{a} \sin \frac{n\pi d}{a} \cdot (1)$$

and thus;

$$w = \frac{Pa^2}{2\pi^3 D} \cdot \Sigma \frac{1}{n^3} (1 + \lambda_n y) e^{-\lambda_n y} \sin \lambda d \sin \lambda x \tag{4.60}$$

1/2/82

4.6.2 Orthotropic Plate Solution

The development of the plate solution for the boundary conditions considered involved the solution of the isotropic plate equation (4.19). If the plate is orthotropic, then the solution of w would involve equation (4.38), where w is still assumed of the form

$$w = \sum Y_n \sin \lambda_n x \tag{4.55}$$

The derivatives of this equation are substituted into equation (4.38):

$$D_x \frac{\partial^4 w}{\partial x^4} + 2H \frac{\partial^4 w}{\partial x^2 \partial y^2} + D_y \frac{\partial^4 w}{\partial y^4} = 0$$

where the load $q = 0$, giving

$$D_x \lambda^4 Y_n \sin \lambda x - 2H\lambda^2 \frac{\partial^2 Y_n}{\partial y^2} \sin \lambda x + D_y \frac{\partial^4 Y_n}{\partial y^4} = 0 \tag{4.61}$$

Assuming, as in the case of the isotropic plate, $Y_n = k_n e^{ky}$, and substituting into the above expression gives

$$k^4 - \left(\frac{2H}{D_y}\right)\lambda^2 k^2 + \lambda^4\left(\frac{D_x}{D_y}\right) = 0$$

The four roots to this equation are

$$k = \pm\lambda\left[\left(\frac{H}{D_y}\right) \pm \frac{1}{D_y}[H^2 - D_x D_y]^{1/2}\right]^{1/2} \quad (4.61a)$$

The solution to equation (4.61) will depend on the conditions of:

Case 1: $H^2 > D_x D_y$ + Closed cell

Case II: $H^2 = D_x D_y$ 0 uniform plate

Case III: $H^2 < D_x D_y$ imaginary open cell

as referenced by the term $[H^2 - D_x D_y]^{1/2}$. Defining $\beta = H/D_y$ and $\alpha = H/\sqrt{D_x D_y}$, equation (4.61a) is

$$k = \pm\lambda[\beta \pm \beta/\alpha(\alpha^2 - 1)^{1/2}]^{1/2} \quad (4.61b)$$

Case I: When $H^2 > D_x D_y$, the term $(\alpha^2 - 1)^{1/2}$ will be positive and therefore, give real roots. The solution of (4.61b), therefore, gives

$$k_1 = -k_2 = +\lambda(\beta/\alpha)^{1/2}\,[\alpha + (\alpha^2 - 1)^{1/2}]^{1/2}$$

$$k_3 = -k_4 = +\lambda(\beta/\alpha)^{1/2}[\alpha - (\alpha^2 - 1)^{1/2}]^{1/2}$$

The solution to Y_n is, therefore,

$$Y_n = C_1 e^{k_1 y} + C_2 e^{-k_1 y} + C_3 e^{k_2 y} + C_4 e^{-k_2 y}$$

The solution for the deflection $w = Y_n \sin \lambda_n x$ is, therefore,

$$w = \sum(C_1 e^{k_1 y} + C_2 e^{-k_1 y} + C_3 e^{k_2 y} + C_4 e^{-k_2 y}) \sin \lambda_n x \quad (3.62a)$$

in hyperbolic form is

$$w = \sum(A \cosh r_1 y + B \sinh r_1 y + C \cosh r_2 y + D \sinh r_2 y) \cdot \sin \lambda_n x \quad (3.62b)$$

where $\lambda_n = n\pi/a$.

Case II: $H^2 = D_x D_y$ Letting $H^2 = D_x D_y$, the term $[\alpha^2 - 1]^{1/2} = 0$ in equation (4.61b), therefore, the roots are

$$k_{1,2} = +\lambda[\beta/\alpha]^{1/2}$$

$$k_{3,4} = -\lambda[\beta/\alpha]^{1/2} = -k_{1,2}$$

The solution to Y_n and thus w is

$$w = \sum (C_1e^{k_1y} + C_2ye^{k_1y} + C_3e^{-k_1y} + C_4ye^{-k_1y}) \sin \lambda_n x \quad (3.63a)$$

or

$$w = \sum(A \sinh r_1y + By \sinh r_1y + C \cosh r_1y + Dy \cosh r_1y) \cdot \sin \lambda_n x \quad (3.63b)$$

Case III: $H^2 < D_xD_y$ For this condition, the term $[H^2 - D_xD_y]^{1/2}$ in equation (4.61b) becomes negative and thus the roots will be imaginary. The equation can thus be transformed into the following:

$$k = \pm\lambda[\beta \pm i(\beta/\alpha)(1 - \alpha^2)^{1/2}]^{1/2}$$

The solution of this equation is

$$k_1 = +\lambda\left[(\beta/\alpha)^{1/2}\left(\frac{1+\alpha}{2}\right)^{1/2} + i(\beta/\alpha)^{1/2}\left(\frac{1-\alpha}{2}\right)^{1/2}\right]$$

$$k_2 = +\lambda\left[(\beta/\alpha)^{1/2}\left(\frac{1+\alpha}{2}\right)^{1/2} - i(\beta/\alpha)^{1/2}\left(\frac{1-\alpha}{2}\right)^{1/2}\right]$$

$$k_3 = -\lambda\left[(\beta/\alpha)^{1/2}\left(\frac{1+\alpha}{2}\right)^{1/2} + i(\beta/\alpha)^{1/2}\left(\frac{1-\alpha}{2}\right)^{1/2}\right]$$

$$k_4 = -\lambda\left[(\beta/\alpha)^{1/2}\left(\frac{1+\alpha}{2}\right)^{1/2} - i(\beta/\alpha)^{1/2}\left(\frac{1-\alpha}{2}\right)^{1/2}\right]$$

Let

$$\phi = \lambda\left[(\beta/\alpha)^{1/2}\left(\frac{1+\alpha}{2}\right)^{1/2}\right]$$

$$\psi = \lambda\left[(\beta/\alpha)^{1/2}\left(\frac{1-\alpha}{2}\right)^{1/2}\right]$$

The roots, therefore, take the form of

$$\text{Roots} = \pm (\psi \pm i\psi)$$

and Y_n, therefore, is

$$Y_n = C_1e^{(\phi+i\psi)y} + C_2e^{(\phi-i\psi)}y + C_3e^{-(\phi+i\psi)y} + C_4e^{-(\phi-i\psi)y}$$

and

$$w = \sum (C_1 \cdot e^{\phi y} \cdot e^{i\psi y} + C_2 e^{\phi y} \cdot e^{-i\psi y} + C_3 e^{-\phi y} \cdot e^{-i\psi y}$$
$$C_4 e^{-\phi y} \cdot e^{+i\psi y}) \sin \lambda_n x \tag{3.64a}$$

However,

$$e^{\pm i\psi y} = \cos \psi y \pm i \sin \psi y$$
$$e^{\pm \phi y} = \cosh \phi y \pm \sinh \phi y$$

Therefore,

$$\begin{aligned} w = [&A(\cosh \phi y + \sinh \phi y) \cdot (\cos \psi y + i \sin \psi y) \\ &+ B(\cosh \phi y + \sinh \phi y) \cdot (\cos \psi y - i \sin \psi y) \\ &+ C(\cosh \phi y - \sinh \phi y) \cdot (\cos \psi y - i \sin \psi y) \\ &+ D(\cosh \phi y - \sinh \phi y) \cdot (\cos \psi y + i \sin \psi y)] \sin \lambda_n x \end{aligned}$$

and collecting terms gives

$$\begin{aligned} w = [&(A + B + C + D) \cdot \cosh \phi y \cdot \cos \psi y \\ &+ (A + B - C - D) \cdot \sinh \phi y \cdot \cos \psi y \\ &+ i(A - B - C + D) \cdot \cosh \phi y \cdot \sin \psi y \\ &+ i(A - B + C - D) \cdot \sinh \phi y \cdot \sin \psi y] \sin\lambda_n x \end{aligned}$$

Defining new coefficients for the combined coefficients gives

$$\begin{aligned} w = \sum (&A \cosh \phi y \cdot \cos \psi y + B \cosh \phi y \cdot \sin \psi y \\ &+ C \sinh \phi y \cdot \cos \psi y + D \sinh \phi y \cdot \sin \psi y) \sin \lambda_n x \end{aligned} \tag{3.64b}$$

The three deflection solutions to the orthotropic plate equation represents three possible material stiffness conditions. In practice, when dealing with variable shapes, the stiffnesses can be equated to practical shapes. For example:

Case I: $H^2 > D_x D_y$ ($\alpha > 1$) represents the condition of a closed cell.

Case II: $H^2 = D_x D_y$ ($\alpha = 1$) represents the case of a uniform thick deck.

Case III: $H^2 < D_x D_y$ ($\alpha < 1$) represents the condition of a stiffened plate or composite girder. Thus selection of the proper equation will permit evaluation of the behavior of a particular type of plate element. Further amplification of these applications will be given in chapter 6. The three deflection equations (4.62), (4.63), and (4.64) can be similarly applied to the plate problem in section 4.6.1.

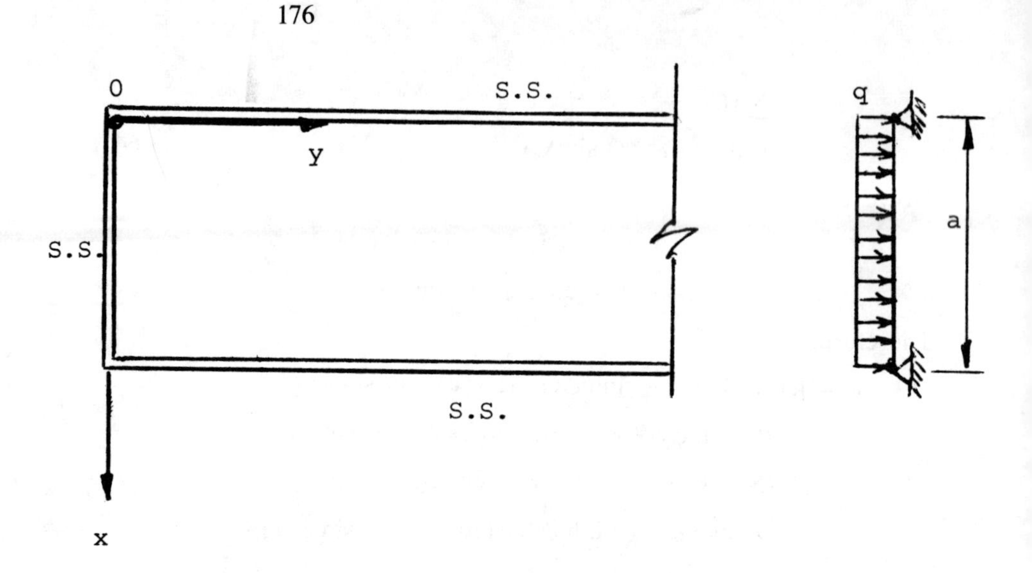

Figure 4-20. Semi-Infinite Simply Supported Plate.

4.7 Semi-Infinite Plate under Uniform Load

4.7.1 Simple Support

In the previous problems, excluding Navier's plate solution, the plate had infinite boundaries in the y direction. Consider now the plate with a finite boundary at $y = 0$, as shown in Figure 4-20. The boundary conditions are:

$$\text{at } y = 0: \quad w = 0, \quad \partial^2 w/\partial y^2 = 0$$

$$\text{at } x = 0, \quad x = a: \quad w = 0, \quad \partial^2 w/\partial x^2 = 0$$

The general loading, given in section 4.4.2, is

$$q(x) = \frac{4q_0}{\pi} \sum \frac{1}{n} \sin \frac{n\pi x}{a}$$

The plate solution will be divided into two parts: (1) the homogeneous solution of the plate equation and (2) particular solution of the plate equation, or

$$\nabla^4 w_H = 0 \tag{1}$$

$$\nabla^4 w_P = q/D \tag{2}$$

The total solution would, therefore, be

$$w_T = w_H + w_P$$

The particular solution has been solved in section 4.4.2 and is

$$w_P = \frac{4qa^4}{D\pi^5} \sum \frac{1}{n^5} \sin \frac{n\pi x}{a} \qquad n = 1, 3, 5, \ldots \tag{4.45}$$

in which only the boundary conditions at $x = 0$ and $x = a$ are satisfied. The homogeneous solution will permit incorporation of the other boundary conditions. Assume:

$$w_H = \sum Y_n \sin \lambda_n x$$

where Y_n is given by equation (4.57a) or

$$Y_n = Ae^{\lambda y} + B\lambda y e^{\lambda y} + Ce^{-\lambda y} + D\lambda e^{-\lambda y}$$

Examining the boundary conditions, if y approaches infinity, $e^{\lambda y}$ would go to infinity, therefore, $A = B = 0$, as in the example given in section 4.6. Therefore,

$$w_H = (Ce^{-\lambda y} + De^{-\lambda y}) \sin \lambda x$$

The total solution is, therefore,

$$w_n = \left\{ \frac{4qa^4}{D\pi^5} \left[\frac{1}{n^5} \right] + \left[(C_n + D_n\lambda_n y)e^{-\lambda_n y} \right] \right\} \sin \lambda_n x$$

where the constant term $4qa^4/D\pi^5$ is arbitrarily placed outside the brackets. Applying the boundary conditions:

$$\text{at } y = 0: w = 0$$

$$\frac{1}{n^5} + C_n = 0$$

therefore

$$C_n = -\frac{1}{n^5}$$

and

$$\text{at } y = 0: \frac{\partial^2 w}{\partial y^2} = 0$$

where

$$\frac{\partial^2 w}{\partial y^2} = \frac{qa^4}{D\pi^5} \cdot \lambda_n^2 [C_n - (2 - \lambda y)D_n] e^{-\lambda_n y} \sin \lambda_n x$$

Therefore,

$$C_n - 2D_n = 0$$

$$C_n = 2D_n$$

but $C_n = -1/n^5$,

Therefore

$$D_n = \frac{1}{2n^5}$$

The deflection equation is, therefore,

$$w_n = \frac{4qa^4}{D\pi^5}\left[\frac{1}{n^5} + (C_n + D_n\lambda_n y)e^{-\lambda_n y}\right] \sin \lambda_n x$$

or

$$w_n = \frac{4qa^4}{D\pi^5}\left[\frac{1}{n^5} + \left(-\frac{1}{n^5} - \frac{1}{2n^5}\lambda_n y\right)e^{-\lambda_n y}\right] \sin \lambda_n x$$

Therefore,

$$w_n = \frac{4qa^4}{D\pi^5 n^5} \cdot \left[1 - \left(1 + \lambda_n \frac{y}{2}\right)e^{-\lambda_n y}\right] \sin \lambda_n x \qquad (4.65a)$$

The general equations are:

$$w = \frac{4qa^4}{D\pi^5} \cdot \Sigma\frac{1}{n^5}\left[1 - \left(1 + \frac{n\pi y}{2a}\right)e^{-n\pi y/a}\right]\sin\frac{n\pi x}{a} \qquad (4.65b)$$

$$M_x = \frac{4qa^2}{\pi^3} : \Sigma\frac{1}{n^3}\left[1 - \left(1 + (1-\mu)\frac{n\pi y}{2a}\right)e^{-n\pi y/a}\right] \sin \lambda_n x \quad (4.66)$$

$$M_y = \frac{4qa^2}{\pi^3} \cdot \Sigma\frac{1}{n^3}\left[(1-\mu)\frac{n\pi y}{2a} - \mu)e^{-n\pi y/a} + \mu\right] \sin \lambda_n x \quad (4.67)$$

Examining the equation when $y \gg 1$ gives

$$M_x = \frac{4qa^2}{\pi^3} \cdot \Sigma\frac{1}{n^3} \sin \lambda_n x$$

$$M_y = \mu M_x \text{ (cylindrical bending)}$$

and for $x = a/2$; $y \to \infty$

$$M_x = \frac{4qa^2}{\pi^3}\left[1 - \frac{1}{27} + \frac{1}{125} - \frac{1}{343} + \frac{1}{729} + \ldots\right] \cong \frac{qa^2}{8}$$

where

$$M_x = qa^2/8$$

for simple beam bending. A plot of the moment distributions are shown in Figure 4-21.

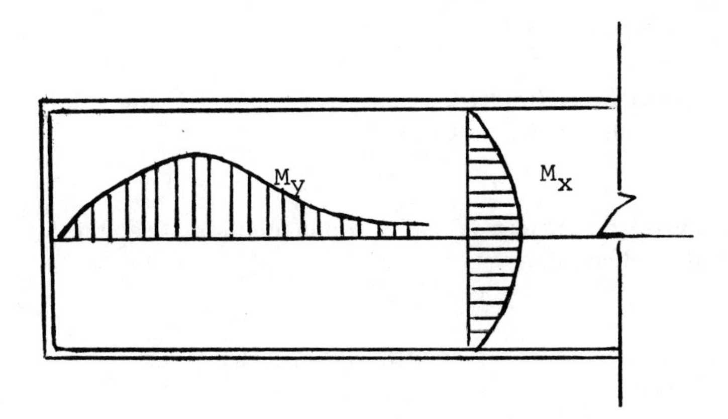

Figure 4-21. Distribution of Moments.

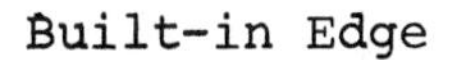

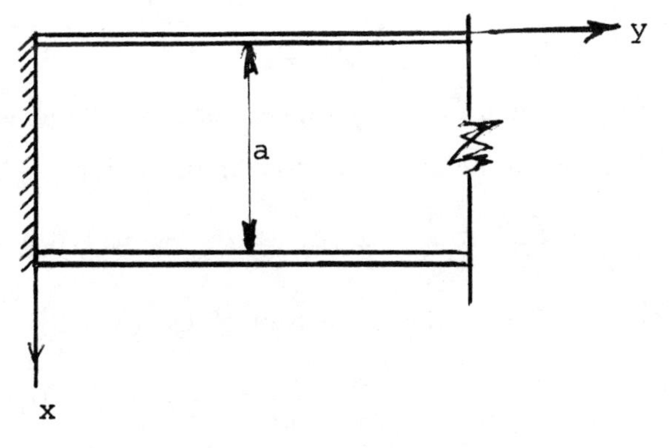

Figure 4-22. Built-In Edged Plate.

4.7.2 Built-in or Fixed Support

If the support at $y = 0$ is fixed, as shown in Figure 4-22, then the boundary conditions would be modified using

$$\text{at } y = 0: \qquad w = \frac{\partial w}{\partial y} = 0$$

The evaluation of the coefficient C and D would then follow.

4.7.3 Free Edge

If the support at $y = 0$ is free, as shown in Figure 4-23, then the following boundary conditions would be used.

$$\text{at } y = 0, \quad M_y = 0, \quad R_y = 0$$

The evaluation of the terms C and D would follow and then the determination of the deflection w.

4.8 Infinite Plate under Constant Load Supported by an Elastic Beam

Consider now the plate, which is infinite in length, subjected to a constant load and supported by an elastic beam, as shown in Figure 4-24. The boundary conditions are:

$$\text{at } y = 0: \quad \frac{\partial w}{\partial y} = 0$$

$$Q_y = -q/2$$

where from beam theory $q/EI = \partial^4 w/\partial x^4$ or $q = EI(\partial^4 w/\partial x^4)$.

The solution of w will be of the form given in section 4.7.1 or

$$w_n = \frac{4qa^4}{D\pi^5}\sum\frac{1}{n^5}[1 + (C_n + D_n\lambda_n y)e^{-\lambda_n y}] \sin \lambda_n x$$

Taking the derivatives of this expression gives

$$\frac{\partial w_n}{\partial y} = -\frac{4qa^3}{D\pi^4} \cdot \frac{1}{n^4}[C_n + D_n\lambda_n y - D_n]e^{-\lambda_n y} \sin \lambda_n x$$

$$\frac{\partial^3 w_n}{\partial y^3} = -\frac{4qa}{D\pi^2} \cdot \frac{1}{n^2} \cdot [C_n(\lambda y - 2)e^{-\lambda_n y}] \sin \lambda_n x$$

$$\frac{\partial^3 w_n}{\partial x^2\,\partial y} = \frac{4qa}{D\pi^2} \cdot \frac{1}{n^2}[C_n(\lambda y)e^{-\lambda_n y}] \sin \lambda_n x$$

The boundary conditions give

$$\text{at } y = 0: \quad \partial w/\partial y = 0$$

therefore,

$$C_n = D_n$$

$$\text{at } y = 0: \quad \frac{EI}{2} \cdot \frac{\partial^4 w}{\partial x^4} = Q_y = -D\frac{\partial}{\partial y}\nabla^2 w$$

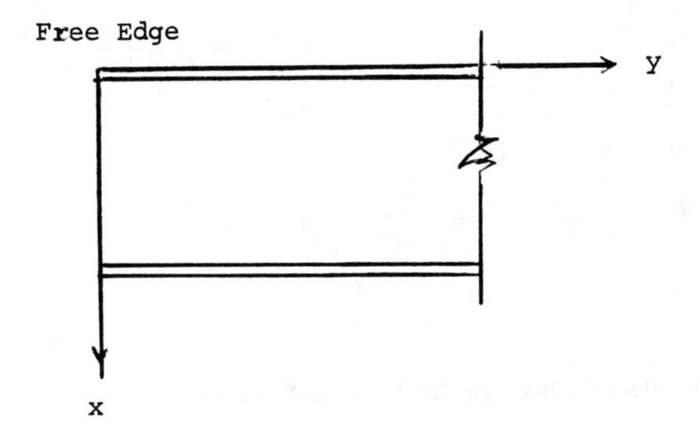

Figure 4-23. Free Edged Plate.

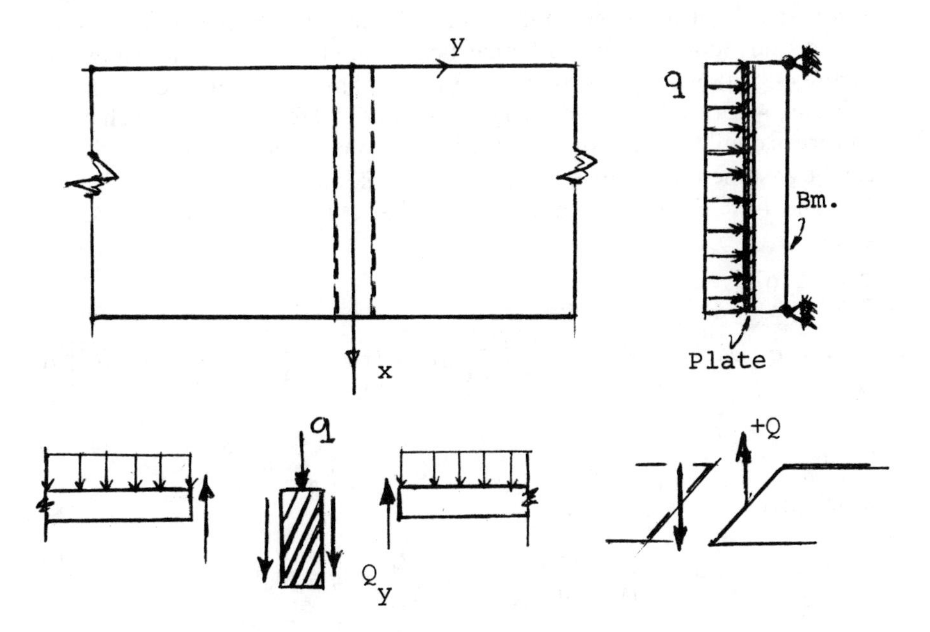

Figure 4-24. Elastically Supported Plate.

Therefore

$$\frac{EI}{2}\,\frac{4q}{D\pi}\cdot\frac{1}{n}(1 + C_n)\sin\lambda_n x = -\frac{4qa}{\pi^2}\,\frac{1}{n^2}2\,C_n\sin\lambda_n x$$

Therefore

$$C_n = -\frac{1}{\dfrac{4Da}{n\pi EI} + 1}$$

The solution of w is then

$$w = \frac{4qa^4}{D\pi^5} \cdot \sum \frac{1}{n^5}\left[1 - \frac{1}{\left(1 + \dfrac{4Da}{n\pi EI}\right)}\left(1 + \frac{n\pi}{a}y\right)e^{-n\pi y/a}\right]\sin \lambda_n x \quad (4.68)$$

4.9 Equivalent Orthotropic Grid System Solution

In the development of the general isotropic plate equation (4.19) and the orthotropic plate equation (4.38), the equilibrium of a small plate element was first examined, as given in Figures 4-2 and 4-3. Suppose the system under study consists of individual beams interconnected, as shown in Figure 4-25(a), where a portion of the system has been removed, as shown in Figure 4-25(b). As shown in Figure 4-25(a), end forces act on each beam and are in equilibrium. If the beams are now considered as plates, the forces can be assumed to act over the plate width $n\lambda$ and λ, as shown in Figures 4-26 and 4-27. Equilibrium of the forces $\Sigma F_z = 0$, $\Sigma M_y = 0$ and $\Sigma M_x = 0$ gives

$\sum F_z = 0$

$$q \cdot dx\,dy + \left[\frac{Q_y}{n\lambda} + \frac{\partial}{\partial y}\left(\frac{Q_y}{n\lambda}\right)dy\right]dx + \left[\frac{Q_x}{\lambda} + \frac{\partial}{\partial x}\left(\frac{Q_x}{\lambda}\right)dx\right]dy - \frac{Q_y}{n\lambda}dx - \frac{Q_x}{\lambda}dy = 0$$

which gives

$$\frac{1}{n\lambda} \cdot \frac{\partial Q_y}{\partial y} + \frac{1}{\lambda} \cdot \frac{\partial Q_x}{\partial x} + q = 0 \quad (4.69)$$

$\sum M_y = 0$

$$\left[\frac{M_{Ty}}{n\lambda} + \frac{\partial}{\partial y}\left(\frac{M_{Ty}}{n\lambda}\right)dy\right]dx - \left[\frac{M_{Ty}}{n\lambda}\right]dx + \left[\frac{M_x}{\lambda} + \frac{\partial}{\partial x}\left(\frac{M_x}{\lambda}\right)dx\right]dy - \frac{M_x}{\lambda}dy - \left[\frac{Q_x}{\lambda} + \frac{\partial}{\partial x}\left(\frac{Q_x}{\lambda}\right)dx\right]dy \cdot \frac{dx}{2} - \left[\frac{Q_x}{\lambda}dy\right] \cdot \frac{dx}{2} = 0$$

which gives

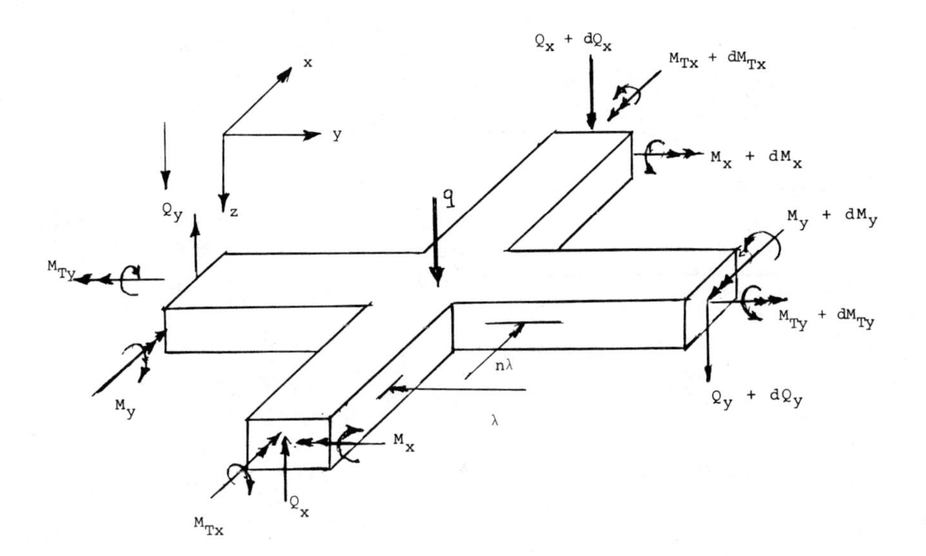

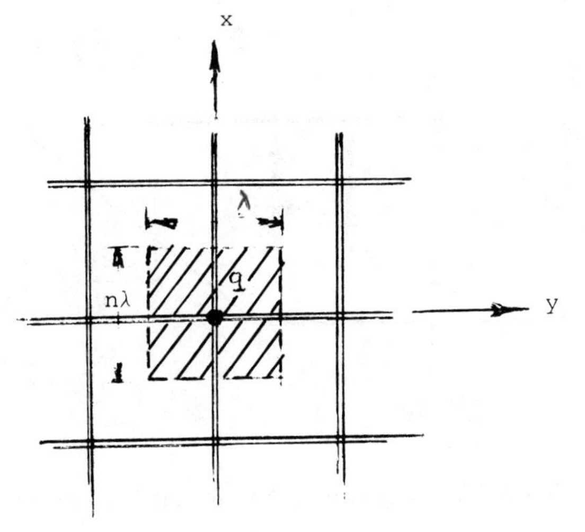

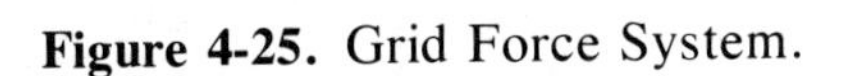

Figure 4-25. Grid Force System.

$$\frac{\partial}{\partial y}\left(\frac{M_{Ty}}{n\lambda}\right) dy\, dx + \frac{\partial}{\partial x}\left(\frac{M_x}{\lambda}\right) dx\, dy - \frac{Q_x}{\lambda}\, dx\, dy = 0$$

or

$$\frac{1}{n\lambda}\,\frac{\partial M_{Ty}}{\partial y} + \frac{1}{\lambda}\,\frac{\partial M_x}{\partial x} - \frac{1}{\lambda} \cdot Q_x = 0 \qquad (4.70)$$

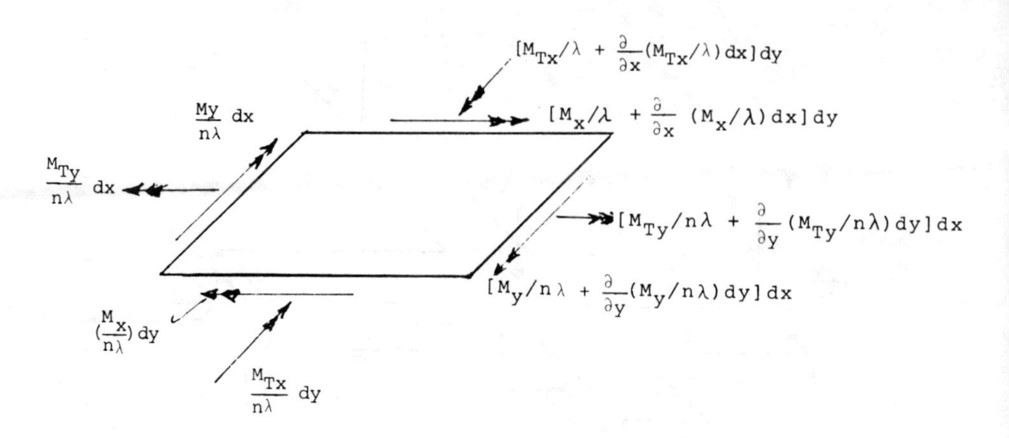

Figure 4-26. Equivalent Moments on Grid System.

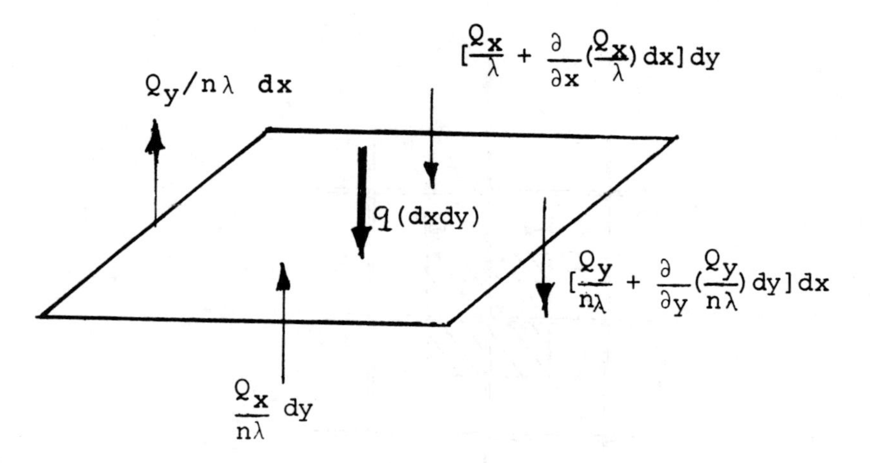

Figure 4-27. Equivalent Shears on Grid System.

$$\sum M_x = 0$$

$$\left[\frac{M_{Tx}}{\lambda} + \frac{\partial}{\partial x}\left(\frac{M_{Tx}}{\lambda}\right) dx\right] dy - \frac{M_{Tx}}{\lambda} \cdot dy + \left[\frac{M_y}{n\lambda} + \frac{\partial}{\partial y}\left(\frac{M_y}{n\lambda}\right) dy\right] dx$$

$$- \frac{M_y}{n\lambda} dy - \left[\frac{Q_y}{n\lambda} + \frac{\partial}{\partial_y}\left(\frac{Q_y}{n\lambda}\right) dy\right] dx \cdot \frac{dy}{2} - \frac{Q_y}{n\lambda} dx \cdot \frac{dy}{2} = 0$$

which gives

$$\frac{1}{\lambda} \cdot \frac{\partial M_{Tx}}{\partial x} + \frac{1}{n\lambda} \frac{\partial M_y}{\partial y} - \frac{1}{n\lambda} Q_y = 0 \tag{4.71}$$

Solving for Q_y and Q_x in equations (4.70) and (4.71), taking the $\partial Q_y / \partial y$ and $\partial Q_x / \partial x$ of these equations, and substituting these relationships into (4.69) gives

$$\left[\frac{1}{\lambda} \frac{\partial^2 M_{Tx}}{\partial x \partial y} + \frac{1}{n\lambda} \frac{\partial^2 M_y}{\partial y^2}\right] + \left[\frac{1}{n\lambda} \frac{\partial^2 M_{Ty}}{\partial x \partial y} + \frac{1}{\lambda} \frac{\partial^2 M_x}{\partial x^2}\right] + q = 0$$

or

$$\frac{1}{\lambda} \frac{\partial^2 M_x}{\partial x^2} + \left[\frac{1}{\lambda} \frac{\partial^2 M_{Tx}}{\partial x \partial y} + \frac{1}{n\lambda} \frac{\partial^2 M_{Ty}}{\partial x \partial y}\right] + \frac{1}{n\lambda} \frac{\partial^2 M_y}{\partial y^2} = -q \tag{4.72}$$

The relationships between bending and torsional moments and the induced vertical deformation w are [29]:

$$M_x = -EI_x \frac{\partial^2 w}{\partial x^2}$$

$$M_y = -EI_y \frac{\partial^2 w}{\partial y^2}. \tag{4.73}$$

$$M_{Tx} = -GK_{Tx} \frac{\partial^2 w}{\partial x \partial y} + EI_{wx} \frac{\partial^4 w}{\partial x^3 \partial y}$$

$$M_{Ty} = -GK_{Ty} \frac{\partial^2 w}{\partial x \partial y} + EI_{wy} \frac{\partial^4 w}{\partial y^3 \partial x}$$

where EI_x = Bending stiffness along x-axis

EI_y = Bending stiffness along y-axis

GK_{Tx} = Pure torsional stiffness along x-axis

GK_{Ty} = Pure torsional stiffness along y-axis

EI_{wx} = Warping torsional stiffness along x-axis

EI_{wy} = Warping torsional stiffness along y-axis

Taking the proper derivatives of equations (4.73), according to the equilibrium equation (4.72) and substituting in these relationships gives

$$\frac{1}{\lambda}\left(-EI_x \frac{\partial^4 w}{\partial x^4}\right) + \left[\frac{1}{\lambda} \cdot \left(-GK_{Tx} \frac{\partial^4 w}{\partial x^2 \partial y^2} + EI_{wx} \frac{\partial^6 w}{\partial x^4 \partial y^2}\right)\right]$$

$$+ \left[\frac{1}{n\lambda} \cdot \left(-GK_{Ty} \frac{\partial^4 w}{\partial y^2 \partial x^2} + EI_{wy} \frac{\partial^6 w}{\partial y^4 \partial x^2}\right)\right] + \frac{1}{n\lambda}\left(-EI_y \frac{\partial^4 w}{\partial y^4}\right) = -q$$

Collecting terms gives

$$-\frac{EI_x}{\lambda}\cdot\frac{\partial^4 w}{\partial x^4}-\left(\frac{GK_{Tx}}{\lambda}+\frac{GK_{Ty}}{n\lambda}\right)\cdot\frac{\partial^4 w}{\partial y^2\,\partial x^2}$$

$$+\frac{EI_{wx}}{\lambda}\cdot\frac{\partial^6 w}{\partial x^4\,\partial y^2}+\frac{EI_{wy}}{n\lambda}\cdot\frac{\partial^6 w}{\partial y^4\,\partial x^2}-\frac{EI_y}{n\lambda}\cdot\frac{\partial^4 w}{\partial y^4}=-q \quad (4.74a)$$

or

$$\frac{\partial^4}{\partial x^4}\left[\frac{EI_x}{\lambda}\cdot w-\frac{EI_{wx}}{\lambda}\cdot\frac{\partial^2 w}{\partial y^2}\right]+\frac{\partial^4 w}{\partial y^2\,\partial x^2}\left[\frac{GK_T}{\lambda}+\frac{GK_{Ty}}{n\lambda}\right]$$

$$+\frac{\partial^4}{\partial y^4}\left[\frac{EI_y}{n\lambda}\cdot w-\frac{EI_{wy}}{n\lambda}\cdot\frac{\partial^2 w}{\partial x^2}\right]=q \quad (4.74b)$$

Defining the stiffness terms as:

$$D_x = EI_x/\lambda$$

$$D_y = EI_y/n\lambda$$

$$2H = G\left[\frac{K_{Tx}}{\lambda}+\frac{K_{Ty}}{n\lambda}\right] \quad (4.75)$$

$$C_x = \frac{EI_{wx}}{\lambda}$$

$$C_y = \frac{EI_{wy}}{n\lambda}$$

Equation (4.74b) becomes

$$D_x\frac{\partial^4 w}{\partial x^4}+\frac{\partial^4 w}{\partial y^2\,\partial x^2}\cdot\left[2H-C_x\frac{\partial^2 w}{\partial x^2}-C_y\frac{\partial^2 w}{\partial y^2}\right]+D_y\frac{\partial^4 w}{\partial y^4}=q \quad (4.76)$$

Note that if the warping stiffness EI_w or $C_x = C_y = 0$, then equation (4.76) reduces to the conventional orthotropic plate equation (4.38).

The shears Q_x and Q_y are found from equations (4.70) and (4.71)

$$\frac{Q_x}{\lambda}=\frac{1}{n\lambda}\cdot\frac{\partial}{\partial y}(M_{Ty})+\frac{1}{\lambda}\,\frac{\partial M_x}{\partial x}$$

therefore

$$\frac{Q_x}{\lambda}=\frac{1}{n\lambda}\cdot\left[-GK_{Ty}\frac{\partial^3 w}{\partial x\,\partial y^2}+EI_{wy}\frac{\partial^5 w}{\partial x\,\partial y^4}\right]-\frac{EI_x}{\lambda}\cdot\frac{\partial^3 w}{\partial x^3} \quad (4.77)$$

$$\frac{Q_y}{n\lambda}=\frac{1}{\lambda}\,\frac{\partial}{\partial x}(M_{Tx})+\frac{1}{n\lambda}\,\frac{\partial M_y}{\partial y}$$

therefore,

$$\frac{Q_y}{n\lambda} = \frac{1}{\lambda} \cdot \left[-GK_{Tx}\frac{\partial^3 w}{\partial y\, \partial x^2} + EI_{wx}\frac{\partial^5 w}{\partial y\, \partial x^4} \right] - \frac{EI_y}{n\lambda} \cdot \frac{\partial^3 w}{\partial y^3} \tag{4.78}$$

The reactions R_x and R_y are also determined by applying the general relationships given by equations (4.24a) and (4.25a):

$$R_x = \frac{Q_x}{\lambda} + \frac{\partial}{\partial y}\left(\frac{M_{Tx}}{\lambda}\right)$$

or

$$R_x = -D_x \cdot \left[\frac{\partial^3 w}{\partial x^3} + \frac{2H}{D_x}\cdot\frac{\partial^3 w}{\partial x\, \partial y^2} - \frac{C_x}{D_x}\cdot\frac{\partial^5 w}{\partial x^3\, \partial y^2} - \frac{C_y}{D_x}\frac{\partial^5 w}{\partial x\, \partial y^4} \right] \tag{4.79}$$

$$R_y = \frac{Q_y}{n\lambda} + \frac{\partial}{\partial x}\left(\frac{M_{Ty}}{n\lambda}\right)$$

$$R_y = -D_y \cdot \left[\frac{\partial^3 w}{\partial y^3} + \frac{2H}{D_y}\cdot\frac{\partial^3 w}{\partial y\, \partial x^2} - \frac{C_x}{D_y}\cdot\frac{\partial^5 w}{\partial y\, \partial x^4} - \frac{C_y}{D_y}\cdot\frac{\partial^5 w}{\partial y\, \partial x^2} \right] \tag{4.80}$$

The solution of these equivalent grid-plate equations will be given in chapter 5, as applied to a bridge structure.

4.10 Circular Plates and Plate Sectors

In all of the previous sections the theories and solutions of certain problems were related to square or rectangular plates. In many engineering applications, the plates may be disks or have circular shapes or be sectors. If such plates require investigation, it is convenient to develop the general plate equations in polar coordinates, shown in Figure 4-28(a). Consider such a plate element and the applied forces acting on one face, as shown in Figure 4-28(b).

The development of these equations would be similar in procedure to those given in section 4-2. Herein only the final equation, however, will be given as presented by Timoshenko [4] and Lekhnitskii [25].

4.10.1 Isotropic Plate [26]

If the plate is isotropic, the general plate equation and force equations are

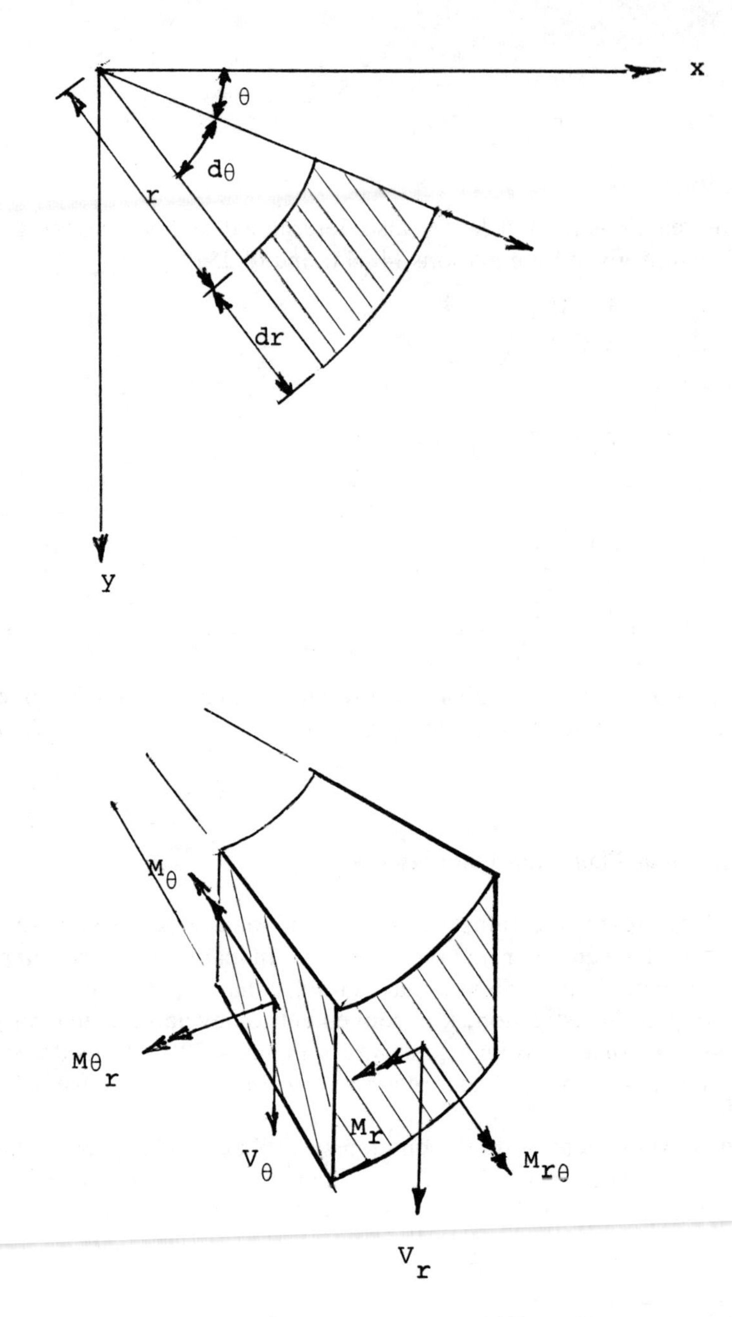

Figure 4-28. Geometry and Forces on Wedge Plate.

$$\nabla^4 w = \left(\frac{\partial^2}{\partial r^2} + \frac{1}{r} \frac{\partial}{\partial r} + \frac{1}{r^2} \frac{\partial^2}{\partial \theta^2} \right)$$

$$\cdot \left(\frac{\partial^2 w}{\partial r^2} + \frac{1}{r} \frac{\partial w}{\partial r} + \frac{1}{r^2} \frac{\partial^2 w}{\partial \theta^2} \right) = \frac{q(r, \theta)}{D}$$

or

$$\nabla^4 w = \frac{\partial^4 w}{\partial r^4} + \frac{2}{r^2} \frac{\partial^4 w}{\partial r^2 \partial \theta^2} + \frac{1}{r^4} \frac{\partial^4 w}{\partial \theta^4} + \frac{2}{r} \frac{\partial^3 w}{\partial r^3} - \frac{2}{r^3} \frac{\partial^3 w}{\partial r \partial \theta^2}$$

$$- \frac{1}{r^2} \frac{\partial^2 w}{\partial r^2} + \frac{4}{r^4} \frac{\partial^2 w}{\partial \theta^2} + \frac{1}{r^3} \frac{\partial w}{\partial r} = \frac{q(r, \theta)}{D} \tag{4.81}$$

$$M_r = -D\left[\frac{\partial^2 w}{\partial r^2} + \mu\left(\frac{1}{r} \frac{\partial w}{\partial r} + \frac{1}{r^2} \frac{\partial^2 w}{\partial \theta^2} \right) \right] \tag{4.82}$$

$$M_\theta = -D\left[\frac{1}{r} \frac{\partial w}{\partial r} + \frac{1}{r^2} \frac{\partial^2 w}{\partial \theta^2} + \mu \frac{\partial^2 w}{\partial r^2} \right] \tag{4.83}$$

$$M_{r\theta} = (1 - \mu)D\left[\frac{1}{r} \frac{\partial^2 w}{\partial r \partial \theta} - \frac{1}{r^2} \frac{\partial w}{\partial \theta} \right] \tag{4.84}$$

4.10.2 Orthotropic Plates [25]

When the material is orthotropic, the above equations take the following form, which includes the properties D_x, D_y, and H:

$$\nabla^4 w = D_r \cdot \left[\frac{\partial^4 w}{\partial r^4} + \frac{2}{r} \cdot \frac{\partial^3 w}{\partial r^3} \right] + 2H \cdot \left[\frac{1}{r^2} \cdot \frac{\partial^4 w}{\partial r^2 \partial \theta^2} - \frac{1}{r^3} \cdot \frac{\partial^3 w}{\partial r \partial \theta^2} \right.$$

$$\left. + \frac{1}{r^4} \frac{\partial^2 w}{\partial \theta^2} \right] + D_\theta \cdot \left[\frac{1}{r^4} \cdot \frac{\partial^4 w}{\partial \theta^4} - \frac{1}{r^2} \frac{\partial^2 w}{\partial r^2} + \frac{2}{r^4} \frac{\partial^2 w}{\partial \theta^2} \right.$$

$$\left. + \frac{1}{r^3} \frac{\partial w}{\partial r} \right] = q(r, \theta) \tag{4.85}$$

$$M_r = -D_r \cdot \left[\frac{\partial^2 w}{\partial r^2} + \mu_\theta \left(\frac{1}{r} \cdot \frac{\partial w}{\partial r} + \frac{1}{r^2} \frac{\partial^2 w}{\partial \theta^2} \right) \right] \tag{4.86}$$

$$M_\theta = -D_\theta \cdot \left[\mu_r \frac{\partial^2 w}{\partial r^2} + \frac{1}{r} \cdot \frac{\partial w}{\partial r} + \frac{1}{r^2} \frac{\partial^2 w}{\partial \theta^2} \right] \tag{4.87}$$

$$M_{r\theta} = + 2D_K \cdot \left[\frac{1}{r} \cdot \frac{\partial^2 w}{\partial r \partial \theta} - \frac{1}{r^2} \frac{\partial w}{\partial \theta} \right] \tag{4.88}$$

$$V_r = -\left[D_r\left(\frac{\partial^3 w}{\partial r^3} + \frac{1}{r}\cdot\frac{\partial^2 w}{\partial r^2}\right) + H\left(\frac{1}{r^2}\cdot\frac{\partial^3 w}{\partial r\,\partial\theta^2} - \frac{1}{r^3}\frac{\partial^2 w}{\partial\theta^2}\right)\right.$$

$$\left. -D_\theta\left(\frac{1}{r^2}\frac{\partial w}{\partial r} + \frac{1}{r^3}\frac{\partial^3 w}{\partial\theta^3}\right)\right] \tag{4.89}$$

$$V_\theta = -\left[H\left(\frac{1}{r}\frac{\partial^3 w}{\partial r^2\,\partial\theta}\right) + D_\theta\left(\frac{1}{r^2}\cdot\frac{\partial^2 w}{\partial\theta\,\partial r} + \frac{1}{r^3}\frac{\partial^3 w}{\partial\theta^3}\right)\right] \tag{4.90}$$

$$R_R = \left(V_r - \frac{1}{r}\cdot\frac{\partial M_{r\theta}}{\partial\theta}\right) \tag{4.91}$$

$$R_\theta = \left(V_\theta - \frac{\partial M_{r\theta}}{\partial r}\right) \tag{4.92}$$

$$R_R = -\left[D_r\left(\frac{\partial^3 w}{\partial r^3} + \frac{1}{r}\frac{\partial^2 w}{\partial r^2}\right) + \bar{H}\left(\frac{1}{r^2}\frac{\partial^3 w}{\partial r\,\partial\theta^2} - \frac{1}{r^3}\frac{\partial^2 w}{\partial\theta^2}\right)\right.$$

$$\left. -D_\theta\left(\frac{1}{r^2}\frac{\partial w}{\partial r} + \frac{1}{r^3}\frac{\partial^2 w}{\partial\theta^2}\right)\right] \tag{4.93}$$

$$R_\theta = -\left[D_\theta\left(\frac{1}{r^2}\cdot\frac{\partial^2}{\partial r\,\partial\theta} + \frac{1}{r^3}\frac{\partial^3 w}{\partial\theta^3}\right)\right.$$

$$\left. + \bar{H}\left(\frac{1}{r}\frac{\partial^3 w}{\partial r^2\,\partial\theta}\right) + 4D_K\left(-\frac{1}{r^2}\frac{\partial^2 w}{\partial r\,\partial\theta} + \frac{1}{r^3}\frac{\partial w}{\partial\theta}\right)\right] \tag{4.94}$$

where:

$$D_r = \frac{E_r h^3}{12(1 - \mu_r\mu_\theta)}$$

$$D_\theta = \frac{E_\theta h^3}{12(1 - \mu_r\mu_\theta)}$$

$$D_K = \frac{G_{r\theta}h^3}{12} \tag{4.95}$$

$$H = D_r\cdot\mu_\theta + 2D_K = D_\theta\mu_r + 2D_K$$

$$\bar{H} = D_r\mu_\theta + 4D_K$$

h = Plate thickness

For the special case when Poisson's Ratio $\mu_\theta = \mu_r = 0$, then $H = 2D_K$ and $\bar{H} = 2H$, and the reactions, in particular, will reduce to:

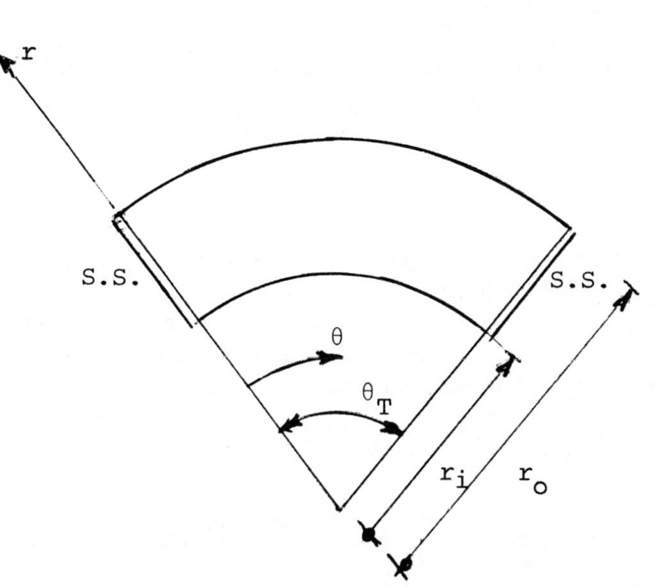

Figure 4-29. Circular Plate Sector Geometry.

$$R_r = -\left[D_r\left(\frac{\partial^3 w}{\partial r^3} + \frac{1}{r}\frac{\partial^2 w}{\partial r^2}\right) + 2H\left(\frac{1}{r^2}\frac{\partial^3 w}{\partial r\,\partial\theta^2} - \frac{1}{r^3}\frac{\partial^2 w}{\partial\theta^2}\right)\right.$$
$$\left. - D_\theta \cdot \left(\frac{1}{r^2}\frac{\partial w}{\partial r} + \frac{1}{r^3}\frac{\partial^2 w}{\partial\theta^2}\right)\right] \tag{4.96}$$

$$R_\theta = -\left[D_\theta\left(\frac{1}{r^2}\frac{\partial^2 w}{\partial r\,\partial\theta} + \frac{1}{r^3}\frac{\partial^3 w}{\partial\theta^3}\right)\right.$$
$$\left. + 2H\left(\frac{1}{r}\frac{\partial^3 w}{\partial r^2\,\partial\theta} - \frac{1}{r^2}\frac{\partial^2 w}{\partial r\,\partial\theta} + \frac{1}{r^3}\frac{\partial w}{\partial\theta}\right)\right] \tag{4.97}$$

If the material is isotropic, then

$$D_{r\theta} = D_r = D_\theta = D = \frac{Eh^3}{(1 - \mu^2)12}$$

and equation (4.85) will reduce to equation (4.81), i.e.,

$$D\nabla^4 w = q(r, \theta)$$

4.10.3 Simply Supported Orthotropic Plate Sector Solutions

The solution of (4.85) will follow the technique presented in section 4.6.2, as applied to rectangular plates. Consider a plate sector, as shown in Figure 4-29, which has simple supports along the radial edges and arbitrary supports along the angular edges. Assume the solution of w to have the form

$$w = \sum R_n \sin \lambda_n \theta \tag{4.98}$$

where

$$\lambda_n = n\pi/\theta_T$$

Equation (4.98) has the unique characteristic that it satisfies the boundary conditions at $\theta = 0$ and $\theta = \theta_T$, i.e., $w = M_\theta = 0$. The evaluation of the variable R_n, which is similar to the term Y_n given in section 4 6.2 for the rectangular plate, is now required. Taking the nth term of the deflection equation (4.98) and substituting the derivatives into equation (4.85) gives

$$D_r R_n'''' - \frac{2H}{r^2} R_n'' \lambda^2 + \frac{D_\theta}{r^4} R_n \lambda^4 + \frac{2D_r}{r} R_n'''$$

$$+ \frac{2H}{r^3} R_n' \lambda^2 - \frac{D_\theta}{r^2} R_n'' - \frac{2(D_\theta + H)}{r^4} R_n \lambda^2 + \frac{D_\theta}{r^3} R_n' = 0 \tag{4.99}$$

where

$$R_n' = \frac{\partial R_n}{\partial r}\,; \qquad R_n'' = \frac{\partial^2 R_n}{\partial r^2}\,; \qquad \text{etc.}$$

This equation is a linear differential equation of variable coefficients [1] and thus the solution of the form $R = e^{mr}$ can not be used directly. Using the technique given by Wang [2], however, will permit the modification of this equation to a constant coefficient differential equation. Therefore, let $r = e^\xi$, thus $\xi = \ln_e r$; and

$$\frac{d\xi}{dr} = \frac{d(\ln_e r)}{dr} = \frac{1}{r}$$

and by chain rule

$$\frac{dR_n}{dr} = \frac{d(R_n)}{d\xi}\,\frac{d\xi}{dr}$$

therefore,

$$\frac{dR_n}{dr} = R_n' \frac{1}{r} \tag{a}$$

where

$$R_n' = \frac{dR_n}{d\xi}$$

Similarly,

$$\frac{d^2R_n}{dr^2} = \frac{1}{r^2}R_n' + \frac{1}{r^2}R_n'' \qquad \text{(b)}$$

$$\frac{d^3R_n}{dr^3} = \frac{1}{r^3}[R_n''' - 3R_n'' + 2R_n'] \qquad \text{(c)}$$

$$\frac{d^4R_n}{dr^4} = \frac{1}{r^4}[R_n'''' - 6R_n''' + 11R_n'' - 6R_n']$$

Now substituting (a) through (d) into (4.99) gives the following solvable fourth order equation:

r^4 cancels out

$$D_rR_n'''' - 4D_rR_n''' + (5D_r - 2H\lambda^2 - D_\theta)R_n'' + (-2D_r + 4H\lambda^2 + 2D_\theta)R_n' + (-2H\lambda^2 - 2D_\theta\lambda^2 + D_\theta\lambda^4)R_n = 0 \qquad (4.100)$$

Assume now that

$$R_n = e^{m\xi} \qquad (4.101)$$

Substituting the derivatives of (4.101) into (4.100) gives the following characteristic equation (4.102):

$$D_rm^4 - 4D_rm^3 + (5D_r - 2H\lambda^2 - D_\theta)m^2 + (-2D_r + 4H\lambda^2 + 2D_\theta)m + (-2H\lambda^2 - 2D_\theta\lambda^2 + D_\theta\lambda^4) = 0 \qquad (4.102)$$

The roots of this equation are found to be

$$m_{1,2,3,4} = \pm \left\{\frac{\alpha + 2\beta\lambda^2 + 1}{2} \pm \left[\frac{1}{4}(1 + \alpha + 2\lambda^2\beta)^2 - \alpha(\lambda^2 - 1)^2\right]^{1/2}\right\}^{1/2} + 1 \qquad (4.103)$$

where:

$$\alpha = D_\theta / D_r$$
$$\beta = H / D_r \qquad (4.104)$$
$$\lambda = n\pi/\theta_T$$

The solution to R_n is, therefore,

$$R_n = C_1e^{m_1\xi} + C_2e^{m_2\xi} + C_3e^{m_3\xi} + C_4e^{m_4\xi}$$

but, as defined previously, $\xi = \ln_e r$ or

$$e^{m\ln_e r} = r^m$$

therefore,

$$R_n = Ar^{m_1} + Br^{m_2} + Cr^{m_3} + Dr^{m_4}$$

and the deflection is

$$w = \sum (Ar^{m_1} + Br^{m_2} + Cr^{m_3} + Dr^{m_4}) \sin \lambda\theta \qquad (4.105)$$

If the plate is isotropic, then $\alpha = \beta = 1$ and the roots are, therefore:

$$m = \pm\left\{\frac{2 + 2\lambda^2}{2} \pm \left[\frac{1}{4}(1 + 1 + 2\lambda^2)^2 - (\lambda^2 - 1)^2\right]^{1/2}\right\}^{1/2} + 1$$

or

$$m = \pm(\lambda \pm 1) + 1$$

The four roots are:

$$m = \lambda + 2$$
$$m = -\lambda$$
$$m = +\lambda$$
$$m = -\lambda + 2$$

where $\lambda = n\pi/\nu_T$.

4.10.4 Orthotropic Plate—Radially Symmetrical

If the plate consists of a complete circular element, then the forces become independent of the coordinate θ and thus $\partial/\partial\theta = 0$, $\partial^2/\partial\theta^2 = 0$, etc. Then equation (4.85) becomes

$$D_r\left[\frac{\partial^4 w}{\partial r^4} + \frac{2}{r}\frac{\partial^3 w}{\partial r^3}\right] + D_\theta\left[-\frac{1}{r^2}\frac{\partial^2 w}{\partial r^2} + \frac{1}{r^3}\frac{\partial w}{\partial r}\right] = q \qquad (4.106)$$

Similarly,

$$M_r = -D_r\left[\frac{\partial^2 w}{\partial r^2} + \mu_\theta\left(\frac{1}{r}\frac{\partial w}{\partial r}\right)\right] \qquad (4.107)$$

$$M_\theta = -D_\theta\left[\mu_r\frac{\partial^2 w}{\partial r^2} + \frac{1}{r}\frac{\partial w}{\partial r}\right] \qquad (4.108)$$

$$M_{r\theta} = 0 \qquad (4.109)$$

$$V_r = -\left[D_r\left(\frac{\partial^3 w}{\partial r^3} + \frac{1}{r}\frac{\partial^2 w}{\partial r^2}\right) - D_\theta\left(\frac{1}{r^2}\frac{\partial w}{\partial r}\right)\right] \qquad (4.110)$$

$$V_\theta = 0 \qquad (4.111)$$

$$R_r = -\left[D_r\left(\frac{\partial^3 w}{\partial r^3} + \frac{1}{r}\frac{\partial^2 w}{\partial r^2}\right) - D_\theta\left(\frac{1}{r^2}\frac{\partial w}{\partial r}\right)\right] \qquad (4.112)$$

$$R_\theta = 0 \tag{4.113}$$

As all of the above expressions are independent of θ, the expressions can be written as d/dr (exact) rather than $\partial/\partial r$ (partial). Therefore, equation (4.106) becomes

$$\frac{d^4w}{dr^4} + \frac{2}{r}\frac{d^3w}{dr^3} + \alpha\left[\frac{1}{r^3}\frac{dw}{dr} - \frac{1}{r^2}\frac{d^2w}{dr^2}\right] = \frac{q}{D_r} \tag{4.114}$$

where $\alpha = D_\theta / D_r$

4.10.5 Isotropic Plate—Radially Symmetrical

Equation (4.114) can be reduced to the following as $D_r = D_\theta = D$:

$$\frac{d^4w}{dr^4} + \frac{2}{r}\frac{d^3w}{dr^3} + \frac{1}{r^3}\frac{dw}{dr} - \frac{1}{r^2}\frac{d^2w}{dr^2} = \frac{q(r)}{D} \tag{4.115}$$

This equation can be written as [26]:

$$\frac{1}{r}\frac{d}{dr}\left[r\frac{d}{dr}\left\{\frac{1}{r}\frac{d}{dr}\left(r\frac{dw}{dr}\right)\right\}\right] = \frac{q(r)}{D} \tag{4.116}$$

Now if $q(r)$ is a known function of r, equation (4.116) can be solved by direct integration.

Example: Consider the case when q is uniform across the entire plate, then,

$$\int d\left[r\frac{d}{dr}\left\{\frac{1}{r}\frac{d}{dr}\left(r\frac{dw}{dr}\right)\right\}\right] = \int\frac{qr}{D}\,dr$$

$$r\frac{d}{dr}\left\{\frac{1}{r}\frac{d}{dr}\left(r\frac{dw}{dr}\right)\right\} = \frac{qr^2}{2D} + C_1$$

$$\int d\left\{\frac{1}{r}\frac{d}{dr}\left(r\frac{dw}{dr}\right)\right\} = \int\left(\frac{qr}{2D} + \frac{C_1}{r}\right)dr$$

$$\frac{1}{r}\frac{d}{dr}\left(r\frac{dw}{dr}\right) = \frac{qr^2}{4D} + C_1\ln r + C_2$$

$$\int d\left(r\frac{dw}{dr}\right) = \int\left(\frac{qr^3}{4D} + C_1 r\ln r + C_2 r\right)dr$$

$$r\frac{dw}{dr} = \frac{qr^4}{16D} + C_1\frac{r^2}{2}\left(\ln r - \frac{1}{2}\right) + C_2\frac{r^2}{2} + C_3$$

$$dw = \int \left(\frac{qr^3}{16D} + C_1 \frac{r}{2}\left(\ln r - \frac{1}{2}\right) + C_2 \frac{r}{2} + \frac{C_3}{r} \right) dr$$

$$w = \frac{qr^4}{64D} + C_1 \frac{r^2}{4}(\ln r - 1) + C_2 \frac{r^2}{4} + C_3 \ln r + C_4 \qquad (4.117)$$

and

$$\frac{dw}{dr} = \frac{qr^3}{16D} + C_1 \frac{r}{2}\left(\ln r - \frac{1}{2}\right) + C_2 \frac{r}{2} + \frac{C_3}{r}$$

$$\frac{d^2w}{dr^2} = \frac{3qr^2}{16D} + \frac{C_1}{2}\left(\ln r + \frac{1}{2}\right) + \frac{C_2}{2} - \frac{C_3}{r^2} \qquad (4.118)$$

$$\frac{d^3w}{dr^3} = \frac{3qr}{8D} + \frac{C_1}{2r} + \frac{2C_3}{r^3}$$

Continuous Plate: If the plate is continuous at $r = 0$, i.e., has no hole in the middle, then $C_1 = C_3 = 0$, as $\ln r$ would go to infinity as r goes to zero. Therefore,

$$w = \frac{qr^4}{64D} + C_2 \frac{r^2}{4} + C_4 \qquad (4.119)$$

and

$$\frac{dw}{dr} = \frac{qr^3}{16D} + C_2 \frac{r}{2}$$

$$\frac{d^2w}{dr^2} = \frac{3qr^2}{16D} + \frac{C_2}{2}$$

Case I: If the plate is simply supported at the rim, as shown in Figure 4-30, then the boundary conditions are $r = R_0$, $w = 0$, and $M_r = 0$ or

$$\text{at } r = R_0: \qquad w = 0$$

$$0 = \frac{qR_0^4}{64D} + C_2 \frac{R_0^2}{4} + C_4$$

$$\text{at } r = R_0: \qquad M_r = 0$$

$$M_r = -D_r \left[\frac{d^2w}{dr^2} + \mu\left(\frac{1}{r}\frac{dw}{dr}\right) \right]$$

$$M_r = -D_r \left[\frac{3qr^2}{16D} + \frac{C_2}{2} + \mu \left(\frac{1}{r}\left(\frac{qr^3}{16D} + C_2 \frac{r}{2} \right) \right) \right]$$

$$0 = \frac{3qR_0^2}{16D} + \frac{C_2}{2} + \mu \frac{qR_0^2}{16D} + \mu C_2 \frac{1}{2}$$

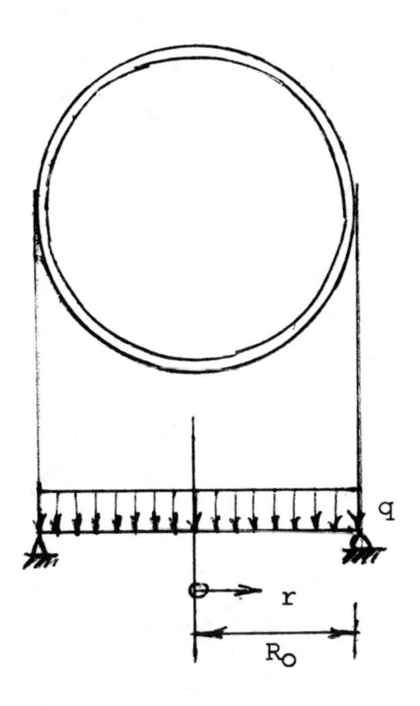

Figure 4-30. Uniformly Loaded Circular Plate.

Solving for C_2 and C_4 gives

$$C_2 = -\frac{qR_0^2}{8D}\left(\frac{3+\mu}{1+\mu}\right)$$

$$C_4 = \frac{qR_0^4}{64D}\left(\frac{5+\mu}{1+\mu}\right)$$

However,

$$w = \frac{qr^4}{64D} + C_2\frac{r}{4} + C_4$$

therefore,

$$w = \frac{qr^4}{64D} - \frac{qR_0^2}{8D}\left(\frac{3+\mu}{1+\mu}\right)\frac{r^2}{4} + \frac{qR_0^4}{64D}\left(\frac{5+\mu}{1+\mu}\right)$$

or

$$w = \frac{q}{64D}\left[\left(\frac{5+\mu}{1+\mu}\right)(R_0^2 - r^2)(R_0^2 - r^2)\right] \tag{4.120}$$

Case II: If the plate is built-in, the boundary conditions are $w = \partial w/\partial r = 0$ at $r = R_0$, therefore,

$$\text{at } r = R_0, \qquad w = 0:$$

$$0 = \frac{qR_0^4}{64D} + C_2\frac{R_0^2}{4} + C_4$$

$$\text{at } r = R_0, \qquad \partial w/\partial r = 0$$

$$0 = \frac{qR_0^3}{16D} + C_2\frac{R_0}{2}$$

Solving for C_2 and C_4 gives

$$C_2 = -\frac{qR_0^2}{8D}$$

$$C_4 = +\frac{qR_0^4}{64D}$$

therefore,

$$w = \frac{q}{64D}(R_0^2 - r^2)^2 \tag{4.121}$$

Plate with Central Hole of Radius R_i: Consider the plate with a hole, as shown in Figure 4-31. This condition has no discontinuity at $r = 0$ as the boundaries are at $r = R_i$ and $r = R_0$. Therefore, the entire equation (4.117) is required:

$$w = \frac{qr^4}{64D} + C_1\frac{r^2}{4}(\ln r - 1) + C_2\frac{r^2}{4} + C_3 \ln r + C_4 \tag{4.117}$$

The boundary conditions are, therefore,

$$\text{at } r = R_0: \quad w = 0, \quad M_r = 0$$

$$\text{at } r = R_i: \quad R_r = 0, \quad M_r = 0$$

where

$$M_r = -D\left[\frac{d^2w}{dr^2} + \mu\left(\frac{1}{r}\frac{dw}{dr}\right)\right]$$

$$R_r = -D\left[\frac{d^3w}{dr^3} + \frac{1}{r}\frac{d^2w}{dr^2} - \frac{1}{r^2}\frac{dw}{dr}\right]$$

Substituting equation (4.121) and its derivatives into M_r and R_r gives

$$M_r = -D\left\{\left[\frac{3qr^2}{16D} + \frac{C_1}{2}\left(\ln r + \frac{1}{2}\right) + \frac{C_2}{2} - \frac{C_3}{r^2}\right] + \frac{\mu}{r}\left[\frac{qr^3}{16D} + \frac{C_1 r}{2}\left(\ln r - \frac{1}{2}\right) + \frac{C_2 r}{2} + \frac{C_3}{r}\right]\right\}$$

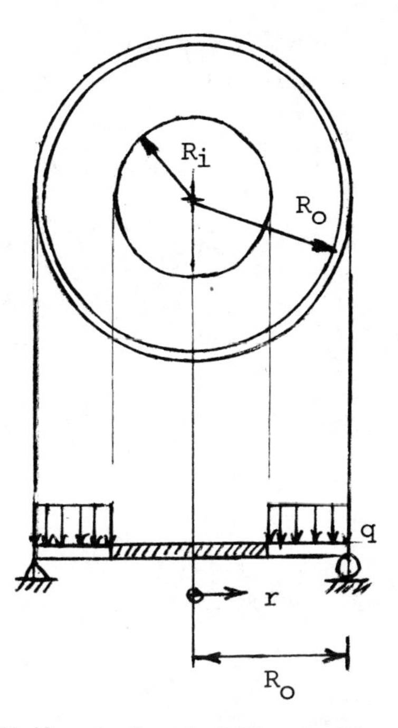

Figure 4-31. Uniformly Loaded Circular Plate with Hole.

$$R_r = -D\left\{\left[\frac{3qr}{8D} + \frac{C_1}{2r} + \frac{2C_3}{r^3}\right] + \frac{1}{r}\left[\frac{3qr^2}{16D} + \frac{C_1}{2}\left(\ln r + \frac{1}{2}\right)\right.\right.$$

$$\left.\left. + \frac{C_2}{2} - \frac{C_3}{r^2}\right] - \frac{1}{r^2}\left[\frac{qr^3}{16D} + C_1\frac{r}{2}\left(\ln r - \frac{1}{2}\right) + \frac{C_2 r}{2} + \frac{C_3}{r}\right]\right\}$$

Applying these boundary conditions and solving for the constants [26] gives

$$C_1 = \frac{qR_i^2}{2D}$$

$$C_2 = -\frac{q}{8D}\left(\frac{3+\mu}{1+\mu}\right)(R_0^2 + R_i^2)$$

$$+ \frac{qR_i^2}{2D}\left\{\frac{R_0^2 \ln R_0 - R_i^2 \ln R_i}{(R_0^2 - R_i^2)} + \frac{(1-\mu)}{2(1+\mu)}\right\}$$

$$C_3 = -\frac{q}{16D}\left\{\left[(3+\mu)(R_0^2 - R_i^2) - 4(1+\mu)R_i^2 \ln\frac{R_0}{R_i}\right]\right\}$$

$$\times \frac{R_0^2 R_i^2}{(1-\mu)(R_0^2 - R_i^2)}$$

$$C_4 = \frac{qR_0^4}{64D}\left(\frac{5+\mu}{1+\mu}\right) - \frac{3qR_0^2R_i^2}{32D}\left(\frac{3+\mu}{1+\mu}\right)$$

$$+ \frac{qR_0^2R_i^2}{16D}\left(\frac{5-\mu}{1-\mu}\right)\ln R_0 - \frac{qR_0^2R_i^2}{8D(R_0^2 - R_i^2)}$$

$$\times \left[R_0^2 \ln R_0 - R_i^2 \ln R_i + 2\left(\frac{1+\mu}{1-\mu}\right)R_i^2 \ln R_0 \ln R_0/R_i\right]$$

Finite Difference and Finite Element Solution of Plates in Bending

5.1 Rectangular Plates

The study and solution of the bending of plates, as given in chapter 4, has been limited to direct integration or Fourier series techniques. These methods were limited in that the boundary conditions were restricted, e.g., simple supports, etc. If the plate has various types of boundary restraints and interacting flexible girders, then these previous techniques may not be applicable. A most general technique that could readily be applied is the finite difference method, [14, 15, 30] which was described in chapter 3 and used in plane stress problems.

Consider now a plate that has a specified grid or nodes on the plate, as shown in Figure 5-1. Assume that the plate is subjected to some load q and thus deforms some amount w_0, w_a, w_b, w_l, and w_r at the respective nodes 0, a, b, l, and r, as shown in Figure 5-1. Assume now that the deflected surface $w(x, y)$ is given by a parabolid:

$$\bar{w} = Ax^2 + Bx + C + Dy + Ey^2 \tag{5.1}$$

such that $\bar{w}$ is identical to w at the five given points (l, r, 0, a, b). Applying now the conditions at the various points gives

$$\left.\begin{array}{l} x = y = 0 \\ \bar{w} = w_0 \end{array}\right\} \bar{w} = w_0 = C \tag{a}$$

$$\left.\begin{array}{ll} x = -n\lambda, & y = 0 \\ \bar{w} = w_l & \end{array}\right\} \begin{array}{l} \bar{w} = w_l = A(n\lambda)^2 - Bn\lambda + C + D(0) + E(0) \\ w_l = A(n\lambda)^2 - Bn\lambda + C \end{array} \tag{b}$$

$$\left.\begin{array}{ll} x = +n\lambda, & y = 0 \\ \bar{w} = w_r & \end{array}\right\} w_r = A(n\lambda)^2 + Bn\lambda + C \tag{c}$$

$$\left.\begin{array}{l} x = 0, \ y = -\lambda \\ \bar{w} = w_b \end{array}\right\} w_b = C - D\lambda + E\lambda^2 \tag{d}$$

$$\left.\begin{array}{ll} x = 0, & y = \lambda \\ \bar{w} = w_a & \end{array}\right\} w_a = C + D\lambda + E\lambda^2 \tag{e}$$

Solving for constants A, B, C, D, E gives

$$A = \frac{1}{2(n\lambda)^2}(w_r - 2w_0 + w_l)$$

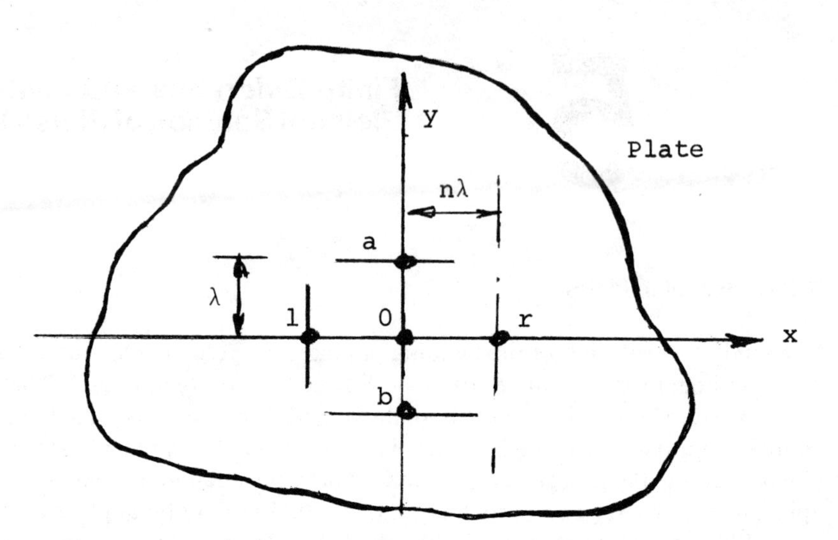

Figure 5-1. General Difference Nodes.

$$
\begin{aligned}
B &= \frac{1}{2n\lambda}(w_r - w_l) \\
C &= w_0 \\
D &= \frac{1}{2\lambda}(w_a - w_b) \\
E &= \frac{1}{2\lambda^2}(w_a - 2w_0 + w_b)
\end{aligned}
\quad \text{(f)}
$$

Substituting equations (f) into (5.1) gives

$$
\bar{w} = \frac{1}{2\lambda^2}\left[(w_r - 2w_0 + w_l)\frac{x^2}{n^2} + (w_r - w_l)\frac{x\lambda}{n} + w_0 2\lambda^2 + (w_a - w_b)y\lambda + (w_a - 2w_0 + w_b)y^2 \right] \quad (5.2)
$$

Equation (5.2) will be used, in part, to evaluate the biharmonic equation:

$$
D_x \frac{\partial^4 w}{\partial x^4} + 2H\frac{\partial^4 w}{\partial x^2\,\partial y^2} + D_y\frac{\partial^4 w}{\partial y^4} = q \quad (4.38)
$$

First extend the mesh pattern to more points, as shown in Figure 5-2. This is required because if the $\partial^4 w/\partial x^4$ etc. were taken of equation (5.2), all terms would vanish.

Take now the partial derivatives of equation (5.2):

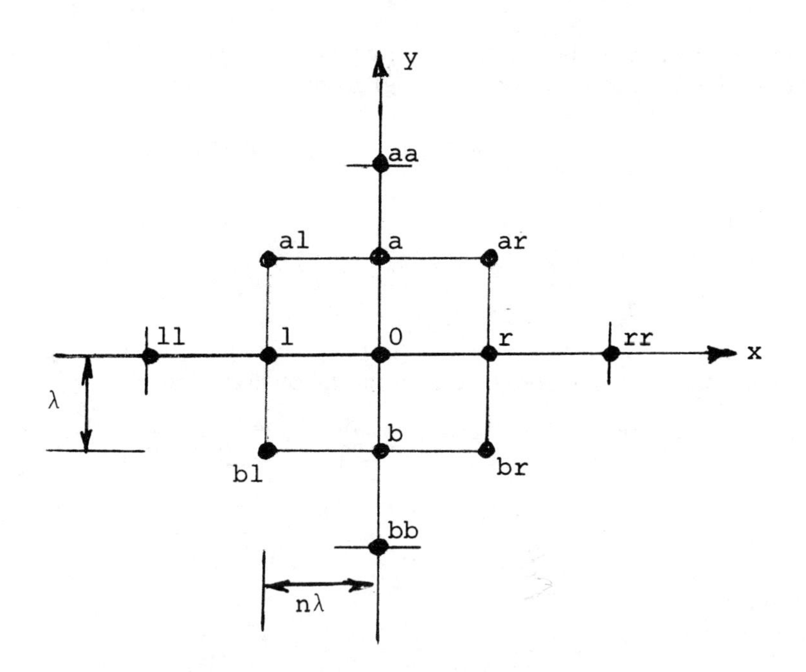

Figure 5-2. General Plate Difference Nodes.

$$\frac{\partial^2 \bar{w}}{\partial x^2} = \frac{1}{(n\lambda)^2}(w_r - 2w_0 + w_l)$$

$$\frac{\partial^2 \bar{w}}{\partial y^2} = \frac{1}{\lambda^2}(w_a - 2w_0 + w_b)$$

The fourth order differential can be written as

$$\frac{\partial^4 \bar{w}}{\partial x^4} = \frac{\partial^2}{\partial x^2}\left(\frac{\partial^2 \bar{w}}{\partial x^2}\right) = \frac{1}{(n\lambda)^2}\left[\left.\frac{\partial^2 \bar{w}}{\partial x^2}\right|_r - 2\left.\frac{\partial^2 \bar{w}}{\partial x^2}\right|_0 + \left.\frac{\partial^2 \bar{w}}{\partial x^2}\right|_l\right] \quad \text{(g)}$$

where expansion is made with respect to nodes r, 0, and l. These expansions are pg. 76-77

$$\left.\frac{\partial^2 \bar{w}}{\partial x^2}\right|_r = \frac{1}{(n\lambda)^2}(w_{rr} - 2w_r + w_0)$$

$$\left.\frac{\partial^2 \bar{w}}{\partial x^2}\right|_0 = \frac{1}{(n\lambda)^2}(w_r - 2w_0 + w_l)$$

$$\left.\frac{\partial^2 \bar{w}}{\partial x^2}\right|_l = \frac{1}{(n\lambda)^2}(w_0 - 2w_l + w_{ll})$$

in accordance with the mesh pattern given in Figure 5-2. Substituting in these relationships into expression (g) gives

$$\frac{\partial^4 \bar{w}}{\partial x^4} = \frac{1}{(n\lambda)^4}(w_{rr} - 4w_r + 6w_0 - 4w_l + w_{ll}) \tag{h}$$

Similarly,

$$\frac{\partial^4 \bar{w}}{\partial y^4} = \frac{1}{\lambda^4}(w_{aa} - 4w_a + 6w_0 - 4w_b + w_{bb}) \tag{i}$$

The mixed partial $(\partial^4 w/\partial x^2 \partial y^2)$ is found by expanding $(\partial^2 w/\partial y^2)$ about nodes r, 0, and l in accordance with the following:

$$\frac{\partial^4 \bar{w}}{\partial x^2 \partial y^2} = \frac{\partial^2}{\partial y^2}\left(\frac{\partial^2 w}{\partial x^2}\right) = \frac{\partial^2}{\partial y^2}\left(\frac{w_r - 2w_0 + w_l}{n^2\lambda^2}\right)$$

$$= \frac{1}{n^2\lambda^2}\left[\frac{\partial^2 w}{\partial y^2}\bigg|_r - 2\frac{\partial^2 w}{\partial y^2}\bigg|_0 + \frac{\partial^2 w}{\partial y^2}\bigg|_l\right]$$

However,

$$\frac{\partial^2 w}{\partial y^2}\bigg|_r = \frac{w_{ar} - 2w_r + w_{br}}{\lambda^2}$$

$$\frac{\partial^2 w}{\partial y^2}\bigg|_0 = \frac{w_a - 2w_0 + w_b}{\lambda^2}$$

$$\frac{\partial^2 w}{\partial y^2}\bigg|_l = \frac{w_{al} - 2w_l + w_{bl}}{\lambda^2}$$

therefore,

$$\frac{\partial^4 w}{\partial x^2 \partial y^2} =$$

$$\frac{1}{n^2\lambda^4}[w_{ar} - 2w_r + w_{br} - 2w_a + 4w_0 - 2w_b + w_{al} - 2w_l + w_{bl}] \tag{j}$$

Let

$$\beta = H/D_y$$

$$\alpha = H/\sqrt{D_x D_y}$$

Substituting (h), (i), (j) into $\nabla^4 w$ of equation (4.38) gives the final general orthotropic plate equation in difference form:

$$\{w_0[6n^4 + 8n^2\beta + 6(\beta/\alpha)^2] - 4[w_r + w_l][(\beta/\alpha)^2 + n^2\beta]$$

$$-4(w_a + w_b)(n^2\beta + n^4) + (w_{rr} + w_{ll})(\beta/\alpha)^2 + (w_{aa} + w_{bb})n^4$$

$$+ (w_{ar} + w_{br} + w_{al} + w_{bl})2n^2\beta\} = qn^4\lambda^4/D_y \tag{5.3}$$

Applying equations (4.31), (4.32), and (4.41), (4.42) in difference form gives the following equations, assuming $\mu_x = \mu_y = 0$:

$$M_x = -\frac{D_x}{(n\lambda)^2}(w_l - 2w_0 + w_r) \tag{5.4}$$

$$M_y = -\frac{D_y}{\lambda^2}(w_a - 2w_0 + w_b) \tag{5.5}$$

$$R_x = -D_x\left[\frac{1}{2n^3\lambda^3}(w_{rr} - 2w_r + 2w_l - w_{ll}) + \frac{2\xi}{2n\lambda^3}(-w_{al} + 2w_l - w_{bl} + w_{ar} - 2w_r + w_{br})\right] \tag{5.6}$$

$$R_y = -D_y\left[\frac{1}{2\lambda^3}(w_{aa} - 2w_a + 2w_b - w_{bb}) + \frac{2\beta}{2n^2\lambda^3}(w_{al} - 2w_a + w_{ar} - w_{bl} + 2w_b - w_{br})\right] \tag{5.7}$$

where $\xi = H/D_x$.

5.2 Rectangular Plates with Interacting Girders

Assume the orthotropic plate is supported on interacting girders, as shown in Figure 5-3. Each girder will have a stiffness EI_x and EI_y and the plate has stiffness D_x, D_y, and H. Equilibrium of the intersecting plate and girders is given by

$$\sum F = 0$$

$$q_T = q_{PL} + q_{Bx} + q_{By} \tag{5.8}$$

where q_T = Externally applied load

q_{PL} = Load resisted by the plate

q_{Bx} = Load resisted by the beam in the x direction

q_{By} = Load resisted by the beam in the y direction

From beam theory:

$$\frac{d^4w}{dy^4} = \frac{P_y}{EI_y}$$

or in difference form,

$$\frac{P_y\lambda^4}{EI_y} = (w_{aa} - 4w_a + 6w_0 - 4w_b + w_{bb}) \tag{5.9}$$

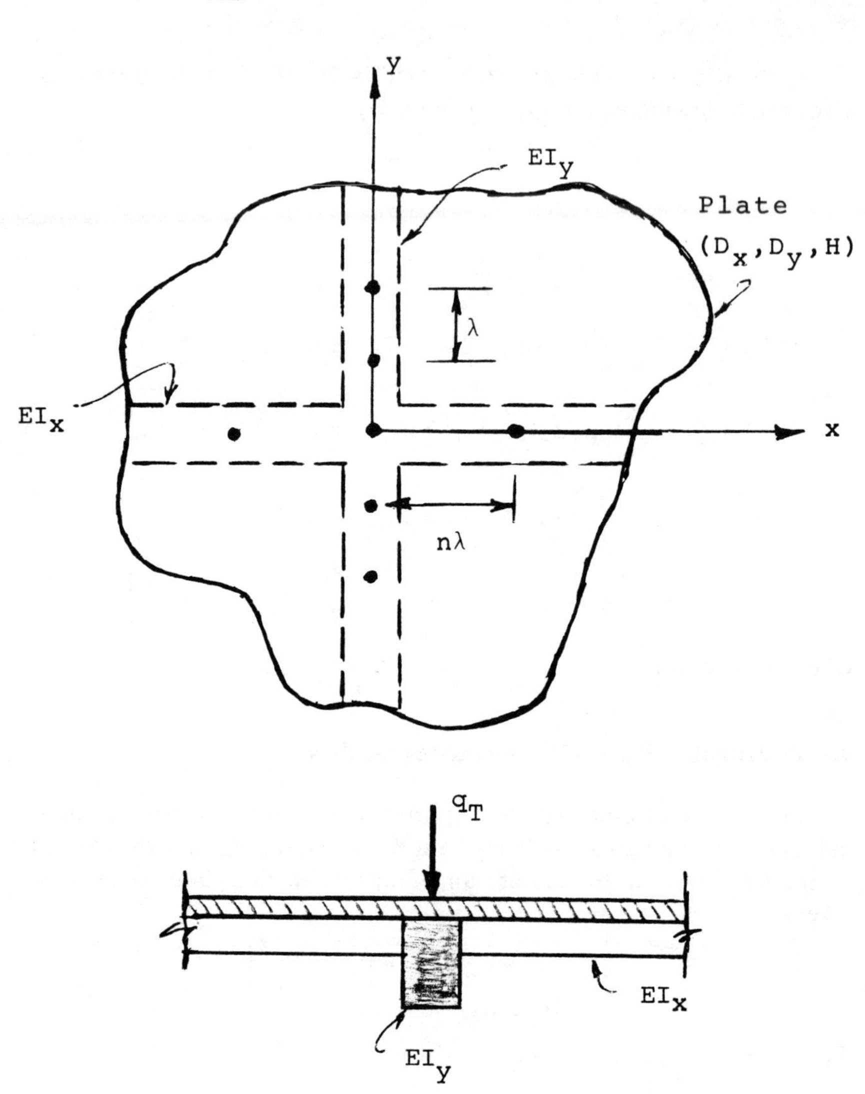

Figure 5-3. Interacting Plate and Girders.

Similarly,

$$\frac{d^4w}{dx^4} = \frac{P_x}{EI_x}$$

$$\frac{P_x(n\lambda)^4}{EI_x} = (w_{ll} - 4w_l + 6w_0 - 4w_r + w_{rr}) \tag{5.10}$$

The forces P_y and P_x given in these equations are per unit length of beam. The forces in the equilibrium plate-girder equation (5.8) are force per unit area. Converting P_x and P_y to a force per area:

$$q_{Bx} = \frac{P_x}{\lambda} \qquad \text{and} \qquad q_{By} = \frac{P_y}{n\lambda}$$

equation (5.8) therefore q_T is

$$\frac{P_{Bx}}{\lambda} + \frac{P_{By}}{n\lambda} + q_{PL} = q_T \tag{5.11}$$

Substituting equations (5.3), (5.9), and (5.10) into (5.11) gives

$$Jw_{Bx}[\ ] + Kw_{By}[\ ] + w_{PL}[\ ] = \frac{qn^4\lambda^4}{D_y}$$

where $J = EI_x/D_y$

$K = (EI_y/\lambda D_y)n^3$

$[\ ] =$ Mesh point parameters

Expansion of this equation gives the following general equation:

$$\begin{aligned} &w_0[6n^4 + 8n^2\beta + 6(\beta/\alpha)^2 + 6K + 6J] - 4(w_r + w_l)[(\beta/\alpha)^2 + n^2\beta + 4J] \\ &- 4(w_a + w_b)(n^2\beta + n^4 + 4K) + (w_{rr} + w_{ll})[(\beta/\alpha)^2 + J] \\ &+ (w_{aa} + w_{bb})(n^4 + K) + (w_{ar} + w_{br} + w_{al} + w_{bl})2n^2\beta = \frac{qn^4\lambda^4}{D_y} \end{aligned} \tag{5.12}$$

This general orthotropic plate equation can also be described in mesh form, as shown in Figure 5-4. Note that if $K = J = 0$, then this general equation (5.12) reduces to equation (5.3), representing an orthotropic plate without girders. If the plate is isotropic i.e., $\alpha = \beta = 1$ and has no girders ($K = J = 0$) and a square mesh ($n = 1$), then equation (5.12) becomes

$$\begin{aligned} &w_0[20] + [w_r + w_l][-8] + [w_a + w_b][-8] + [w_{rr} + w_{ll}][1] \\ &+ [w_{aa} + w_{bb}][1] + [w_{ar} + w_{br} + w_{al} + w_{bl}][2] = \frac{q\lambda^4}{D} \end{aligned} \tag{5.13}$$

This equation is shown in mesh form in Figure 5-5.

With the general plate equation described in difference form, the solution to a particular problem can readily be determined by applying the appropriate boundary conditions and accordingly varying the general mesh pattern. The general plate equation is then written for each mesh point on the plate. The solution of the resulting system of equations gives the final deformations of the plate. The forces are then determined by using these deformations in conjunction with the difference force equations.

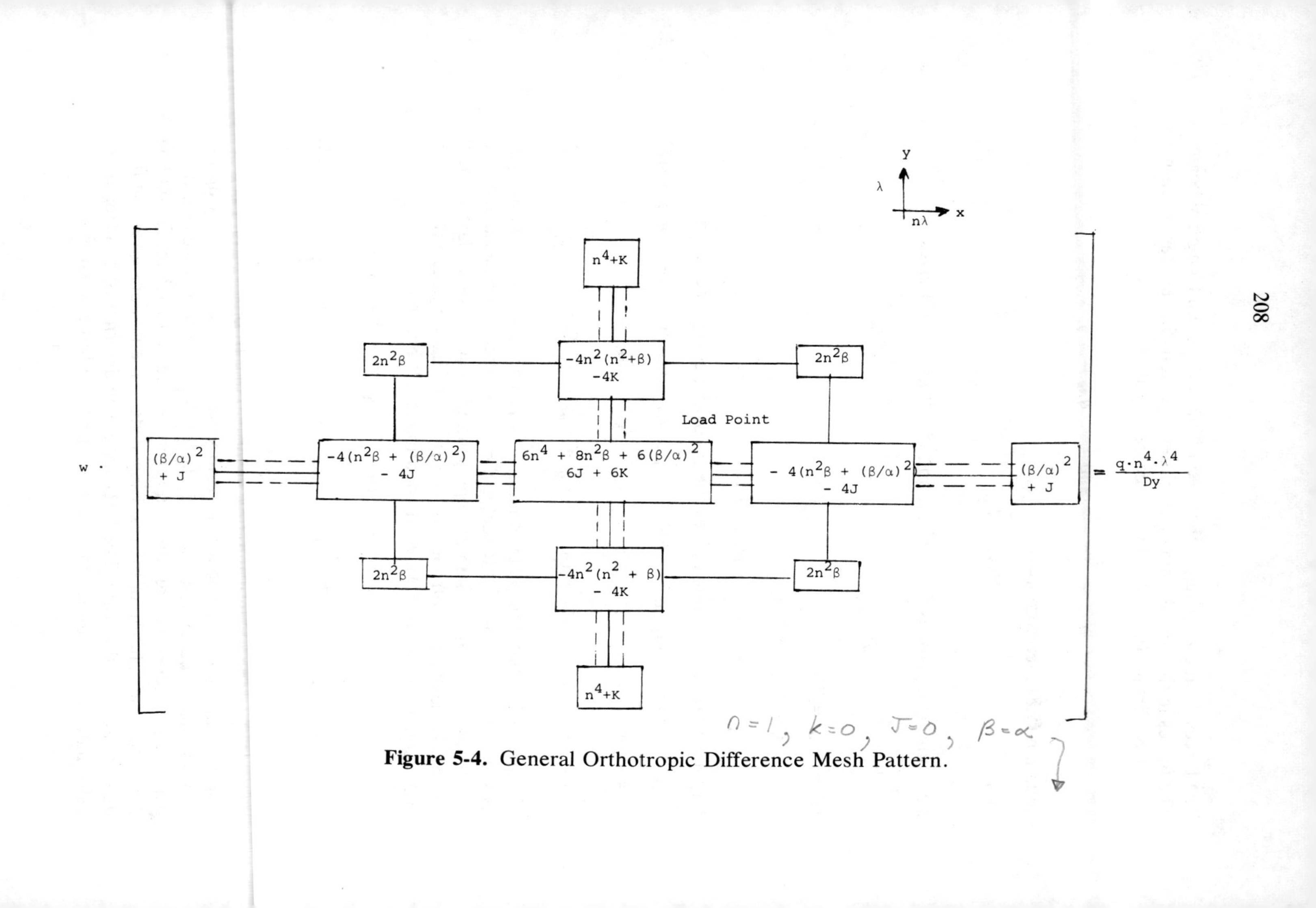

Figure 5-4. General Orthotropic Difference Mesh Pattern.

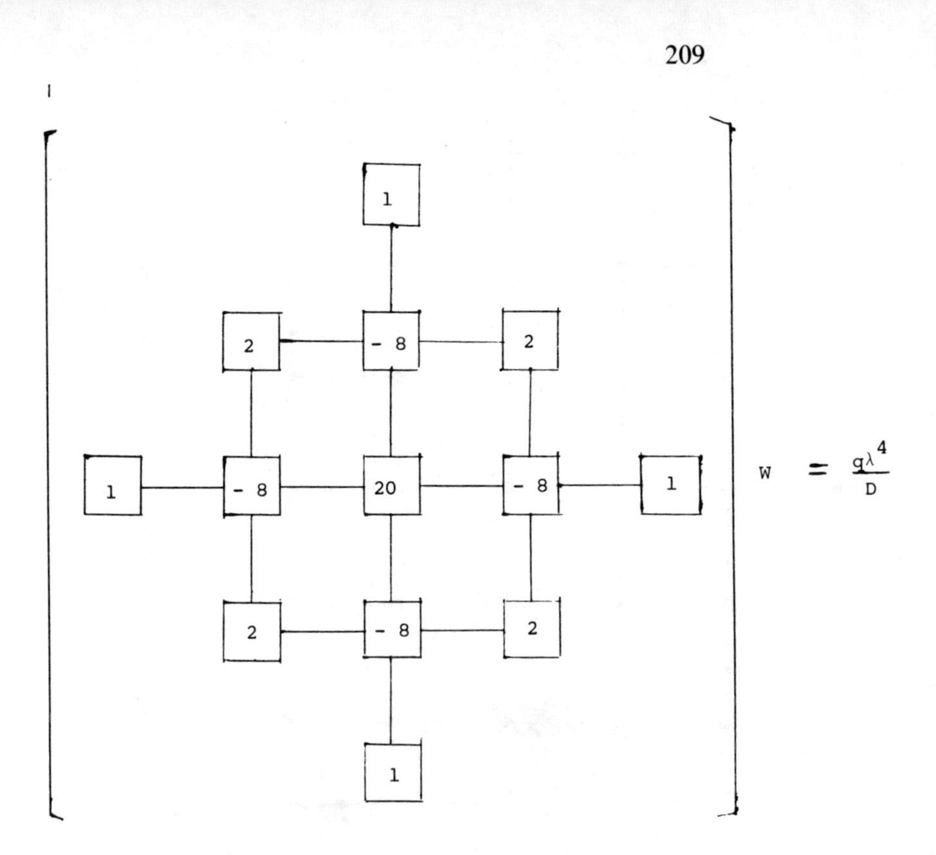

Figure 5-5. General Isotropic Difference Mesh Pattern.

The following example will illustrate the application of this technique. Chapter 6 will detail the application relative to bridges and the computer program used to generate and solve the required difference equations.

5.3 Example—Difference Plate Solution

A plate which has two built-in edges and two simple supports is subdivided into equally spaced nodes, as shown in Figure 5-6. It is desired to determine the deflections and moments at all points when the plate is subjected to (a) a uniform load q (lb/ft^2) and (b) a concentrated load P applied at mesh point 13.

Part (a) Uniform Load on Entire Plate

Due to symmetry of load + boundary

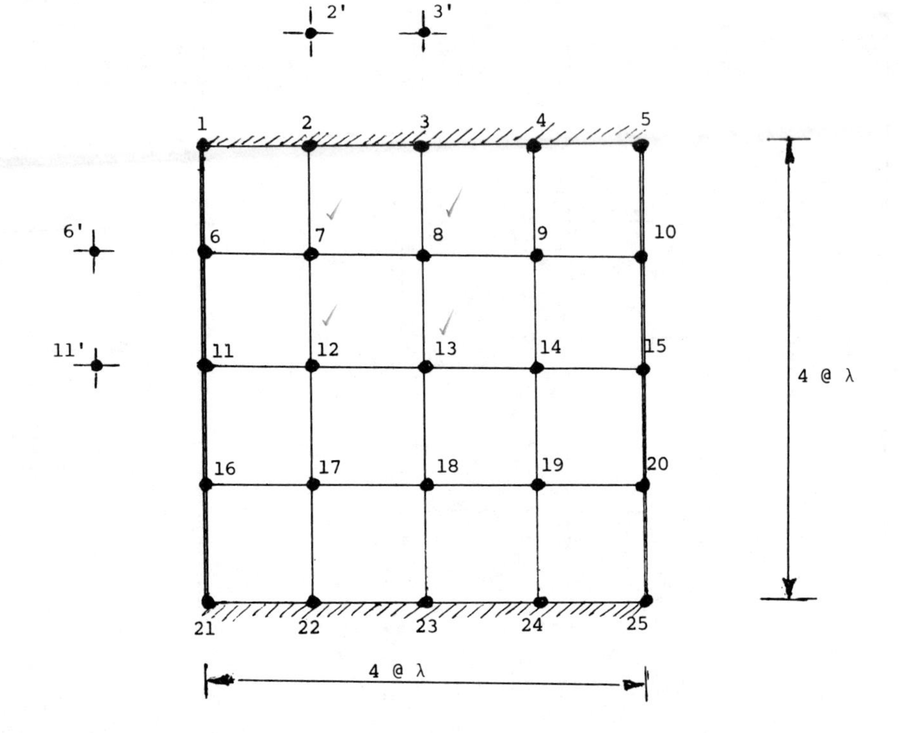

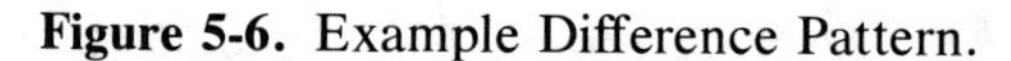

Figure 5-6. Example Difference Pattern.

$$w_7 = w_9 = w_{17} = w_{19}$$

$$w_8 = w_{18}$$

$$w_{12} = w_{14}$$

Therefore, there are a total of four unknowns—w_7, w_8, w_{12}, and w_{13}. Now, applying the mesh plate equation (5.13) as given in Figure 5-5 at the four nodes 7, 8, 12, and 13 gives

Load at Point 7

$$w_6' - 8w_6 + 20w_7 - 8w_8 + w_9 + 2w_{11} - 8w_{12} + 2w_{13}$$

$$+ 2w_1 - 8w_2 + 2w_3 + w_2' + w_{17} = q/D \qquad \text{(a)}$$

Load at Point 8

$$w_6 - 8w_7 + 20w_8 - 8w_9 + w_{10} + 2w_{12} - 8w_{13} + 2w_{14}$$

$$+ 2w_2 - 8w_3 + 2w_4 + w_3' + w_{18} = q/D \qquad \text{(b)}$$

Load at Point 12

$$w'_{11} - 8w_{11} + 20w_{12} - 8w_{13} + w_{14} + 2w_{16} - 8w_{17}$$
$$+ 2w_{18} + 2w_6 - 8w_7 + 2w_8 + w_2 + w_{22} = q/D \qquad \text{(c)}$$

Load at Point 13

$$w_{11} - 8w_{12} + 20w_{13} - 8w_{14} + w_{15} + 2w_{17} - 8w_{18}$$
$$+ 2w_{19} + 2w_7 - 8w_8 + 2w_9 + w_3 + w_{23} = q/D \qquad \text{(d)}$$

The four equations can now be modified by applying the boundary conditions, which are:

$$w_1 = w_2 = w_3 = w_4 = w_5 = 0$$
$$w_{21} = w_{22} = w_{23} = w_{24} = w_{25} = 0$$
$$w_6 = w_{11} = w_{16} = w_{10} = w_{15} = w_{20} = 0$$

Along the simple supports $M = 0$, therefore,

$$M_6 = 0 = w'_6 - 2w_6 + w_7$$

However, $w_6 = 0$; therefore, $w'_6 = -w_7$.

$$M_{11} = 0 = w'_{11} - 2w_{11} + w_{12}$$

However $w_{11} = 0$; therefore $w'_{11} = -w_{12}$.

$$\text{Slope at 2} = 0 = (w'_2 - w_7)/2\lambda \qquad \text{or} \qquad w'_2 = w_7$$
$$\text{Slope at 3} = 0 = (w'_3 - w_8)2\lambda \qquad \text{or} \qquad w'_3 = w_8$$

Substituting all of these relationships into equations (a), (b), (c), and (d) and noting symmetry gives

Load 7

$$22w_7 - 8w_8 - 8w_{12} + 2w_{13} = \frac{q\lambda^4}{D}$$

Load 8

$$-16w_7 + 22w_8 + 4w_{12} - 8w_{13} = \frac{q\lambda^4}{D}$$

Load 12

$$-16w_7 + 4w_8 + 20w_{12} - 8w_{13} = \frac{q\lambda^4}{D}$$

Load 13

$$8w_7 - 16w_8 - 16w_{12} + 20w_{13} = \frac{q\lambda^4}{D}$$

The solution is

$$w_7 = 0.308\frac{q\lambda^4}{D}$$

$$w_8 = 0.414\frac{q\lambda^4}{D}$$

$$w_{12} = 0.466\frac{q\lambda^4}{D}$$

$$w_{13} = 0.631\frac{q\lambda^4}{D}$$

using these deflections and applying the moment equations (5.4) and (5.5) gives

Moments

Point	$M_x/q\lambda^2$	$M_y/q\lambda^2$
2	0	−0.6167
3	0	−0.829
7	+0.2022	+0.1504
8	+0.2122	+0.1977
12	+0.3013	+0.3159
13	+0.3299	+0.4336

Part (b): Concentrated Load P at Point 13

The previous equations are the same except for the loading; noting that P is converted to an equivalent uniform load $q = P/\lambda^2$:

$$22w_7 - 8w_8 - 8w_{12} + 2w_{13} = 0$$

$$-16w_7 + 22w_8 + 4w_{12} - 8w_{13} = 0$$

$$-16w_7 + 4w_8 + 20w_{12} - 8w_{13} = 0$$

$$8w_7 - 16w_8 - 16w_{12} + 20w_{13} = P\lambda^2/D$$

Solution of above gives

$$w_7 = 0.0417\frac{P\lambda^2}{D}$$

$$w_8 = 0.0724\frac{P\lambda^2}{D}$$

$$w_{12} = 0.0815\frac{P\lambda^2}{D}$$

$$w_{13} = 0.1564\frac{P\lambda^2}{D}$$

Solving for moments:

Moments

Point	M_x/P	M_y/P
2	0	−0.0835
3	0	−0.1449
7	+0.01107	+0.00201
8	+0.6137	−0.01157
12	+0.00654	+0.07948
13	+0.1499	+0.16801

5.4 Errors in Finite Differences

Because a parabolid was used to approximate the actual deflected shape, some error will be introduced, depending on the mesh spacing. The order of the error, as influenced by the mesh spacing can be found by using Taylor's theorem [1]. Applying this theorem, the exact difference equations can be determined, including error terms.

Noting that a given function $f(m)$ can be written as

$$f(m) = f(n) + (m - n)\frac{d(n)}{dx} + \frac{(m - n)^2}{2!}\frac{d^2(n)}{dx^2} + \frac{(m - n)^3}{3!}\frac{d^3(n)}{dx^3} + \dots \tag{5.14}$$

which was previously given in chapter 2, in which $f(m)$ represents the function of x at $x = m$ and $f(n)$ equals the function of x at $x = n$, or

$$y_m = y_n + (m - n)\frac{dy}{dx_n} + \frac{(m - n)^2}{2!}\frac{d^2y}{dx_n^2} + \frac{(m - n)^3}{3!}\frac{d^3y}{dx_n^3} + \frac{(m - n)^4}{4!}\frac{d^4y}{dx_n^4} + \dots \tag{5.15}$$

Let m assume successive values of $n + 2s$, $n + s$, n, $n - s$, $n - 2s$, where these represent the node points rr, r, 0, l, ll of the mesh previously described. Then writing the function y at these points gives

$$y_{n+2s} = y_n + 2s\frac{dy}{dx_n} + \frac{(2s)^2}{2!}\frac{d^2y}{dx_n^2} + \frac{(2s)^3}{3!}\frac{d^3y}{dx_n^3} + \frac{(2s)^4}{4!}\frac{d^4y}{dx_n^4} + \ldots \tag{5.16}$$

$$y_{n+s} = y_n + s\frac{dy}{dx_n} + \frac{s^2}{2!}\frac{d^2y}{dx_n^2} + \frac{s^3}{3!}\frac{d^3y}{dx_n^3} + \frac{s^4}{4!}\frac{d^4y}{dx_n^4} + \ldots \tag{5.17}$$

$$y_n = y_n \tag{5.18}$$

$$y_{n-s} = y_n - s\frac{dy}{dx_n} + \frac{s^2}{2!}\frac{d^2y}{dx_n^2} - \frac{s^3}{3!}\frac{d^3y}{dx_n^3} + \frac{s^4}{4!}\frac{d^4y}{dx_n^4} + \ldots \tag{5.19}$$

$$y_{n-2s} = y_n - 2s\frac{dy}{dx_n} + \frac{(2s)^2}{2!}\frac{d^2y}{dx_n^2} - \frac{(2s)^3}{3!}\frac{d^3y}{dx_n^3} + \frac{(2s)^4}{4!}\frac{d^4y}{dx_n^4} + \ldots \tag{5.20}$$

Adding algebraically [30] the appropriate equations (5.16) through (5.20) gives

$$\frac{y_{n+s} - y_{n-s}}{2s} = \frac{dy}{dx} + \frac{s^2}{3!}\frac{d^3y}{dx^3} + \frac{s^4}{5!}\frac{d^5y}{dx^5} + \frac{s^6}{7!}\frac{d^2y}{dx^7} + \ldots \tag{5.21}$$

$$\frac{y_{n+s} - 2y_n + y_{n-s}}{s^2} = \frac{d^2y}{dx^2} + \frac{2s^2d^4y}{4!dx^4} + \frac{2s^4}{6!}\frac{d^6y}{dx^6} + \frac{2s^6}{8!}\frac{d^8y}{dx^8} + \ldots \tag{5.22}$$

$$\frac{y_{n+2s} - 2y_{n+s} + 2y_{n-s} - y_{n-2s}}{2s^3} = \frac{d^3y}{dx^3} + s^2\frac{(2^5 - 2)}{5!}\frac{d^5y}{dx^5} + s^4\frac{(2^7 - 2)}{7!}\frac{d^7y}{dx^7} + \ldots \tag{5.23}$$

$$\frac{y_{n+2s} - 4y_{n+s} + 6y_n - 4y_{n-s} + y_{n-2s}}{s^4} = \frac{d^4y}{dx^4} + s^2\frac{(2^7 - 8)}{6!}\frac{d^6y}{dx^6} + s^4\frac{(2^9 - 8)}{8!}\frac{d^8y}{dx^8} + \ldots \tag{5.24}$$

Let the finite differences be δ; then equations (5.21) through (5.24) become

$$\frac{dy}{dx} = \delta^1 - \frac{s^2\delta^3}{6} + \frac{s^4\delta^5}{120} - \ldots \tag{5.25}$$

$$\frac{d^2y}{dx^2} = \delta^2 - \frac{s^2\delta^4}{12} + \frac{s^4\delta^6}{360} - \ldots \tag{5.26}$$

$$\frac{d^3y}{dx^3} = \delta^3 - \frac{s^2\delta^5}{4} + \frac{7s^4\delta^7}{280} - \ldots \tag{5.27}$$

$$\frac{d^4y}{dx^4} = \delta^4 - \frac{s^2\delta^6}{6} + \frac{7s^4\delta^8}{80} - \ldots \tag{5.28}$$

From equations (5.25) through (5.28) it is seen that the error can be reduced by decreasing mesh spacing s or by taking more terms. Note that the error is approximately equal to the power of s^2.

5.5 Curved Plates

As shown in section 5.1, the general orthotropic differential plate equation (4.38) representing a rectangular plate was written in difference form. The procedure used to develop the rectangular coordinate difference relationships can also be employed in developing polar difference equations [31]. These relationships will be incorporated into the curved orthotropic plate equation (4.85).

$$D_r\left[\frac{\partial^4 w}{\partial r^4} + \frac{2}{r}\frac{\partial^3 w}{\partial r^3}\right] + 2H\left[\frac{1}{r^2}\frac{\partial^4 w}{\partial r^2\,\partial\theta^2} - \frac{1}{r^3}\frac{\partial^3 w}{\partial r\,\partial\theta^2} + \frac{1}{r^4}\frac{\partial^2 w}{\partial\theta^2}\right]$$

$$+ D_\theta\left[\frac{1}{r^4}\frac{\partial^4 w}{\partial\theta^4} - \frac{1}{r^2}\frac{\partial^2 w}{\partial r^2} + \frac{2}{r^4}\frac{\partial^2 w}{\partial\theta^2} + \frac{1}{r^3}\frac{\partial w}{\partial r}\right] = q(r, \theta)$$

Examine now a segment of a deflected plate that has given deflections of magnitude w_0, w_r, w_l, w_b, as shown in Figure 5-7. Through these points a surface will be prescribed by the equation $\bar{w} = Ar^2 + Br + C + Dr\theta + Er\theta^2$ such that at each location (r, θ) the surface will coincide with these given deflections. Taking then the partial derivatives of the equation of this surface with respect to r and θ and expanding the number of mesh points, these derivatives may be substituted into equation (4.85), thus describing the deflection of a loaded sectoral plate as functions of the deflections at discrete points.

Now examine the polar coordinate mesh pattern, Figure 5-7, which has points on the mesh which are arbitrarily assigned letters. The actual deflection surface $w(r, \theta)$ of the plate is assumed to have the following form:

$$\bar{w} = Ar^2 + Br + C + Dr\theta + Er\theta^2$$

such that $\bar{w}$ is identical to actual deflections w at points $(a, 0, b, l, r)$. Applying the conditions:

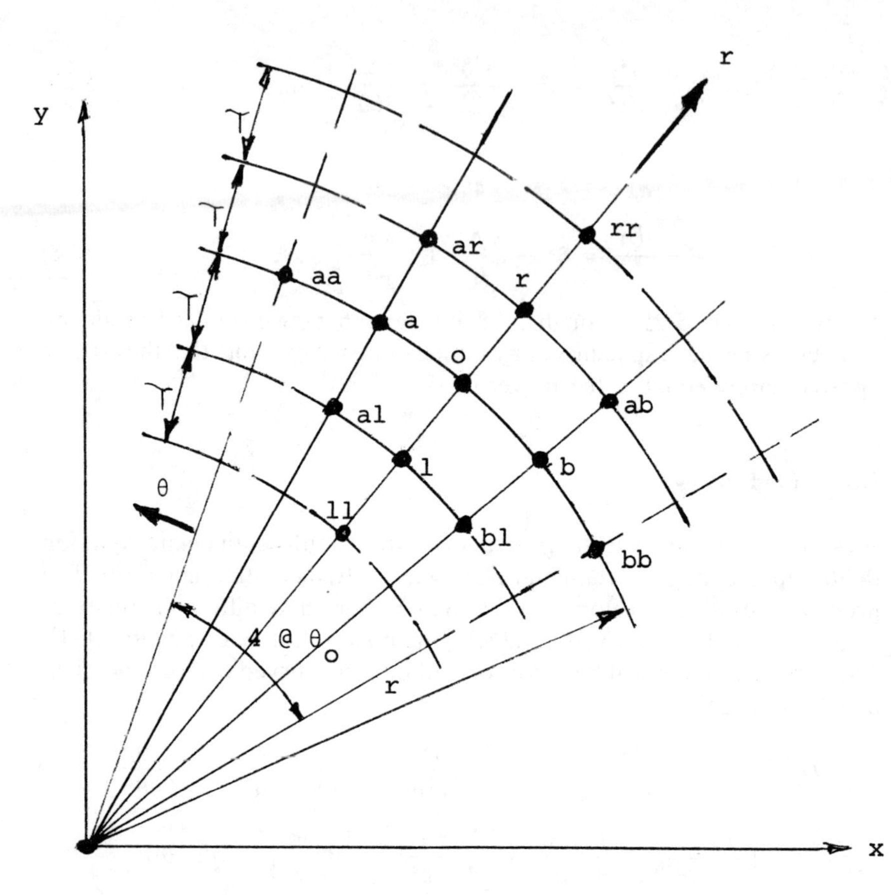

Figure 5-7. Angular Difference Nodes.

at $r = r$, $\theta = 0$: $\bar{w} = w_0$

$$w_0 = Ar^2 + Br + C \tag{a}$$

at $r = r - \lambda$, $\theta = 0$: $\bar{w} = w_l$

$$w_1 = A(r - \lambda)^2 + B(r - \lambda) + C \tag{b}$$

at $r = r + \lambda$, $\theta = 0$: $\bar{w} = w_r$

$$w_r = A(r + \lambda)^2 + B(r + \lambda) + C \tag{c}$$

at $r = r$, $\theta = \theta_0$: $\bar{w} = w_a$

$$w_a = Ar^2 + Br + C + Dr\theta_0 + Er\theta_0^2 \tag{d}$$

at $r = r$, $\theta = -\theta_0$: $\bar{w} = w_b$

$$w_b = Ar^2 + Br + C - Dr\theta_0 + Er\theta_0^2 \tag{e}$$

Solving for constants A, B, C, D, E from equations (a) through (e) yields

$$A = \frac{w_l - 2w_0 + w_r}{2\lambda^2}$$

$$B = \frac{-w_l(2r + \lambda) + 4rw_0 - w_r(2r - \lambda)}{2\lambda^2}$$

$$C = \frac{w_l r(\lambda + r) + 2(\lambda^2 - r^2)w_0 + w_r r(-\lambda + r)}{2\lambda^2}$$

$$D = \frac{w_a - w_b}{2r\theta_0}$$

$$E = \frac{w_a - 2w_0 + w_b}{2r\theta_0^2}$$

Substituting the constants A, B, C, D, and E into the equation $\bar{w} = Ar^2 + Br + C + Dr\theta + Er\theta^2$ results in the following:

$$\bar{w} = \frac{(w_l - 2w_0 + w_r)}{2\lambda^2} r^2 + \frac{[-w_l(2r + \lambda) + 4rw_0 - w_r(2r - \lambda)]}{2\lambda^2} r + \frac{[w_l r(\lambda + r) + 2(\lambda^2 - r^2)w_0 + w_r r(-\lambda + r)]}{2\lambda^2} + \frac{(w_a - w_b)}{2\theta_0}\theta + \frac{(w_a - 2w_0 + w_b)}{2\theta_0^2}\theta^2 \qquad \text{(f)}$$

Equation (f) now represents the surface of the plate through points $0, r, l, a$, and b. In order to use this equation in solving the differential equation (4.85), partial derivatives must be evaluated at the various mesh points. These partials are as follows:

$$\left.\frac{\partial w}{\partial r}\right|_{r=r,\theta=0} = \frac{w_r - w_l}{2\lambda} \qquad \text{(g)}$$

$$\left.\frac{\partial w}{\partial \theta}\right|_{r=r,\theta=0} = \frac{w_a - w_b}{2\theta_0} \qquad \text{(h)}$$

$$\left.\frac{\partial^2 w}{\partial r^2}\right|_{r=r,\theta=0} = \frac{w_l - 2w_0 + w_r}{\lambda^2} \qquad \text{(i)}$$

$$\left.\frac{\partial^2 w}{\partial \theta^2}\right|_{r=r,\theta=0} = \frac{w_a - 2w_0 + w_b}{\theta_0^2} \qquad \text{(j)}$$

$$\left.\frac{\partial^3 w}{\partial r^3}\right|_{r=r,\theta=0} = \frac{1}{2\lambda^3}(w_{rr} - 2w_r + 2w_l - w_{ll}) \tag{k}$$

$$\left.\frac{\partial^3 w}{\partial \theta^3}\right|_{r=r,\theta=0} = \frac{1}{2\theta_0^3}(w_{aa} - 2w_a + 2w_b - w_{bb}) \tag{l}$$

$$\left.\frac{\partial^4 w}{\partial r^4}\right|_{r=r,\theta=0} = \frac{1}{\lambda^4}(w_{rr} - 4w_r + 6w_0 - 4w_l + w_{ll}) \tag{m}$$

$$\left.\frac{\partial^4 w}{\partial \theta^4}\right|_{r=r,\theta=0} = \frac{1}{\theta_0^4}(w_{aa} - 4w_a + 6w_0 - 4w_b + w_{bb}) \tag{n}$$

The mixed partial derivatives take on the following values:

$$\left.\frac{\partial^3 w}{\partial r\,\partial \theta^2}\right|_{r=r,\theta=0} = \frac{1}{2\lambda\theta_0^2}(w_{ar} - w_{al} - 2w_r + 2w_l + w_{br} - w_{bl}) \tag{o}$$

$$\left.\frac{\partial^4 w}{\partial r^2\,\partial \theta^2}\right|_{r=r,\theta=0} = \frac{1}{\lambda^2\theta_0^2}(w_{ar} - 2w_r + w_{br} - 2w_a + 4w_0 - 2w_b + w_{al} - 2w_l + w_{bl}) \tag{p}$$

$$\left.\frac{\partial^2 w}{\partial r\,\partial \theta}\right|_{r=r,\theta=0} = \frac{1}{4\theta_0\lambda}(w_{ar} - w_{al} - w_{br} + w_{bl}) \tag{q}$$

$$\left.\frac{\partial^3 w}{\partial r^2\,\partial \theta}\right|_{r=r,\theta=0} = \frac{1}{2\theta_0\lambda^2}\cdot(w_{ar} - w_{br} - 2w_a + 2w_b + w_{al} - w_{bl}) \tag{r}$$

The above partial derivatives may now be substituted into the differential plate equation (4.85), giving the solution in finite difference form. Letting $\alpha = D_\theta/D_r$; $\beta = H/D_r$, the following equation (5.29) is obtained:

$$\begin{aligned}
& r^4\theta_0^4[w_{rr} - 4w_r + 6w_0 - 4w_l + w_{ll}] \\
& + 2\beta r^2\lambda^2\theta_0^2[w_{ar} - 2w_r + w_{br} - 2w_a + 4w_0 - 2w_b + w_{al} \\
& - 2w_l + w_{bl}] + \alpha\lambda^4[w_{aa} - 4w_a + 6w_0 - 4w_b + w_{bb}] \\
& + \lambda r^3\theta_0^4[w_{rr} - 2w_r + 2w_l - w_{ll}] \\
& - \beta\lambda^3\theta_0^2 r[w_{ar} - w_{al} - 2w_r + 2w_l + w_{br} - w_{bl}] \\
& - \alpha r^2\lambda^2\theta_0^4[w_l - 2w_0 + w_r] + 2(\alpha + \beta)\theta_0^2\lambda^4\,[w_a - 2w_0 + w_b] \\
& + \alpha/2r\theta_0^4\lambda^3[w_r - w_l] \\
& = \frac{q\lambda^4 r^4\theta_0^4}{D_r}
\end{aligned} \tag{5.29}$$

Equation (5.29) is now rearranged and each specific deflection term collected, yielding the following equation (5.30):

$$
\begin{aligned}
& w_0[6r^4\theta_0^4 + 8\beta r^2\lambda^2\theta_0^2 + 6\alpha\lambda^4 + 2\alpha r^2\lambda^2\,\theta_0^2 \\
& - 4(\alpha + \beta)\theta_0^2\lambda^4] + w_a[-4\beta r^2\lambda^2\theta_0^2 - 4\alpha\lambda^4 \\
& + 2(\alpha + \beta)\theta_0^2\lambda^4] + w_b[-4\beta r^2\lambda^2\theta_0^2 - 4\alpha\lambda^4 \\
& + 2(\alpha + \beta)\theta_0^2\lambda^4] + w_r[-4r^2\theta_0^4 - 4\beta r^2\lambda^2\theta_0^2 \\
& - 2\lambda r^3\theta_0^4 + 2\beta\lambda^3\theta_0^2 r - \alpha r^2\lambda^2\theta_0^4 + \alpha/2r\theta_0^4\lambda^3] \\
& + w_l[-4r^4\theta_0^4 - 4\beta r^2\lambda^2\theta_0^2 + 2\lambda r^3\theta_0^4 - 2\beta\lambda^3\theta_0^2 r \\
& - \alpha r^2\lambda^2\theta_0^4 - \alpha/2r\theta_0^4\lambda^3] \\
& + w_{rr}[r^4\theta_0^4 + \lambda r^3\theta_0^4] + w_{ll}[r^4\theta_0^4 - \lambda r^3\theta_0^4] \\
& + w_{al}[2\beta r^2\lambda^2\theta_0^2 + \beta\lambda^3\theta_0^2 r] + w_{ar}[2\beta r^2 \\
& \lambda^2\theta_0^2 - \beta\lambda^3\theta_0^2 r] + w_{bl}[2\beta r^2\lambda^2\theta_0^2 + \beta\lambda^3\theta_0^2 r] \\
& + w_{br}[2\beta r^2\lambda^2\theta_0^2 - \beta\lambda^3\theta_0^2 r] + w_{aa}[\alpha\lambda^4] + w_{bb}[\alpha\lambda^4] \\
& = \frac{q\lambda^4 r^4\theta_0^4}{D_r}
\end{aligned}
\tag{5.30}
$$

Equation (5.30) is the final solution of the differential plate equation (4.85) in finite difference form. Equation (5.30) may be described in a more convenient form by assigning the coefficients to the corresponding node points in a graphical format similar to Figure 5-7. Figure 5-8 describes equation (5.30) in this manner.

The relationship between the specified deflections, moments, and reactions at point 0, with the Poisson's ratio effect equal to zero, as per equations (4.86), (4.87), (4.93), and (4.94) are

$$
M_r\Big|_0 = -D_r\left[\frac{\partial^2 w}{\partial r^2}\right] = \frac{-D_r}{\lambda^2}(w_l - 2w_0 + w_r) \tag{5.31}
$$

$$
\begin{aligned}
M_\theta\Big|_0 &= -D_\theta\left(\frac{1}{r}\,\frac{\partial w}{\partial r} + \frac{1}{r^2}\,\frac{\partial^2 w}{\partial \theta^2}\right) \\
&= -D_\theta\left[\frac{(w_r - w_l)}{2r\lambda} + \frac{(w_a - 2w_0 + w_b}{r^2\theta_0^2}\right]
\end{aligned}
\tag{5.32}
$$

$$
\begin{aligned}
R_r\Big|_0 = \; & -D_r\left[\left(\frac{\partial^3 w}{\partial r^3} + \frac{1}{r}\,\frac{\partial^2 w}{\partial r^2}\right) + 2\beta\left(\frac{1}{r^2}\,\frac{\partial^3 w}{\partial r\,\partial\theta^2} - \frac{1}{r^3}\,\frac{\partial^2 w}{\partial\theta^2}\right)\right. \\
& \left. - \alpha\left(\frac{1}{r^2}\,\frac{\partial w}{\partial r} + \frac{1}{r^3}\,\frac{\partial^2 w}{\partial\theta^2}\right)\right]
\end{aligned}
$$

or

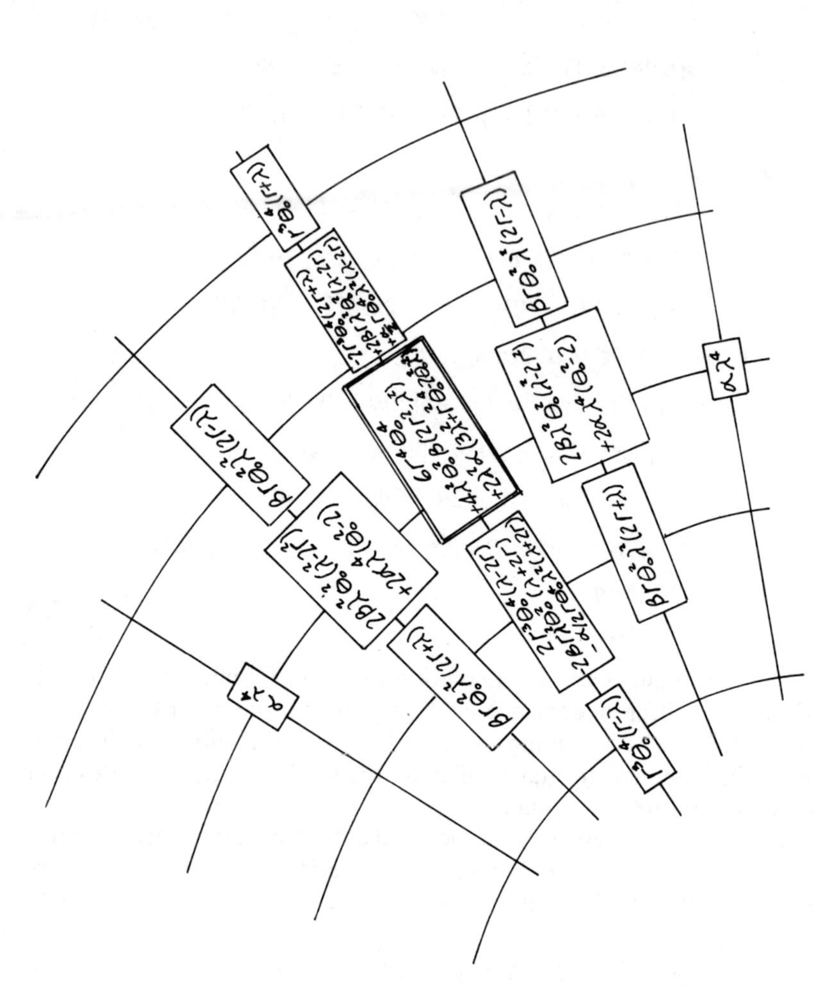

Figure 5-8. General Orthotropic Angular Difference Mesh Pattern.

$$
\begin{aligned}
R_r \Big|_0 = & -D_r\Big\{(w_{rr} - 2w_r + 2w_l - w_{ll})\frac{1}{2\lambda^3} + \frac{1}{\lambda^2 r}(w_l - 2w_0 \\
& + w_r) + 2\beta\Big[(w_{ar} - w_{al} - 2w_r + 2w_l + w_{br} - w_{bl})\frac{1}{2r^2\lambda\theta_0^2} \\
& - (w_a - 2w_0 + w_b)\frac{1}{r^3\theta_0^2}\Big] \\
& - \alpha\Big[\frac{1}{2r^2\lambda}(w_r - w_1) + \frac{1}{r^3\theta_0^2}(w_a - 2w_0 + w_b)\Big]\Big\} \qquad (5.33)
\end{aligned}
$$

$$R_\theta\Big|_0 = -\Big[D_\theta\Big(\frac{1}{r^2}\frac{\partial^2 w}{\partial r\,\partial\theta} + \frac{1}{r^3}\frac{\partial^3 w}{\partial\theta^3}\Big)$$

$$+ H\Big(\frac{2}{r}\frac{\partial^3 w}{\partial r^2\,\partial\theta} - \frac{2}{r^2}\frac{\partial^2 w}{\partial r\,\partial\theta} + \frac{2}{r^3}\frac{\partial w}{\partial\theta}\Big)\Big]$$

or

$$R_\theta\Big|_0 = -\Big\{D_\theta\Big[\frac{1}{r^2}\Big(\frac{w_{ar} - w_{al} - w_{br} + w_{bl}}{4\theta_0\lambda}\Big)$$

$$+ \frac{1}{r^3}\Big(\frac{w_{aa} - 2w_a + 2w_b - w_{bb}}{2\theta_0^3}\Big)\Big] + H\Big[\frac{2}{r^3}\Big(\frac{w_a - w_b}{2\theta_0}\Big)$$

$$+ \frac{2}{r}\Big(\frac{w_{ar} - w_{br} - 2w_a + 2w_b + w_{al} - w_{bl}}{2\theta_0\lambda^2}\Big)$$

$$- \frac{2}{r^2}\Big(\frac{w_{ar} - w_{al} - w_{br} + w_{bl}}{4w_0\lambda}\Big)\Big]\Big\} \qquad (5.34)$$

The solution of the general orthotropic curved plate equation (4.85) is now determined in finite difference form by equation (5.30). The moments and reactions in the angular (θ) and the radial (r) directions are also known and are represented by equations (5.31) through (5.34).

In order to determine plate deflections due to some loading, a mesh spacing, as described by Figure 5-7, must be determined. Then the proper boundary conditions must be evaluated by using equations (5.31) through (5.34). The condition at each mesh point may be described by equation (5.30) in terms of the load at the point and the surrounding deflections. Thus, a set of simultaneous equations will evolve, the solution of which will yield the deflections at each prescribed mesh point. Additional forces at the boundary may then be determined by using equations (5.31) through (5.34). Application of these equations, relative to curved orthotropic bridges, will be described in chapter 5.

5.6 Curved Plates with Interacting Girders

As in the case of the rectangular orthotropic plate interacting with flexible members, given in section 5.2, the curved orthotropic plate may also be supported by curved and radial girders [31, 32, 33]. Assuming that the total load to any mesh point is q_T and that this load is equal to the sum of each particular load taken by the plate, radial girder, and angular girder, gives the following equation:

$$q_T = q_{PL} + q_{rG} + q_{AG} \qquad \text{(a)}$$

where $q_{PL} = [\text{Eq. } (5.30)] \times D_r/(\lambda^4 r^4 \theta_0^4)$

$$q_{rG} = [w_{ll} - 4w_l + 6w_0 - 4w_r + w_{rr}] \cdot \frac{EI_r}{\lambda^4} \cdot \frac{1}{r\theta_0} \tag{5.35}$$

$$q_{AG} = [w_{aa} + w_a(\theta_0^2 - 4) + w_0(6 - 2\theta_0^2) + w_b(\theta_0^2 - 4) + w_{bb}] \frac{EI_\theta}{r^4\theta_0^4} \frac{1}{\psi\lambda} \tag{5.36}$$

where $\psi = [\cos \phi_x + \tan (\phi/2) \sin \phi_x](1 + k) - k$

$k = EI_\theta/GK_T$

$\phi_x =$ Angle to load point

$\phi =$ Total central angle

Now substituting the respective finite difference equations, representing the load deformation equations for q_{PL}, equation (5.30); q_{rG}, equation (5.35); and q_{AG}, equation (5.36) into the above equation (a) gives equation (b). The brackets, [ ], indicate the general finite difference mesh terms.

$$q_T = w_{PL}[\ ]\frac{D_r}{\lambda^4 r^4 \theta_0^4} + w_{rG}[\ ]\frac{EI_r}{\lambda^4 r\theta_0} + w_{AG}[\ ]\frac{EI_\theta}{r^4\theta_0^4}\frac{1}{\psi\lambda} \tag{b}$$

Now dividing this equation (b) through by the factor $D_r/(\lambda^4 r^4 \theta_0^4)$ gives the following:

$$\frac{q_T\lambda^4 r^4\theta_0^4}{D_r} = w_s[\ ] + w_{rG}[\ ]\frac{EI_r r^3\theta_0^3}{D_r} + w_{AG}[\ ]\frac{EI_\theta\lambda^3}{D_r\psi} \tag{c}$$

Let

$$J = \frac{EI_r}{D_r} r^3\theta_0^3 \qquad K = \frac{EI_\theta}{D_r}\frac{\lambda^3}{\psi}$$

reducing equation (c) to the general equation (5.37).

$$w_s[\ ] + w_{rG}[\ ]J + w_{AG}[\ ]K = \frac{q_T\lambda^4 r^4\theta_0^4}{D_r} \tag{5.37}$$

This plate equation (5.30), as represented in Figure 5-8, would then contain the various J and K terms, as listed in Figure 5-9.

5.7 Equivalent Orthotropic Grid System Difference Equations

In section 4.9, the equations of equilibrium of a system of interacting

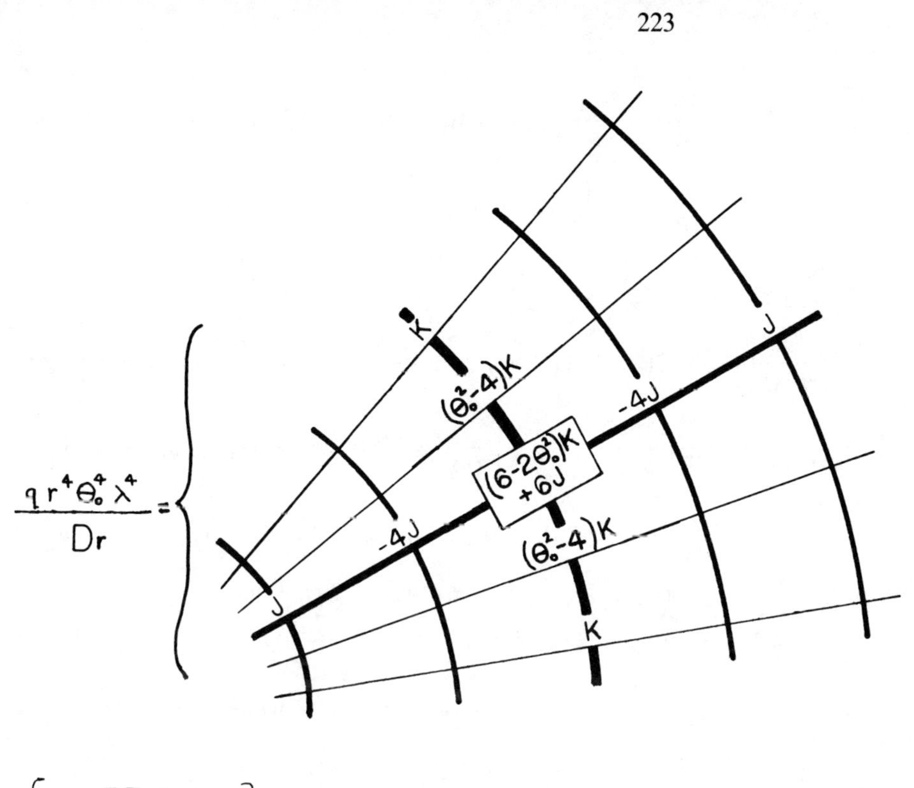

Figure 5-9. Angular Girder Grid Pattern.

girders were developed. The resulting differential equation describing the load deformation of this system was found to be

$$D_x \frac{\partial^4 w}{\partial x^4} - C_x \frac{\partial^6 w}{\partial x^4 \partial y^2} + D_y \frac{\partial^4 w}{\partial y^4} - C_y \frac{\partial^6 w}{\partial x^2 \partial y^4} + 2H \frac{\partial^4 w}{\partial x^2 \partial y^2} = q \quad (4.76)$$

where
$$\begin{aligned} D_x &= EI_x/\lambda \\ D_y &= EI_y/n\lambda \\ C_x &= EI_{wx}/\lambda \\ C_y &= EI_{wy}/n\lambda \\ 2H &= G(K_{Tx}/\lambda + K_{Ty}/n\lambda) \end{aligned} \quad (4.75)$$

Using the procedure described in section 5.1 the difference equations for the partial differentials in equation (4.76) have been determined as [34]:

$$\frac{\partial^4 w}{\partial x^4} = \frac{1}{(n\lambda)^4}(w_{ll} - 4w_l + 6w_0 - 4w_r + w_{rr}) \quad \text{(a)}$$

$$\frac{\partial^4 w}{\partial y^4} = \frac{1}{\lambda^4}(w_{aa} - 4w_a + 6w_0 - 4w_b + w_{bb}) \quad \text{(b)}$$

$$\frac{\partial^4 w}{\partial x^2\,\partial y^2} =$$

$$\frac{1}{n^2\lambda^4}[4w_0 - 2(w_r + w_l + w_a + w_b) + (w_{ar} + w_{br} + w_{al} + w_{bl})] \quad \text{(c)}$$

$$\frac{\partial^6 w}{\partial x^4\,\partial y^2} = \frac{\partial^2}{\partial y^2}\left(\frac{\partial^4 w}{\partial x^4}\right) = \frac{\partial^2}{\partial y^2}\left[\frac{1}{(n\lambda)^4}(w_{ll} - 4w_l + 6w_0 - 4w_r + w_{rr})\right]$$

$$= \frac{1}{n^4\lambda^6}[w_{all} - 2w_{ll} + w_{bll} - 4(w_{al} - 2w_l + w_{bl})$$

$$+ 6(w_a - 2w_0 + w_b) - 4(w_{ar} - 2w_r + w_{br})$$

$$+ (w_{arr} - 2w_{rr} + w_{brr})]$$

$$= \frac{1}{n^4\lambda^6}[(w_{all} + w_{bll} + w_{arr} + w_{brr}) - 2(w_{ll} + w_{rr})$$

$$- 4(w_{al} + w_{bl} + w_{ar} + w_{br}) + 8(w_l + w_r)$$

$$+ 6(w_a + w_b) - 12w_0] \quad \text{(d)}$$

similarly:

$$\frac{\partial^6 w}{\partial x^2\,\partial y^4} = \frac{1}{n^2\lambda^6}[(w_{aar} + w_{aal} + w_{bbr} + w_{bbl}) - 2(w_{aa} + w_{bb})$$

$$- 4(w_{ar} + w_{al} + w_{br} + w_{bl}) + 8(w_a + w_b)$$

$$+ 6(w_r + w_l) - 12w_0] \quad \text{(e)}$$

These difference relationships are referred to the nodes given in Figure 5-10. Substituting equations (a) through (e) into equation (4.76) gives the resulting difference equation as follows and as shown in Figure 5-11.

$$w_0[6n^4 + 6(\beta/\alpha)^2 + 8n^2\beta + 12A + 12n^2B]$$

$$-4(w_r + w_l)[n^2\beta + (\beta/\alpha)^2 + 2A + 6/4n^2B]$$

$$-4(w_a + w_b)[n^2\beta + n^4 + 6/4A + 2n^2B]$$

$$+ (w_{rr} + w_{ll})[(\beta/\alpha)^2 + 2A] + (w_{aa} + w_{bb})[n^4 + 2n^2B]$$

$$+ (w_{ar} + w_{br} + w_{al} + w_{bl})[2n^2\beta + 4A + 4n^2B]$$

$$- A(w_{all} + w_{bll} + w_{arr} + w_{brr}) - n^2B(w_{aar} + w_{aal} - w_{bbr}$$

$$+ w_{bbl}) = qn^4\lambda^4/D_y \quad (5.38)$$

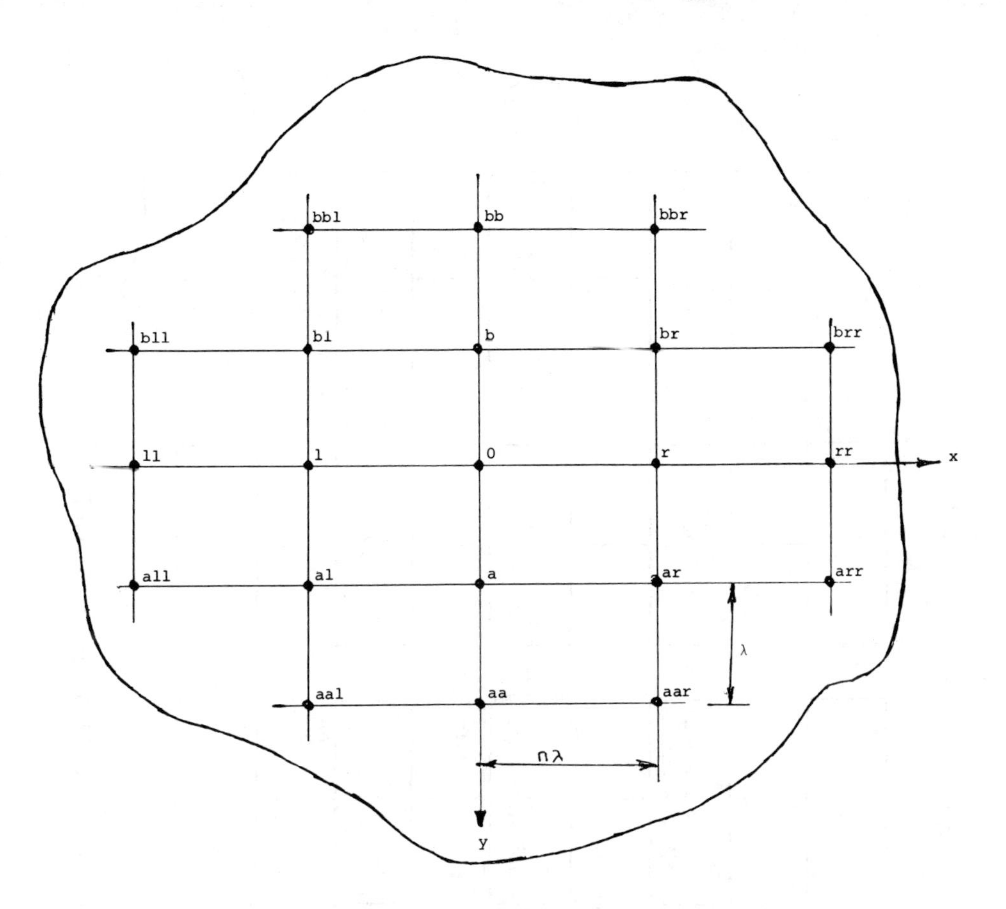

Figure 5-10. General Grid Difference Nodes.

where $\alpha = H/\sqrt{D_x D_y}$

$\beta = H/D_y$

$A = C_x/\lambda^2 D_y$

$B = C_y/\lambda^2 D_y$

5.8 Finite Element Solution

5.8.1 General Equations–Triangular Element

As described in chapter 3, section 3.10, a plate can be divided into a series of interconnecting plate elements to represent the actual structure. The solution of such a system required the evaluation of the plate stiffness matrix [23]:

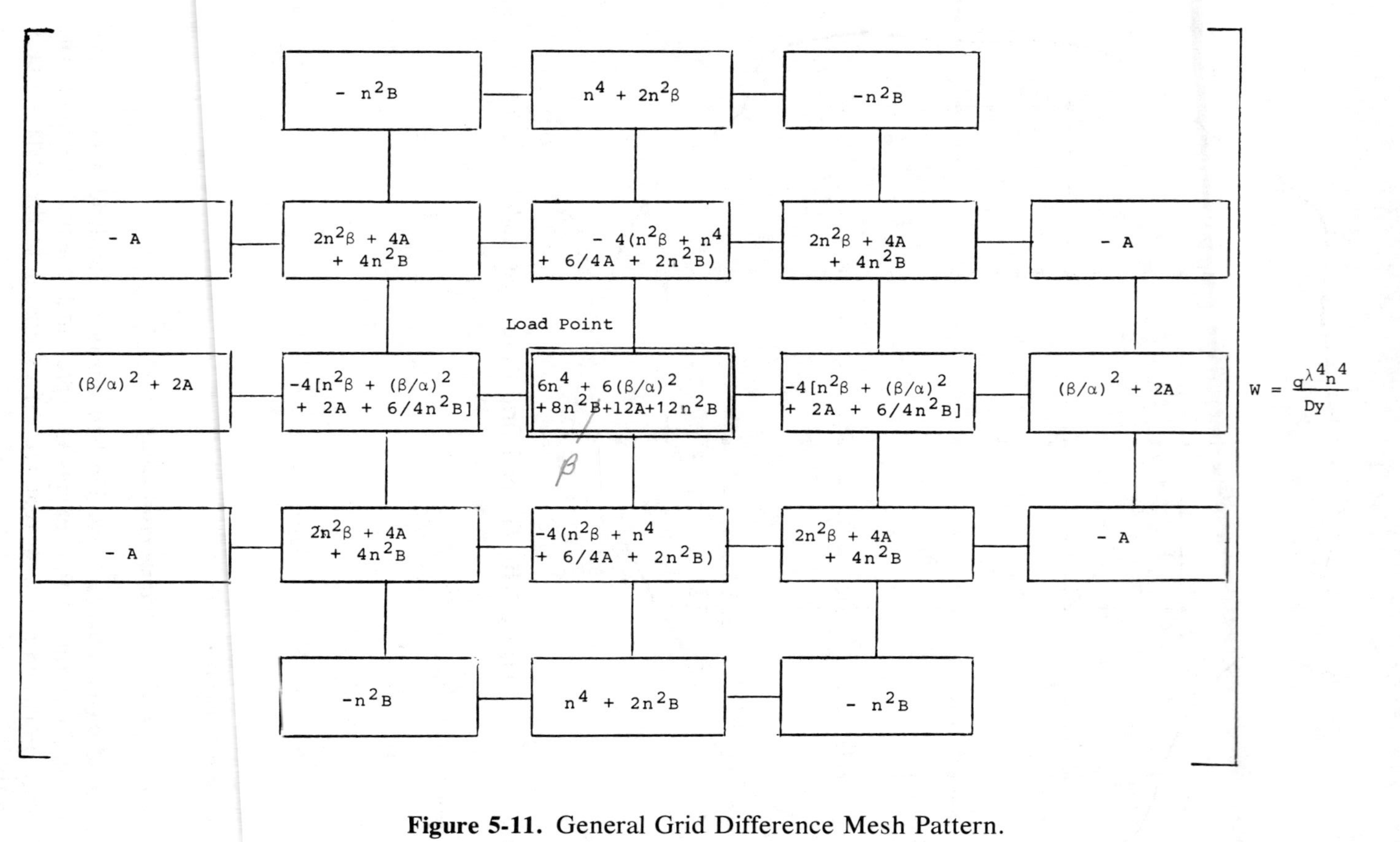

Figure 5-11. General Grid Difference Mesh Pattern.

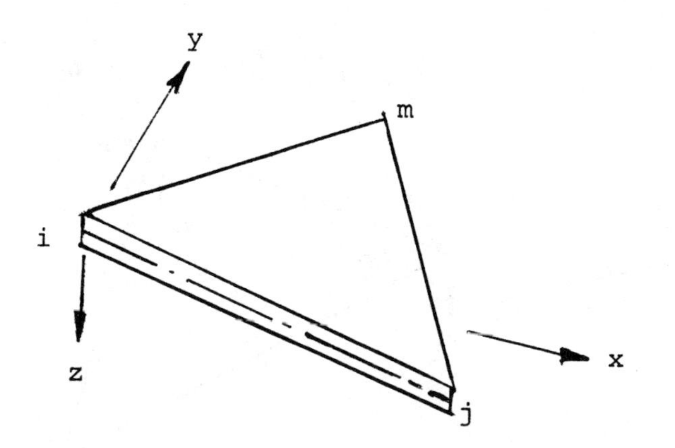

Figure 5-12. Triangular Finite Element Nodes.

$$K = \int_v B^T \chi \, B \, dV \tag{3.119}$$

This equation will now be used to develop the required stiffness of a triangular plate element in bending [23]. The solution of this K matrix, in conjunction with equation (3.119) or

$$P = \int_v B^T B \, dV \Delta$$

gives the resulting displacements at each node of the plate. In considering plate bending, the classical plate equations as given in chapter 4 will apply. These equations were developed based on the assumptions that:

1. Plates are thin and undergo small displacements.
2. There is no deformation of the middle surface.
3. Points on the plate that are normal to the middle surface before bending remain on the normal after bending.
4. Normal stresses in the direction transverse to the plate are zero (i.e., no in-plane or membrane forces).

Consider now a triangular plate element lying in the x-y plane, as shown in Figure 5-12, with nodes i, j, and m numbered positively in a counterclockwise manner. This element has 9 degrees of freedom as associated with nodes i, j, and m, as shown in Figure 5-13. These are the vertical displacements w normal to the plate surface (coincident with z-axis) and rotations θ_x and θ_y with respect to the x- and y-axes, respectively.

Since the displacements of the middle surface throughout the element is defined by w, then the rotations about the x- and y-axes, can be assumed as:

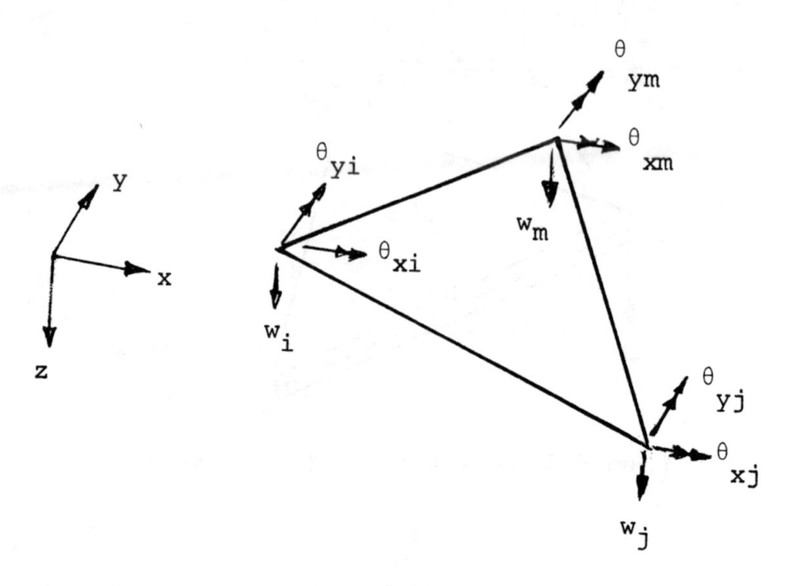

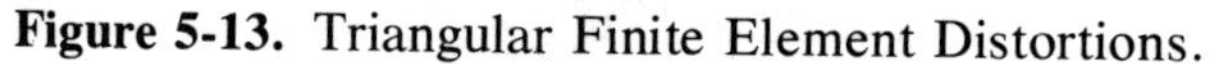

Figure 5-13. Triangular Finite Element Distortions.

$$\theta_x = -\frac{\partial w}{\partial y}\ ; \qquad \theta_y = \frac{\partial w}{\partial x}$$

As in the case of the plane stress finite element solution, section 3.10, a displacement or shape function $[N]$, which will uniquely relate the displacements w to the deformations in the plate, must be obtained. In general, such a relationship is

$$[w] = [N][\delta]^* \tag{5.39}$$

where $[\delta^*] = \begin{vmatrix} \delta_i \\ \delta_j \\ \delta_m \end{vmatrix}$;

and $[\delta_i] = \begin{vmatrix} w \\ \theta_x \\ \theta_y \end{vmatrix}_i$; $[\delta_j] = \begin{vmatrix} w \\ \theta_x \\ \theta_y \end{vmatrix}_j$; $[\delta_m] = \begin{vmatrix} w \\ \theta_x \\ \theta_y \end{vmatrix}_m$ (5.40)

and w, θ_x and θ_y are the translational displacements in the z direction and the rotational displacement about the x and y axes respectively, as shown positively in Figure 5-13.

As noted in Figure 5-13, there are nine degrees of freedom, therefore,

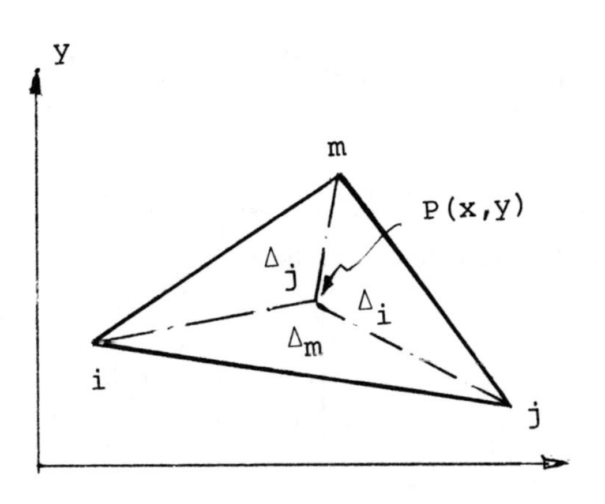

Figure 5-14. Equivalent Triangular Finite Element.

the function to be chosen must contain nine terms in the expansion. There are basically two classifications of displacement functions, compatible and noncompatible [23]. Both types of functions impose continuity at the nodes, but the compatible function also imposes complete continuity between the element interfaces, whereas the noncompatible element imposes only transverse displacement continuity. At first glance, one would likely choose a compatible function, however, Zienkiewicz [23] found that the noncompatible function often yields better results.

The general displacement function, which meets the above conditions, is assumed to be a cubic expansion [23], as given by equation (5.41):

$$w = \alpha_1 + \alpha_2 x + \alpha_3 y + \alpha_4 x^2 + \alpha_5 xy + \alpha_6 y^2 + \alpha_7 x^3 + \alpha_8 (x^2 y + xy^2) + \alpha_9 y^3 \tag{5.41}$$

This displacement function gives good results providing two sides of the element are not parallel to the x- and y-axes (in which the matrix relating the coefficients α_i and w becomes singular). In this latter case, therefore, another method must be employed known as the *area coordinate method* [23].

The area coordinates, for a triangular element, are shown in Figure 5-14. Any point $P(x, y)$ lying in the element divides the element into three subtriangles such that:

$$\text{Area } \Delta_i + \text{Area } \Delta_j + \text{Area } \Delta_m = \text{Area } \Delta$$

denoted as

$$\Delta_i + \Delta_j + \Delta_m = \Delta \tag{5.42}$$

Define the new area coordinates as

$$L_i = \frac{\Delta_i}{\Delta}\,; \qquad L_j = \frac{\Delta_j}{\Delta}\,; \qquad L_m = \frac{\Delta_m}{\Delta} \tag{5.43}$$

or

$$L_i + L_j + L_m = \Delta_i/\Delta + \Delta_j/\Delta + \Delta_m/\Delta = 1 \tag{5.44}$$

The total area of the triangle can be taken as

$$\Delta = \left(\frac{1}{2}\right) \det \begin{vmatrix} 1 & x_i & y_i \\ 1 & x_j & y_j \\ 1 & x_m & y_m \end{vmatrix} \tag{5.45}$$

and

$$\Delta_i = \left(\frac{1}{2}\right) \det \begin{vmatrix} 1 & x & y \\ 1 & x_j & y_j \\ 1 & x_m & y_m \end{vmatrix}\,; \qquad \Delta_j = \left(\frac{1}{2}\right) \det \begin{vmatrix} 1 & x_i & x_j \\ 1 & x & y \\ 1 & x_m & y_m \end{vmatrix}$$

$$\Delta_m = \left(\frac{1}{2}\right) \det \begin{vmatrix} 1 & x_i & y_i \\ 1 & x_j & x_i \\ 1 & x & y \end{vmatrix} \tag{5.46}$$

or, expanding these determinants

$$\begin{aligned} \Delta &= 1/2(a_i + a_j + a_m) \\ \Delta_i &= 1/2(a_i + b_i x + c_i y) \\ \Delta_j &= 1/2(a_j + b_j x + c_j y) \\ \Delta_m &= 1/2(a_m + b_m x + c_m y) \end{aligned} \tag{5.47}$$

where,

$$\begin{aligned} a_i &= x_j y_m - x_m y_j \\ b_i &= y_j - y_m \\ c_i &= x_m - x_j \\ a_j &= x_m y_i - x_i y_m \\ b_j &= y_m - y_i \\ c_j &= x_i - x_m \\ a_m &= x_i y_j - x_j y_i \\ b_m &= y_i - y_j \\ c_m &= x_j - x_i \end{aligned} \tag{5.48}$$

These functions can be combined to yield

$$\begin{bmatrix} L_i \\ L_j \\ L_m \end{bmatrix} = \frac{1}{2\Delta} \begin{bmatrix} a_i & b_i & c_i \\ a_j & b_j & c_j \\ a_m & b_m & c_m \end{bmatrix} \begin{bmatrix} 1 \\ x \\ y \end{bmatrix} \tag{5.49}$$

Solving equation (5.49) for x and y yields

$$\begin{aligned} x &= L_i x_i + L_j x_j + L_m x_m \\ y &= L_i y_i + L_j y_j + L_m y_m \end{aligned} \tag{5.50}$$

At this point, the displacement function w will be divided in two parts or,

$$w = w^* + w^R \tag{5.51}$$

where w^* is the displacement function for a simply supported element, dealing only with rotations, and w^R, which is rigid body displacements, containing only translations.

The rigid body function w^R is defined in terms of a linear function of x and y using area coordinates, i.e.,

$$w^R = w_i L_i + w_j L_j + w_m L_m \tag{5.52}$$

The solution of the simply supported case is

$$[\delta_i^*] = \begin{bmatrix} \theta_x^* \\ \theta_y^* \end{bmatrix}_i \quad ; \quad [\delta_j^*] = \begin{bmatrix} \theta_x^* \\ \theta_y^* \end{bmatrix}_j \quad ; \quad [\delta_m^*] = \begin{bmatrix} \theta_x^* \\ \theta_y^* \end{bmatrix}_m \tag{5.53}$$

such that using equation (5.51) gives

$$\begin{aligned} \theta_x &= -\frac{\partial}{\partial y}(w^* + w^R) = \theta_x^* - \frac{\partial w^R}{\partial y} \\ \theta_y &= \frac{\partial}{\partial x}(w^* + w^R) = \theta_y^* + \frac{\partial w^R}{\partial x} \end{aligned} \tag{5.54}$$

The relative slopes are

$$\begin{aligned} \theta_x^* &= \theta_x + \frac{\partial w^R}{\partial y} \\ \theta_y^* &= \theta_y - \frac{\partial w^R}{\partial x} \end{aligned} \tag{5.55}$$

Using equations (5.49) and (5.52):

$$\begin{aligned} \frac{\partial w^R}{\partial y} &= (c_i w_i + c_j w_j + c_m w_m)/2\Delta \\ \frac{\partial w^R}{\partial x} &= (b_i w_i + b_j w_j + b_m w_m/2\Delta \end{aligned} \tag{5.56}$$

Substituting equations (5.56) into (5.55) gives

$$\begin{aligned}
\theta^*_{xi} &= \theta_{xi} + (c_i w_i + c_j w_j + c_m w_m)/2\Delta \\
\theta^*_{yi} &= \theta_{yi} + (b_i w_i + b_j w_j + b_m w_m)/2\Delta \\
\theta^*_{xj} &= \theta_{xj} + (c_i w_i + c_j w_j + c_m w_m)/2\Delta \\
\theta^*_{yj} &= \theta_{yj} + (b_i w_i + b_j w_j + b_m w_m)/2\Delta \\
\theta^*_{xm} &= \theta_{xm} + (c_i w_i + c_j w_j + c_m w_m)/2\Delta \\
\theta^*_{ym} &= \theta_{ym} + (b_i w_i + b_j w_j + b_m w_m)/2\Delta
\end{aligned} \qquad (5.57)$$

Using equation (5.57), the relative displacement vector can be expressed in terms of the total vector:

$$[\delta^*]^* = [T][\delta]^* \qquad (5.58)$$

where $[T]$ is defined as

$$[T] = \frac{1}{2\Delta}\begin{bmatrix} c_i & 2\Delta & 0 & c_j & 0 & 0 & c_m & 0 & 0 \\ -b_i & 0 & 2\Delta & -b_j & 0 & 0 & -b_m & 0 & 0 \\ c_i & 0 & 0 & c_j & 2\Delta & 0 & c_m & 0 & 0 \\ -b_i & 0 & 0 & -b_j & 0 & 2\Delta & -b_m & 0 & 0 \\ c_i & 0 & 0 & c_j & 0 & 0 & c_m & 2\Delta & 0 \\ -b_i & 0 & 0 & -b_j & 0 & 0 & -b_m & 0 & 2\Delta \end{bmatrix} \qquad (5.59)$$

Here a relative shape function must be developed in terms of the relative displacement vector:

$$w^* = [N][\delta^*]^* \qquad (5.60)$$

where $$w^* = [N_i \; N_j \; N_m]\begin{bmatrix} \delta^*_i \\ \delta^*_j \\ \delta^*_m \end{bmatrix} \qquad (5.61)$$

and

$$\begin{aligned}
[N_i] &= [N_{xi} \; N_{yi}] \\
[N_j] &= [N_{xj} \; N_{yj}] \\
[N_m] &= [N_{xm} \; N_{ym}]
\end{aligned} \qquad (5.62)$$

This expanded gives

$$w^* = N_{xi}\theta^*_{xi} + N_{yi}\theta^*_{yi} + N_{xj}\theta^*_{xj} + N_{yj}\theta^*_{yj} + N_{xm}\theta^*_{xm} + N_{ym}\theta^*_{ym} \qquad (5.63)$$

The individual relative displacement functions, N_{xi}, etc., must impose the boundary conditions, e.g., simply supported along with continuity at the nodes.

Equation (5.57) yields

$$\frac{\partial N_{xi}}{\partial y} = 0 \text{ at nodes } j \text{ and } m, \text{ and } -1 \text{ at node } i$$

$$\frac{\partial N_{xi}}{\partial x} - \frac{\partial N_{yi}}{\partial y} = 0 \text{ at all nodes}$$

$$\frac{\partial N_{yi}}{\partial x} = 0 \text{ at nodes } j \text{ and } m \text{ and } +1 \text{ at node } i \tag{5.64}$$

Similar conditions exist also at nodes j and m.

In addition to the above boundary conditions, the displacement functions must satisfy the condition of constant strain for convergence. Functions that satisfy this condition are

$$\begin{aligned}
N_{xi} &= b_m(L_i^2L_j + \alpha L_iL_jL_m) - b_j(L_i^2L_m + \alpha L_iL_jL_m)\\
N_{yi} &= c_m(L_i^2L_j + \alpha L_iL_jL_m) - c_j(L_i^2L_m + \alpha L_iL_jL_m)\\
N_{xj} &= b_i(L_j^2L_m + \alpha L_iL_jL_m) - b_m(L_j^2L_m + \alpha L_iL_jL_m)\\
N_{yj} &= c_i(L_j^2L_m + \alpha L_iL_jL_m) - c_m(L_j^2L_i + \alpha L_iL_jL_m)\\
N_{xm} &= b_j(L_m^2L_i + \alpha L_iL_jL_m) - b_i(L_m^2L_j + \alpha L_iL_jL_m)\\
N_{ym} &= c_j(L_m^2L_i + \alpha L_iL_jL_m) - c_i(L_m^2L_j + \alpha L_iL_jL_m)
\end{aligned} \tag{5.65}$$

where α is a constant which needs to be evaluated in such a manner that constant strain is obtained. To accomplish this, a general quadratic of w^* is assumed to be

$$w^* = \lambda_iL_jL_m + \lambda_jL_mL_i + \lambda_mL_iL_j \tag{5.66}$$

where λ_i, λ_j, and λ_m are constants.

From equation (5.57)

$$\begin{aligned}
\theta_{xi}^* &= -\frac{\partial w^*}{\partial y} = -\lambda_jc_m - \lambda_mc_j\\
\theta_{yi}^* &= \frac{\partial w^*}{\partial x} = \lambda_jb_m + \lambda_mb_j\\
\theta_{xj}^* &= -\lambda_mc_i - \lambda_ic_m\\
\theta_{yj}^* &= \lambda_mb_i + \lambda_ib_m\\
\theta_{xm}^* &= -\lambda_ic_j - \lambda_jc_i\\
\theta_{ym}^* &= \lambda_ib_j + \lambda_jb_i
\end{aligned} \tag{5.67}$$

such that

$$w^* = \frac{1}{2\Delta}[\lambda_i(-c_m N_{xj} - c_j N_{xm} + b_m N_{yj} + b_j N_{ym}) + \lambda_j(-c_i N_{xm} - c_m N_{xi} + b_i N_{ym} + b_m N_{yi}) + \lambda_m(-c_j N_{xi} - c_i N_{xj} + b_j N_{yi} + b_i N_{yj})] \quad (5.68)$$

Substituting equation (5.65) into (5.68) and equating this to (5.66) will yield $\alpha = 1/2$, which will define the necessary displacement function as follows:

$$\begin{aligned} w^* = {} & [b_m(L_i^2 L_j + \tfrac{1}{2}L_iL_jL_m) - b_j(L_i^2 L_m + \tfrac{1}{2}L_jL_m)]\theta^*_{xi} \\ & + [c_m(L_i^2 L_j + \tfrac{1}{2}L_iL_jL_m) - c_j(L_i^2 L_m + \tfrac{1}{2}L_iL_jL_m)]\theta^*_{yi} \\ & + [b_i[L_j^2 L_m + \tfrac{1}{2}L_iL_jL_m) - b_m(L_j^2 L_i + \tfrac{1}{2}L_iL_jL_m)]\theta^*_{xj} \\ & + [c_i[L_j^2 L_m + \tfrac{1}{2}L_iL_jL_m) - c_m(L_j^2 L_i + \tfrac{1}{2}L_iL_jL_m)]\theta^*_{yj} \\ & + [b_j(L_m^2 L_i + \tfrac{1}{2}L_iL_jL_m) - b_i(L_m^2 L_j + \tfrac{1}{2}L_iL_jL_m)]\theta^*_{xm} \\ & + [c_j(L_m^2 L_i + \tfrac{1}{2}L_iL_jL_m) - c_i(L_m^2 L_j + \tfrac{1}{2}L_iL_jL_m)]\theta^*_{ym} \end{aligned} \quad (5.69)$$

where L_K, b_K, and c_K ($K = i, j, m$) are defined by equations (5.48) and (5.49).

As is the case of classical plate theory, stresses in the plate are determined by internal moments M_x, M_y, and M_{xy}, as given previously in equations (4.16) through (4.18). For an isotropic plate these moments are

$$\begin{aligned} M_x &= \frac{Et^3}{12(1-\mu^2)}\left(\frac{\partial^2 w^*}{\partial x^2} + \mu\frac{\partial^2 w^*}{\partial y^2}\right) \\ M_y &= \frac{Et^3}{12(1-\mu^2)}\left(\frac{\partial^2 w^*}{\partial y^2} + \mu\frac{\partial^2 w^*}{\partial x^2}\right) \\ M_{xy} &= \frac{Et^3}{12(1+\mu)}\frac{\partial^2 w^*}{\partial x\,\partial y} \end{aligned} \quad (5.70)$$

The strain vector is denoted as

$$[\varepsilon] = \begin{bmatrix} -\dfrac{\partial^2 w^*}{\partial x^2} \\ -\dfrac{\partial^2 w^*}{\partial y^2} \\ \dfrac{2\partial^2 w^*}{\partial x\,\partial y} \end{bmatrix} \quad (5.71)$$

in terms of stress

$$[\sigma] = [D][\varepsilon] \tag{5.72}$$

Strain is related to the displacements by

$$[\varepsilon] = [B][\delta]^* \tag{5.73}$$

Performing the operation indicated in equation (5.71) with respect to (5.69) yields

$$[B^*] = [B_i^* \; B_j^* \; B_m^*] \tag{5.74}$$

where:

$$[B_i]^* = \begin{bmatrix} -\dfrac{\partial^2 N_{xi}}{\partial x^2} & -\dfrac{\partial^2 N_{yi}}{\partial x^2} \\ -\dfrac{\partial^2 N_{xi}}{\partial y^2} & -\dfrac{\partial^2 N_{yi}}{\partial y^2} \\ \dfrac{2\partial^2 N_{xi}}{\partial x\,\partial y} & \dfrac{2\partial^2 N_{yi}}{\partial x\,\partial y} \end{bmatrix} \tag{5.75}$$

and similar relationships for $[B_j]^*$ and $[B_m]^*$. Substituting equations (5.74) and (5.58) into (5.72) gives

$$[\sigma] = [D][B^*][\delta^*]^* = [D][B^*][T][\delta]^* \tag{5.76}$$

or

$$[\sigma] = [s][T][\delta]^* \tag{5.77}$$

where $[s]$ is a 3×6 matrix with components

$$\begin{aligned} s_{1n} &= \frac{1}{8\Delta^3}[D_x(A_n + C_n x + B_n y) + D_1(E_n + G_n x + F_n y)] \\ s_{2n} &= \frac{1}{8\Delta^3}[D_1(A_n + C_n x + B_n y) + D_y(E_n + G_n x + F_n y)] \\ s_{3n} &= \frac{2D_{xy}}{8\Delta^3}[H_n + B_n x + G_n y] \qquad n = 1, 2, \ldots, 6 \end{aligned} \tag{5.78}$$

for an isotropic material

$$[D] = \begin{bmatrix} D_x & D_1 & 0 \\ D_1 & D_y & 0 \\ 0 & 0 & D_{xy} \end{bmatrix} \tag{5.79}$$

Equation (5.79) is valid for any position x or y in the element. If the centroid of the element is used as the element coordinate axes, then (5.77) reduces to

$$s_{1n} = \frac{1}{8\Delta^3}[D_x A_n + D_1 E_n]$$

$$s_{2n} = \frac{1}{8\Delta^3}[D_1 A_n + D_y E_n] \tag{5.80}$$

$$s_{3n} = \frac{2D_{xy}}{8\Delta^3}[H_n]$$

The element stiffness matrix $[K^*]$ can be obtained by the principle of virtual displacements, as given previously by equation (3.119), in section 3.10, or

$$[K^*] = \int_v [B^*]^T [D][B^*]\,dv \tag{5.81}$$

If the element coordinate axis is located at the element centroid, the stiffness matrix can be expressed explicitly as

$$\begin{aligned}
K_{ij}^* = {} & \frac{D_x}{64\Delta^5}\left[A_iA_j + \frac{C_iC_j}{12}x_T^2 + \frac{C_iB_j + B_iC_j}{12}(xy)_T + \frac{B_iB_j}{12}y_T^2\right] \\
& + \frac{D_1}{64\Delta^5}\left[A_jE_i + A_iE_j + \frac{G_iC_j + C_iG_j}{12}x_T^2\right. \\
& \left. + \frac{G_iB_j + F_iC_j + C_iF_j + B_iG_j}{12}(xy)_T + \frac{F_iB_j + B_iF_j}{12}y_T^2\right] \\
& + \frac{D_y}{64\Delta^5}\left[E_iE_j + \frac{G_iG_j}{12}x_T^2 + \frac{G_iF_j + F_iG_j}{12}(xy)_T + \frac{F_iF_j}{12}y_T^2\right] \\
& + \frac{D_{xy}}{64\Delta^5}\left[H_iH_j + \frac{B_iB_j}{12}x_T^2 + \frac{G_iB_j + B_iG_j}{12}(xy)_T + \frac{G_iG_j}{12}x_T^2\right]
\end{aligned} \tag{5.82}$$

$$i = 1, 2, \ldots, 6$$

$$j = 1, 2, \ldots, 6$$

The total element stiffness matrix is given by

$$[K] = [T]^T[K^*][T] \tag{5.83}$$

Performing these matrix operations yields the required 9×9 stiffness matrix to define the triangular plate element in bending.

The terms A, B, C, D, E, F, G, and H required to describe the matrix elements are shown in Table 5-1, where the other terms are defined as follows:

Table 5-1
Finite Element Bending Coefficients

```
ALPHA=AK*BI*BJ+AJ*BI*BK+AI*BJ*BK
BETA=AK*CI*CJ+AJ*CI*CK+AI*CJ*CK
GAMMA=.5*(AJ*BI*CK+AI*BJ*CK+AJ*BK*CI+AI*BK*CJ+AK*BJ*CI+AK*BI*CJ)
LAMBDA=BI*BJ*CK+BJ*BK*CI+BI*BK*CJ
PHI=BJ*CI*CK+BI*CJ*CK+BK*CI*CJ
XT2=XI*XI+XJ*XJ+XK*XK
XYT=XI*YI+XJ*YJ+XK*YK
YT2=YI*YI+YJ*YJ+YK*YK
A(1)=2.*BI*BI*(AJ*BK-AK*BJ)+ALPHA*(BK-BJ)
A(2)=2.*BI*BI*(AJ*CK-AK*CJ)+4.*AI*BI*(CK*BJ-CJ*BK)+ALPHA*(CK-CJ)
A(3)=2.*BJ**2*(AK*BI-AI*BK)+ALPHA*(BI-BK)
A(4)=2.*BJ**2*(AK*CI-AI*CK)+4.*AJ*BJ*(CI*BK-CK*BI)+ALPHA*(CI-CK)
A(5)=2.*BK**2*(AI*BJ-AJ*BI)+ALPHA*(BJ-BI)
A(6)=2.*BK**2*(AI*CJ-AJ*CI)+4.*AK*BK*(CJ*BI-CI*BJ)+ALPHA*(CJ-CI)
B(1)=2.*BI**2*(BK*CJ-BJ*CK)+LAMBDA*(BK-BJ)
B(2)=4.*BI*CI*(CK*BJ-CJ*BK)+LAMBDA*(CK-CJ)
B(3)=2.*BJ**2*(BI*CK-BK*CI)+LAMBDA*(BI-BK)
B(4)=4.*BJ*CJ*(CI*BK-CK*BI)+LAMBDA*(CI-CK)
B(5)=2.*BK**2*(BJ*CI-BI*CJ)+LAMBDA*(BJ-BI)
B(6)=4.*BK*CK*(CJ*BI-CI*BJ)+LAMBDA*(CJ-CI)
C(1)=3.*BI*BJ*BK*(BK-BJ)
C(2)=3.*BI*BJ*BK*(CK-CJ)+6.*BI**2*(CK*BJ-CJ*BK)
C(3)=3.*BI*BJ*BK*(BI-BK)
C(4)=3.*BI*BJ*BK*(CI-CK)+6.*BJ**2*(CI*BK-CK*BI)
C(5)=3.*BI*BJ*BK*(BJ-BI)
C(6)=3.*BI*BJ*BK*(CJ-CI)+6.*BK**2*(CJ*BI-CI*BJ)
E(1)=2.*CI**2*(AJ*BK-AK*BJ)+4.*AI*CI*(BK*CJ-BJ*CK)+BETA*(BK-BJ)
E(2)=2.*CI**2*(AJ*CK-AK*CJ)+BETA*(CK-CJ)
E(3)=2.*CJ**2*(AK*BI-AI*BK)+4.*AJ*CJ*(BI*CK-BK*CI) +BETA*(BI-BK)
E(4)=2.*CJ**2*(AK*CI-AI*CK)+BETA*(CI-CK)
E(5)=2.*CK**2*(AI*BJ-AJ*BI)+4.*AK*CK*(BJ*CI-BI*CJ)+BETA*(BJ-BI)
E(6)=2.*CK**2*(AI*CJ-AJ*CI)+BETA*(CJ-CI)
F(1)=3.*CI*CJ*CK*(BK-BJ)+6.*CI**2*(BK*CJ-BJ*CK)
F(2)=3.*CI*CJ*CK*(CK-CJ)
F(3)=3.*CI*CJ*CK*(BK)+6.*CJ**
F(3)=3.*CI*CJ*CK*(BI-BK)+6.*CJ**2*(BI*CK-BK*CI)
F(4)=3.*CI*CJ*CK*(CI-CK)
F(5)=3.*CI*CJ*CK*(BJ-BI)+6.*CK**2*(BJ*CI-BI*CJ)
F(6)=3.*CI*CJ*CK*(CJ-CI)
G(1)=4.*CI*BI*(BK*CJ-BJ*CK)+PHI*(BK-BJ)
G(2)=2.*CI**2*(CK*BJ-CJ*BK)+PHI*(CK-CJ)
G(3)=4.*CJ*BJ*(BI*CK-BK*CI)+PHI*(BI-BK)
G(4)=2.*CJ**2*(CI*BK-CK*BI)+PHI*(CI-CK)
G(5)=4.*CK*BK*(BJ*CI-BI*CJ)+PHI*(BJ-BI)
G(6)=2.*CK**2*(CJ*BI-CI*BJ)+PHI*(CJ-CI)
H(1)=2.*BI*CI*(BK*AJ-BJ*AK)+2.*AI*BI*(BK*CJ-BJ*CK)+GAMMA*(BK-BJ)
H(2)=2.*BI*CI*(CK*AJ-CJ*AK)+2.*AI*CI*(CK*BJ-CJ*BK)+GAMMA*(CK-CJ)
H(3)=2.*BJ*CJ*(BI*AK-BK*AI)+2.*AJ*BJ*(BI*CK-BK*CI)+GAMMA*(BI-BK)
H(4)=2.*BJ*CJ*(CI*AK-CK*AI)+2.*AJ*CJ*(CI*BK-CK*BI)+GAMMA*(CI-CK)
H(5)=2.*BK*CK*(BJ*AI-BI*AJ)+2.*AK*BK*(BJ*CI-BI*CJ)+GAMMA*(BJ-BI)
H(6)=2.*BK*CK*(CJ*AI-CI*AJ)+2.*AK*CK*(CJ*BI-CI*BJ)+GAMMA*(CJ-CI)
 K(I,J)=DX/(64.*DELT**5)*(A(I)*A(J)+C(I)*C(J)/12.*XT2+(C(I)*B(J)+B(
1I)*C(J))/12.*XYT+B(I)*B(J)/12.*YT2)+D1/(64.*DELT**5)*(E(I)*A(J)+A(
2I)*E(J)+(G(I)*C(J)+C(I)*G(J))/12.*XT2+(G(I)*B(J)+F(I)*C(J)+
4 C(I)*F(J)+B(I)*G(J))/12.*XYT+(F(I)*B(J)+B(I)*F(J))/12.*YT2)
5 +DY/(64.*DELT**5)*(E(I)*E(J)+G(I)*G(J)/12.*XT2+(G(I)*F(J)+F(I)
6 *G(J))/12.*XYT +F(I)*F(J)/12.*YT2) +DXY/(64.*DELT**5)*(H(I)*H(J)
8 +B(I)*B(J)/12.*XT2 +(G(I)*B(J)+B(I)*G(J))/12.*XYT+G(I)*G(J)/
1 12.*YT2)
```

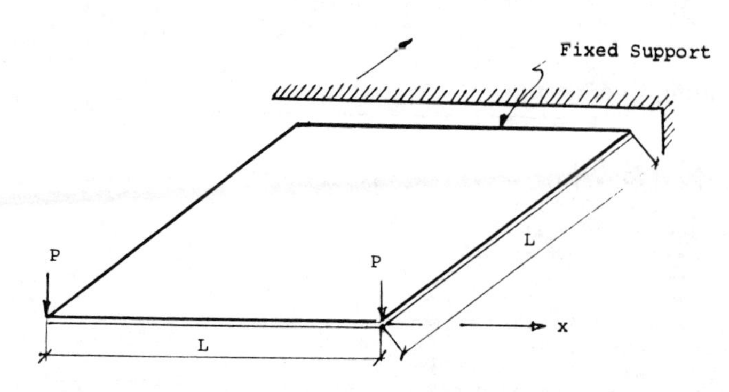

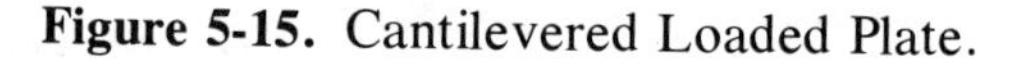

Figure 5-15. Cantilevered Loaded Plate.

$$x_T^2 = x_i^2 + x_j^2 + x_k^2$$

$$(xy)_T = x_i y_i + x_j y_j + x_k y_k$$

$$y_T^2 = y_i^2 + y_j^2 + y_k^2$$

$$D_x = \frac{Et^3}{12(1 - \mu^2)}$$

$$D_y = D_x$$

$$D_1 = \mu D_x$$

$$D_{xy} = \frac{Et^3}{24(1 + \mu)}$$

5.8.2 Example Problem

As shown in Figure 5-15, a plate of dimension $L \times L$ is cantilevered from a fixed support and is subjected to an end load of magnitude P. This plate will be idealized or modeled into two finite elements, as shown in Figure 5-16 (a). A solution to this problem will involve a total of twelve unknown displacements or (w, θ_x, θ_y) at each of the four nodes 1, 2, 3 and 4, as shown in Figure 5-16 (b). However, due to the end fixity at nodes 3 and 4, six of these displacements are zero. The plate geometry and stiffness will be assumed as follows:

$$L = 0.25, \quad P = 1, \quad t = 1, \quad E = 10.92, \quad \mu = 0.3, \quad D_x = D_y = 1$$

Element Stiffness Matrix—Element 1: The element stiffness can be developed by substituting the proper element properties into equations (5.59), (5.82), and (5.83). Recall that the element relative stiffness matrix, $[K^*]$, of equation (5.82) was developed for the element axes coinciding with the element centroid. Therefore, the element node coordinates must be com-

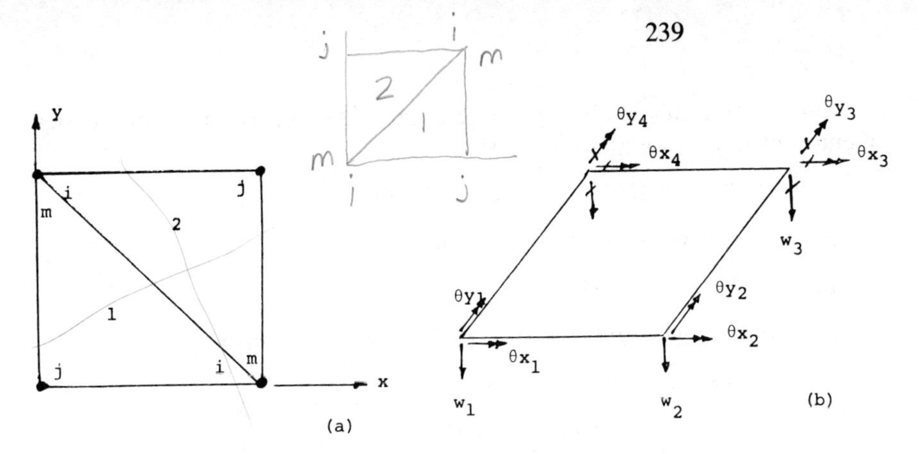

Figure 5-16. Finite Element Idealization.

puted with respect to the element centroid. This operation can be accomplished by the following matrix equation [23].

$$\begin{bmatrix} x \\ y \end{bmatrix}_c = \begin{bmatrix} x \\ y \end{bmatrix} - \tfrac{1}{3}\begin{bmatrix} x_i + x_j + x_m \\ y_i + y_j + y_m \end{bmatrix}$$

or for element 1

$$\begin{bmatrix} x_i \\ x_j \\ x_m \end{bmatrix}_c = \begin{bmatrix} x_i \\ x_j \\ x_m \end{bmatrix} - \frac{x_i + x_j + x_m}{3} \qquad \text{(a)}$$

$$= \begin{bmatrix} 0.0 \\ 0.25 \\ 0.25 \end{bmatrix} - 0.1667 = \begin{bmatrix} -0.1667 \\ 0.0833 \\ 0.0833 \end{bmatrix}$$

likewise,

$$\begin{bmatrix} y_i \\ y_j \\ y_m \end{bmatrix}_c = \begin{bmatrix} y_i \\ y_j \\ y_m \end{bmatrix} - \frac{y_i + y_j + y_m}{3} \qquad \text{(b)}$$

$$= \begin{bmatrix} 0.0 \\ 0.0 \\ 0.25 \end{bmatrix} - 0.0833 = \begin{bmatrix} -0.0833 \\ -0.0833 \\ 0.1667 \end{bmatrix}$$

Substitution of the element node coordinates with respect to the element centroid into equation (5.48) gives

$$a_i = (0.0833)(0.1667) - (0.0833)(-0.0833) = 0.020833$$

$$b_i = (-0.0833 - 0.1667) = -0.25$$

$$c_i = (0.0833 - 0.0833) = 0$$

$$a_j = (0.0833)(-0.0833) - (-0.1667)(0.1667) = 0.020833$$

$$b_j = (0.1667 + 0.0833) = 0.25$$

$$c_j = (-0.1667 - 0.0833) = -0.25$$

$$a_m = (-0.1667)(-0.0833) - (0.0833)(-0.0833) = 0.020833$$

$$b_m = (-0.833 + 0.0833) = 0$$

$$c_m = (0.0833 + 0.1667) = 0.25$$

The total area of the triangle Δ is

$$\Delta = 0.03125$$

The element transformation matrix, $[T]$, of equaton (5.59) multiplied through by $1/2\Delta$ is

$$[T]_1 = \begin{bmatrix} 0 & 1 & 0 & -4 & 0 & 0 & +4 & 0 & 0 \\ +4 & 0 & 1 & -4 & 0 & 0 & 0 & 0 & 0 \\ 0 & 0 & 0 & -4 & 1 & 0 & +4 & 0 & 0 \\ +4 & 0 & 0 & -4 & 0 & 1 & 0 & 0 & 0 \\ 0 & 0 & 0 & -4 & 0 & 0 & +4 & 1 & 0 \\ +4 & 0 & 0 & -4 & 0 & 0 & 0 & 0 & 1 \end{bmatrix}$$

Substitution of the constants as given in the computer listing Table 5-1 into equation (5.82) for the relative stiffness matrix $[k^*]$ gives

$$[K^*] = \begin{bmatrix} 0.433 & -0.017 & 0.050 & -0.167 & 0.250 & -0.083 \\ -0.017 & 1.867 & -0.133 & 0.750 & 0.717 & 0.250 \\ 0.050 & -0.133 & 1.600 & 0 & 0.750 & -0.167 \\ -0.167 & 0.750 & 0 & 1.600 & -0.133 & 0.050 \\ 0.250 & 0.717 & 0.750 & -0.133 & 1.867 & -0.017 \\ -0.083 & 0.250 & -0.167 & 0.050 & -0.017 & 0.433 \end{bmatrix}$$

The total element stiffness matrix, $[k]$ can be computed using equation (5.83), or $[K]_1 = [T]^T[K^*][T]_1$ where the $[T]_1^T$ matrix is found by exchanging the rows and columns of the $[T]$, matrix or:

$$[T]^T = \begin{bmatrix} 0 & 4 & 0 & 4 & 0 & 4 \\ 1 & 0 & 0 & 0 & 0 & 0 \\ 0 & 1 & 0 & 0 & 0 & 0 \\ -4 & -4 & -4 & -4 & -4 & -4 \\ 0 & 0 & 1 & 0 & 0 & 0 \\ 0 & 0 & 0 & 1 & 0 & 0 \\ 4 & 0 & 4 & 0 & 4 & 0 \\ 0 & 0 & 0 & 0 & 1 & 0 \\ 0 & 0 & 0 & 0 & 0 & 1 \end{bmatrix}$$

Multiplying $[T]^T[K^*]$, gives

$$[T]_1^T[K^*] = \begin{bmatrix} -1.068 & 11.468 & -1.200 & 9.6 & 2.268 & 2.932 \\ 0.433 & -0.017 & 0.050 & -0.167 & 0.250 & -0.083 \\ -0.017 & 1.867 & -0.133 & 0.750 & 0.717 & 0.250 \\ -1.864 & -13.736 & -8.400 & -8.400 & -13.736 & -1.864 \\ 0.050 & -0.133 & 1.600 & 0 & 0.750 & -0.167 \\ -0.167 & 0.750 & 0 & 1.600 & -0.133 & 0.050 \\ 2.932 & 2.268 & 9.600 & -1.200 & 11.468 & -1.068 \\ 0.250 & 0.717 & 0.750 & -0.133 & 1.867 & -0.017 \\ -0.083 & 0.250 & -0.167 & 0.050 & -0.017 & 0.433 \end{bmatrix}$$

The total element stiffness matrix is now found by multiplying the previous matrix $[T]_1^T[K^*]$, times $[T]$, which gives $[K]_1$, or Matrix 5-1.

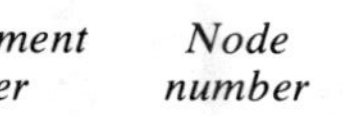

Matrix 5-1

$$
[K]_1 = \begin{bmatrix}
96 & -1.07 & 11.47 & -96.0 & -1.2 & 9.6 & 0 & 2.27 & 2.93 \\
-1.07 & 0.433 & -0.017 & -1.87 & 0.05 & -0.167 & 2.93 & 0.25 & -0.0833 \\
11.47 & -0.017 & 1.86 & -13.73 & -0.133 & 0.75 & 2.27 & 0.717 & 0.25 \\
-96.0 & -1.87 & -13.73 & 192 & -8.4 & -8.4 & -96.0 & 13.73 & -1.87 \\
-1.2 & 0.05 & -0.133 & -8.4 & 1.6 & 0 & 9.6 & 0.75 & -0.167 \\
9.6 & -0.167 & 0.75 & -8.4 & 0 & 1.6 & -1.2 & -0.133 & 0.05 \\
0 & 2.93 & 2.27 & -96.0 & 9.6 & -1.2 & 96.0 & 11.47 & -1.07 \\
2.27 & 0.25 & 0.717 & -13.73 & 0.75 & -0.133 & 11.47 & 1.86 & -0.017 \\
2.93 & -0.088 & 0.25 & -1.87 & -0.167 & 0.05 & -1.07 & -0.017 & 0.433
\end{bmatrix}
\begin{bmatrix} w_1 \\ \theta_{x1} \\ \theta_{y1} \\ w_2 \\ \theta_{x2} \\ \theta_{y2} \\ w_3 \\ \theta_{x3} \\ \theta_{y3} \end{bmatrix}
$$

Column groups (node numbers) below the matrix: 1 (columns 1–3), 2 (columns 4–6), 3 (columns 7–9).

Displacement number	Node number
w_1	
θ_{x1}	1
θ_{y1}	
w_2	
θ_{x2}	2
θ_{y2}	
w_3	
θ_{x3}	3
θ_{y3}	

Element Stiffness Matrix—Element 2: The member stiffness matrix for element 2 is developed in the same manner as element 1. Substituting the node coordinate values for element 2 into equations (a) and (b) gives

$$\begin{bmatrix} x_i \\ x_j \\ x_m \end{bmatrix}_c = \begin{bmatrix} 0.25 \\ 0.0 \\ 0.0 \end{bmatrix} - 0.0833 = \begin{bmatrix} 0.1667 \\ -0.0833 \\ -0.0833 \end{bmatrix} \tag{c}$$

$$\begin{bmatrix} y_i \\ y_j \\ y_m \end{bmatrix}_c = \begin{bmatrix} 0.25 \\ 0.25 \\ 0.0 \end{bmatrix} - 0.1667 = \begin{bmatrix} 0.0833 \\ 0.0833 \\ -0.1667 \end{bmatrix} \tag{d}$$

Notice that equations (c) and (d) are the same as equations (a) and (b), except for the sign. Therefore, upon substituting these values into equation (5.48), the same values for the constants are obtained as for element 1 except for the sign.

Substitution of the element node coordinates with respect to the element centroid into equation (5.48) gives

$$\begin{array}{lll} a_i = a_j & = a_m = 0.02083 & \\ b_i = 0.25 & b_j = -0.25 & b_m = 0 \\ c_i = 0 & c_j = 0.25 & c_m = -0.25 \end{array}$$

The total area of the element is

$$\Delta = 0.03125$$

The element transformation matrix of equation (5.59) is

$$[T]_2 = \begin{bmatrix} 0 & 1 & 0 & 4 & 0 & 0 & -4 & 0 & 0 \\ -4 & 0 & 1 & 4 & 0 & 0 & 0 & 0 & 0 \\ 0 & 0 & 0 & 4 & 1 & 0 & -4 & 0 & 0 \\ -4 & 0 & 0 & 4 & 0 & 1 & 0 & 0 & 0 \\ 0 & 0 & 0 & 4 & 0 & 0 & -4 & 1 & 0 \\ -4 & 0 & 0 & 4 & 0 & 0 & 0 & 0 & 1 \end{bmatrix}$$

The transpose matrix, using the $[T]_2$ matrix is

$$[T]_2^T = \begin{bmatrix} 0 & -4 & 0 & -4 & 0 & -4 \\ 1 & 0 & 0 & 0 & 0 & 0 \\ 0 & 1 & 0 & 0 & 0 & 0 \\ 4 & 4 & 4 & 4 & 4 & 4 \\ 0 & 0 & 1 & 0 & 0 & 0 \\ 0 & 0 & 0 & 1 & 0 & 0 \\ -4 & 0 & -4 & 0 & -4 & 0 \\ 0 & 0 & 0 & 0 & 1 & 0 \\ 0 & 0 & 0 & 0 & 0 & 1 \end{bmatrix}$$

The element stiffness matrix is now computed using (5.83), or

$$[K]_2 = [T]_2^T[K^*]_2[T]_2$$

multiplying $[T]_2^T[K^*]_2$ gives

$$[T]_2^T\,[k^*]_2 = \begin{bmatrix} 1.068 & -11.468 & 1.200 & -9.600 & -2.268 & -2.932 \\ 0.433 & -0.017 & 0.050 & -0.167 & 0.250 & -0.083 \\ -0.017 & 1.867 & -0.133 & 0.750 & 0.717 & 0.250 \\ 1.864 & 13.736 & 8.400 & 8.400 & 13.736 & 1.864 \\ 0.050 & -0.133 & 1.600 & 0 & 0.750 & -0.167 \\ -0.167 & 0.750 & 0 & 1.60 & -0.133 & 0.050 \\ -2.932 & -2.268 & -9.600 & 1.200 & -11.468 & 1.068 \\ 0.250 & 0.717 & 0.750 & -0.133 & 1.867 & -0.017 \\ -0.083 & -0.250 & -0.167 & 0.050 & -0.017 & 0.433 \end{bmatrix}$$

The total finite element stiffness matrix is now found by multiplying the previous matrix $[T]_2^T\,[k^*]_2$ times $[T]$ which gives $[k]_2$ or Matrix 5-2.

Matrix 5-2

$$[k]_2 = \begin{bmatrix} 96.0 & 1.07 & -11.47 & -96.0 & 1.2 & -9.6 & 0 & -2.27 & -2.93 \\ 1.07 & 0.433 & -0.017 & 1.87 & 0.05 & -0.167 & -2.93 & 0.25 & -0.0833 \\ -11.47 & -.017 & 1.87 & 13.73 & -0.133 & 0.75 & -2.27 & 0.717 & 0.25 \\ -96.0 & 1.87 & 13.73 & 192 & 8.4 & 8.4 & -96.0 & 13.73 & 1.87 \\ 1.2 & 0.05 & -0.133 & 8.4 & 1.6 & 0 & -9.6 & 0.75 & -0.167 \\ -9.6 & -0.167 & 0.75 & 8.4 & 0 & 1.6 & 1.2 & -0.133 & 0.05 \\ 0 & -2.93 & -2.27 & -96.0 & -9.6 & 1.2 & 96.0 & -11.47 & 1.07 \\ -2.27 & 0.25 & 0.717 & 13.73 & 0.75 & -0.133 & -11.47 & 1.87 & -0.017 \\ -2.93 & -0.088 & 0.25 & 1.87 & -0.167 & 0.05 & 1.07 & -0.017 & 0.433 \end{bmatrix} \begin{bmatrix} w_3 \\ \theta_{x3} \\ \theta_{y3} \\ w_4 \\ \theta_{x4} \\ \theta_{y4} \\ w_1 \\ \theta_{x1} \\ \theta_{y1} \end{bmatrix}$$

Column groups (node numbers): columns 1–3: 3; columns 4–6: 4; columns 7–9: 1.

Displacement number	Node number
w_3	
θ_{x3}	3
θ_{y3}	
w_4	
θ_{x4}	4
θ_{y4}	
w_1	
θ_{x1}	1
θ_{y1}	

The two resulting element stiffness matrixes are related to forces, and are written in the form of:

$$[k_1] \cdot \begin{bmatrix} w_1 \\ \theta_{x1} \\ \theta_{y1} \\ w_2 \\ \theta_{x2} \\ \theta_{y2} \\ w_3 \\ \theta_{x3} \\ \theta_{y3} \end{bmatrix} = \begin{bmatrix} +1/2 \\ 0 \\ 0 \\ +1 \\ 0 \\ 0 \\ V_3 \\ M_{x3} \\ M_{y3} \end{bmatrix} \quad \text{and } [k_2] \cdot \begin{bmatrix} w_3 \\ \theta_{x3} \\ \theta_{y3} \\ w_4 \\ \theta_{x4} \\ \theta_{y4} \\ w_1 \\ \theta_{x1} \\ \theta_{y1} \end{bmatrix} = \begin{bmatrix} V_3 \\ M_{x3} \\ M_{y3} \\ V_4 \\ M_{x4} \\ M_{y4} \\ 1/2 \\ 0 \\ 0 \end{bmatrix}$$

These two resulting matrices are now combined to give the system Matrix 5-3.

The solution of the first 6×6 matrix gives the unknown displacements at nodes 1 and 2. The solution of the forces at nodes 3 and 4 can then be found by solving the remaining matrix. These results are as follows;

Deflections	*Forces*
$w_1 = 0.0462$	$V_3 = -1.489$
$\theta_{x1} = 0.2807$	$M_{x3} = -0.265$
$\theta_{y1} = 0.0380$	$M_{y3} = 0.078$
$w_2 = 0.0433$	$V_4 = -0.506$
$\theta_{x2} = 0.2566$	$M_{x4} = -0.226$
$\theta_{y2} = 0.0378$	$M_{y4} = 0.020$

Matrix 5-3

$$
\left[\begin{array}{cccccc|cccccc}
192.00 & -12.54 & 12.54 & -96.00 & -1.20 & 9.60 & 0 & -0.66 & 0.66 & -96.0 & -9.6 & 1.20 \\
-12.54 & 2.29 & -0.034 & -1.87 & 0.05 & -0.17 & 0.66 & 0.50 & 0.637 & 13.73 & 0.75 & -0.133 \\
12.54 & -0.034 & 2.303 & -13.73 & -0.133 & 0.75 & -0.66 & 0.637 & 0.50 & 1.87 & -0.167 & 0.05 \\
-96.99 & -1.87 & -13.73 & 192.00 & -8.40 & -8.40 & -96.0 & -13.73 & -1.87 & 0 & 0 & 0 \\
-1.20 & 0.05 & -0.133 & -8.40 & 1.60 & 0 & 9.60 & 0.75 & -0.17 & 0 & 0 & 0 \\
9.60 & -0.17 & 0.75 & -8.40 & 0 & 1.60 & -1.20 & -0.133 & 0.05 & 0 & 0 & 0 \\
\hline
0 & 0.66 & -0.66 & -96.00 & 9.60 & -1.20 & 192.0 & 12.54 & -12.54 & -96.0 & 1.20 & -9.6 \\
-0.66 & 0.50 & 0.637 & -13.73 & 0.75 & -0.133 & 12.54 & 2.303 & -0.034 & 1.87 & 0.05 & -0.167 \\
0.66 & 0.637 & 0.500 & -1.87 & -0.17 & 0.05 & -12.54 & -0.034 & 2.29 & 13.73 & -0.133 & 0.75 \\
-96.00 & 13.73 & 1.87 & 0 & 0 & 0 & -96.0 & 1.87 & 13.73 & 192. & 8.4 & 8.4 \\
-9.60 & 0.75 & -0.167 & 0 & 0 & 0 & 1.20 & 0.05 & -0.133 & 8.4 & 1.6 & 0 \\
1.20 & -0.133 & 0.05 & 0 & 0 & 0 & -9.60 & -0.167 & 0.75 & 8.4 & 0 & 1.6
\end{array}\right]
\left[\begin{array}{c}
w_1 \\ \theta_{x1} \\ \theta_{y1} \\ w_2 \\ \theta_{x2} \\ \theta_{y2} \\ w_3 \\ \theta_{x3} \\ \theta_{y3} \\ w_4 \\ \theta_{x4} \\ \theta_{y4}
\end{array}\right]
=
\left[\begin{array}{c}
1 \\ 0 \\ 0 \\ 1 \\ 0 \\ 0 \\ V_3 \\ M_{x3} \\ M_{y3} \\ V_4 \\ M_{x4} \\ M_{y4}
\end{array}\right]
$$

Bridge and Building System Solutions

In this chapter the analytical determination of the responses of bridge and building floor systems will be presented. These methods will employ the equations outlined in the previous chapter. In some instances, the results of these system solutions have been applied in the design and field studies of actual structures [35, 36, 37]. The need for employing these various techniques is dictated by the complexity of the structure. Depending on the type of structure and on computer availability, each technique will have its advantage. The selection of the method is, therefore, left to the discretion of the engineer.

6.1 Rectangular Orthotropic Bridges

The behavior of orthotropic bridges can be determined by incorporating the interaction of the plate equation (4.38) with flexible girders. Such interaction has been determined by two analytical schemes [38, 39], designated as: (1) the slope-deflection technique and (2) the finite difference technique. Both of these techniques will now be explained. A computer program to evaluate the system equations is given in chapter 7.

6.1.1 Slope-Deflection Method

This method incorporates the ideas used in the familiar plane-frame slope deflection technique [40]. In developing the general slope-deflection equations, the Fourier series solution of the expressions will be used. In part, the equations given in section 4.6.2. will be used. Case I equation (3.62b), Case II equation (3.63b), and Case III equation (3.64b) are the deflection relationships for different categories of plate stiffness for a plate that has two simple supports with the other two edges arbitrary. These plate equations will be used in the following slope-deflection developments.

Examining now Figure 6-1, we have a continuous plate on a series of flexible supports subjected to some loading. If we now remove the deflected plate *a*-*b* of the distorted system of Figure 6-1, the final end slopes and deflections must be retained. In order to retain these geometrical

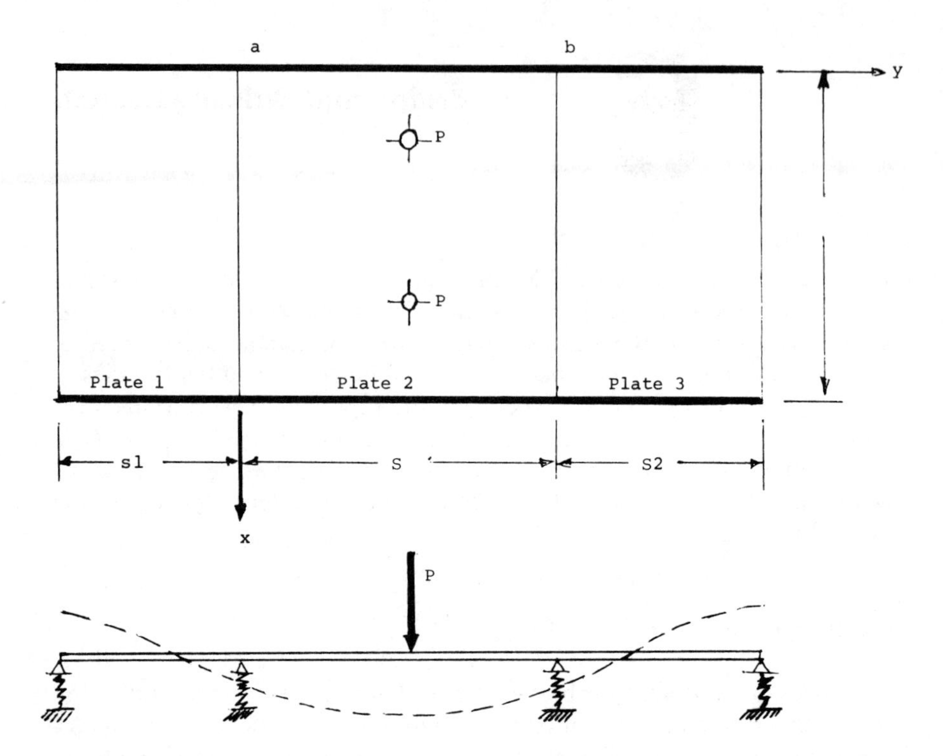

Figure 6-1. Distorted Continuous Plate on Flexible Supports.

conditions external forces consisting of end moments and reactions must be present. These forces, as well as the end slopes, deflections, and loads, will all be expressed as functions of sin λx in order to conform to the initial plate solution, equations (3.62), (3.63), and (3.64).

Case I

$$w = \sum(A \cosh r_1 y + B \sinh r_1 y + C \cosh r_2 y + D \sinh r_2 y)\sin \lambda x$$

Case II

$$w = \sum(A \sinh r_1 y + By \sinh r_1 y + C \cosh r_1 y + Dy \cosh r_1 y)\sin \lambda x$$

Case III

$$w = \sum(A \cosh \phi y \cos \psi y + B \cosh \phi y \sin \psi y + C \sinh \phi y \cos \psi y) + D \sinh \phi y \sin \psi y)\sin \lambda x$$

Figure 6-2 describes the displaced plate *a-b*. In order to determine the magnitude of the indeterminate moments and reactions, as mentioned previously, a method similar to the planar slope-deflection technique will

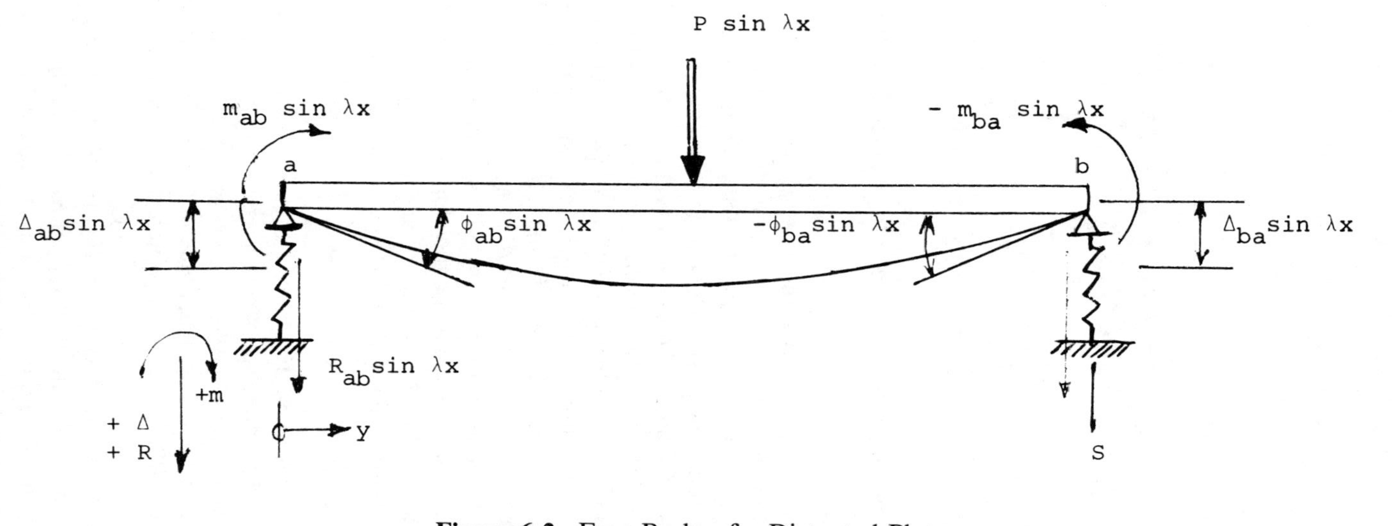

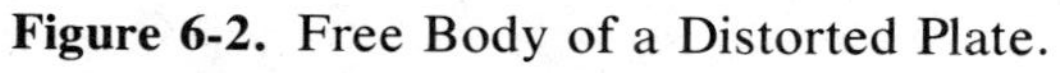
Figure 6-2. Free Body of a Distorted Plate.

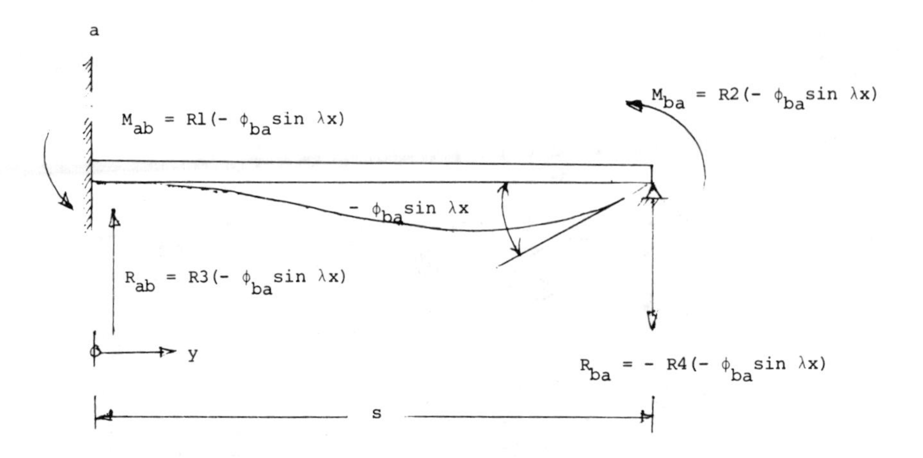

Figure 6-3. Induced Plate Forces due to Applied End Rotation.

be developed. It will be recalled that the slope-deflection method requires the expression of the indeterminate end moments in terms of end slopes, deflections, and fixed end moments (FEM). These expressions are written for all joints of a system. Then a series of equilibrium equations, $\Sigma m = 0$, at each joint is expressed in terms of the geometrical conditions and the FEM. The simultaneous solution of these equations then yields the actual end slopes and deflections. The substitution of these values into the moment, slope, deflection relationship equation will then evaluate the indeterminate moments. A similar analogy may be developed for the indeterminate reactions, utilizing a $\Sigma F = 0$ equation at each joint.

In order to use this idea for the analysis of continuous plates on flexible supports, a series of equations must be developed relating the end slopes and deflections and fixed end moments. The following development will consider individually the effects of end slopes, end deflections, and FEM on a plate.

Rotation Only, No Deflection: Examining Figure 6-3, the plate a-b has a fixed support at a and a simple support at b with a given applied slope $\phi_{ba} \cdot \sin \lambda x$. In order to retain the slope $\phi_{ba} \cdot \sin \lambda x$ certain end moments and end reactions must be imposed. The magnitude of these forces are designated by the constants $R1$, $R2$, $R3$, and $R4$, which will be defined later.

Deflection Only, No Rotation: Examining Figure 6-4, the plate a-b will have fixed supports at a and b. A deflection $\Delta_{ba} \cdot \sin \lambda x$ will be imposed at support b, inducing end moments and end reactions designated by the constants $D1$, $D2$, $D3$, and $D4$, which will be defined later.

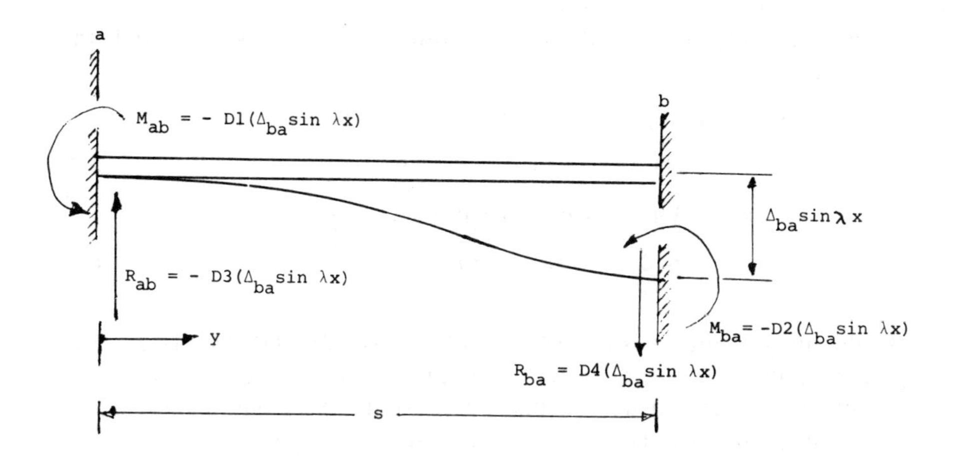

Figure 6-4. Induced Plate Forces due to Applied End Deflection.

In addition to the above solutions, two other cases are required. One case is similar to that described by Figure 6-3, but all boundaries are reversed and the slope is now applied at support a as $\phi_{ab} \cdot \sin \lambda x$. The other case is similar to the one described by Figure 6-4, except the deflection is now applied at support a as $\Delta_{ab} \cdot \sin \lambda x$. For both of these cases, similar R and D constants will be developed.

Fixed End Moments and Reactions: The effects of the end slopes and deflections on the end moments and reactions have just been examined. However, the effects of external loading must also be considered. At present, these loads cause fixed end moments and reactions at supports a and b designated by the terms, *FEMA, FEMB, FERA,* and *FERB*.

Moment Equations: The final moments, in terms of slopes, deflections, and fixed end moments may now be written. Summing up the individual effects of moment ΣMab and ΣMba as expressed in Figures 6-3 and 6-4 and the other cases discussed, yields the following equations.

$$M_{ab} = R2\ \phi_{ab} \sin \lambda x + R1\ \phi_{ba} \sin \lambda x + D2\ \Delta_{ab} \sin \lambda x - D1\ \Delta_{ba} \sin \lambda x - FEMAB \tag{6.1}$$

$$M_{ba} = R1\ \phi_{ab} \sin \lambda x + R2\ \phi_{ba} \sin \lambda x + D1\ \Delta_{ab} \sin \lambda x - D2\ \Delta_{ba} \sin \lambda x + FEMBA \tag{6.2}$$

Reaction Equations: To determine the final reactions in terms of slopes, deflections, and fixed end reactions, the sum of each part ΣRab and ΣRba

described by Figures 6-3 and 6-4 and the other cases discussed will yield the following equations.

$$R_{ab} = R4\ \phi_{ab} \sin \lambda x + R3\ \phi_{ba} \sin \lambda x + D4\ \Delta_{ab} \sin \lambda x - D3\ \Delta_{ba} \sin \lambda x - FERAB \quad (6.3)$$

$$R_{ba} = -R3\ \phi_{ab} \sin \lambda x - R4\ \phi_{ba} \sin \lambda x - D3\ \Delta_{ab} \sin \lambda x + D4\ \Delta_{ba} \sin \lambda x - FERBA \quad (6.4)$$

Evaluation of Plate Constants R, $R2$, $R3$, $R4$, $D1$, $D2$, $D3$, $D4$**:** In order to use equations (6.1) through (6.4), the plate constants R and D must be evaluated. These constants actually represent the induced moments and reactions at the various ends of a plate (Figures 6-3 and 6-4), due to either the applied slopes ϕ or deflections Δ at these ends.

The following approach will be used in determining these moments and reactions for the various orthotropic plate deflection solutions, Cases I, II, and III.

Moments and Reactions on Plate Due to Induced End Slopes ϕ: Examining Figure 6-3, the plate has a fixed support at $y = 0$, $(w = w' = 0)$ and a simple support at $y = s$, $(w = 0)$ and with a given sinusoidal rotation at $y = s$, $(w' = -\phi_{ba} \sin \lambda x)$. These boundary conditions will now permit us to evaluate the constants A, B, C, and D in all of the general deflection equations (3.62), (3.63), and (3.64). These equations represent the deflection of a plate simply supported along the y-axis at $x = 0$ and $x = 1$, and a fixed and simple support at $y = 0$ and $y = s$, with an applied rotation $-\phi_{ba} \sin \lambda x$ at $y = s$.

The evaluation of the moments and reactions at the supports requires the use of equations (4.32) and (4.42). Therefore, differentiating the various general case deflection equations (3.62), (3.63), and (3.64) with the constants A, B, C, and D now determined, and using equations (4.32) and (4.42), the moments and reactions, and thus the R constants, are evaluated at $y = 0$ and $y = s$. For Case I, the R constants are represented by equations (6.5) through (6.8); for Case II, equations (6.9) through (6.12); and for Case III, equations (6.13) through (6.16). All of these exprcssions are listed on the following pages.

Moments and Reactions on Plate Due to Induced End Deflections Δ: Examining Figure 6-4, the plate has a fixed support at $y = 0$ $(w = w' = 0)$ and a fixed support at $y = s$, $(w' = 0)$ and with a given sinusoidal deflection at $y = s$, $(w = \Delta_{ba} \sin \lambda x)$. These boundary conditions will permit us to evaluate the constants A, B, C, and D in all of the general deflection equations (3.62), (3.63), and (3.64). These equations will now represent the deflection of a plate simply supported along the y-axis at $x = 0$ and $x = 1$, and with fixed supports at $y = 0$ and $y = s$ with an applied deflection Δ_{ba} sin λx at $y = s$.

The determination of the moments and reactions at the supports may be determined by using equations (4.32) and (4.42). Applying these equations to the various general deflection equations (3.62), (3.63), and (3.64) with the constants A, B, C, and D now evaluated, gives the D term at $y = 0$ and $y = s$. The D values for Case I are given by equations (6.17) through (6.20), for Case II we have equations (6.21) through (6.24), and for Case III values given by equations (6.25) through (6.28). The summation Σ is for integer values of $n = 1$ to infinity.

Case I

$$R1 = D_y \sum (r_1^2 - r_2^2)(r_1 \sinh r_2 s - r_2 \sinh r_1 s)k \tag{6.5}$$

$$\begin{aligned} R2 = D_y \sum (r_1^2 - r_2^2)(r_1 \cosh r_1 s \sinh r_2 s \\ - r_2 \cosh r_2 s \sinh r_1 s)k \end{aligned} \tag{6.6}$$

$$R3 = D_y \sum (r_1 r_2 (r_2^2 - r_1^2)(\cosh r_2 s - \cosh r_1 s)k \tag{6.7}$$

$$\begin{aligned} R4 = D_y \sum [r_1 r_2 (r_1^2 + r_2^2 - 4\beta\lambda^2)(1 - \cosh r_1 s \cosh r_2 s) \\ + (r_1^4 + r_2^4 - 2\beta\lambda^2(r_1^2 + r_2^2) \sinh r_1 s \sinh r_2 s]k \end{aligned} \tag{6.8}$$

$$k = 1/[2r_1 r_2 (1 - \cosh r_1 s \cosh r_2 s) + (r_1^2 + r_2^2) \sinh r_1 s \sinh r_2 s]$$

Case II

$$R1 = D_y \sum 2r_1(\sinh r_1 s - r_1 s \cosh r_1 s)k \tag{6.9}$$

$$R2 = D_y \sum 2r_1(\cosh r_1 s \sinh r_1 s - r_1 s)k \tag{6.10}$$

$$R3 = D_y \sum r_1^2 (2r_1 s \sinh r_1 s)k \tag{6.11}$$

$$\begin{aligned} R4 = D_y \sum \{r_1^2 [3 \sinh^2 r_1 s - (r_1 s)^2] \\ + 2\lambda^2 \beta [(r_1 s)^2 - \sinh^2 r_1 s]\}k \end{aligned} \tag{6.12}$$

$$k = 1/[\sinh^2 r_1 s - (r_1 s)^2]$$

Case III

$$R1 = D_y \sum 2\phi\psi(\psi \sinh \phi s \cos \psi s - \phi \cosh \phi s \sin \psi s)k \tag{6.13}$$

$$R2 = D_y \sum \phi\psi(\psi \sinh 2\phi s - \phi \sin 2\psi s)k \tag{6.14}$$

$$R3 = D_y \sum 2\phi\psi(\phi^2 + \psi^2)(\sinh \phi s \sin \psi s)k \tag{6.15}$$

$$\begin{aligned} R4 = D_y \sum [\psi^2 (3\phi^2 - \psi^2 - 2\beta\lambda^2)\sinh^2 \phi s \\ + \phi^2 (3\psi^2 - \phi^2 - 2\beta\lambda^2)\sin^2 \psi s]k \end{aligned} \tag{6.16}$$

$$k = 1/[\psi^2 \sinh^2 \phi s - \phi^2 \sin^2 \psi s]$$

Case I

$$D1 = D_y \sum r_1r_2(r_1^2 - r_2^2)(\cosh r_2s - \cosh r_1s)k \tag{6.17}$$

$$D2 = D_y \sum r_1r_2[(r_1^2 + r_2^2)(\cosh r_1s \cosh r_2s - 1) - 2r_1r_2 \sinh r_1s \sinh r_2s]k \tag{6.18}$$

$$D3 = D_y \sum r_1r_2(r_2^2 - r_1^2)(r_2 \sinh r_2s - r_1 \sinh r_1s)k \tag{6.19}$$

$$D4 = D_y \sum r_1r_2(r_2^2 - r_1^2)(r_2 \sinh r_2s \cosh r_1s - r_1 \sinh r_1s \cosh r_2s)k \tag{6.20}$$

$$k = 1/[2r_1r_2(1 - \cosh r_1s \cosh r_2s) + (r_2^2 + r_1^2) \sinh r_1s \sinh r_2s]$$

Case II

$$D1 = D_y \sum -r_1^2(2r_1s \sinh r_1s)k \tag{6.21}$$

$$D2 = D_y \sum r_1^2[\sinh^2 r_1s + (r_1s)^2]k \tag{6.22}$$

$$D3 = D_y \sum 2r_1^3(r_1s \cosh r_1s + \sinh r_1s)k \tag{6.23}$$

$$D4 = D_y \sum 2r_1^3(r_1s + \sinh r_1s \cosh r_1s)k \tag{6.24}$$

$$k = 1/[\sinh^2 r_1s - (r_1s)^2]$$

Case III

$$D1 = D_y \sum -2\phi\psi(\phi^2 + \psi^2)(\sinh \phi s \sin \psi s)k \tag{6.25}$$

$$D2 = D_y \sum (\phi^2 + \psi^2)(\psi^2 \sinh^2 \phi s + \phi^2 \sin^2 \psi s)k \tag{6.26}$$

$$D3 = D_y \sum 2\phi\psi(\phi^2 + \psi^2)(\phi \cosh \phi s \sin \psi s + \psi \sinh \phi s \cos \psi s)k \tag{6.27}$$

$$D4 = D_y \sum \phi\psi(\phi^2 + \psi^2)(\psi \sinh 2\phi s + \phi \sin 2\psi s)k \tag{6.28}$$

$$k = 1/[\psi^2 \sinh^2 \phi s - \phi^2 \sin^2 \psi s]$$

Fixed End Moments and Reactions on a Plate due to Line Loads: The equations representing the plate fixed end moments (FEM) and fixed end reactions (FER), as listed in the general slope-deflection equations (6.1) through (6.4), may be evaluated by using the Muller-Breslau Principle [40] in conjunction with Maxwell's Theorem of Reciprocal Deflections [40].

As explained in reference [39], the fixed end moments and reactions due to a unit line loading located at some distance $y = v$ on the plate may be easily expressed by using the previously derived deflection equations (3.62), (3.63), and (3.64) with the constants A, B, C, and D evaluated according to the boundary conditions described in Figures 6-3 and 6-4. These final deflection equations were also used to determine the plate constants R and D. The fixed end moments and reactions, due to a unit line loading according to the Case I condition, are listed as equations (6.29) and (6.30). The solutions for Case II are equations (6.31) and (6.32), and for Case III equations (6.33) and (6.34). All of these equations are listed below.

These solutions would be multiplied by the actual value of the line loading to give the proper value of moments and reactions. Also, the fixed end moments M_{ab} and reactions R_{ab} at support a may be obtained by substituting $(s - v)$ for v in the listed equations.

Case I

$$m^f_{ba} = \sum \frac{\sin \lambda x}{k}[(\cosh r_2 s - \cosh r_1 s)(r_1 \sinh r_2 v - r_2 \sinh r_1 v) + (r_1 \sinh r_2 s - r_2 \sinh r_1 s)(\cosh r_1 v - \cosh r_2 v)] \quad (6.29)$$

$$R^f_{ba} = \sum \frac{\sin \lambda x}{k}[r_1 r_2(\cosh r_2 s - \cosh r_1 s)(\cosh r_2 v - \cosh r_1 v) + (r_2 \sinh r_2 s - r_1 \sinh r_1 s)(r_2 \sinh r_1 v - r_1 \sinh r_2 v)] \quad (6.30)$$

$$k = [2r_1 r_2(1 - \cosh r_1 s \cosh r_2 s) + (r_1^2 + r_2^2)(\sinh r_1 s \sinh r_2 s)]$$

Case II

$$m^f_{ba} = \sum \frac{\sin \lambda x}{k}[s \sinh r_1 s \sinh r_1 v + (r_1 s \cosh r_1 s - \sinh r_1 s)\, v \sinh r_1 v - (r_1 s \sinh r_1 s)\, v \cosh r_1 v] \quad (6.31)$$

$$R^f_{ba} = \sum \frac{\sin \lambda x}{k}[(r_1 s \cosh r_1 s + \sinh r_1 s)\sinh r_1 v + (r_1 s \sinh r_1 s)(r_1 v \sinh r_1 v) - (r_1 s \cosh r_1 s + \sinh r_1 s)\, r_1 v \cosh r_1 v] \quad (6.32)$$

$$k = [\sinh^2 r_1 s - (r_1 s)^2]$$

Case III

$$m^f_{ba} = \sum \frac{\sin \lambda x}{k}[(\phi \sinh \phi s \sin \psi s)\cosh \phi v \sin \psi v - (\psi \sinh \phi s \sin \psi s)\sinh \phi v \cos \psi v + \sinh \phi v \sin \psi v\, (\psi \sinh \phi s \cos \psi s - \phi \cosh \phi s \sin \psi s)] \quad (6.33)$$

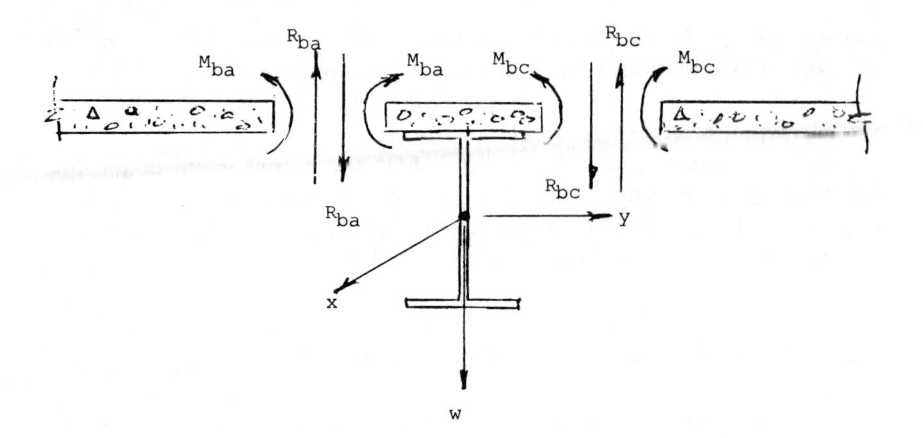

Figure 6-5. Joint Equilibrium.

$$
\begin{aligned}
R^f_{ba} = \sum \frac{\sin \lambda x}{k} [&\phi \cosh \phi v \sin \psi v (\psi \sinh \phi s \cos \psi s \\
&+ \phi \cosh \phi s \sin \psi s) - \psi(\psi \sinh \phi s \cos \psi s \\
&+ \phi \cosh \phi s \sin \psi s) \sinh \phi v \cos \psi v \\
&- (\phi^2 + \psi^2) \sinh \phi s \sin \psi s \sinh \phi v \sin \psi v]
\end{aligned}
\tag{6.34}
$$

$$k = [\phi^2 \sin^2 \psi s - \psi^2 \sinh^2 \phi s]$$

Consideration of Elastic Supports: The previous development would be sufficient to describe the behavior of the bridge if the supports (floor beams) were rigid. The displacements at the ends of the plate at $y = 0$ and $y = s$ would be zero, and the moments, reactions, and slopes could be readily determined. The floor beams, however, are elastic and they deflect and rotate with the plates that they support.

Figure 6-5 shows a typical plate and floor beam junction with the plate moments and reactions interacting with the floor beam. These plate reactions and moments are not concentrated, but are distributed along the length of the junction in the x direction, resulting in a floor beam torque m_t and a vertical reaction R_b which equals q.

The differential equation describing the deflection of an elastic beam due to a distributed load [29] is

$$\frac{d^4 \Delta_x}{dx^4} = \frac{q}{EI} \tag{6.35}$$

if the deflection is assumed to vary sinusoidally with x as

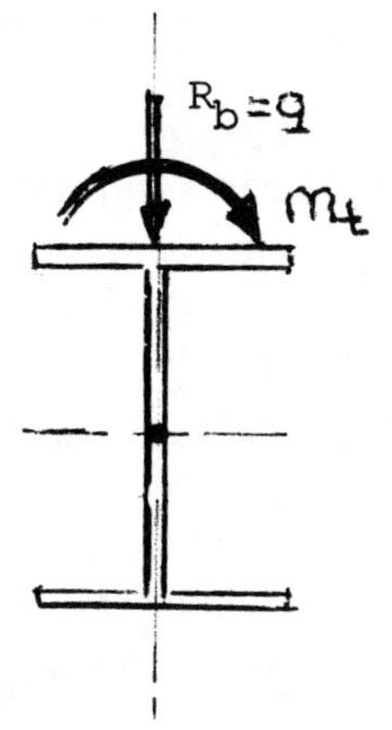

Figure 6-6. Forces on Flexible Girder.

$$\Delta_x = \Delta_n \sin \lambda x \qquad (6.36)$$

Equation (6.35) may be rewritten to express the uniform load as a function of deflection.

$$q = \lambda^4 EI\Delta_n \sin \lambda x \qquad (6.37)$$

Likewise, the differential equation describing the angle of twist of a beam due to a uniform torque [29] is

$$GJ \frac{d^2\Phi_x}{dx^2} - EI_w \frac{d^4\Phi_x}{dx^4} = -m_{tx} \qquad (6.38)$$

where both pure and warping torsions are considered. By assuming that the angle of twist varies sinusoidally in x as represented by equation (6.39),

$$\Phi_x = \Phi_n \sin \lambda x \qquad (6.39)$$

equation (6.38) may be rewritten to express the uniform torque as a function of rotation.

$$m_{tx} = \lambda^2 GJ\Phi_n \sin \lambda x + \lambda^4 EI_w\Phi_n \sin \lambda x \qquad (6.40)$$

Final Slope Deflection Equations: The force equilibrium condition for a typical plate-floor beam junction b, as shown in Figure 6-5 and Figure 6-6, is determined by statics. That is, the separate sums of the moments and reactions of the plate and girder forces are equal to zero. The results of these summations are given as equations (6.41) and (6.42), using equations (6.1) through (6.4), changing M_{ab} to M_{bc} and R_{ab} to R_{bc}, and equations (6.37) and (6.40), changing Δ_n and Φ_n to Δ_b and Φ_b.

$$\sum M = M_{ba} + M_{bc} + m_t = 0$$

$$\begin{aligned}
&\sum[R1(\Phi_{ab} \sin \lambda x) + R2(\Phi_{ba} \sin \lambda x) + D1(\Delta_{ab} \sin \lambda x) \\
&\quad - D2(\Delta_{ba} \sin \lambda x) - FEMBA] + \sum[R2(\Phi_{bc} \sin \lambda x) \\
&\quad + R1(\Phi_{cb} \sin \lambda x) + D2(\Delta_{bc} \sin \lambda x) - D1(\Delta_{cb} \sin \lambda x) \\
&\quad + FEMBC] + \sum[\lambda^2 GJ\Phi_b \sin \lambda x + \lambda^4 EI_w \Phi_b \sin \lambda x] = 0
\end{aligned} \tag{6.41}$$

$$\sum F = R_{ba} + R_{bc} + q = 0$$

$$\begin{aligned}
&\sum[-R3(\Phi_{ab} \sin \lambda x) - R4(\Phi_{ba} \sin \lambda x) - D3(\Delta_{ab} \sin \lambda x) \\
&\quad + D4(\Delta_{ba} \sin \lambda x) + FERBA] + \sum[R4(\Phi_{bc} \sin \lambda x) \\
&\quad + R3(\Phi_{cb} \sin \lambda x) + D4(\Delta_{bc} \sin \lambda x) - D3(\Delta_{cb} \sin \lambda x) \\
&\qquad + FERBC] + \sum \lambda^4 EI\Delta_b \sin \lambda x = 0
\end{aligned} \tag{6.42}$$

Floor Beam Shear Forces, Moments, and Torsional Stresses. There are two additional differential equations similar to equation (6.35) relating the deflection of a beam to its shear force and bending moment [29].

$$\frac{d^3\Delta_x}{dx^3} = \frac{V}{EI} \tag{6.43}$$

and

$$\frac{d^2\Delta_x}{dx^2} = \frac{M}{EI} \tag{6.44}$$

Using equation (6.36), equations (6.43) and (6.44) are rearranged as

$$V = -\lambda^3 EI\Delta_n \cos \lambda x \tag{6.45}$$

and

$$M = -\lambda^2 EI\Delta_n \sin \lambda x \tag{6.46}$$

In addition to the stresses caused by this shear force and bending moment, there are stresses caused by the rotation of the beam, Φ. These are the pure torsion shear, warping shear, and warping normal stresses [29] given as equations (6.47), (6.48), and (6.49), respectively.

$$\tau_t = Gt\frac{d\Phi_x}{dx} = GT\,\lambda\phi_n \cos \lambda x \tag{6.47}$$

$$\tau_{ws} = -ES_w\frac{d^3\Phi_x}{dx^3} = ES_{ws}\lambda^3\phi_n \cos \lambda x \tag{6.48}$$

$$\sigma_{ws} = EW_n\frac{d^2\Phi_x}{dx^2} = -EW_n\lambda^2\phi_n \sin \lambda x \tag{6.49}$$

The solution of the slope deflection equations (6.41) and (6.42) will yield the numerical values of Δ_x and Φ_x for each floor beam. The values of Δ_x are used directly in equations (6.45) and (6.46) to find the shears and moments. Using equations (6.47) through (6.49), the respective stresses can then be determined.

Slope Deflection Equation Solution: Equations (6.41) and (6.42) are two simultaneous equations for each floor beam in which are contained six unknowns: three slopes and three deflections. These unknowns are at the floor beam in question, and at those adjacent to this member. If an effort is made to provide additional equations by adding additional plates, it is seen that the number of unknowns will increase by four, while the number of equations will increase by four. There are then six equations and ten unknowns. The number of unknowns will always exceed the number of equations by four until the end floor beams are reached, at which point four equations and zero unknowns will be added, resulting in a matrix of $2N$ equations, where N is the number of floor beams.

The solution of the system of slope deflection equations will yield deformations that will completely describe the elastic behavior of the transverse floor beams and deck plate. A computer program to evaluate such a system is given in chapter 7 and Appendix A-1.

The technique just described has recently been extended in the study of single and continuous curved orthotropic bridge systems [41, 42]. These results have permitted the development of various design aids [43, 44] and have yielded excellent comparisons with experimental model and prototype tests [45, 46, 47].

6.1.2 Finite Difference Method

It was shown in section 5.1 that the orthotropic plate equation (4.38) could be expressed in terms of some predetermined coordinate points according to the Figure 5-4 mesh pattern. In order to apply this to a system of plates and girders with certain boundary conditions, additional relationships are necessary. For the various boundaries that may be encountered in bridge structures, consisting of simple supports and free edges, the general plate equation (5.12), Figure 5-4, has been modified [38, 39] and these new plate equations are as listed in Figures 6-7 through 6-11. It will be noticed that girders or beams are included in the system, as described previously, as well as their stiffnesses J and K.

To determine the deflections of a system of flexible girders with a top plate and edges consisting of simple supports and free edges, the given system must first be divided into mesh spacings. With these spacings and

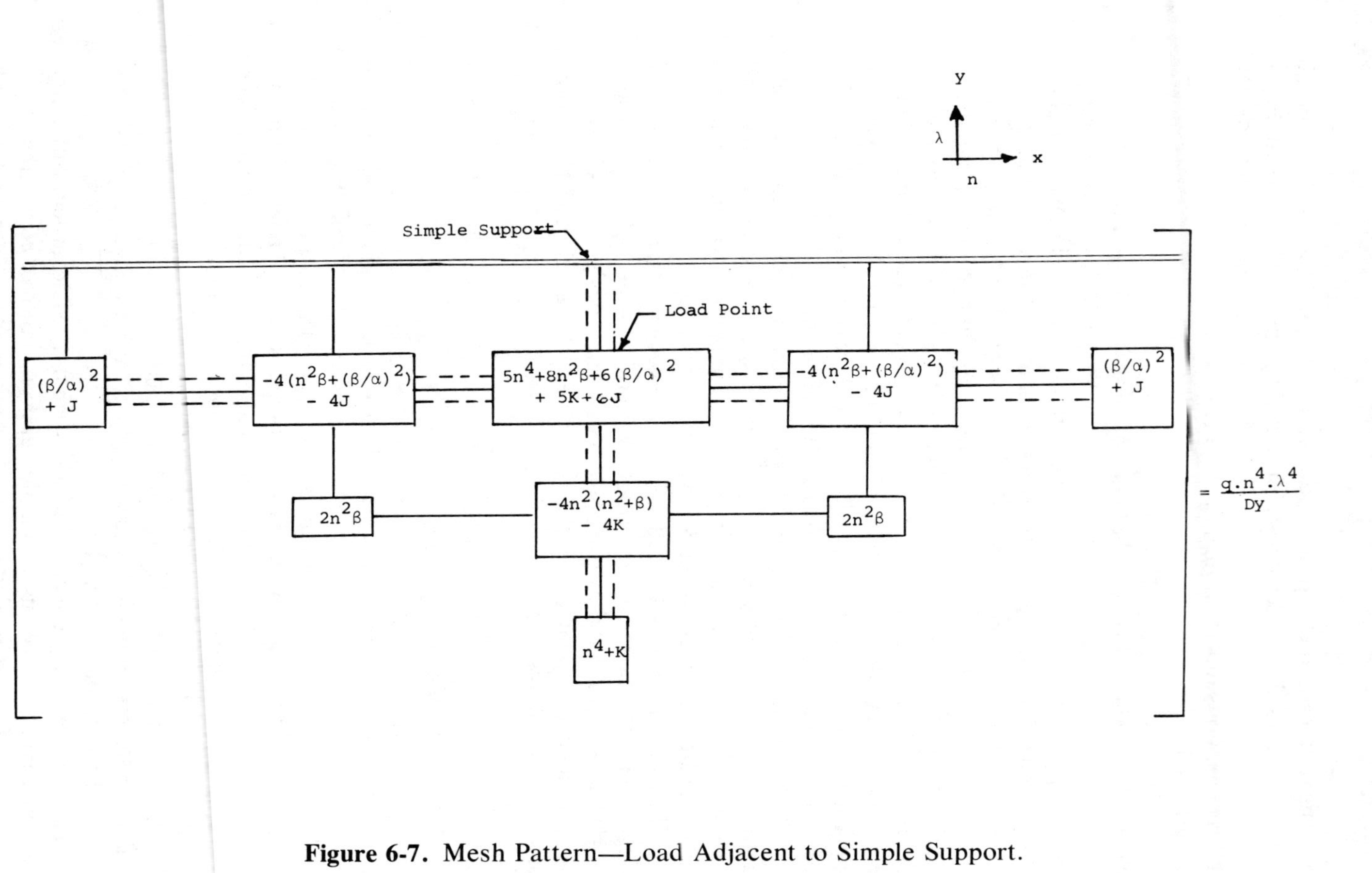

Figure 6-7. Mesh Pattern—Load Adjacent to Simple Support.

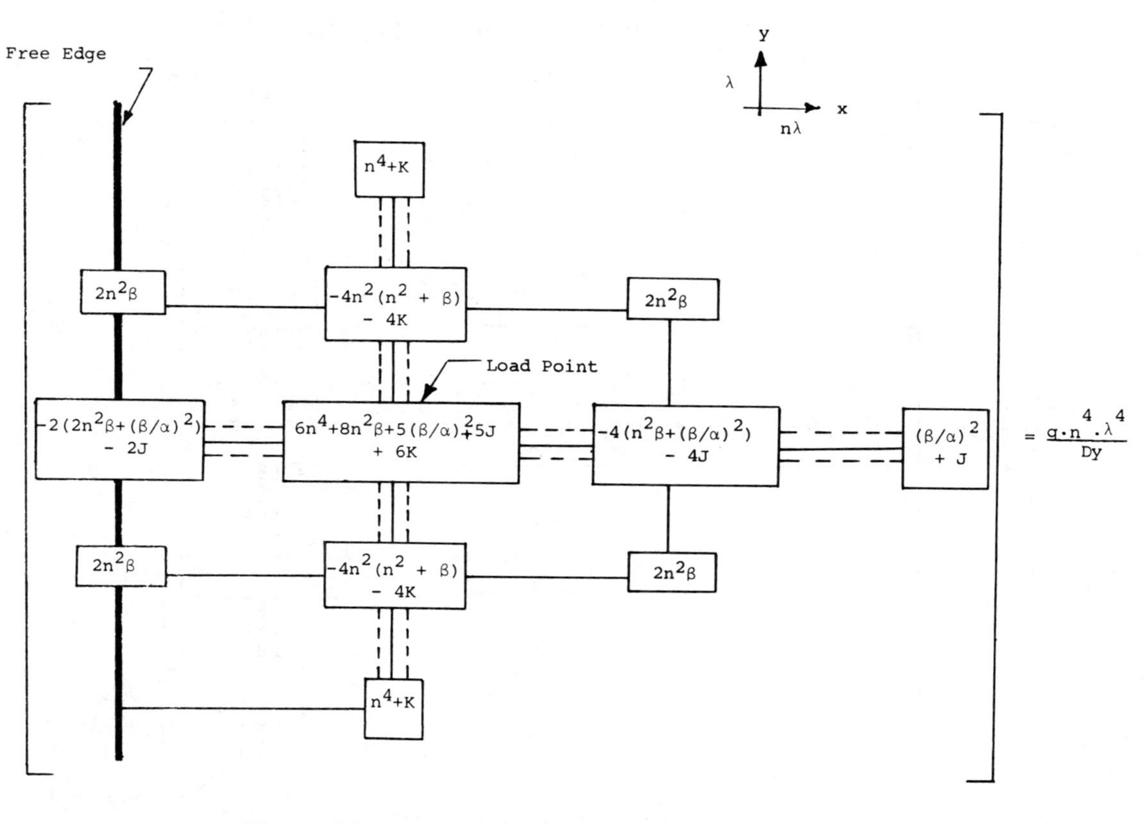

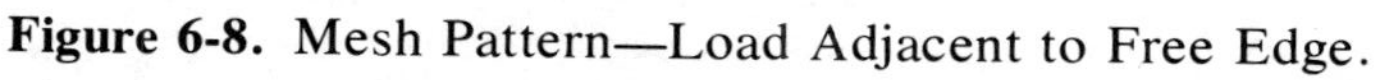

Figure 6-8. Mesh Pattern—Load Adjacent to Free Edge.

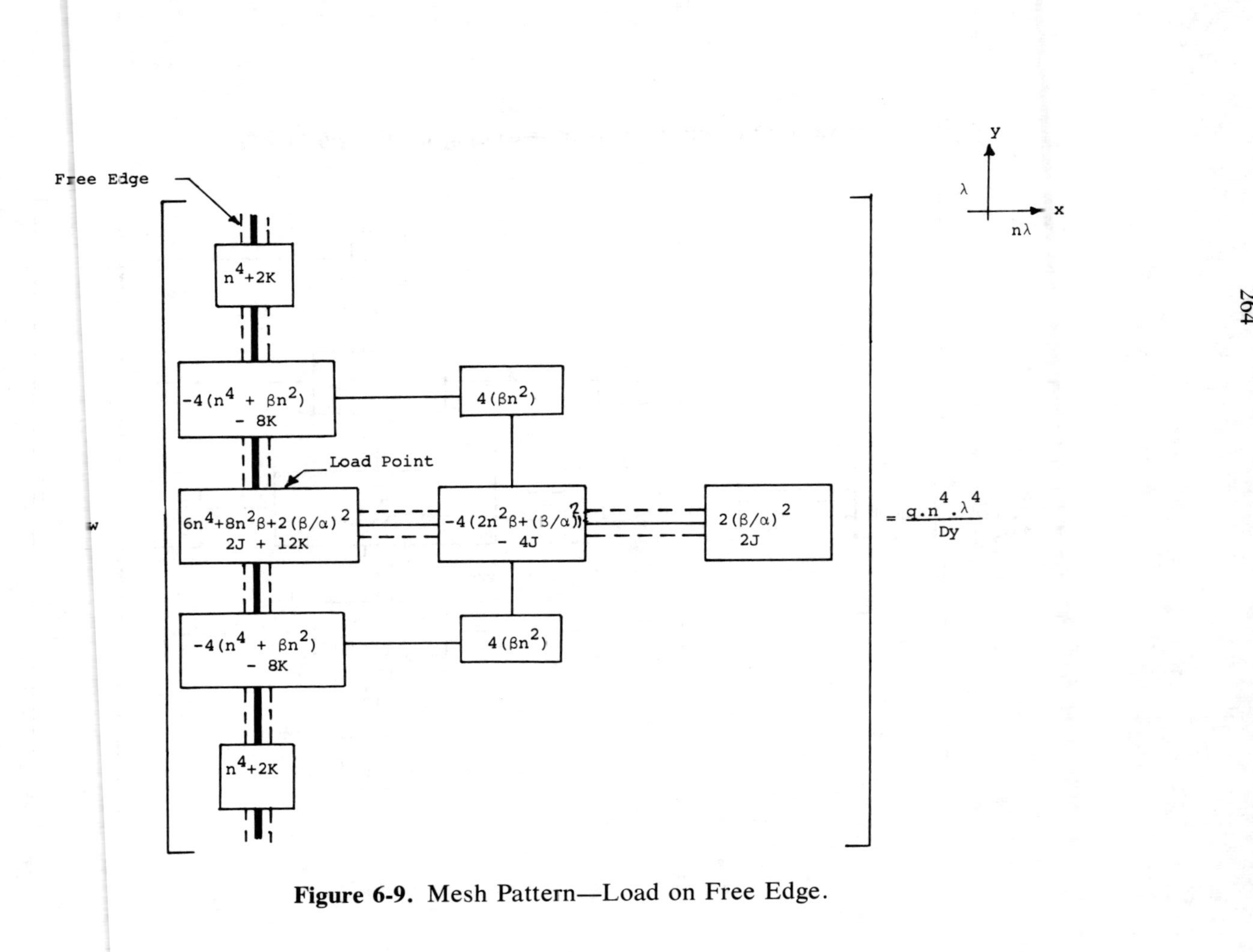

Figure 6-9. Mesh Pattern—Load on Free Edge.

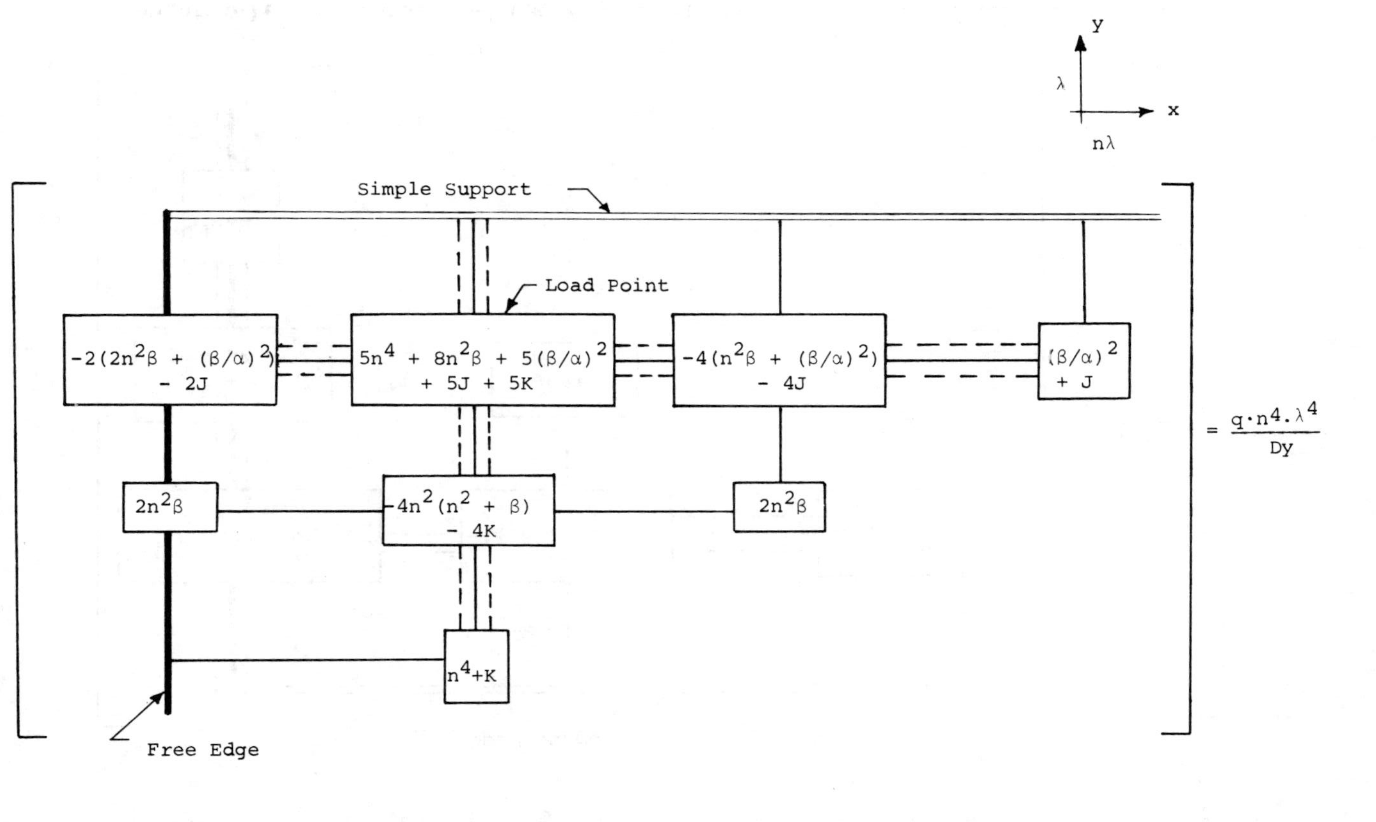

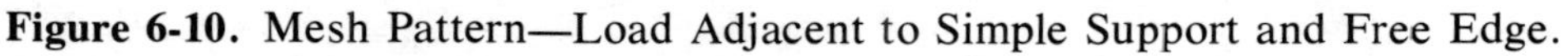

Figure 6-10. Mesh Pattern—Load Adjacent to Simple Support and Free Edge.

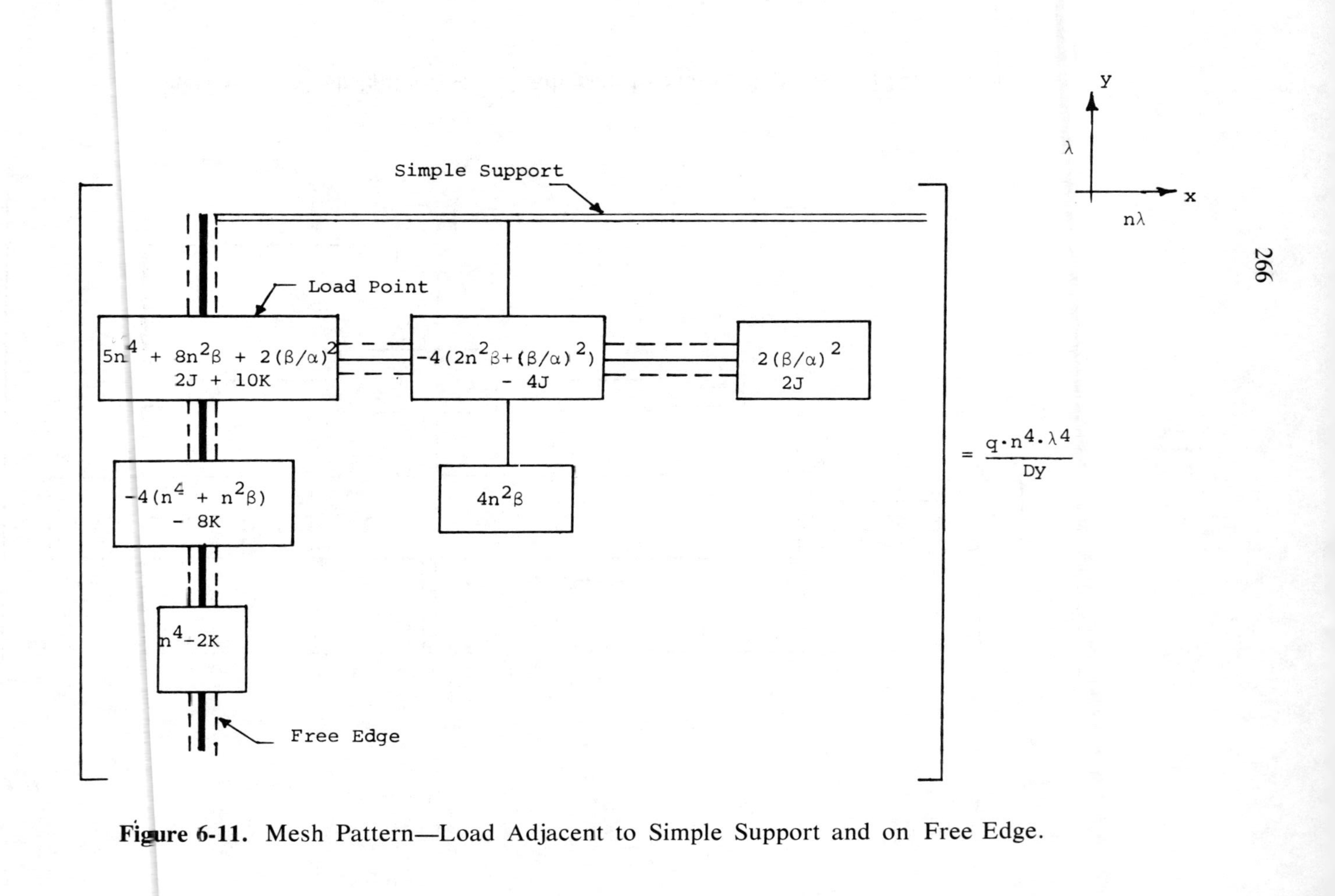

Figure 6-11. Mesh Pattern—Load Adjacent to Simple Support and on Free Edge.

the corresponding girder and plate stiffnesses and the loading, a series of equations may be generated. The solution of these equations gives the deflection at each preselected mesh point. Using this data and equations (5.4) or (5.5):

$$m_{x,y} = -\frac{EI}{(\text{spacing})^2} \cdot (w_{l_a} - 2w_0 + w_{r_b})$$

The corresponding moments may then be evaluated at the various mesh points. The computer program to perform the solution of the general difference equations is given in reference [39]. The application of this technique, relative to prediction of actual bridge response, has been excellent [35, 36].

6.2 Curved Orthotropic Bridges

Section 5.5 described the development of the curved orthotropic plate equation (4.85) in difference form as equation (5.30). In order to use this equation in the study of curved orthotropic bridges appropriate boundary conditions must be employed to modify the general equation given in Figure 5-8.

The specific boundary conditions that will be applicable are those associated with Figure 6-12. This figure shows a plate sector with simple supports and free boundaries. Thus the boundary conditions are

$$M_r = R_r = 0 \qquad \text{at} \qquad r = r_i, \qquad r = r_o$$

$$w = M_\theta = 0 \qquad \text{at } \theta = 0, \qquad \theta = \theta_T$$

where the difference equations (5.31), (5.32), and (5.33) are used.

With the bridge defined, the general plate equation is modified at fifteen locations described in Figure 6-13. Figures 6-14 through 6-18 show the specific equations for conditions $2L$, $3B$, $4BL$, $6B$, and $7BL$, respectively. The remaining nine conditions required for a complete solution are given in references [31, 32]. Also exterior girder stiffness may be included in the system, as given in reference [31].

With all boundaries defined, a system of equations may be generated that describes the load-deformation relationship of the entire plate. The solution of this system of equations will give the deflections at each mesh point. The generation of these equations and their solution are accomplished by a computer program. Using the deflection output data, the internal bending moments are evaluated, as are the torsional or twisting moments. The evaluation of these moments is performed by a computer program [32].

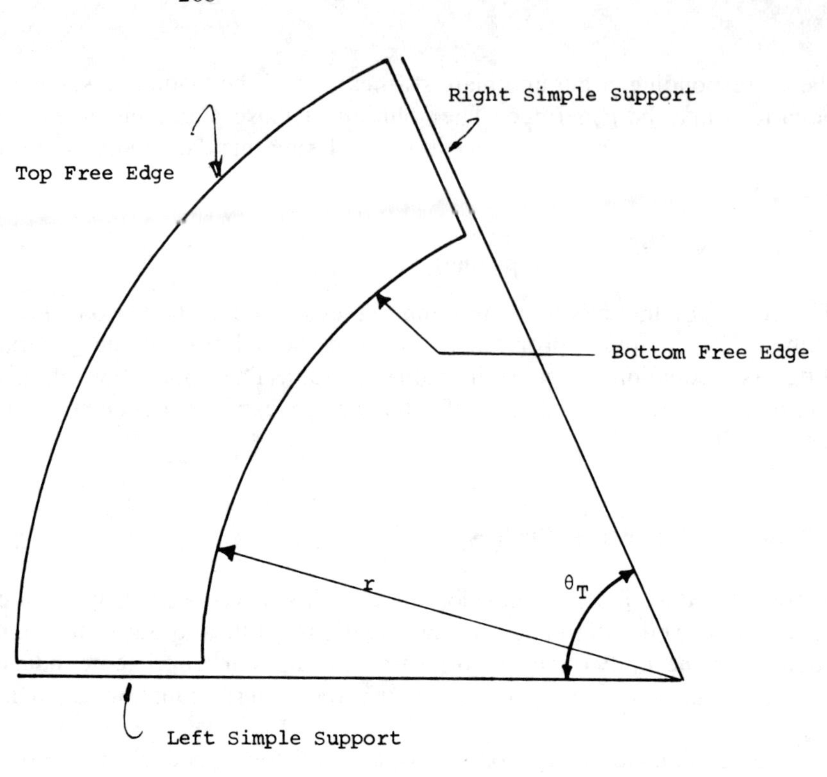

Figure 6-12. Circular Plate Segment Boundaries.

6.3 Bridge Grid System

The difference representation of the plate equation that represents intersecting beams is given by equation (5.38). The force equations given in section 4.9 have also been written in difference form [34]. These relationships have been used to modify the general plate equation (5.38) to accommodate the various boundary conditions for a bridge as described in Figure 6-19. This bridge is simply supported at two ends and has two free supports. Six general finite difference mesh patterns are required to accommodate this model. The following will describe the required modifications and resulting mesh patterns.

Mesh Condition 1—General Interior Load Point
The general mesh pattern for any load point completely within the boundaries is described in Figure 5-11 and is equation (5.38), written in mesh pattern form.

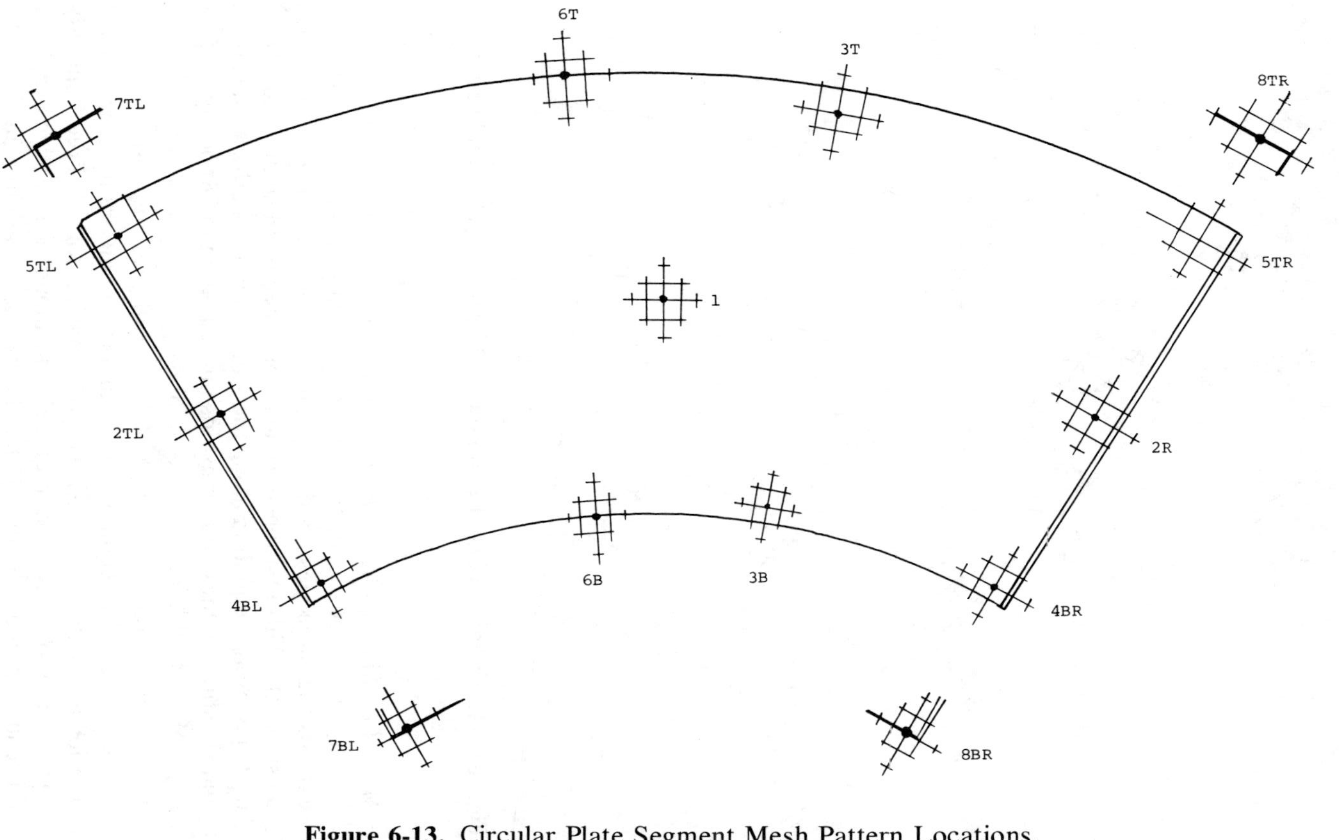

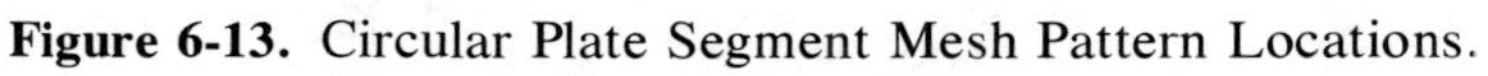

Figure 6-13. Circular Plate Segment Mesh Pattern Locations.

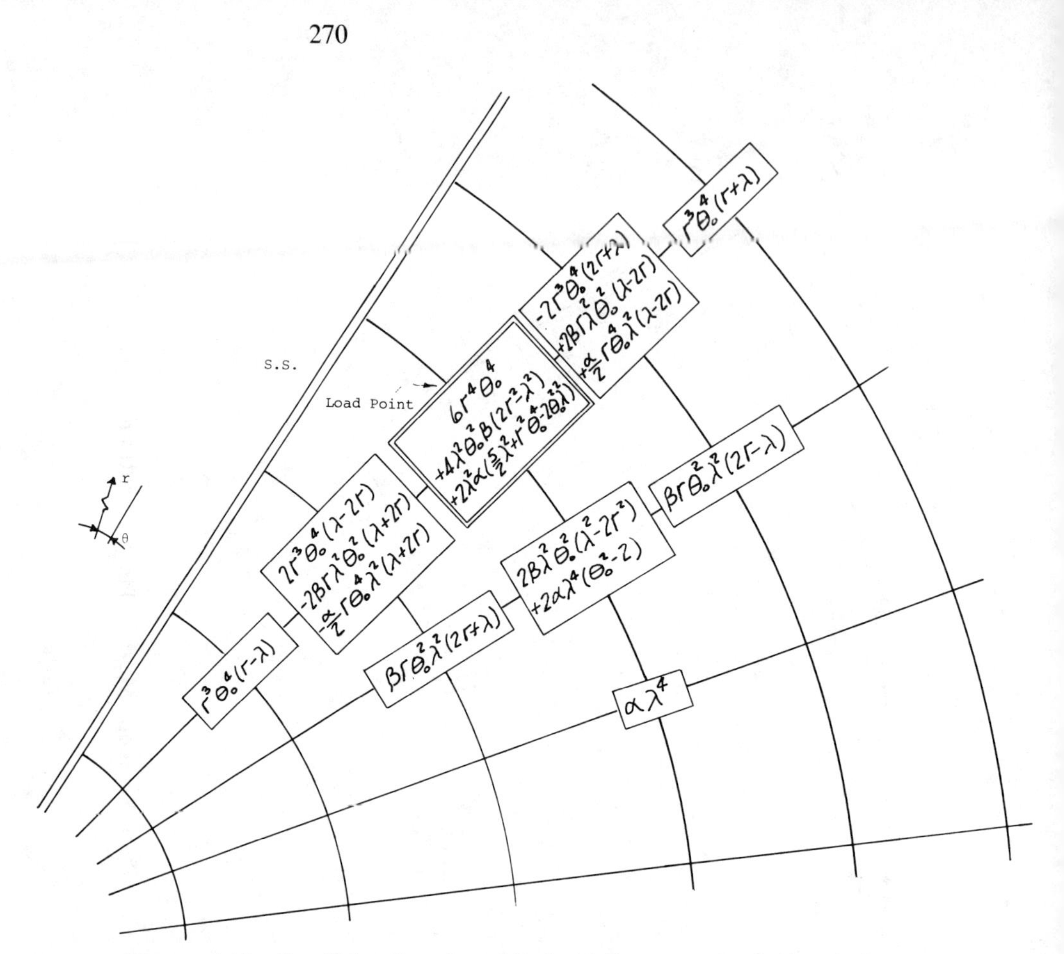

Figure 6-14. Condition 2*L*—Load Point Adjacent to Left Simple Support.

Mesh Condition 2—Load Point Adjacent to Simple Support
This type of loading requires three boundary conditions in order to express the three mesh points beyond the boundary relative to known mesh points. This relationship can be obtained by the expression $M_y = 0$. Applying this equation along the simple supports at three locations and substituting these relationships into equation (5.38) or Figure 5-11 gives Mesh Condition 2, Figure 6-20.

Mesh Condition 3—Load Point Adjacent to Free Support
This condition also requires three boundary conditions in order to relate interior and exterior points. The relationship for all three boundary conditions is $M_x = 0$ at each required point along the free edge. Applying this equation and substituting the results into equation (5.38) gives the final Mesh Condition 3, Figure 6-21.

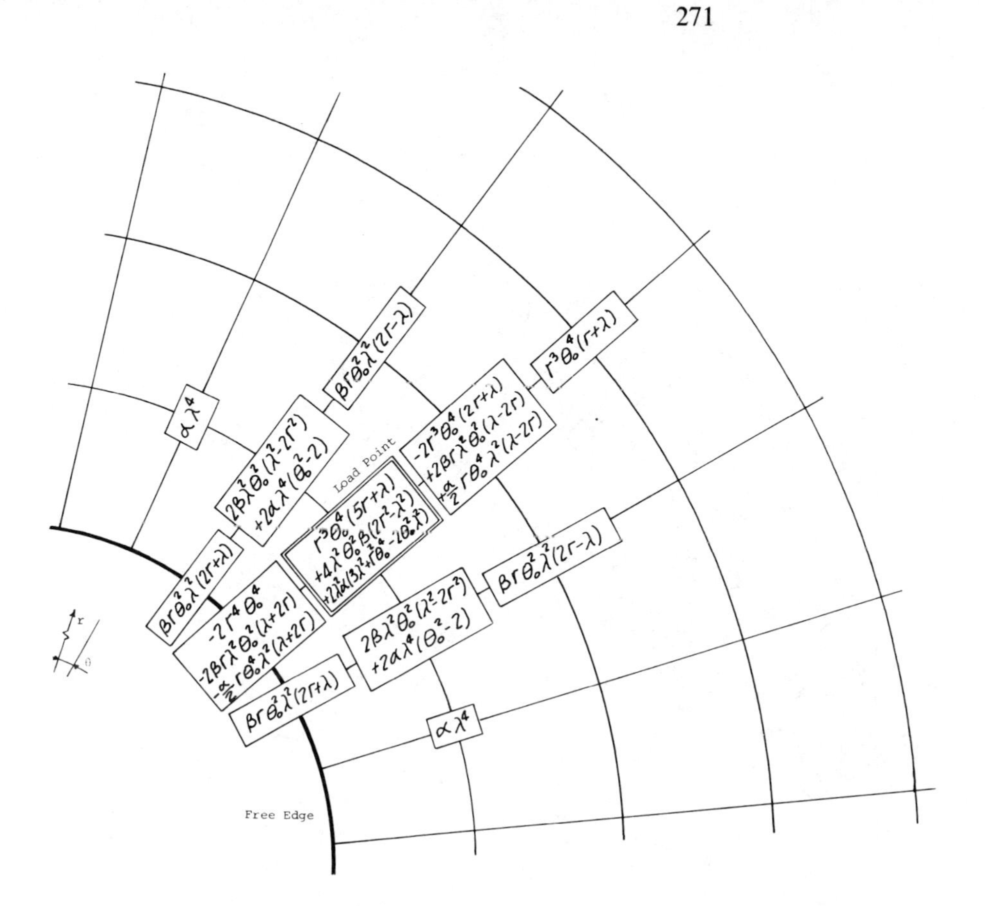

Figure 6-15. Condition 3*B*—Load Point Adjacent to Bottom Free Edge.

Mesh Condition 4—Load Point on a Free Support
This loading condition requires eight boundary conditions. The boundary conditions that were used are:

(1) $M_x = 0$ for five mesh points

(2) $R_x = 0$ for one mesh point

(3) $\left(\dfrac{\partial w}{\partial x}\right)_1 = \dfrac{1}{2}\left[\left(\dfrac{\partial w}{\partial x}\right)_{al} + \left(\dfrac{\partial w}{\partial x}\right)_{bl}\right]$ one mesh point

(4) $\left(\dfrac{\partial w}{\partial y}\right)_1 = \dfrac{1}{2}\left[\left(\dfrac{\partial w}{\partial y}\right)_{o} + \left(\dfrac{\partial w}{\partial y}\right)_{ll}\right]$ one mesh point

The first two types of boundary conditions, for a free edge, are self-explanatory. The last two conditions are relationships relating the slope

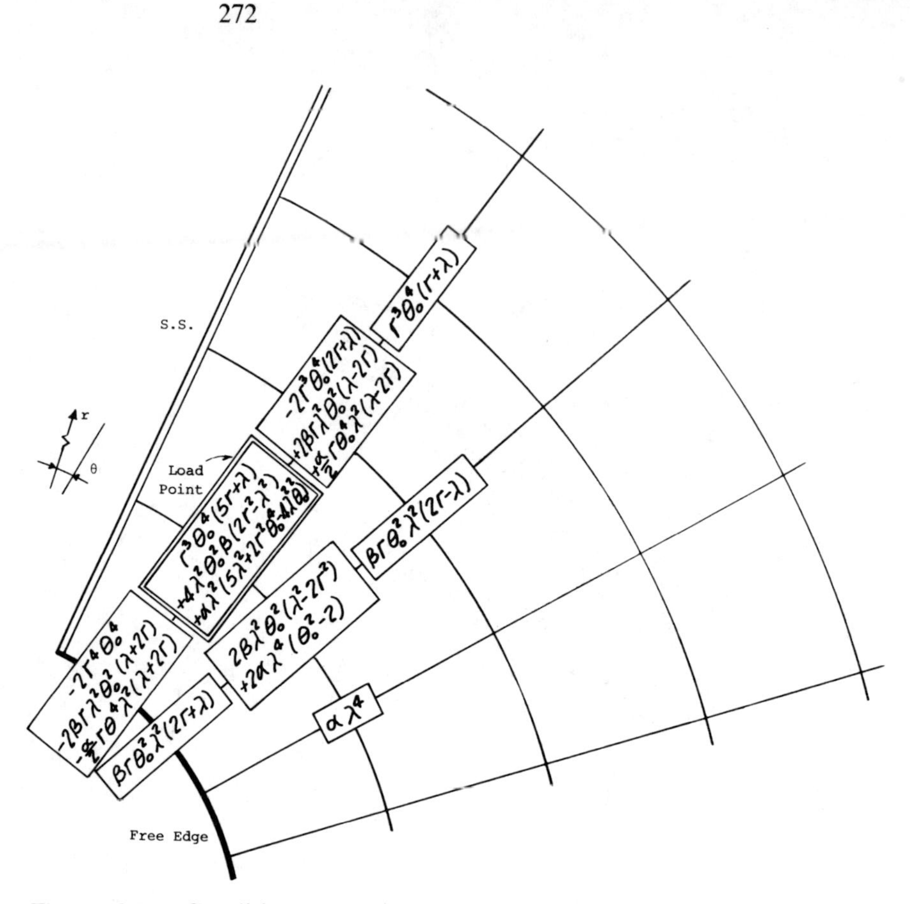

Figure 6-16. Condition 4*BL*—Load Point Adjacent to Left Simple Support and Free Edges.

at a given point to the sum of the average slopes surrounding that point. These slope equations were required in order to express *all* exterior points relative to the interior points. The results from all of the eight boundary conditions and their substitution gives Mesh Condition 4, Figure 6-22.

Mesh Condition 5—Load Point Adjacent to a Simple and Free Support
This load condition requires six boundary conditions. As utilized in Mesh Conditions 2 and 3, the moments $M_y = 0$ along the simple support and $M_n = 0$ along the free support are used. The solution of these conditions, when substituted into equation (5.38), results in mesh Figure 6-23.

Mesh Condition 6—Load Point Adjacent to Simple Support on Free Support
There are ten boundary conditions that are required for this loading condition. The required relationships are:

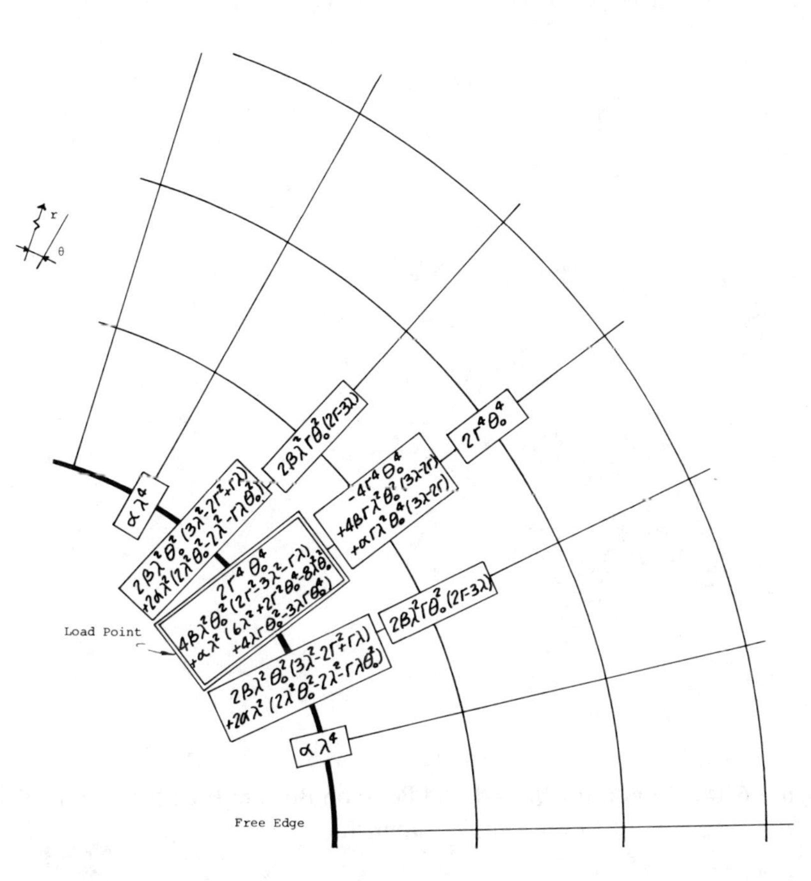

Figure 6-17. Condition 6*B*—Load Point on Bottom Free Edge.

(1) $M_y = 0$ at two points

(2) $M_x = 0$ at four points

(3) $$\left(\frac{\partial w}{\partial y}\right)_a = \frac{1}{2}\left[\left(\frac{\partial w}{\partial y}\right)_{ar} + \left(\frac{\partial w}{\partial y}\right)_{al}\right]$$

(4) $$\left(\frac{\partial w}{\partial y}\right)_l = \frac{1}{2}\left[\left(\frac{\partial w}{\partial y}\right)_0 + \left(\frac{\partial w}{\partial y}\right)_{ll}\right]$$

(5) $$\left(\frac{\partial w}{\partial x}\right)_l = \frac{1}{2}\left[\left(\frac{\partial w}{\partial x}\right)_{al} + \left(\frac{\partial w}{\partial x}\right)_{bl}\right]$$

The solution of these nine equations and their relationships results in Mesh Condition 6, Figure 6-24.

With all of the required mesh patterns defined and developed, the solution of any given loaded bridge grid system can be solved. The genera-

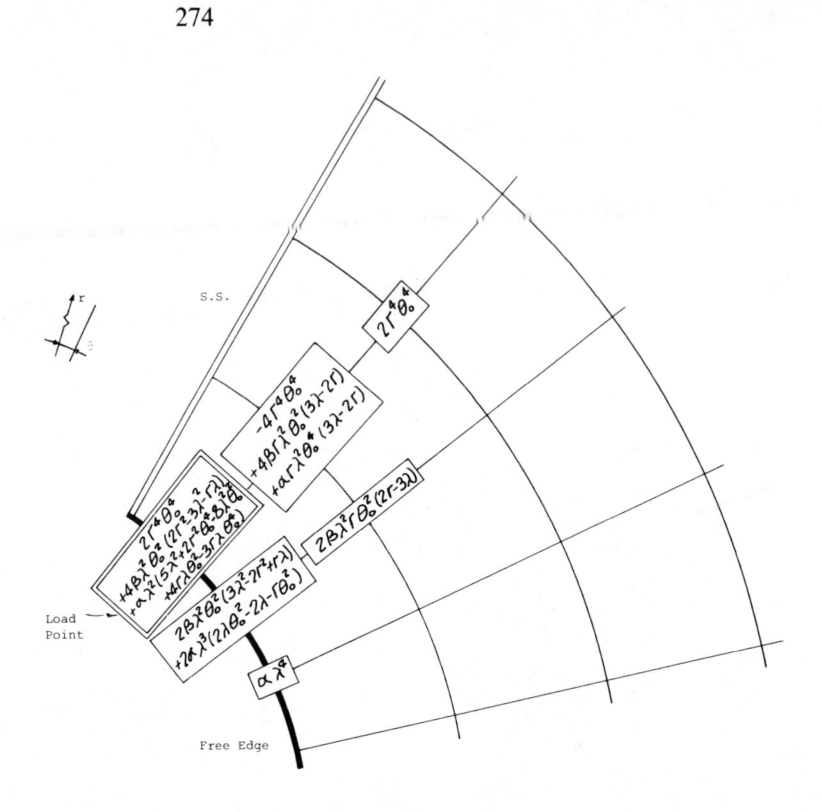

Figure 6-18. Condition *7BL*—Load Point on Bottom Free Edge and Adjacent to Left Simple Support.

tion of the required equations and their solutions has been accomplished by a computer program. The evaluation of the deflection at each grid point will then permit evaluation of each force at that grid point. Equations (4.73) and (4.77) through (4.80) have been derived in finite difference form, and the forces then evaluated by a computer program [34]. The two basic computer programs that have been written will evaluate deflections and then forces. The results from these equations and programs have been presented in the analysis of a hypothetical bridge [34].

6.4 Building Grid

Many times the structural engineer is required to design a grid framework. The analysis of such a system is often performed by a computer-oriented

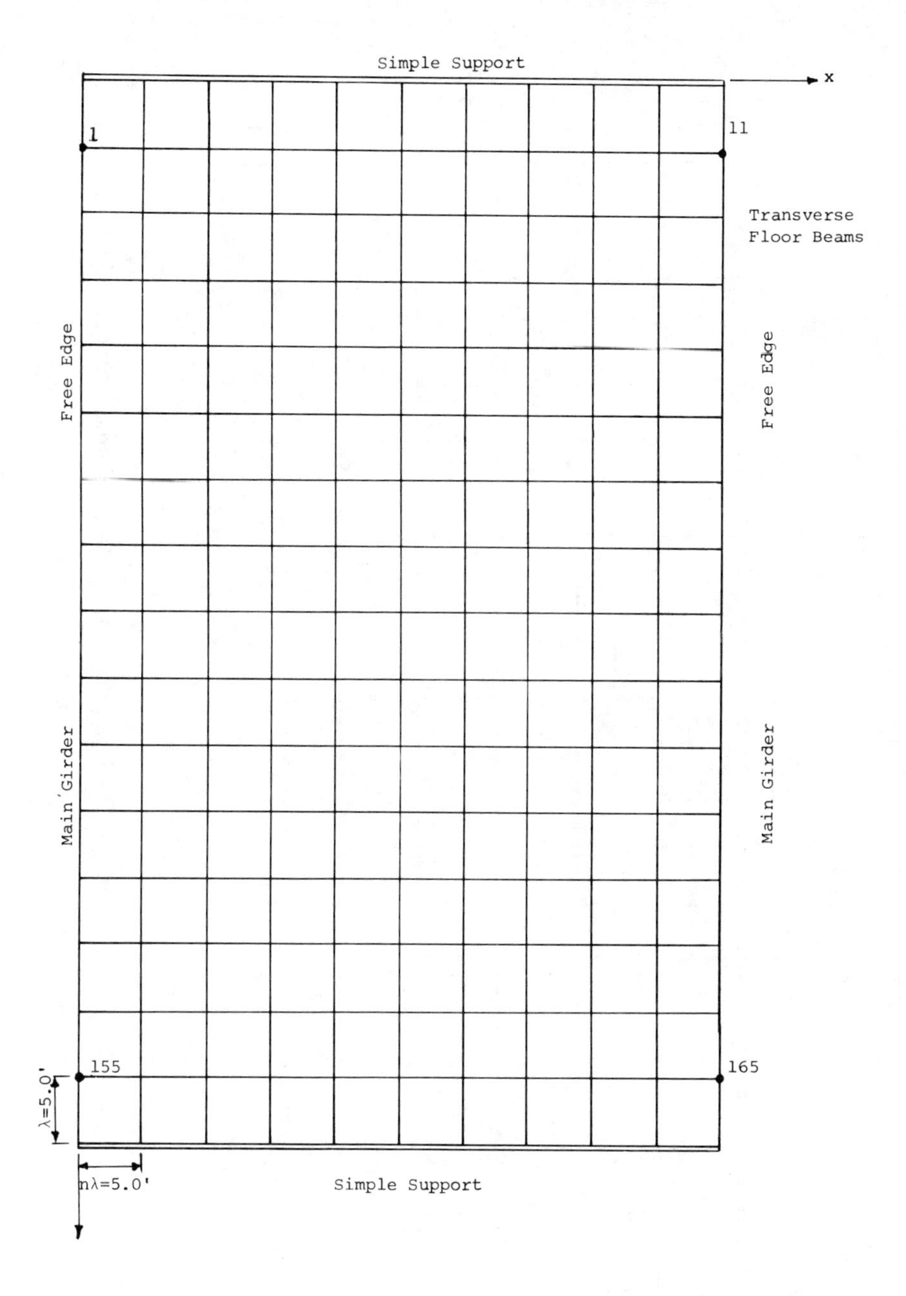

Figure 6-19. Bridge Idealization as a Grid.

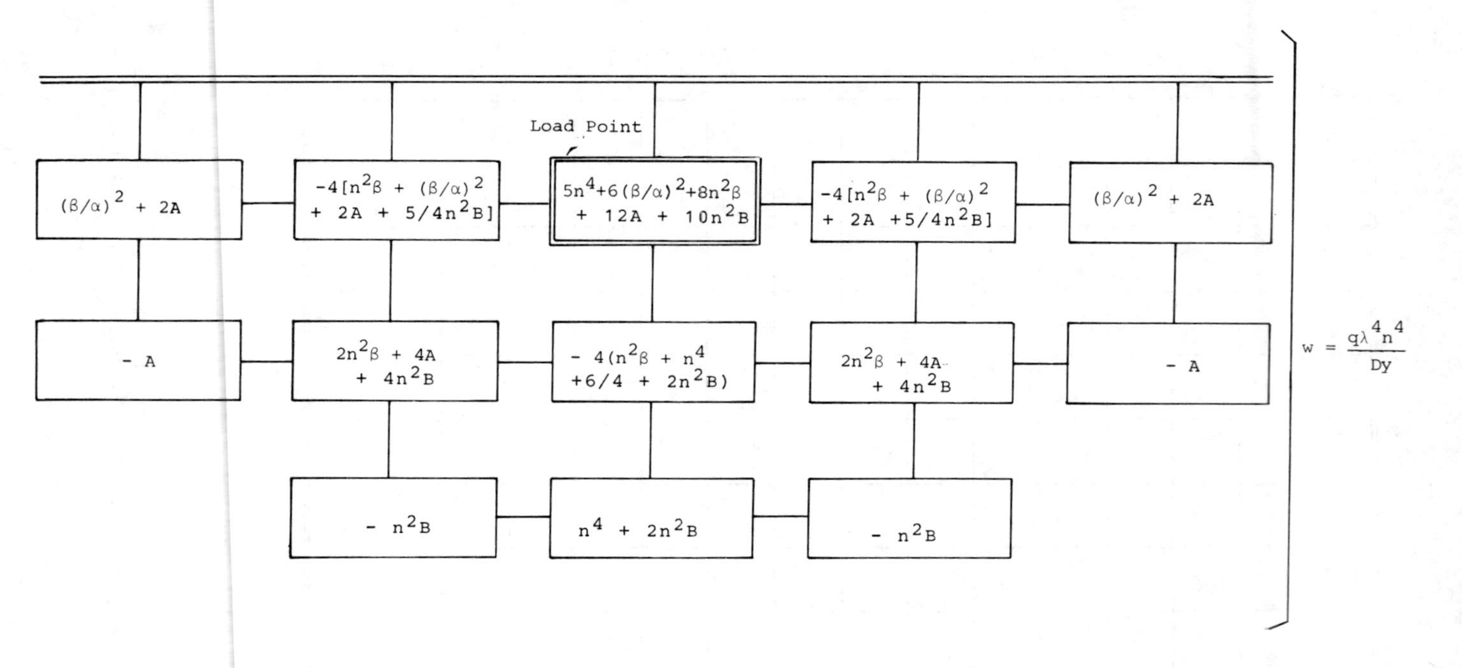

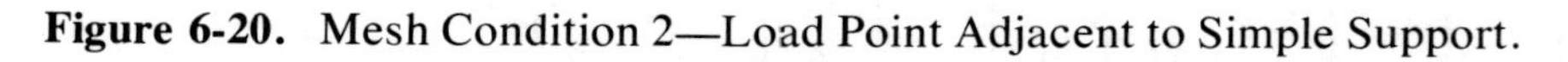

Figure 6-20. Mesh Condition 2—Load Point Adjacent to Simple Support.

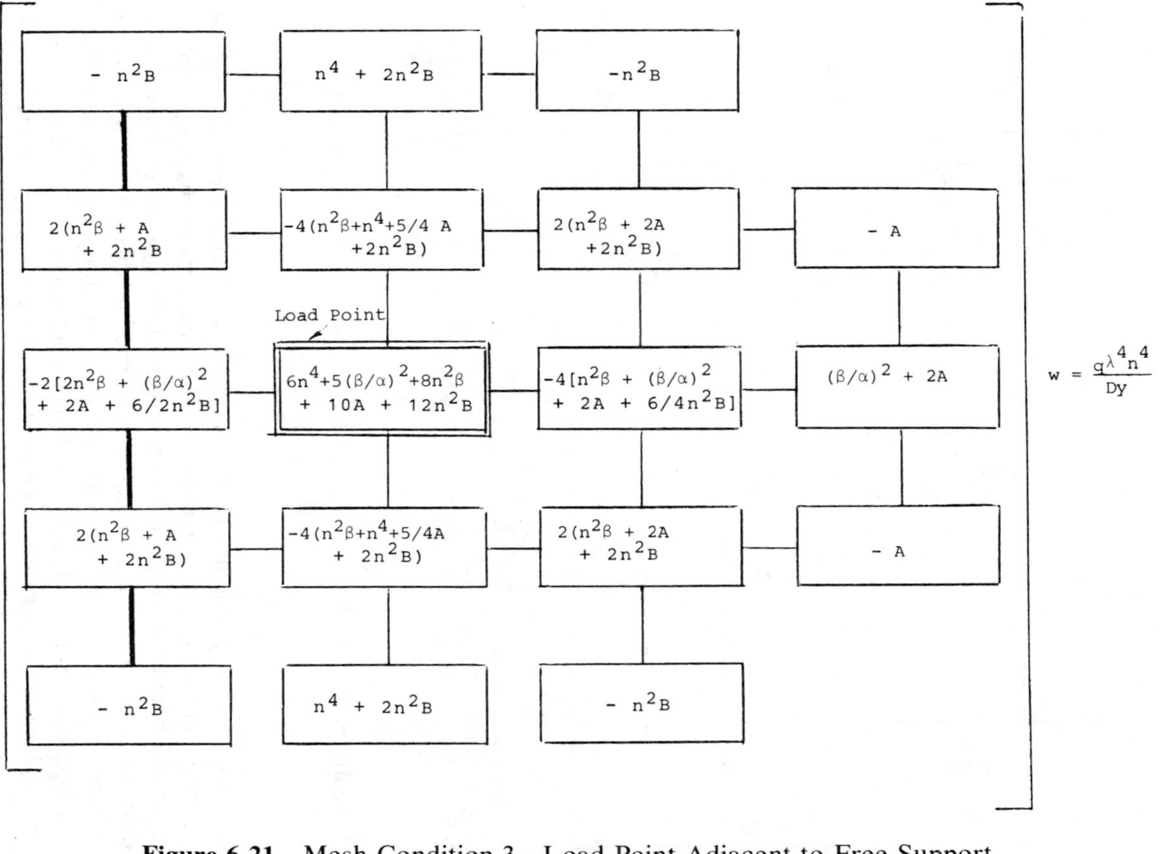

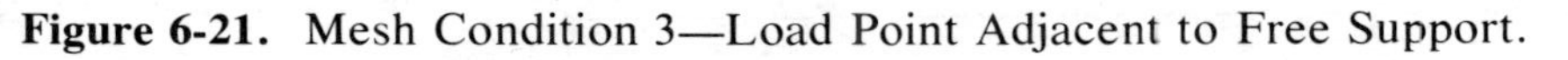

Figure 6-21. Mesh Condition 3—Load Point Adjacent to Free Support.

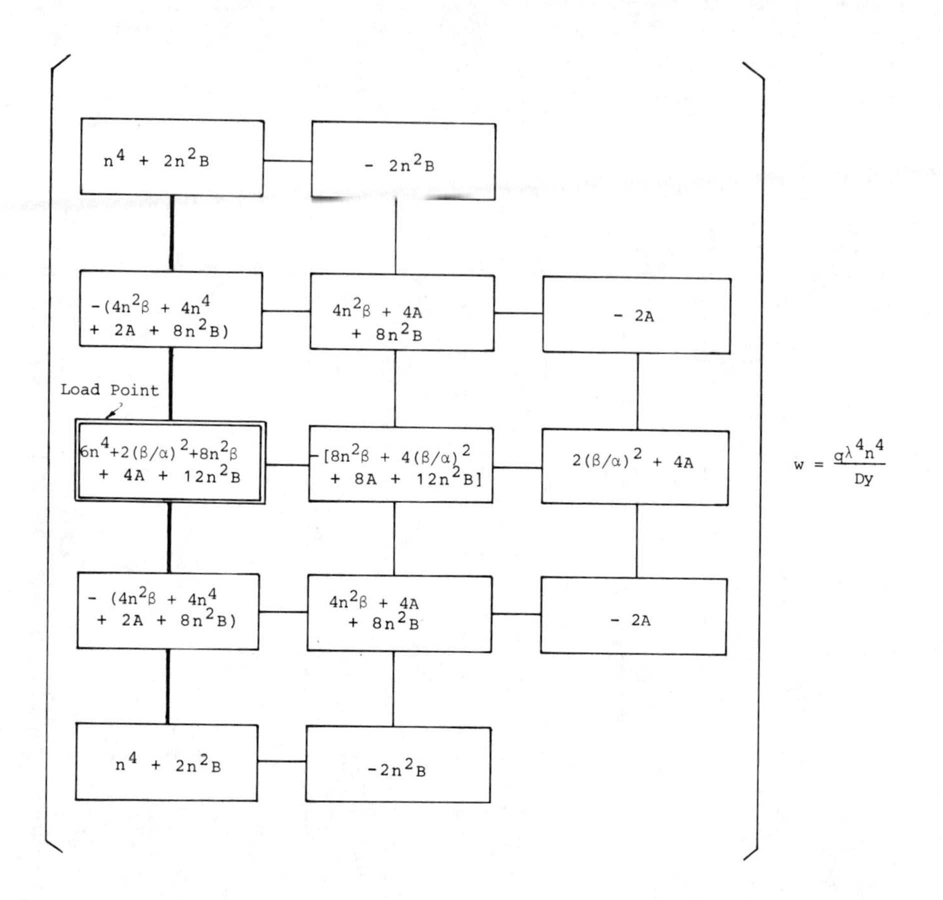

Figure 6-22. Mesh Condition 4—Load Point on a Free Support.

stiffness matrix technique. However, the use of such methods requires extensive computer programming, or, if existing programs are available, extensive computer time and storage. It is, therefore, often judicious to develop a system of linear equations governed by difference relationships to represent the loaded grid. The system of equations can be solved by a small desk computer and requires minimal analysis time. The development and application of such difference equations, as given in part in section 5.3, will be demonstrated by the design and analysis of a steel grid restaurant and lounge, as shown in Figures 6-25 and 6-26 [37]. The technique and results will now be given.

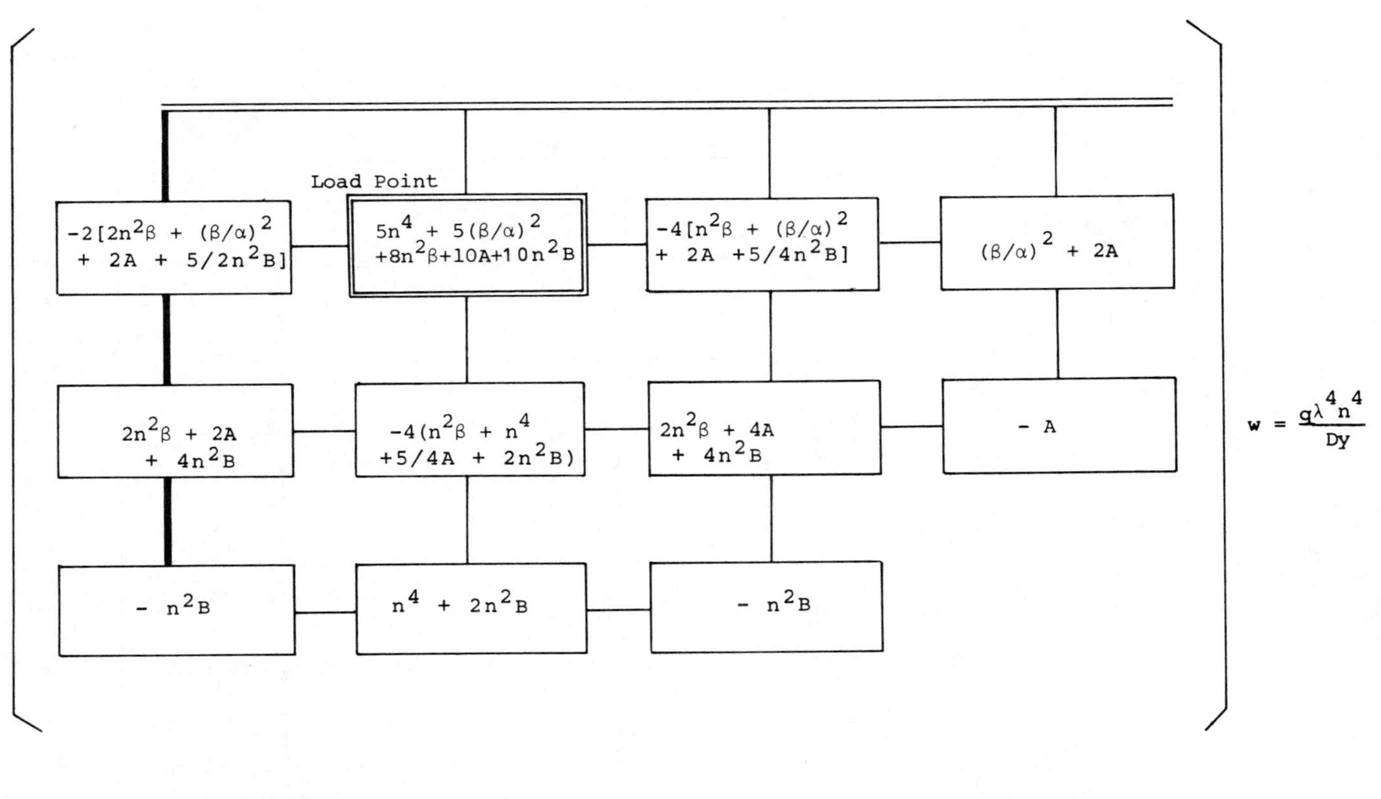

Figure 6-23. Mesh Condition 5—Load Point Adjacent to Simple and Free Support.

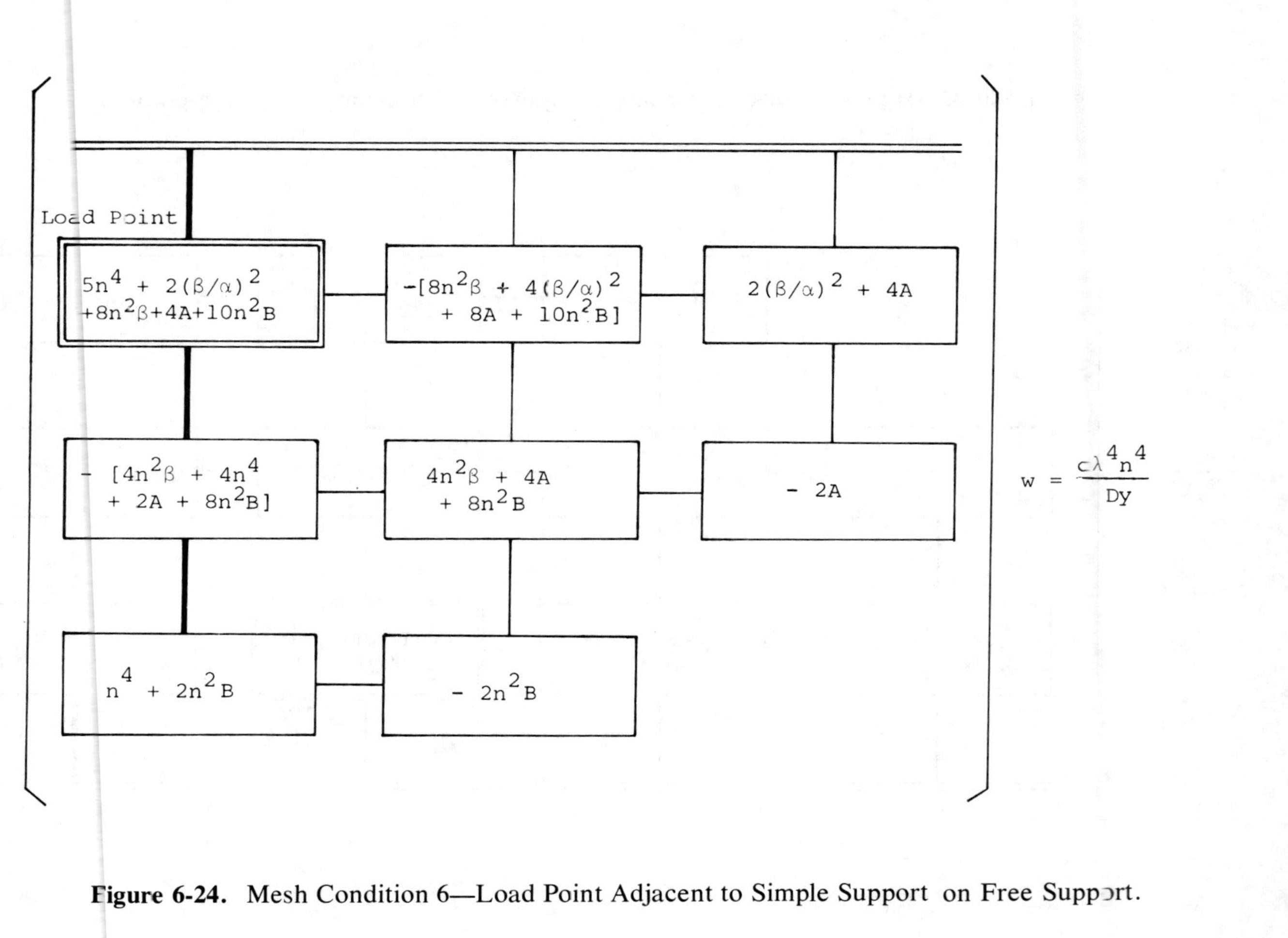

Figure 6-24. Mesh Condition 6—Load Point Adjacent to Simple Support on Free Support.

Figure 6-25. Building System—Interior Girder Grid.

Figure 6-26. Exterior Columns.

6.4.1 *Basic Equations*

As will be recalled, the load-deformation relationship of a girder subjected to a load P in pounds per foot is given by the conventional differential equation:

$$\frac{d^4w}{dx^4} = \frac{P_x}{EI_x} \tag{6.50}$$

The shear and moment relationships are

$$\frac{d^3w}{dx^3} = \frac{V_x}{EI_x} \tag{6.51}$$

$$\frac{d^2w}{dx^2} = \frac{-M_x}{EI_x} \tag{6.52}$$

These equations can be transformed into central difference form, as given in part in section 5.2, using the relationships:

$$\frac{d^4w}{dx^4} = (w_{n-2} - 4w_{n-1} + 6w_n - 4w_{n+1} + w_{n+2})/\lambda^4 \tag{6.53}$$

$$\frac{d^3w}{dx^4} = (-w_{n-2} + 2w_{n-1} - 2w_{n+1} + w_{n-2})/2\lambda^3 \tag{6.54}$$

$$\frac{d^2w}{dx^2} = (w_{n-1} - 2w_n + w_{n+1})/\lambda^2 \tag{6.55}$$

Substituting equations (6.53) through (6.55) into equations (6.50) through (6.52) gives

$$P_x = \frac{EI_x}{\lambda^4}(w_{n-2} - 4w_{n-1} + 6w_n - 4w_{n+1} + w_{n+2}) \tag{6.56}$$

$$V_x = \frac{EI_x}{2\lambda^3}(-w_{n-2} + 2w_{n-1} - 2w_{n+1} + w_{n+2}) \tag{6.57}$$

$$M_x = -\frac{EI_x}{\lambda^2}(w_{n-1} - 2w_n + w_{n+1}) \tag{6.58}$$

where the integer n represents a given mesh point surrounded by adjacent points spaced at distances λ, as shown in Figure 6-27. Considering two intersecting beams, as shown in Figure 6-28, and assuming that the deflection at midpoint (n,m) is the same, the total applied load at the intersection can be equated to the resistance of each individual girder:

$$P_x + P_y = P_T$$

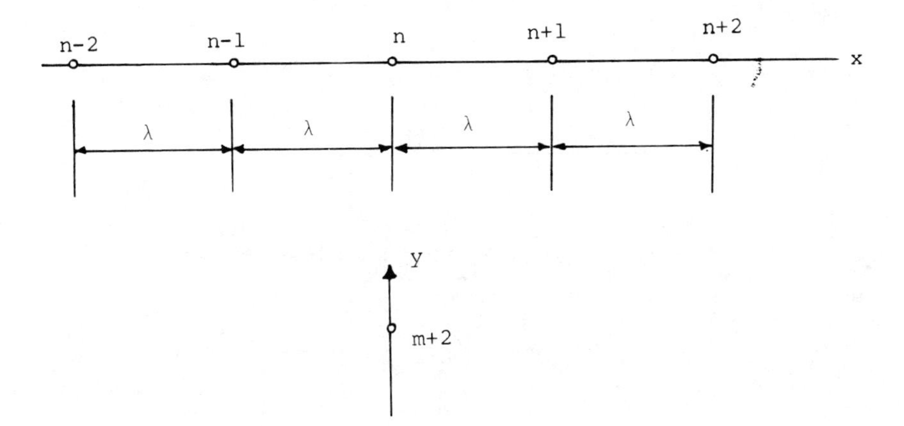

Figure 6-27. General Mesh Layout.

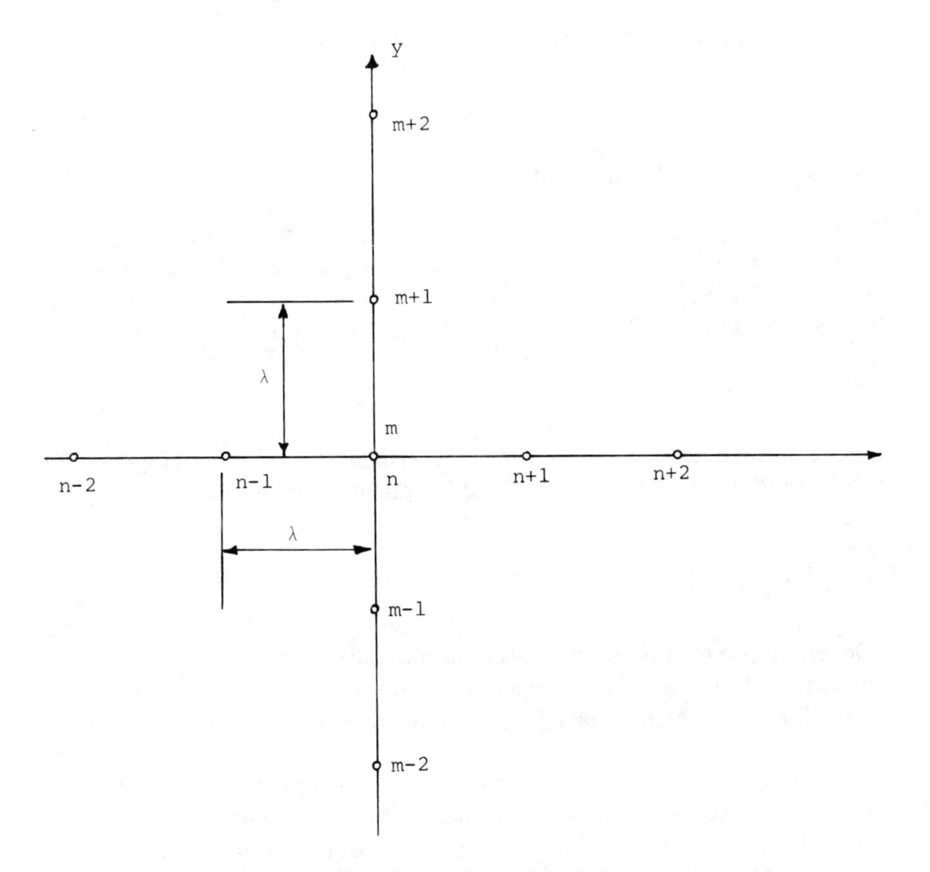

Figure 6-28. Two Girder Mesh Layout.

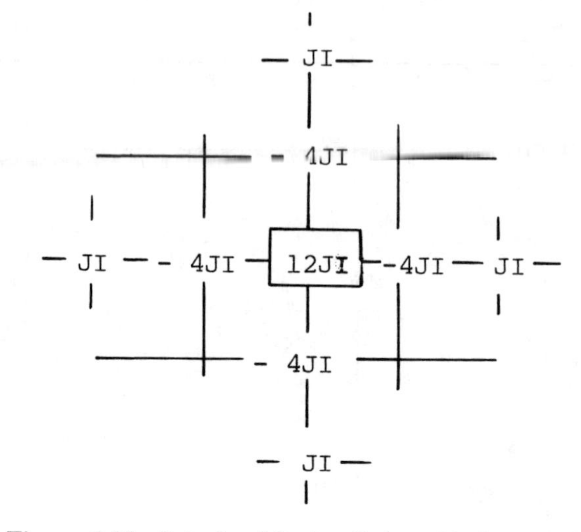

Figure 6-29. Interior Mesh—Points 11 through 20.

or

$$\frac{EI_x}{\lambda^4}(w_{n-2} - 4w_{n-1} + 6w_n - 4w_{n+1} + w_{n+2})$$

$$+ \frac{EI_y}{\lambda^4}(w_{m-2} - 4w_{m-1} + 6w_m - 4w_{m+1} + w_{m+2}) = q_T \quad (6.59a)$$

or, assuming $EI_x = EI_y = JI$, the equation can be represented in mesh form as:

$$JI[\ ] \cdot w = P \cdot \lambda^4 \quad (6.59b)$$

where the bracket [ ] is represented by coefficients, as shown in Figure 6-29. The term $E = 30 \times 10^3$ ksi, and (I_x, I_y) represents the moment of inertia of the girders.

6.4.2 Modification of Equations

Depending on the grid boundary conditions, the basic equation (6.59) must be modified. The boundaries may be simple supports ($w = M = 0$). Therefore, for the particular problem, equations (6.50), (6.51) or $w' = (w_{n+1} - w_{n-1})/2\lambda$ may be applied.

The structure that was analyzed using the difference method is shown in Figure 6-30. The structure is to be analyzed under a uniform dead and live load, therefore, only one eighth of the structure need be analyzed, as shown in Figure 6-31. The structure is supported on six fixed columns, and has an

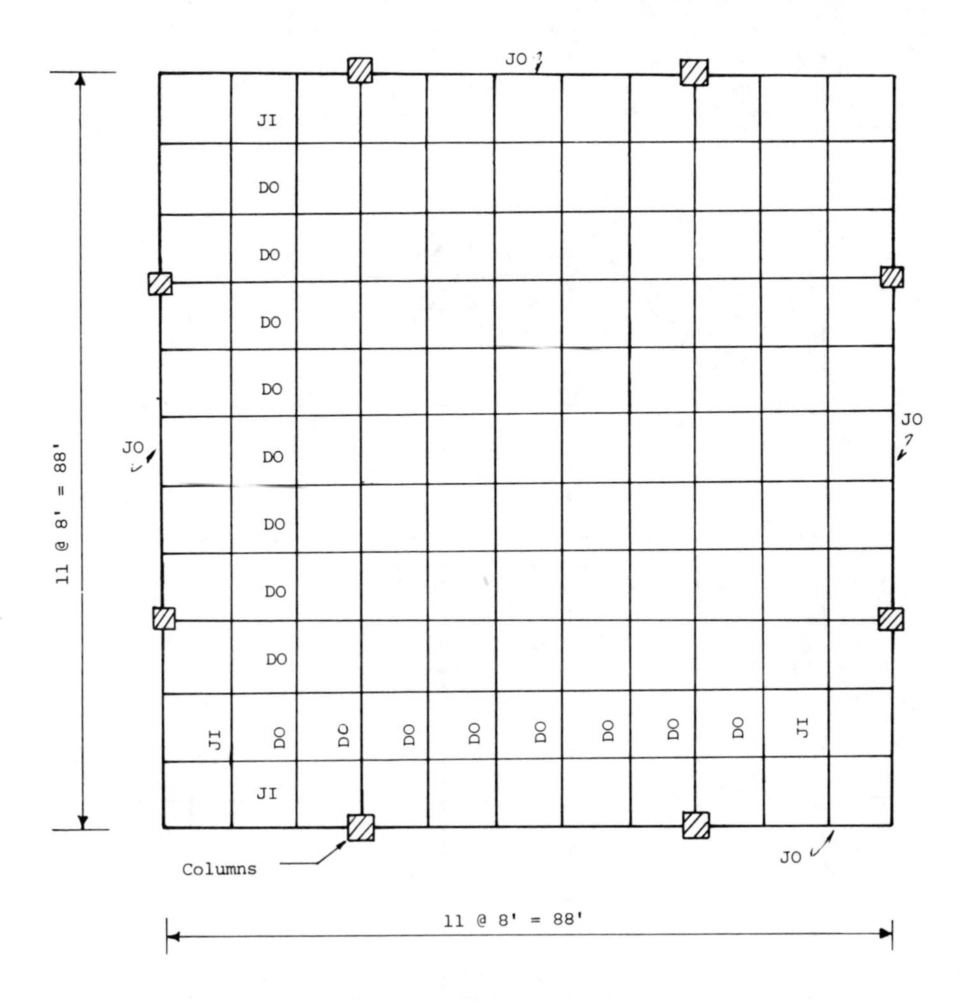

Figure 6-30. General Building Plan.

outside girder of stiffness *JO* and ten interior cross girders *JI* at spacings of 11.0 feet. The preliminary designs indicated that the outside girders were to be fabricated into steel w shapes (F_y = 36.0 ksi) with a 5/16-in web, 8 × 5/8-in flange plates, and total depth of 51.0 in. The bending stiffness of the girder is I = 9520.0 in^4. The inside cross girders had 5/16-in web plate, 8 × 5/8-in flange plates and a total depth of 42.0 in. The bending stiffness was computed as I = 6010.0 in^4.

In order properly to represent the load-deformation response of each mesh point of the system, as shown in Figure 6-31, the general mesh pattern, Figure 6-29, must be modified accordingly:

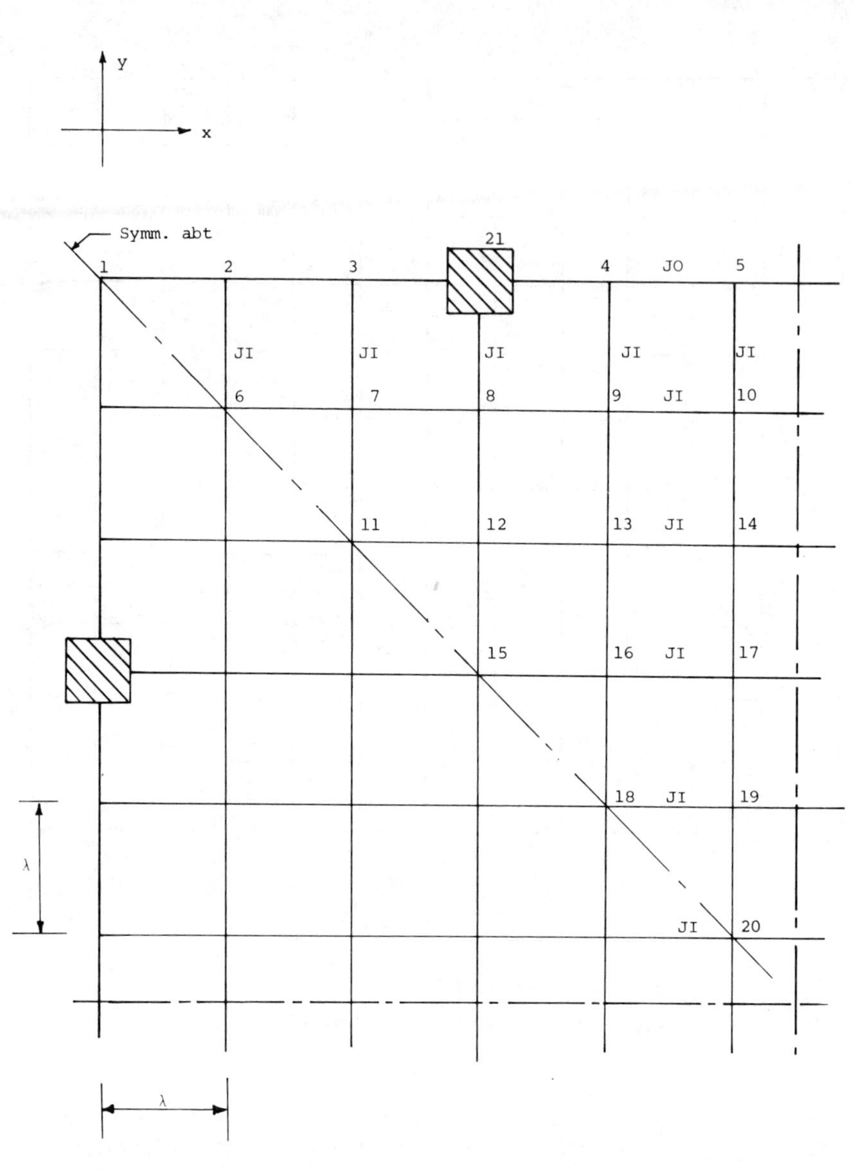

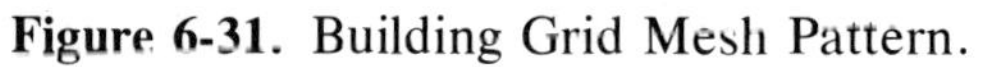
Figure 6-31. Building Grid Mesh Pattern.

Load Point 1: Positioning the general pattern, Figure 6-29, over the mesh point 1 indicates that deflection points $(n - 2)$, $(n - 1)$, $(m + 1)$, and $(m + 2)$ are beyond the boundary. These points can be related to the interior mesh points by applying the boundary condition $M = V = 0$, as given by

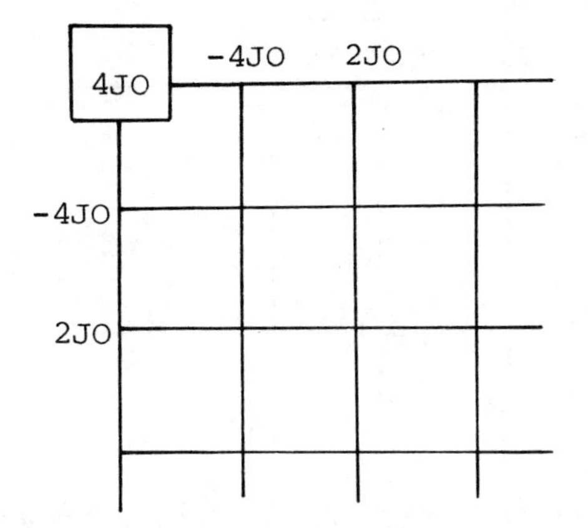

Figure 6-32. Corner Mesh Pattern Point 1.

equations (6.57) and (6.58). Consider first the modification of the girder spanning in the x direction.

$$M = 0 : \qquad w_{n-1} = 2w_n - w_{n+1} \tag{6.60}$$

$$V = 0 : \qquad w_{n-2} = -2w_{n+1} + 2w_{n-1} + w_{n+2} \tag{6.61a}$$

Substituting equation (6.60) into equation (6.61a) gives

$$w_{n-2} = -2w_{n+1} + 4w_n - 2w_{n+1} + w_{n+2}$$

or

$$w_{n-2} = +4w_n - 4w_{n+1} + w_{n+2} \tag{6.61b}$$

where n = point 1, $n + 1$ = point 2, $n + 2$ = point 3.

Substituting equations (6.60) and (6.61b) into the general load equation (6.56) gives

$$P \cdot \lambda^4 = EI_x(+4w_n - 4w_{n+1} + w_{n+2} - 8w_n + 4w_{n+1} + 6w_n - 4w_{n+1} + w_{n+2})$$

$$P \cdot \lambda^4 = EI_x(+2w_n - 4w_{n+1} + 2w_{n+2}) \tag{6.62a}$$

Letting $JO = EI_x$, equation (6.62a) can be written in mesh form as:

$$P\lambda^4 = \left\{ \boxed{2\ JO}\boxed{-\ 4\ JO}\boxed{+\ 2\ JO} \right\} w \tag{6.62b}$$

Similarly, the outside girder spacing in the y direction can be written as:

$$P\lambda^4 = \left\{ \boxed{2\ JO}\boxed{-\ 4\ JO}\boxed{+\ 2\ JO} \right\} w \tag{6.62c}$$

where $JO = EI_y$. Equations (6.62b) and (6.62c) are represented in mesh form as shown in Figure 6-32. This mesh pattern represents the load-deformation response of the outside girders at corner point 1.

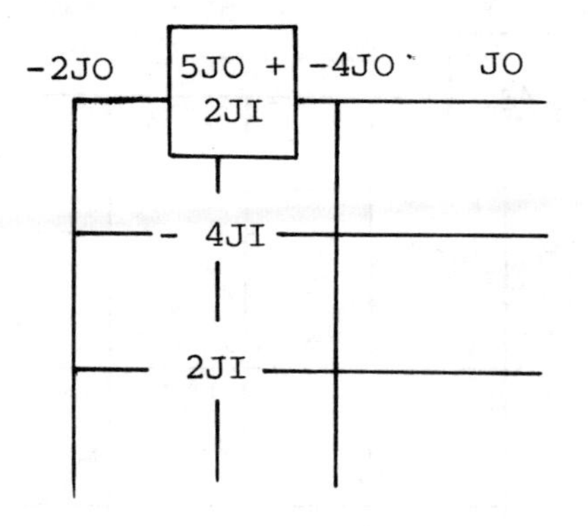

Figure 6-33. Corner Edge Mesh Pattern Point 2.

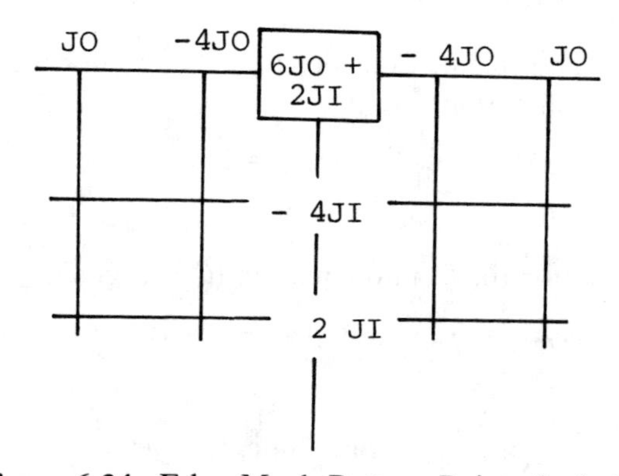

Figure 6-34. Edge Mesh Pattern Points 3, 4, 5.

Load Point 2: Positioning the general pattern, Figure 6-29, over mesh point 2 indicates that deflections points $(n - 2)$, $(m + 1)$, and $(m + 2)$ are beyond the boundary. The mesh point $(n - 2)$ is related to the interior parts by applying the $M = 0$ equation. Mesh points $(m + 1)$ and $(m + 2)$ are related to the interior points by application of the equation $M = V = 0$, as described previously. The resulting mesh pattern is shown in Figure 6-33.

Load Points 3, 4, 5: The resulting mesh pattern for these points would require modification of deflection modes $(m + 2)$ and $(m + 1)$ only. The boundary conditions required are, therefore, $V = M = 0$, as discussed under load point 1. The resulting mesh pattern for these load points is given in Figure 6-34.

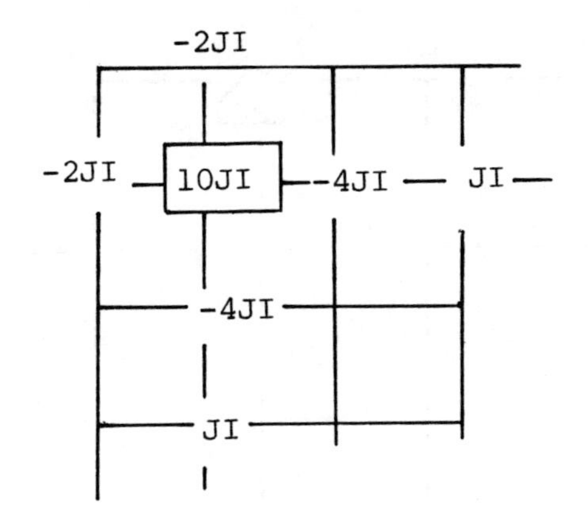

Figure 6-35. Adjacent to Corner Mesh Pattern Point 6.

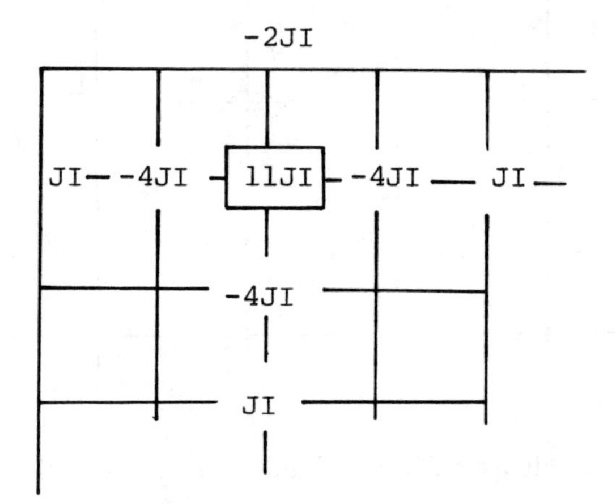

Figure 6-36. Adjacent to Edge Mesh Pattern Points 7, 8, 9, 10.

Load Point 6: The mesh pattern for this point is given in Figure 6-35. The boundary condition for the relationship of points $(m + 2)$ and $(n - 2)$ is $M = 0$.

Load Points 7, 8, 9, and 10: The mesh pattern for these points is shown in Figure 6-36. The boundary condition required to modify the general pattern is $M = 0$ for point $(m + 2)$.

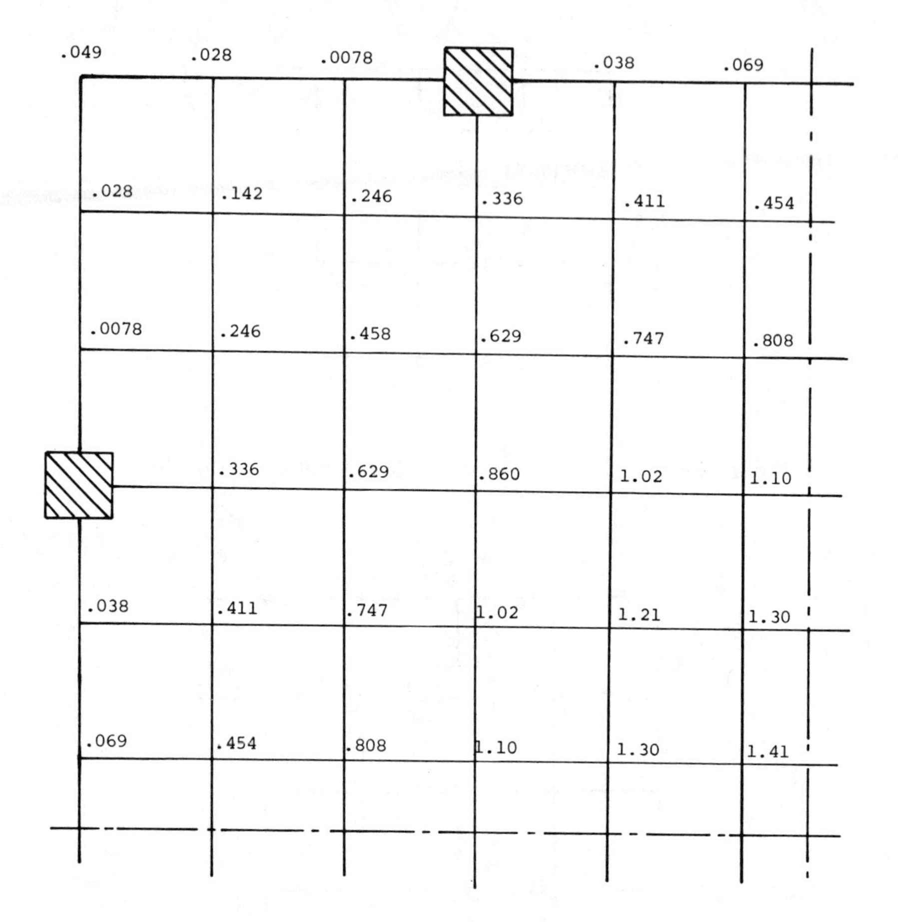

Figure 6-37. Resultant Deflections (inches).

6.4.3 Solution

With the mesh patterns developed for each respective point in the system, Figure 6-31, a set of equations were written which are of the form:

$$[\ ]\cdot w = P\lambda^4 \tag{6.63}$$

The load term P is in pounds per foot and is computed using a dead load (17.7 pounds per square foot) plus the weight of steel and the live load (30 pounds per square foot). The loads are computed for each mesh point 1 through 20. The system of equations is then solved, resulting in the displacement at each of these points, shown in Table 6-1. The moments are

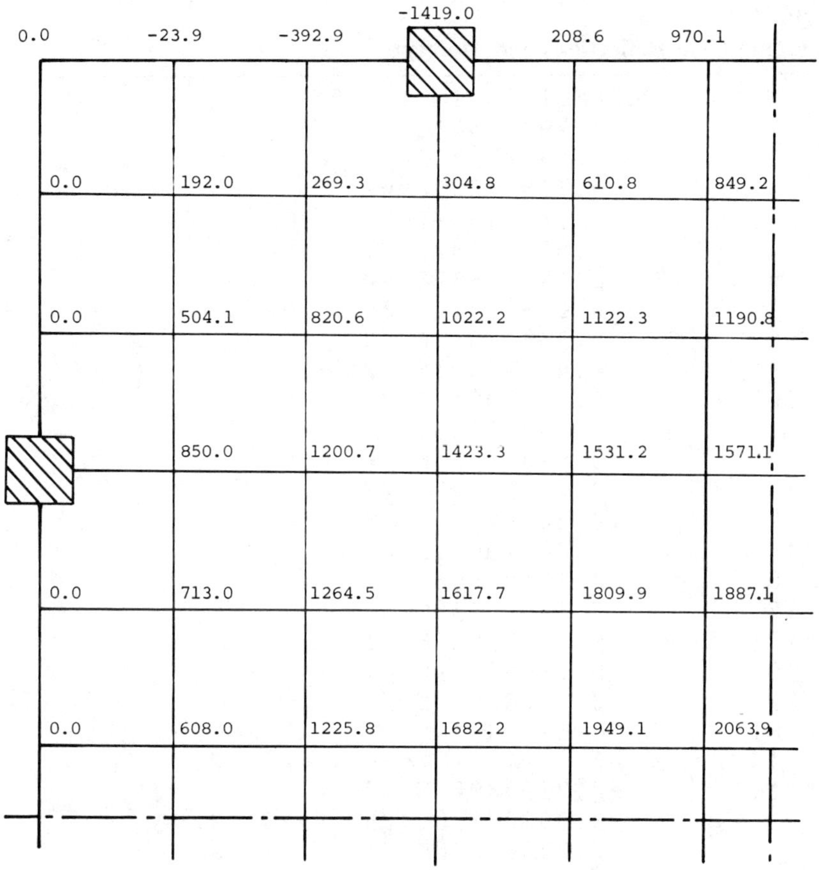

Figure 6-38. Moments M_x(k-in).

computed at each respective point using equation (6.58), resulting in the values given in Table 6-1. The deflection and moments are also drawn on Figures 6-37, 6-38, and 6-39. Examination of these data give the following maximums:

Stress

$$\sigma_{b_{20}} = \frac{2063.9}{6010} \times 21 = 7.2 \text{ ksi}$$

$$\sigma_{b_{21}} = -\frac{1419.0}{9520} \times 25.5 = -3.80 \text{ ksi}$$

Table 6-1
Building Grid Deflections and Moments

POINT	DEFLECTION IN INCHES
1	Ø.49388383E-Ø1
2	Ø.28184945E-Ø1
3	Ø.77536365E-Ø2
4	Ø.38Ø35657E-Ø1
5	Ø.69338782E-Ø1
6	Ø.141975ØØE ØØ
7	Ø.24595124E ØØ
8	Ø.33616372E ØØ
9	Ø.41Ø79441E ØØ
1Ø	Ø.4542Ø258E ØØ
11	Ø.45837965E ØØ
12	Ø.62886553E ØØ
13	Ø.747Ø9947E ØØ
14	Ø.8Ø796771E ØØ
15	Ø.86Ø19491E ØØ
16	Ø.1Ø187698E Ø1
17	Ø.1Ø99Ø776E Ø1
18	Ø.12Ø77412E Ø1
19	Ø.13Ø42Ø1ØE Ø1
2Ø	Ø.14Ø96947E Ø1

POINT	X MOMENT K IN.	Y MOMENT K IN.
1	Ø.	Ø.
2	-Ø.239280Ø8E Ø2	Ø.
3	-Ø.39287576E Ø3	Ø.
4	Ø.2Ø863836E Ø3	Ø.
5	Ø.97ØØ7Ø79E Ø3	Ø.
6	Ø.19199569E Ø3	Ø.19199569E Ø3
7	Ø.26927142E Ø3	Ø.5Ø41436ØE Ø3
8	Ø.3Ø48391ØE Ø3	Ø.85Ø28Ø37E Ø3
9	Ø.61Ø831Ø9E Ø3	Ø.71317294E Ø3
1Ø	Ø.84922892E Ø3	Ø.6Ø84Ø842E Ø3
11	Ø.82Ø55515E Ø3	Ø.82Ø55515E Ø3
12	Ø.1Ø222466E Ø4	Ø.12ØØ6956E Ø4
13	Ø.11222912E Ø4	Ø.12645ØØ2E Ø4
14	Ø.119Ø8142E Ø4	Ø.12257742E Ø4
15	Ø.14233182E Ø4	Ø.14233183E Ø4
16	Ø.153122Ø6E Ø4	Ø.16179Ø83E Ø4
17	Ø.15711255E Ø4	Ø.16822235E Ø4
18	Ø.18Ø98753E Ø4	Ø.18Ø98753E Ø4
19	Ø.18871216E Ø4	Ø.19491366E Ø4
2Ø	Ø.2Ø638565E Ø4	Ø.2Ø638565E Ø4
21	-Ø.14189911E Ø4	Ø.

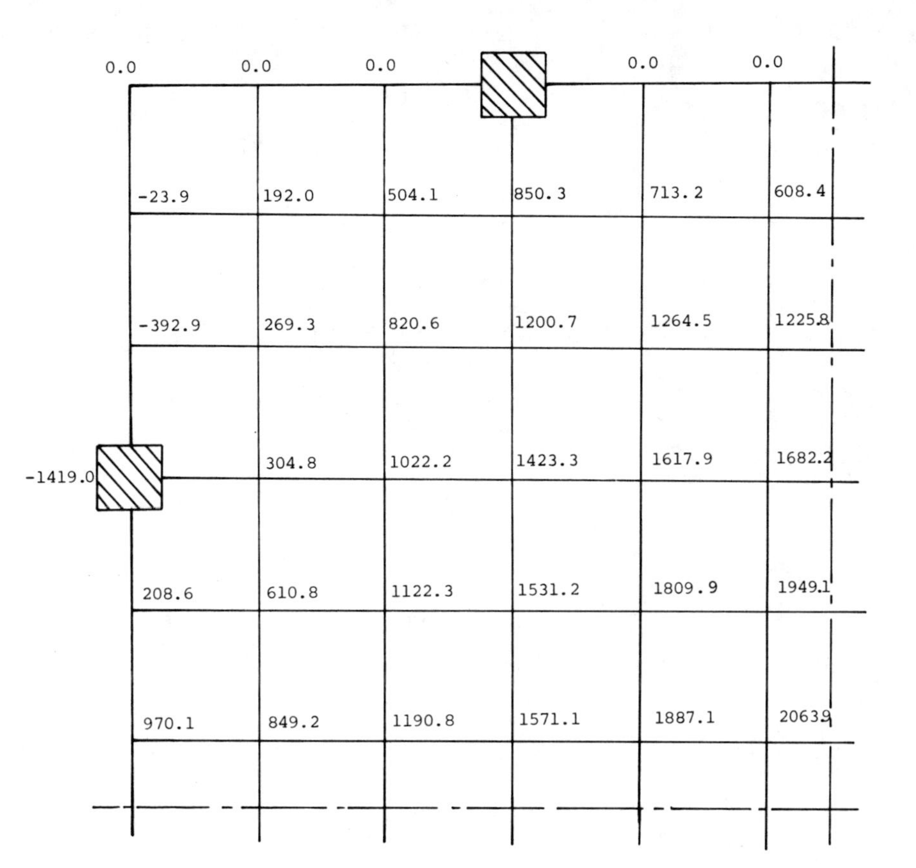

Figure 6-39. Moments M_y(k-in).

The allowable bending stress is 20.0 ksi; therefore, the design stress is adequate and certainly conservative.

Deflection

The maximum deflection, as listed in Figure 6-37, is

$$\Delta_{20} = 1.41 \text{ in}$$

The allowable deflection is $\Delta = L/800 = (88 \times 12)/800 = 1.32$ in. This deflection controls the design and the system is adequate.

7 Computer Programs

7.1 Rectangular Orthotropic Bridges

7.1.1 Slope-Deflection Method

General: As described previously, the deformations and forces throughout a series of continuous orthotropic plates on flexible supports when subjected to some forces can be determined by solving a set of plate-girder interaction equations (6.41) and (6.42). The solution of these equations and the types of bridges that may be considered will now be presented.

Generally, two basic types of structural arrangements for bridges are currently being designed. Figures 7-1 and 7-2 describe these general bridge types, when subjected to the standard HS20 AASTO design truck.

Figure 7-1 describes longitudinal girders, spanning between the supports, of length l and with a transverse spacing s. This type of system is exactly the same as the one described in Figure 6-1, which was enlisted during the development of the slope-deflection method. To apply the slope-deflection technique to this type of structure, a series of equations would be written relating the slopes, deflections, and fixed end moments (FEM) and reactions (FER) for each girder a through f. Only plates cd and de would require FEM and FER as the truck is located only on these plates. There are six slopes ϕ and six deflections Δ to be determined, thus requiring twelve equations of statics. These equations would be $\Sigma m = 0$ and $\Sigma F = 0$ about each joint a through f. In order to perform the development and evaluation of these or any other sets of equations describing the behavior of this bridge, a computer program has been written [39] and will be given herein. This program considers the three types of slab stiffness:

Case I $\quad H^2 > D_x \cdot D_y$ (Closed Rib)

Case II $\quad H^2 = D_x \cdot D_y$ (Uniform Plate)

Case III $\quad H^2 < D_x \cdot D_y$ (Open Rib)

By the inclusion of all three of these cases in the slope-deflection analysis, any type of slab stiffness and configuration may be considered. The computer program will require the bridge dimensions and stiffness parameters, as well as the location of wheel loads, for input data. Consideration of the loads on the girders rather than on the plates is also included. The program

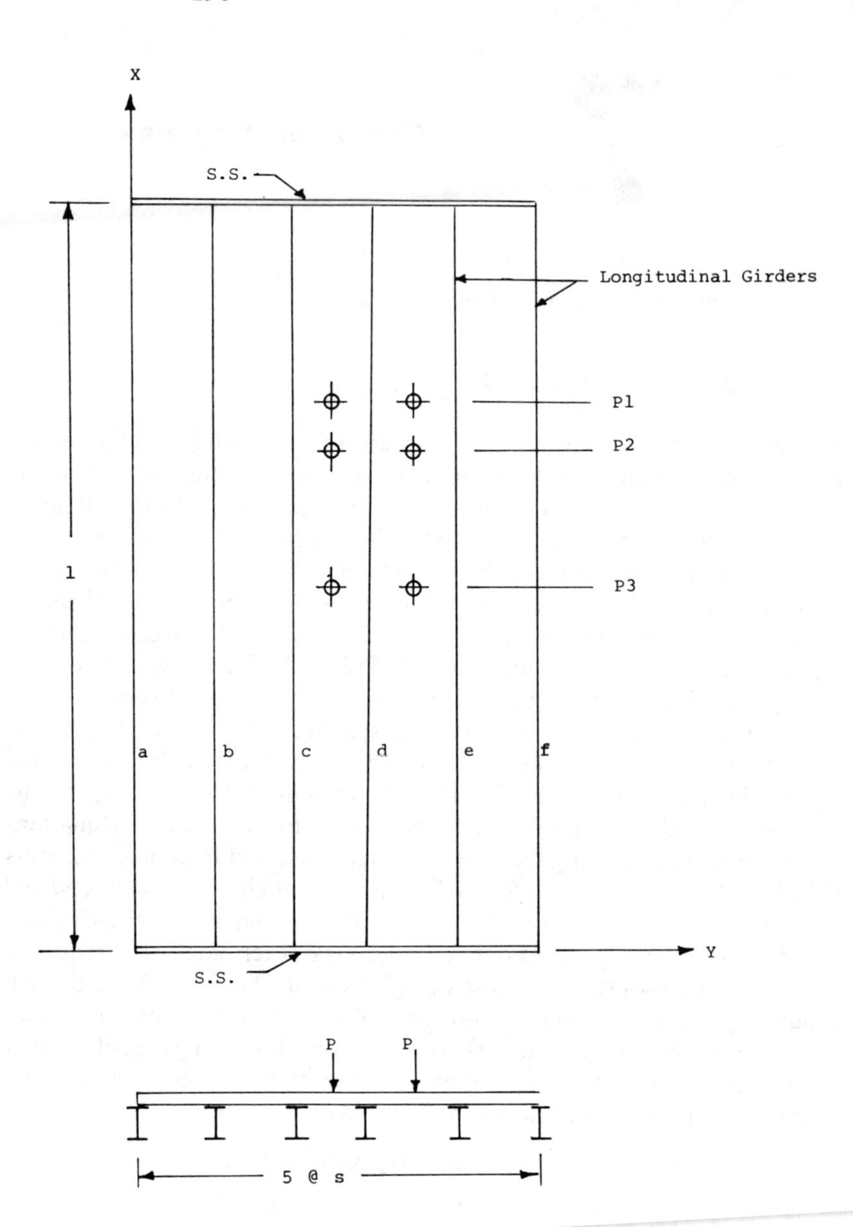

Figure 7.1. Main Longitudinal Girder Type Bridge.

will set up the required equations, solve the system, and then print out the unknown slopes, deflections, and forces along each girder. This analysis is performed for *n* terms, depending on the number of terms desired.

Other programs [39] have also been written to determine the internal deflections and forces, using the slope-deflection data. For instance, the

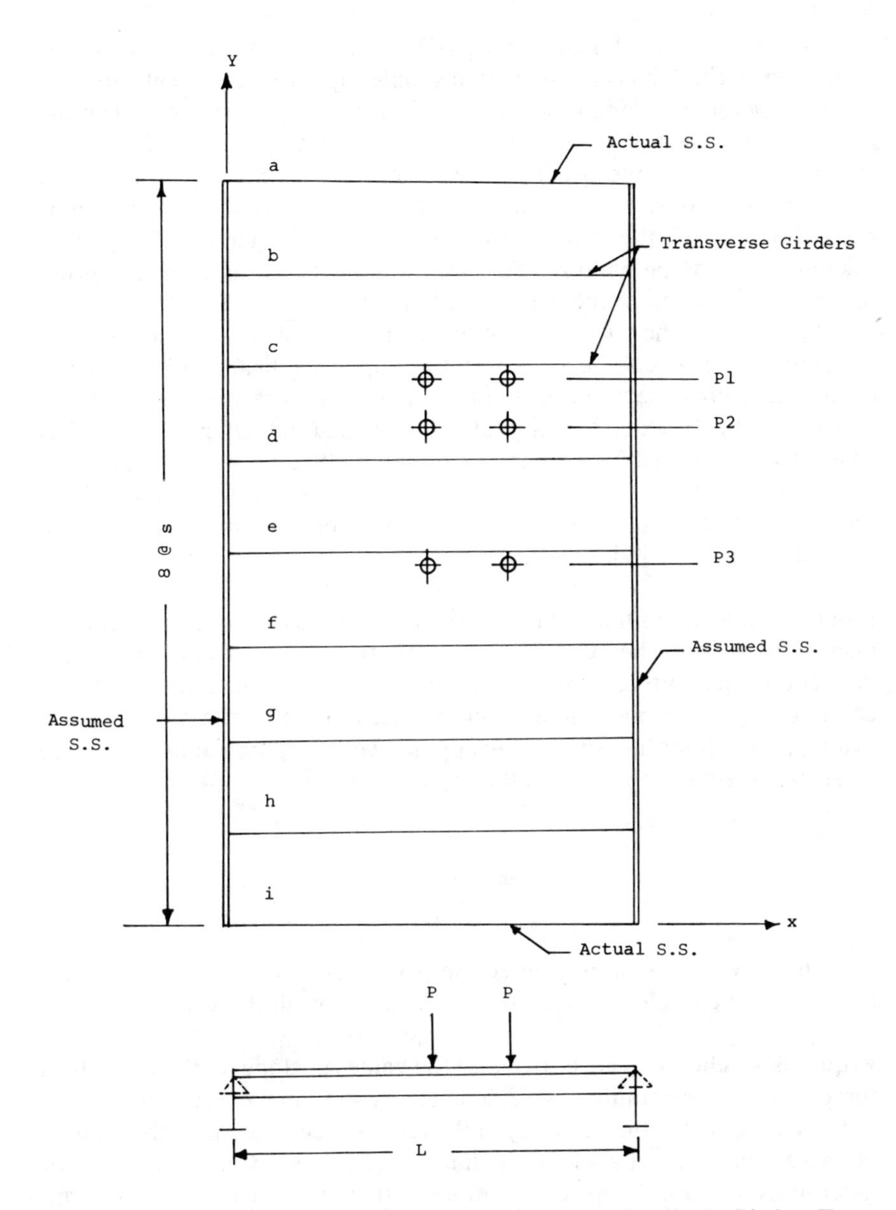

Figure 7-2. Transverse and Main Outside Longitudinal Girder Type Bridge.

complete analysis of plate *cd* for any type of slab stiffness is made possible by these programs.

Another structural system to be solved by the slope-deflection method would be that shown in Figure 7-2. This system has transverse girders framing into flexible longitudinal girders. In order to apply the slope-deflection method, it is necessary that simple supports be present along the y-axis or perpendicular to the girders. This was predetermined when the plate solution was assumed to be of the form $w = \Sigma Y_n \cdot \sin \lambda x$. Therefore, the longitudinal flexible girder will be replaced by simple supports to meet the plate conditions. To account for the actual simple support at the a and i ends of the bridge, the girder stiffnesses will be set equal to infinity. This system may now be analyzed for slopes, deflections, and finally, girder forces, as discussed in the previous paragraph.

The effects of the outside longitudinal girders must now be considered. Assuming the bridge to be a large, simply supported beam with the inertia of the entire cross section, beam deflections and moments may be determined for the truck axle loads at various longitudinal positions. This analysis is performed by a computer program [39] and then the computer output is added to the values determined by the slope-deflected procedure. The addition of the plate and beam solutions are also accomplished by a computer program [39].

Limiting Plate Conditions: Since the program was written in a general manner to solve either Case I, II, or III, the correctness of the slope-deflection equations and the programming was checked for a limiting condition by examining three different cases in which the structure and loading, and all stiffness values except α, were held constant. For these three cases α was set equal to the following values:

$$\alpha = 0.999$$

$$\alpha = 1.000$$

$$\alpha = 1.001$$

The output values of slope, deflection, shear force, and bending moment for the transverse floor beams were the same for all three cases.

Required Number of Terms: Because the values of slope, deflection, shear force, and bending moment are calculated by summing the results for each Fourier series term, the accuracy of the results is dependent on the number of terms. Short of using an infinite number of terms, the accuracy may be determined by examining computations with increasing numbers of terms to note any convergence of the results. Such computations have been performed [48], the results for which are shown in Figures 7-3 through 7-5. The bridge configuration, stiffnesses, and loading that were used are shown in Figure 7-6. The number of terms used was from five to thirty and increased in increments of five. The results show the deflection, shear

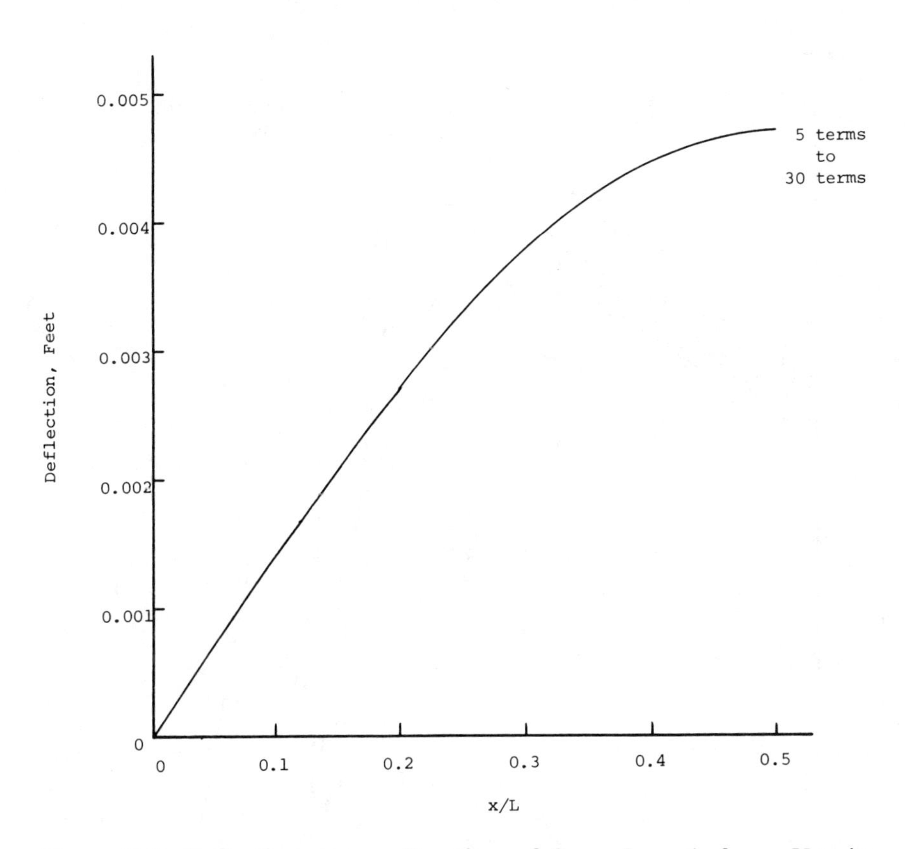

Figure 7-3. Deflection versus Fraction of Span Length for a Varying Number of Fourier Series Terms.

force, and bending moment for floor beam number 13 at tenth points along the beam.

In Figure 7-3 it is seen that the deflection curve is essentially unaffected by the number of terms. Likewise, the bending moments in Figure 7-5 display little variation as the number of terms is increased. If concern should arise as to the minor differences in bending moment, the error at $x/L = 0.5$ is less than five percent.

The results for shear force, however, are greatly affected by the number of terms. As Figure 7-4 shows, not only is the magnitude affected, but the position of the largest magnitude varies.

These results are not surprising, nor are those for deflection and bending moment. From elementary strength of materials, it is known that for a simply-supported load, a sinusoidal function is a good approximation for

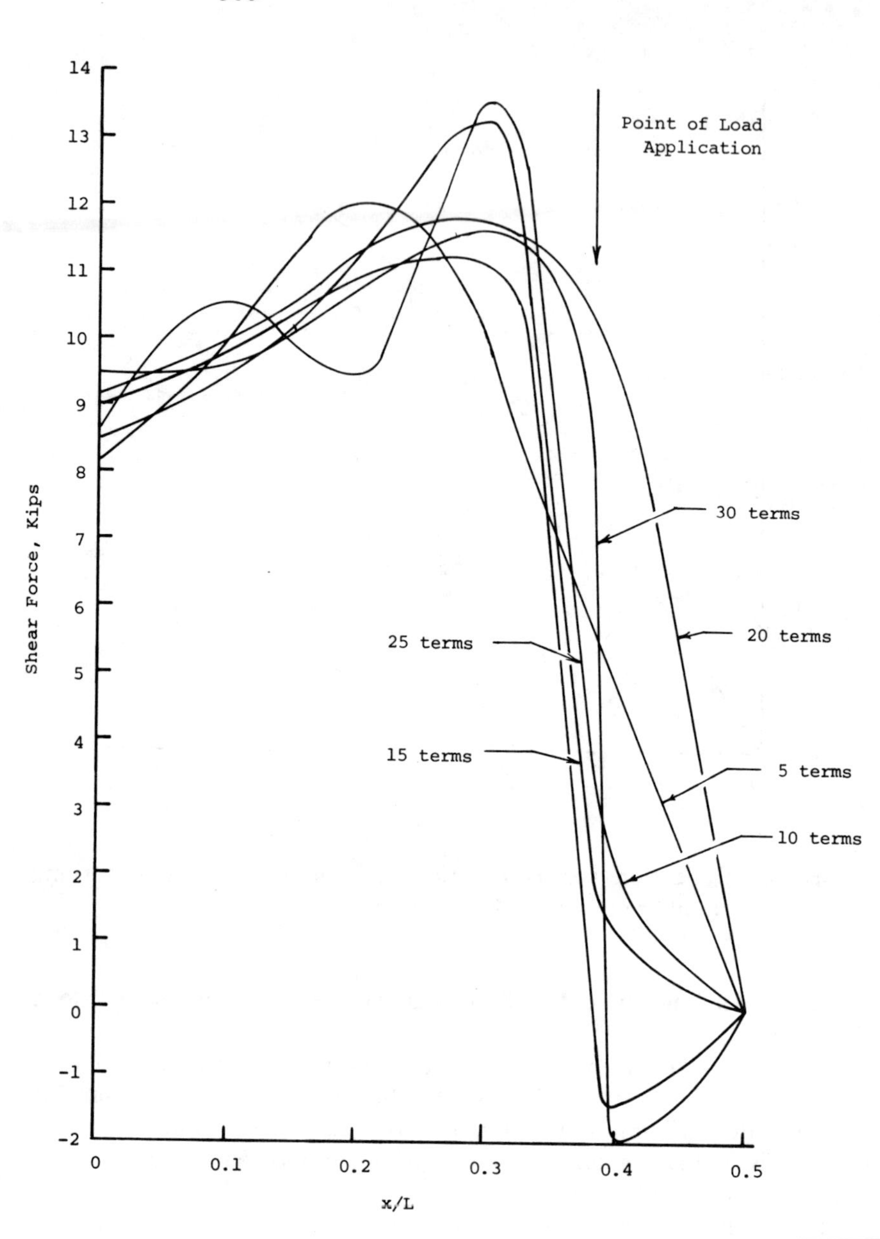

Figure 7-4. Shear Force versus Fraction of Span Length for a Varying Number of Fourier Series Terms.

the deflection curve. The bending moment is a ramp function and the shear force is a step function. Obviously, the deflections herein obtained agree very well. The bending moment variation is approximated reasonably well;

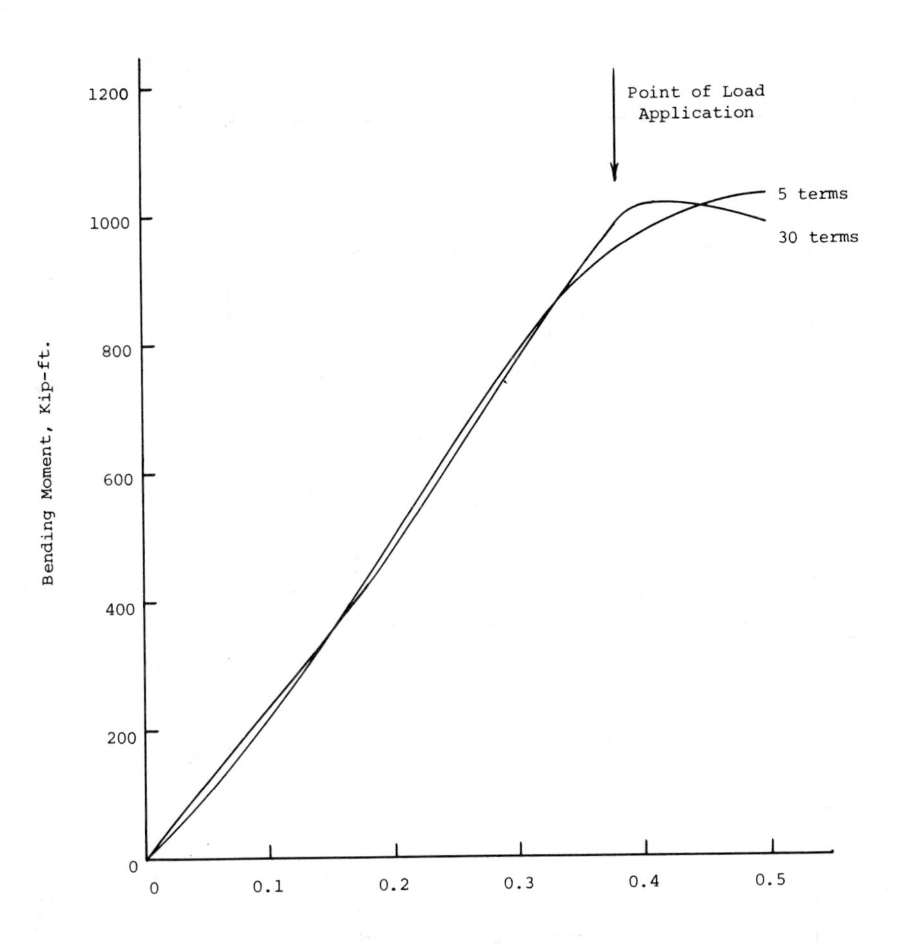

Figure 7-5. Bending Moment versus Fraction of Span Length for a Varying Number of Fourier Series Terms.

however, a step function is difficult to approximate with a small number of sine waves. Since the shear forces is a minor consideration in floor beam design and conservative results are desired, fifteen terms are deemed sufficient.

Floor Beam Studies: In order to verify the mathematical correctness of the computer program, a simple model as shown in Figure 7-7 was studied. Since the stiffnesses of the deck plate can be considered negligible, the resulting center floor beam deflection, shear force, and bending moment should agree with those for a simply supported beam with a concentrated load at its midspan.

The maximum floor beam deflection, shear force, and bending moment

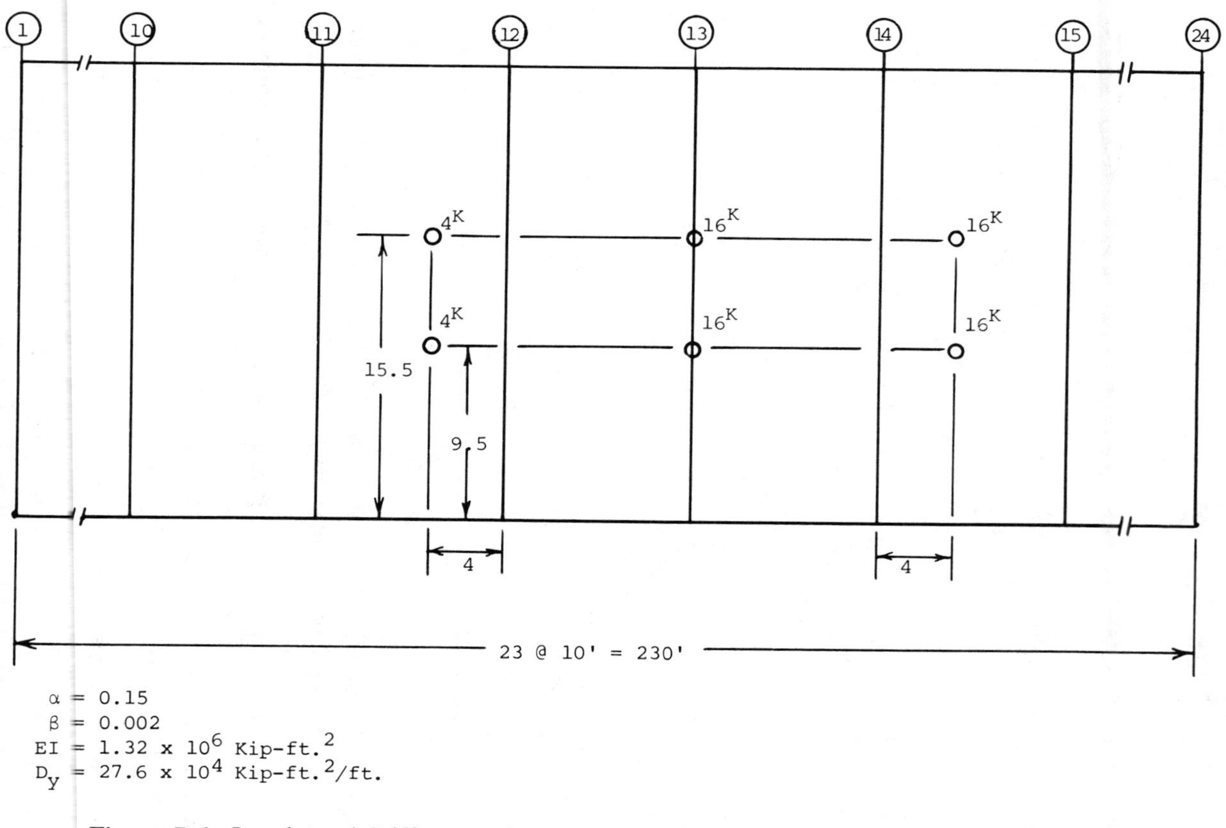

Figure 7-6. Loads and Stiffnesses for Determining Effect of Fourier Series Terms.

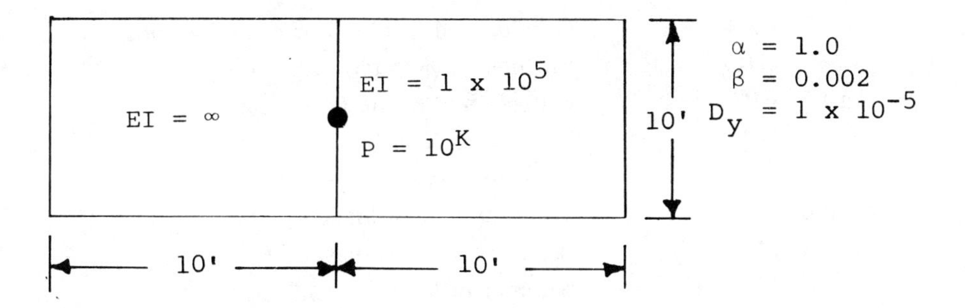

Figure 7-7. Simple Plate Model.

compared with simply supported beam results are given as follows. A total of fifteen Fourier series terms were used in obtaining the results.

Function	*Slope Deflection*	*Simple Beam*
Deflection	0.2083×10^{-2}	0.2083×10^{-2}
Shear Force	4.802	5.000
Bending Moment	24.37	25.00

These results are in very good agreement and serve to verify the technique.

Computer Program Input: The computer program that performed the preliminary studies first presented, and which will be illustrated in a complete bridge design study, has the following input requirements. The source statement listing of the program, written for a UNIVAC 1108 computer, appears in the Appendix A-1. A description of the input cards to be used with the program is included below.

Card 1	Description	
Card 2	ALPHA	$= H/\sqrt{D_x D_y}$
	BETA	$= H/D_y$
	S	Floor beam spacing (ft)
	CL	Span width (ft)
	N	Number of floor beams × 2
Card 3	NDGJ	Number of divisions along the floor beam for which deflections, shears, moments, and slopes are desired
Card 4	EIWO, EIWI	Outside and inside floor beam warping Stiffnesses (Kip-ft^2)

	GJO GJI	Outside and inside floor beam bending Stiffnesses (Kip-ft²)
Card 5	EIO EII	Outside and inside floor beam bending Stiffnesses (Kip-ft²)
	DY	Plate Bending Stiffness (Kip-ft²/ft)
Card 6	NUM	Total number of loaded plates
Card 7	M	Plate number (Actual × 2)
	ML(M)	Number of loads on each plate
Card 8	Q(M,J1)	Magnitude of load (Kips)
	U(M,J1)	Distance to load in x direction (ft)
	VL(M,J1)	Distance to load in y direction from left-hand floor beam (ft)
	VR(M,J1)	Distance to load in y direction from right-hand floor beam (ft)

(Since J1 = ML(M), Card 8 is repeated for each load.)

(Cards 7 and 8 are repeated for each loaded plate.)

Card 9	NT	Number of Fourier series terms to be used

Example Input Data: In order to illustrate the use of this program, a hypothetical orthotropic deck system will be studied. The system consists of nineteen transverse floor beams, spaced at 15.0 ft, as shown in Figure 7-8. Two additional floor beams with infinite stiffness are also included in the analysis, in order to simulate the simple supports. The top 3/8 in deck plate is stiffened by 3/8 × 10-in plates, spaced at 16 in. The complete details of the design of these elements is given in chapter 8. The computed stiffnesses of these members are.

Plate

$$D_y = 1.43 \times 10^4 \text{ k-ft}^2/\text{ft}$$

$$D_x = 11.0 \text{ k-ft}^2/\text{ft}$$

$$H = 28.6 \text{ k-ft}^2/\text{ft}$$

Floor Beam

$$EI = 3.19 \times 10^6 \text{ k-ft}^2$$

The plate parameters are computed as:

$$\alpha = H/\sqrt{D_x D_y}$$

$$\alpha = 28.6/\sqrt{11.0 \times 1.43 \times 10^4} = 0.07$$

$$\beta = H/D_y = 28.6/1.43 \times 10^4 = 0.002$$

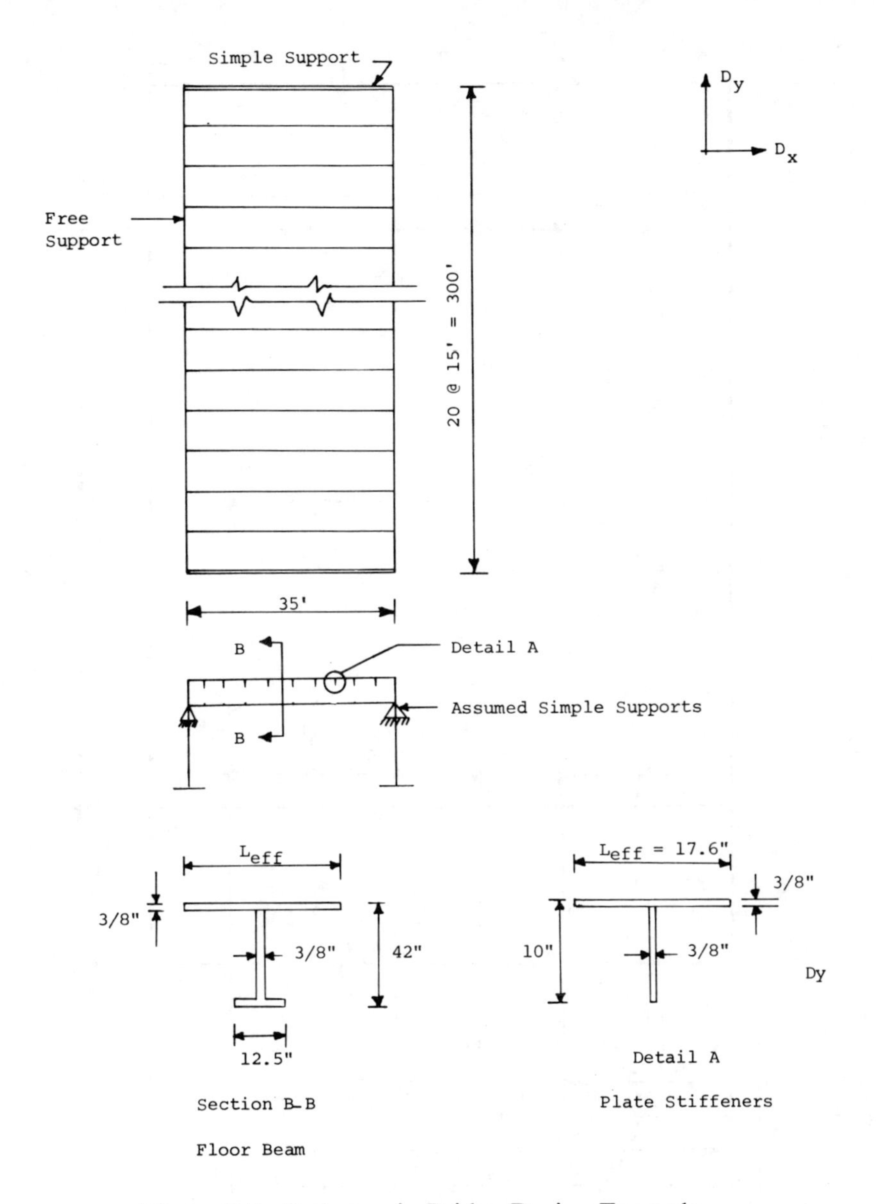

Figure 7-8. Orthotropic Bridge Design Example.

For this case, however, it will be assumed that $\alpha = 0.35$. The loads to be applied to the structure consist of three trucks, as shown in Figure 7-9. Figure 7-9 also shows the parameters required for the input data. The general scheme required for computer input data is shown in Figure 7-10.

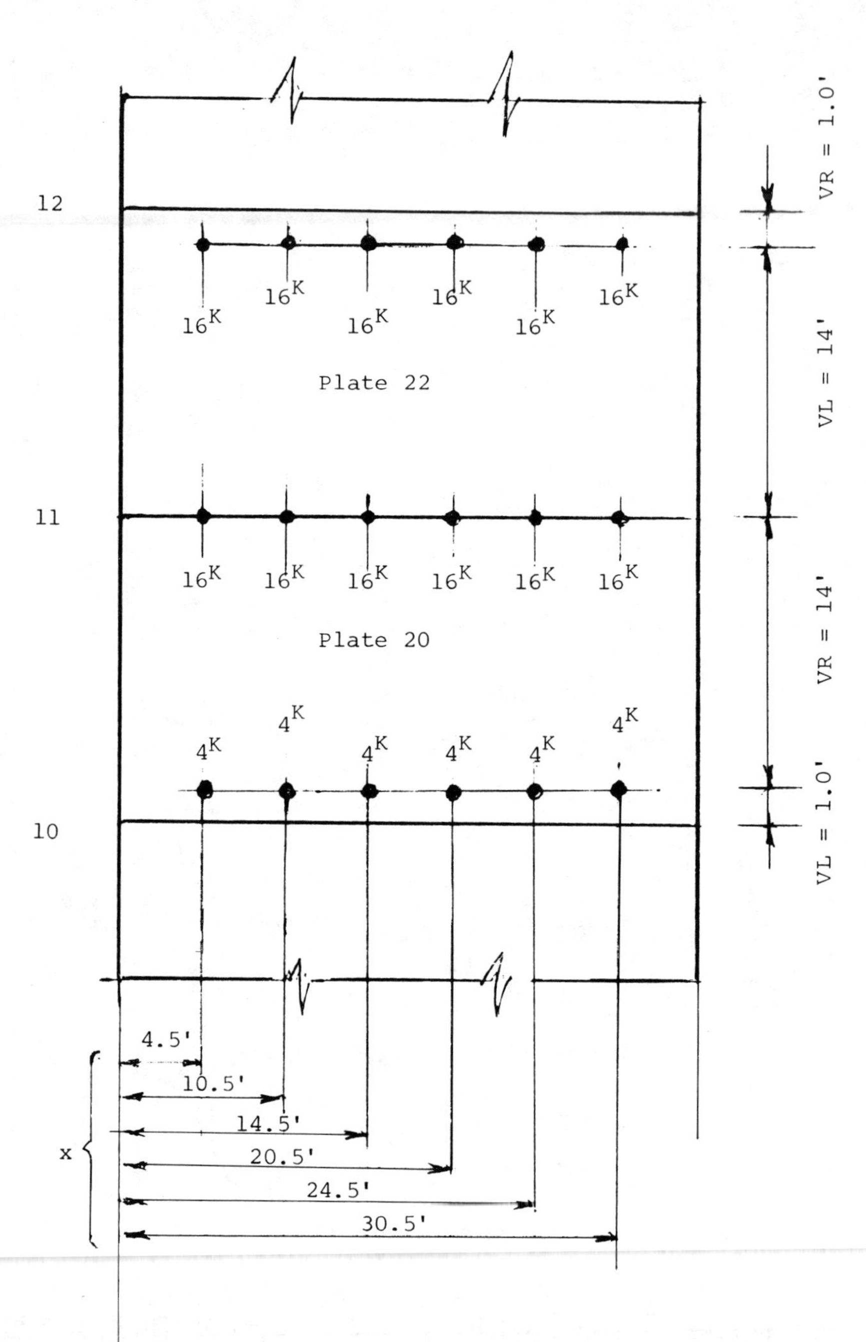

Figure 7-9. Bridge Deck Loading.

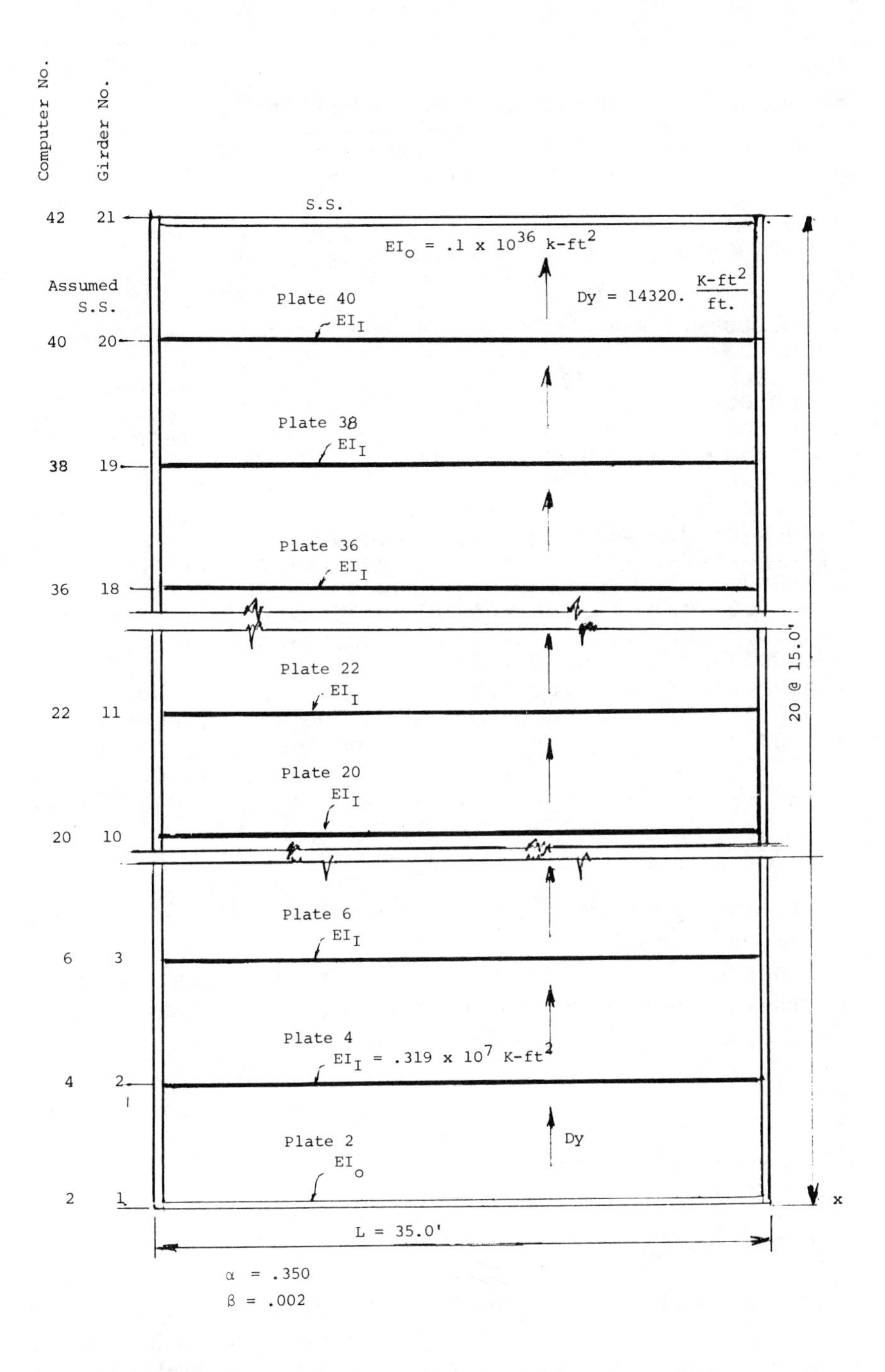

Figure 7-10. Bridge Deck Analysis Idealization.

Table 7-1
Rectangular Orthotropic Bridge Computer Input Data

```
      3 LANES, 15-FOOT FLOOR BEAM SPACING
 .35            .20000-002   .15000+002   .35000+002 42
10
 1.995      D5   .1000000+001   .1000000+001   2.86      D2.
 .1000000+036    .319       D7   .1432      D5
2
20  6
 .4000000+001   .4500000+001   .1000000+001   .1400000+002
 .4000000+001   .1050000+002   .1000000+001   .1400000+002
 .4000000+001   .1450000+002   .1000000+001   .1400000+002
 .4000000+001   .2050000+002   .1000000+001   .1400000+002
 .4000000+001   .2450000+002   .1000000+001   .1400000+002
 .4000000+001   .3050000+002   .1000000+001   .1400000+002
22  12
 .1600000+002   .4500000+001   .0000000+000   .1500000+002
 .1600000+002   .1050000+002   .0000000+000   .1500000+002
 .1600000+002   .1450000+002   .0000000+000   .1500000+002
 .1600000+002   .2050000+002   .0000000+000   .1500000+002
 .1600000+002   .2450000+002   .0000000+000   .1500000+002
 .1600000+002   .3050000+002   .0000000+000   .1500000+002
 .1600000+002   .4500000+001   .1400000+002   .1000000+001
 .1600000+002   .1050000+002   .1400000+002   .1000000+001
 .1600000+002   .1450000+002   .1400000+002   .1000000+001
 .1600000+002   .2050000+002   .1400000+002   .1000000+001
 .1600000+002   .2450000+002   .1400000+002   .1000000+001
 .1600000+002   .3050000+002   .1400000+002   .1000000+001
15
```

The resulting input data required is shown in Table 7-1. The output data is shown in Table 7-2.

If the effects of the main longitudinal girders are to be considered, then a secondary analysis of the structure is required. This is performed, as outlined previously, by considering the bridge to be a large beam, as illustrated in Figure 7-11.

7.2 Curved Orthotropic Bridges

7.2.1 Finite Difference Method

Utilizing the finite difference equations given in section 6.2 and in references [31, 32], the mathematical model can be constructed such that it will represent a physical bridge structure to be analyzed. The bridge deck, as shown in Figure 7-12, is divided into a polar grid mesh pattern consisting of a minimum and maximum radius, a total enclosed angle, and subdivisions of these radii and angles, which are listed as "LAMBDA," λ, and "THETA SUB 0", θ_0. At each intersection of the grid, there are present the three stiffnesses of the bridge and the point of application of the vertical loading. Thus, if the grid system is fine enough, it can be visualized that any

Table 7-2
Rectangular Orthotropic Bridge Computer Output Data

SINGLE SPAN ORTHOTROPIC BRIDGE ANALYSIS

3 LANES, 15-FOOT FLOOR BEAM SPACING

BRIDGE WIDTH	.3500000+002
BRIDGE LENGTH	.3000000+003
NUMBER OF GIRDERS	21
GIRDER SPACING	.1500000+002

EIWI	EIWO	GJI	GJO
.10000+001	.19950+006	.28600+003	.10000+001

EII	EIO
.31900+007	.10000+036

ALPHA	.3500000+000
BETA	.2000000-002
DY	.1432000+005

PLATE LOAD DATA

PLATE	LOAD NO	MAGNITUDE	X-COORDINATE	VL	VR
20	1	4.0	4.5	1.0	14.0
	2	4.0	10.5	1.0	14.0
	3	4.0	14.5	1.0	14.0
	4	4.0	20.5	1.0	14.0
	5	4.0	24.5	1.0	14.0
	6	4.0	30.5	1.0	14.0

PLATE	LOAD NO	MAGNITUDE	X-COORDINATE	VL	VR
22	1	16.0	4.5	.0	15.0
	2	16.0	10.5	.0	15.0
	3	16.0	14.5	.0	15.0
	4	16.0	20.5	.0	15.0
	5	16.0	24.5	.0	15.0
	6	16.0	30.5	.0	15.0
	7	16.0	4.5	14.0	1.0
	8	16.0	10.5	14.0	1.0
	9	16.0	14.5	14.0	1.0
	10	16.0	20.5	14.0	1.0
	11	16.0	24.5	14.0	1.0
	12	16.0	30.5	14.0	1.0

Table 7-2 (cont.)

PROGRAM OUTPUT FOR 15 FOURIER SERIES SUM

GIRDER 1

X/L	DEFLECTION	SHEAR FORCE	BENDING MOMENT	SLOPE
.00	.0000	.2822-006	.0000	.0000
.10	.3372-039	.1621-006	-.1246-005	-.2645-009
.20	.7833-039	-.6283-006	.6636-006	-.4774-009
.30	.1187-038	.4699-007	.6264-006	-.2477-009
.40	.1495-038	-.2816-006	.1177-005	.1259-009
.50	.1640-038	.4703-011	.2808-005	.2321-009
.60	.1495-038	.2816-006	.1177-005	.1259-009
.70	.1187-038	-.4700-007	.6264-006	-.2477-009
.80	.7833-039	.6283-006	.6636-006	-.4774-009
.90	.3372-039	-.1621-006	-.1246-005	-.2645-009
1.00	-.5743-046	-.2822-006	.1812-012	.4919-016

GIRDER 2

X/L	DEFLECTION	SHEAR FORCE	BENDING MOMENT	SLOPE
.00	.0000	.2898-005	.0000	.0000
.10	-.6912-009	.3278-005	-.8836-005	.1276-008
.20	-.1343-008	.6877-005	-.3127-004	-.9086-009
.30	-.1879-008	.3843-005	-.4562-004	.6979-009
.40	-.2239-008	.3484-005	-.5954-004	.1492-009
.50	-.2370-008	-.6242-011	-.7076-004	-.2165-008
.60	-.2239-008	-.3484-005	-.5954-004	.1492-009
.70	-.1879-008	-.3843-005	-.4562-004	.6979-009
.80	-.1343-008	-.6877-005	-.3127-004	-.9086-009
.90	-.6912-009	-.3278-005	-.8836-005	.1276-008
1.00	.1279-015	-.2898-005	.1860-011	.2180-015

GIRDER 3

X/L	DEFLECTION	SHEAR FORCE	BENDING MOMENT	SLOPE
.00	.0000	-.1774-004	.0000	.0000
.10	.3587-008	-.1830-004	.5586-004	-.4297-008
.20	.6939-008	-.3224-004	.1621-003	.5946-008
.30	.9687-008	-.2057-004	.2371-003	-.1012-008
.40	.1152-007	-.1595-004	.3062-003	.1399-008
.50	.1218-007	.3173-011	.3546-003	.1146-007
.60	.1152-007	.1595-004	.3062-003	.1399-008
.70	.9687-008	.2057-004	.2371-003	-.1012-008
.80	.6939-008	.3224-004	.1621-003	.5946-008
.90	.3587-008	.1830-004	.5586-004	-.4297-008
1.00	-.6648-015	.1774-004	-.1139-010	-.1198-014

GIRDER 4

X/L	DEFLECTION	SHEAR FORCE	BENDING MOMENT	SLOPE
.00	.0000	-.1250-003	.0000	.0000
.10	.1227-007	-.1165-003	.4557-003	.1922-007
.20	.2290-007	-.3226-004	.6494-003	-.2003-007
.30	.3099-007	-.3410-004	.8441-003	.1179-007
.40	.3592-007	-.8969-006	.8847-003	.2710-008
.50	.3751-007	-.1254-011	.8018-003	-.3839-007
.60	.3592-007	.8969-006	.8847-003	.2710-008
.70	.3099-007	.3410-004	.8441-003	.1179-007
.80	.2290-007	.3226-004	.6494-003	-.2003-007
.90	.1227-007	.1165-003	.4557-003	.1922-007
1.00	-.2303-014	.1250-003	-.8023-010	.4697-014

Table 7-2 (cont.)

GIRDER 5

X/L	DEFLECTION	SHEAR FORCE	BENDING MOMENT	SLOPE
.00	.0000	.1547-002	.0000	.0000
.10	-.1909-006	.1476-002	-.5447-002	-.1142-006
.20	-.3615-006	.9869-003	-.9514-002	.8994-008
.30	-.4955-006	.7583-003	-.1293-001	-.1554-006
.40	-.5806-006	.3453-003	-.1478-001	-.1342-006
.50	-.6095-006	-.2659-010	-.1503-001	.2943-007
.60	-.5806-006	-.3453-003	-.1478-001	-.1342-006
.70	-.4955-006	-.7583-003	-.1293-001	-.1554-006
.80	-.3615-006	-.9869-003	-.9514-002	.8994-008
.90	-.1909-006	-.1476-002	-.5447-002	-.1142-006
1.00	.3566-013	-.1547-002	.9934-009	-.1271-013

GIRDER 6

X/L	DEFLECTION	SHEAR FORCE	BENDING MOMENT	SLOPE
.00	.0000	-.2822-002	.0000	.0000
.10	.3186-006	-.2754-002	.1027-001	.5090-006
.20	.5995-006	-.1261-002	.1622-001	.6363-007
.30	.8175-006	-.1075-002	.2177-001	.8036-006
.40	.9536-006	-.4145-003	.2398-001	.7328-006
.50	.9989-006	.8977-011	.2321-001	.6259-007
.60	.9536-006	.4145-003	.2398-001	.7328-006
.70	.8175-006	.1075-002	.2177-001	.8036-006
.80	.5995-006	.1261-002	.1622-001	.6363-007
.90	.3186-006	.2754-002	.1027-001	.5090-006
1.00	-.5962-013	.2822-002	-.1812-008	.4391-013

GIRDER 7

X/L	DEFLECTION	SHEAR FORCE	BENDING MOMENT	SLOPE
.00	.0000	-.4266-001	.0000	.0000
.10	.5930-005	-.4008-001	.1444+000	-.7337-006
.20	.1130-004	-.3847-001	.2876+000	.2221-005
.30	.1559-004	-.2734-001	.3975+000	-.1758-008
.40	.1835-004	-.1471-001	.4736+000	.9256-006
.50	.1931-004	.1523-008	.5060+000	.3924-005
.60	.1835-004	.1471-001	.4736+000	.9256-006
.70	.1559-004	.2734-001	.3975+000	-.1758-008
.80	.1130-004	.3847-001	.2876+000	.2221-005
.90	.5930-005	.4008-001	.1444+000	-.7337-006
1.00	-.1105-011	.4266-001	-.2738-007	-.4260-012

GIRDER 8

X/L	DEFLECTION	SHEAR FORCE	BENDING MOMENT	SLOPE
.00	.0000	.2609+000	.0000	.0000
.10	-.3572-004	.2453+000	-.8874+000	-.4664-005
.20	-.6805-004	.2265+000	-.1738+001	-.2333-004
.30	-.9377-004	.1622+000	-.2398+001	-.1902-004
.40	-.1103-003	.8549-001	-.2843+001	-.2643-004
.50	-.1161-003	-.8723-008	-.3021+001	-.4015-004
.60	-.1103-003	-.8549-001	-.2843+001	-.2643-004
.70	-.9377-004	-.1622+000	-.2398+001	-.1902-004
.80	-.6805-004	-.2265+000	-.1738+001	-.2333-004
.90	-.3572-004	-.2453+000	-.8874+000	-.4664-005
1.00	.6658-011	-.2609+000	.1675-006	.3134-011

Table 7-2 (cont.)

GIRDER 9

X/L	DEFLECTION	SHEAR FORCE	BENDING MOMENT	SLOPE
.00	.0000	.6454+000	.0000	.0000
.10	-.8352-004	.6278+000	-.2268+001	.1126-005
.20	-.1585-003	.4608+000	-.4114+001	.5991-004
.30	-.2177-003	.3430+000	-.5646+001	.2641-004
.40	-.2555-003	.1838+000	-.6538+001	.4896-004
.50	-.2684-003	-.1306-007	-.6746+001	.1034-003
.60	-.2555-003	-.1838+000	-.6538+001	.4896-004
.70	-.2177-003	-.3430+000	-.5646+001	.2641-004
.80	-.1585-003	-.4608+000	-.4114+001	.2641-004
.90	-.8352-004	-.6278+000	-.2268+001	.5991-004
1.00	.1559-010	-.6454+000	.4143-006	-.9236-011

GIRDER 10

X/L	DEFLECTION	SHEAR FORCE	BENDING MOMENT	SLOPE
.00	.0000	-.1343+002	.0000	.0000
.10	.1713-002	-.1321+002	.4764+002	.3511-003
.20	.3248-002	-.9026+001	.8462+002	.4352-003
.30	.4457-002	-.6821+001	.1161+003	.8372-003
.40	.5227-002	-.3724+001	.1335+003	.9030-003
.50	.5489-002	.2285-006	.1365+003	.7294-003
.60	.5227-002	.3724+001	.1335+003	.9030-003
.70	.4457-002	.6821+001	.1161+003	.8372-003
.80	.3248-002	.9026+001	.8462+002	.4352-003
.90	.1713-002	.1321+002	.4764+002	.3511-003
1.00	-.3198-009	.1343+002	-.8621-005	-.2993-010

GIRDER 11

X/L	DEFLECTION	SHEAR FORCE	BENDING MOMENT	SLOPE
.00	.0000	-.4897+002	.0000	.0000
.10	.6162-002	-.4854+002	.1750+003	.3423-003
.20	.1167-001	-.3116+002	.3052+003	.3296-003
.30	.1600-001	-.2386+002	.4183+003	.7488-003
.40	.1875-001	-.1309+002	.4782+003	.7833-003
.50	.1969-001	.6904-006	.5080+003	.5429-003
.60	.1875-001	.1309+002	.4782+003	.7833-003
.70	.1600-001	.2386+002	.4183+003	.7488-003
.80	.1167-001	.3116+002	.3052+003	.3296-003
.90	.6162-002	.4854+002	.1750+003	.3423-003
1.00	-.1151-008	.4897+002	-.3144-004	-.1436-010

GIRDER 12

X/L	DEFLECTION	SHEAR FORCE	BENDING MOMENT	SLOPE
.00	.0000	-.4402+002	.0000	.0000
.10	.5533-002	-.4369+002	.1575+003	-.8569-003
.20	.1048-001	-.2784+002	.2741+003	-.3048-003
.30	.1436-001	-.2135+002	.3757+003	-.1725-002
.40	.1683-001	-.1176+002	.4292+003	-.1601-002
.50	.1767-001	.6061-006	.4350+003	-.4730-003
.60	.1683-001	.1176+002	.4292+003	-.1601-002
.70	.1436-001	.2135+002	.3757+003	-.1725-002
.80	.1048-001	.2784+002	.2741+003	-.3048-003
.90	.5533-002	.4369+002	.1575+003	-.8569-003
1.00	-.1033-008	.4402+002	-.2826-004	-.4436-010

Table 7-2 (cont.)

X/L	DEFLECTION	SHEAR FORCE	BENDING MOMENT	SLOPE
		GIRDER 13		
.00	.0000	-.1999+001	.0000	.0000
.10	.2849-003	-.1832+001	.6630+001	-.7384-004
.20	.5441-003	-.1966+001	.1376+002	-.4703-003
.30	.7514-003	-.1373+001	.1903+002	-.3403-003
.40	.8858-003	-.7192+000	.2292+002	-.4948-003
.50	.9326-003	.8456-007	.2482+002	-.8069-003
.60	.8858-003	.7192+000	.2292+002	-.4948-003
.70	.7514-003	.1373+001	.1903+002	-.3403-003
.80	.5441-003	.1966+001	.1376+002	-.4703-003
.90	.2849-003	.1832+001	.6630+001	-.7384-004
1.00	-.5306-010	.1999+001	-.1283-005	.6523-010
		GIRDER 14		
X/L	DEFLECTION	SHEAR FORCE	BENDING MOMENT	SLOPE
.00	.0000	.1260+001	.0000	.0000
.10	-.1734-003	.1185+001	-.4279+001	.3239-005
.20	-.3303-003	.1107+001	-.8430+001	.8868-004
.30	-.4553-003	.7894+000	-.1163+002	.5029-004
.40	-.5358-003	.4205+000	-.1381+002	.8065-004
.50	-.5637-003	-.4229-007	-.1470+002	.1532-003
.60	-.5358-003	-.4205+000	-.1381+002	.8065-004
.70	-.4553-003	-.7894+000	-.1163+002	.5029-004
.80	-.3303-003	-.1107+001	-.8430+001	.8868-004
.90	-.1734-003	-.1185+001	-.4279+001	.3239-005
1.00	.3231-010	-.1260+001	.8087-006	-.1344-010
		GIRDER 15		
X/L	DEFLECTION	SHEAR FORCE	BENDING MOMENT	SLOPE
.00	.0000	-.6998-001	.0000	.0000
.10	.1065-004	-.6453-001	.2279+000	.8534-005
.20	.2041-004	-.7939-001	.5078+000	-.4410-005
.30	.2825-004	-.5373-001	.7091+000	.1073-004
.40	.3338-004	-.3116-001	.8688+000	.7713-005
.50	.3518-004	.3481-008	.9588+000	-.8278-005
.60	.3338-004	.3116-001	.8688+000	.7713-005
.70	.2825-004	.5373-001	.7091+000	.1073-004
.80	.2041-004	.7939-001	.5078+000	-.4410-005
.90	.1065-004	.6453-001	.2279+000	.8534-005
1.00	-.1982-011	.6998-001	-.4493-007	.1583-011
		GIRDER 16		
X/L	DEFLECTION	SHEAR FORCE	BENDING MOMENT	SLOPE
.00	.0000	-.2849-001	.0000	.0000
.10	.3463-005	-.2730-001	.1009+000	-.2776-005
.20	.6550-005	-.1715-001	.1730+000	-.1171-006
.30	.8972-005	-.1348-001	.2348+000	-.4100-005
.40	.1050-004	-.6000-002	.2670+000	-.3701-005
.50	.1102-004	.4594-009	.2693+000	.5134-007
.60	.1050-004	.6000-002	.2670+000	-.3701-005
.70	.8972-005	.1348-001	.2348+000	-.4100-005
.80	.6550-005	.1715-001	.1730+000	-.1171-006
.90	.3463-005	.2730-001	.1009+000	-.2776-005
1.00	-.6470-012	.2849-001	-.1829-007	-.2618-012

Table 7-2 (cont.)

GIRDER 17

X/L	DEFLECTION	SHEAR FORCE	BENDING MOMENT	SLOPE
.00	.0000	.5614-002	.0000	.0000
.10	-.6552-006	.5314-002	-.1994-001	.5302-006
.20	-.1237-005	.3011-002	-.3305-001	-.2834-006
.30	-.1691-005	.2452-002	-.4451-001	.5344-006
.40	-.1976-005	.9311-003	-.4999-001	.3723-006
.50	-.2072-005	-.7151-010	-.4967-001	-.5608-006
.60	-.1976-005	-.9311-003	-.4999-001	.3723-006
.70	-.1691-005	-.2452-002	-.4451-001	.5344-006
.80	-.1237-005	-.3011-002	-.3305-001	-.2834-006
.90	-.6552-006	-.5314-002	-.1994-001	.5302-006
1.00	.1225-012	-.5614-002	.3604-008	.9135-013

GIRDER 18

X/L	DEFLECTION	SHEAR FORCE	BENDING MOMENT	SLOPE
.00	.0000	-.5584-004	.0000	.0000
.10	-.1892-007	-.2757-004	.3197-003	-.1016-006
.20	-.3853-007	.3527-003	-.6438-003	.1300-006
.30	-.5604-007	.1881-003	-.1174-002	-.3828-007
.40	-.6877-007	.2087-003	-.1985-002	.1376-007
.50	-.7377-007	-.2314-010	-.2820-002	.2454-006
.60	-.6877-007	-.2087-003	-.1985-002	.1376-007
.70	-.5604-007	-.1881-003	-.1174-002	-.3828-007
.80	-.3853-007	-.3527-003	-.6438-003	.1300-006
.90	-.1892-007	.2757-004	.3197-003	-.1016-006
1.00	.3438-014	.5584-004	-.3585-010	-.2720-013

GIRDER 19

X/L	DEFLECTION	SHEAR FORCE	BENDING MOMENT	SLOPE
.00	.0000	-.8440-004	.0000	.0000
.10	.1845-007	-.9146-004	.2654-003	.2718-007
.20	.3575-007	-.1718-003	.8266-003	-.2899-007
.30	.4999-007	-.1067-003	.1218-002	.1093-007
.40	.5952-007	-.8835-004	.1587-002	-.7963-009
.50	.6296-007	.1719-010	.1855-002	-.5690-007
.60	.5952-007	.8835-004	.1587-002	-.7963-009
.70	.4999-007	.1067-003	.1218-002	.1093-007
.80	.3575-007	.1718-003	.8266-003	-.2899-007
.90	.1845-007	.9146-004	.2654-003	.2718-007
1.00	-.3416-014	.8440-004	-.5418-010	.6037-014

GIRDER 20

X/L	DEFLECTION	SHEAR FORCE	BENDING MOMENT	SLOPE
.00	.0000	-.3266-005	.0000	.0000
.10	-.1356-008	.6919-006	.1651-004	-.7828-008
.20	-.2739-008	.2365-004	-.5204-004	.4172-008
.30	-.3948-008	.1009-004	-.8653-004	-.5028-008
.40	-.4810-008	.1440-004	-.1348-003	-.2328-008
.50	-.5146-008	-.3605-010	-.1884-003	.1062-007
.60	-.4810-008	-.1440-004	-.1348-003	-.2328-008
.70	-.3948-008	-.1009-004	-.8653-004	-.5028-008
.80	-.2739-008	-.2365-004	-.5204-004	.4172-008
.90	-.1356-008	-.6920-006	.1651-004	-.7828-008
1.00	.2468-015	.3266-005	-.2096-011	-.1066-014

Table 7-2 (cont.)

		GIRDER 21		
X/L	DEFLECTION	SHEAR FORCE	BENDING MOMENT	SLOPE
.00	.0000	.3849-005	.0000	.0000
.10	-.7121-038	.2853-005	-.1451-004	.1475-008
.20	-.1272-037	-.1891-005	-.1003-004	.2692-008
.30	-.1689-037	.1625-005	-.1541-004	.1371-008
.40	-.1929-037	-.1062-005	-.1575-004	-.7801-009
.50	-.1990-037	.2767-010	-.7561-005	-.1387-008
.60	-.1929-037	.1062-005	-.1575-004	-.7801-009
.70	-.1689-037	-.1625-005	-.1541-004	.1371-008
.80	-.1272-037	.1891-005	-.1002-004	.2692-008
.90	-.7121-038	-.2853-005	-.1451-004	.1475-008
1.00	.1360-044	-.3849-005	.2471-011	-.2710-015

loading can be described and any change of stiffness can be incorporated into the analysis. At present, the system of the simultaneous equations, as developed in the subroutine, has a maximum size of 140 × 140. Hence, for this program, a mesh pattern of 140 points could be entertained. The finite difference equations have been computer programmed for the solution of deflections and for the solution of the moments or strains, as given in Appendix A-2 and A-3, respectively. The size and stiffnesses are input to the deflection program, along with the desired mesh grid and the particular cases of loading. The program then generates a set of simultaneous equations, one for every point on the grid, incorporating the effects of the boundary and edge conditions. These equations are set up in matrix form and solved internally using a Gaussian reduction method. The solution is a set of deflections prescribing the deformed shape of the model. These deflections are printed out, in addition to the stiffnesses of the bridge, with an accuracy check of the equation. This program is written in FORTRAN IV and is computed by the IBM 7090/7094 system. The output is given in such a format as to interface with the input data to the moment program. This program takes the given deflections and, by means of finite differences, calculates the curvatures and moments of the bridge. These moments, the angular moment, the radial moment, and the torsional moment, are given for each point of the grid.

Finite Difference Deflection Program—Input Parameters: The following describes the necessary input data for this program.

Card 1	Bridge name	General description
Card 2	Loading description	

Transverse Girders:

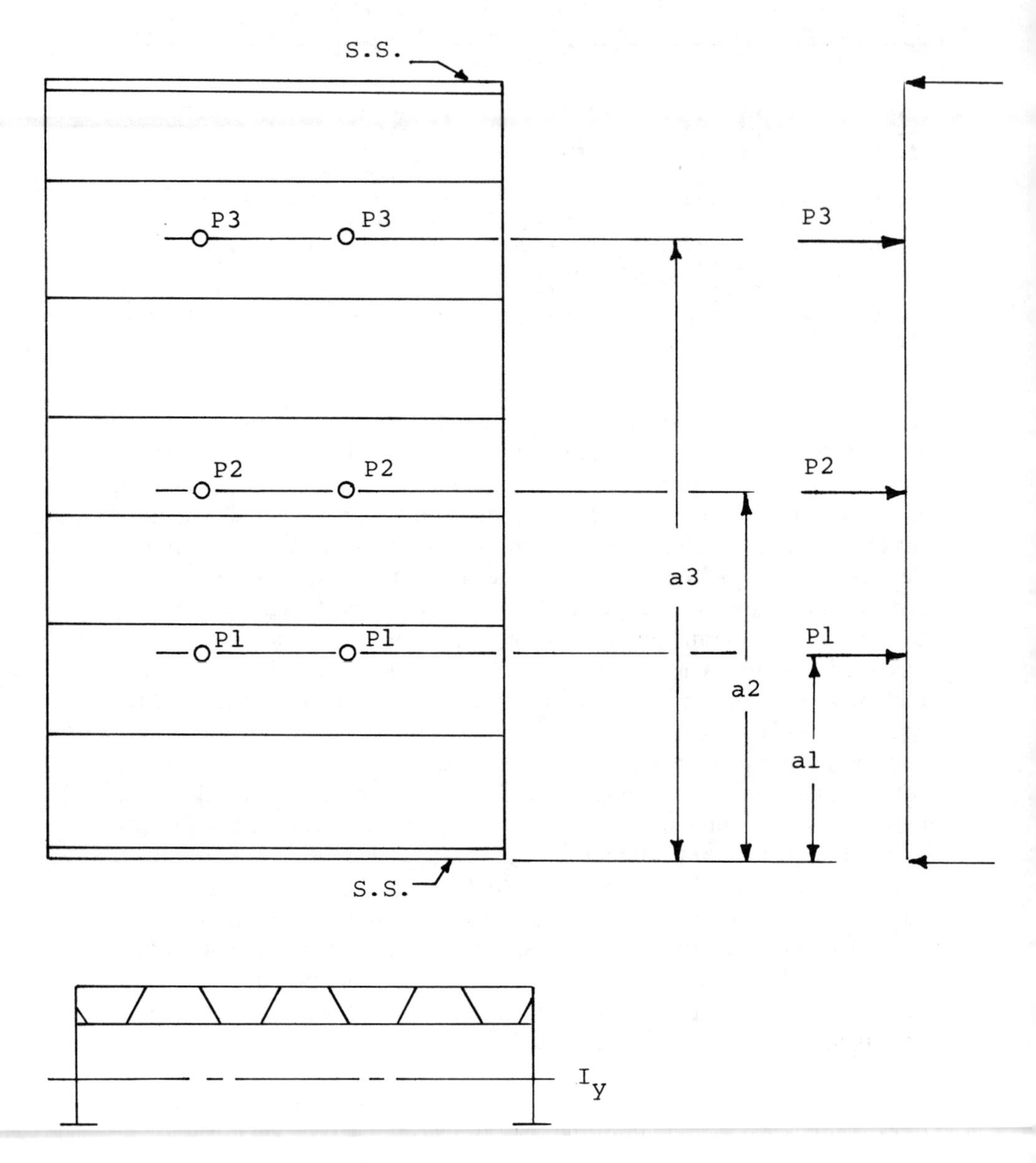

Figure 7-11. Main Girder Solution.

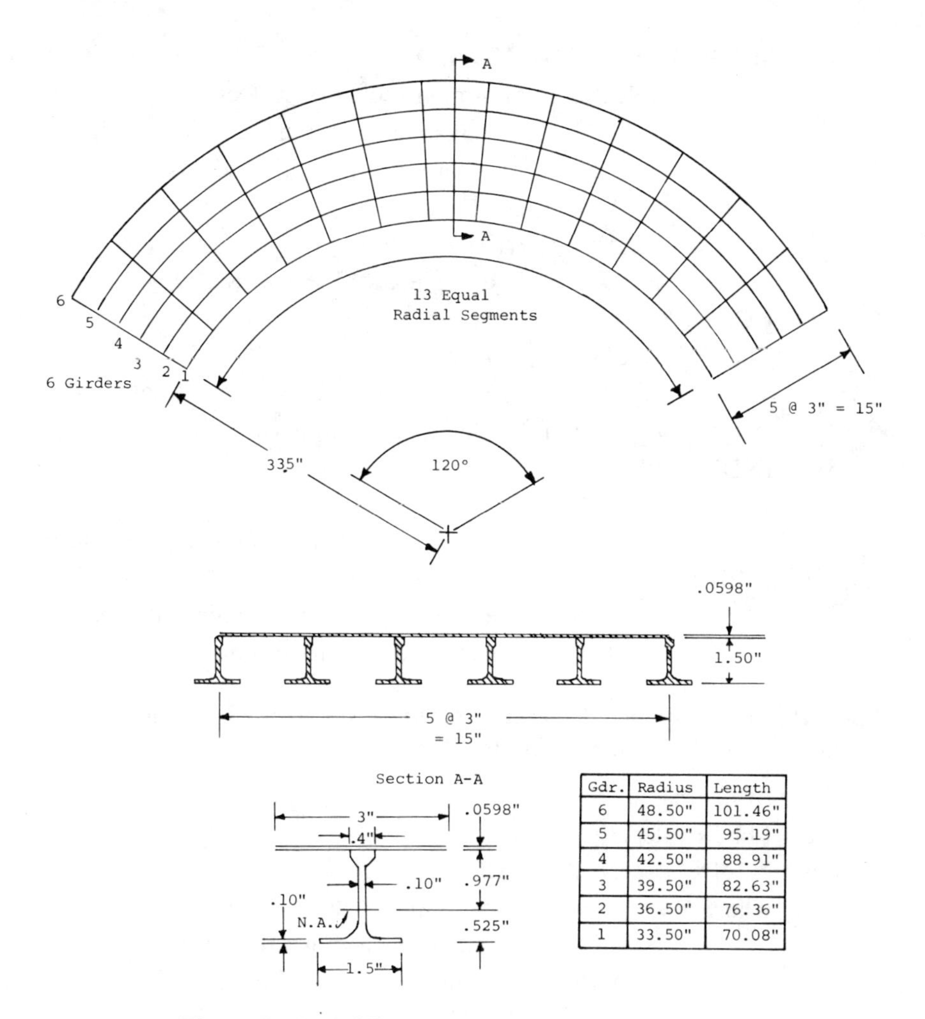

Gdr.	Radius	Length
6	48.50"	101.46"
5	45.50"	95.19"
4	42.50"	88.91"
3	39.50"	82.63"
2	36.50"	76.36"
1	33.50"	70.08"

Figure 7-12. Stiffened Plate Model Geometry.

Card 3 N, NN, MM, ICASE, IQUA, IORTHO

where:
- N = Total number of mesh points
- NN = Number of mesh points per row
- MM = Number of iterations desired
- ICASE = Number of loadings being considered
- IQUA = Coeff. of equations output (1, No; 2, Yes)
- IORTHO = Stiffnesses output (1, Orthotropic; $\neq 1$, Deck and Girder)

Card 4 MINRAD, LAM, OTHETA, ALP, BET, DR

where: MINRAD = Minimum radius (r) in feet

LAM = Radial spacing (λ_r) in feet

OTHETA = Angular spacing (θ_0) in radians

ALP = D_θ / D_r (α)

BET = H / D_r (β)

DR = Radial plate stiffness in K-ft²/ft

Card 5 KCON, KSTIF

where: KCON = $E_{i\theta}/JG$

KSTIF = EI_θ(K-ft²)—Angular girder stiffness

Card 6 JSTIF = EI_r (K-ft²)—Radial girder stiffness

Card 7 INDEX, LL

where: INDEX = Case load number

LL = Number of loaded points in a given case

Card 8 I, BB(I)

where: I = Location of load, mesh point number

BB(I) = Value of load at mesh point I

Finite Difference Moment Program—*Input Parameters:* The following describes the necessary input data for this program.

Card 1 Bridge name
Card 2 Loading description } General description

Card 3 N, NN, RMIN, AL, THO, ICASE

where N = Total number of mesh points

NN = Number of mesh points per row

RMIN = Minimum radius—(r) in feet

AL = Radial spacing—(λ_r) in feet

THO = Angular spacing—(θ_0) in radians

ICASE = Number of loadings being considered

Card 4 EIA(I) = EI = Angular girder stiffness in K-ft²

Card 5 HIA(I) = GK = Torsion stiffness in K-ft²

Card 6 EIR(I) = EI_r = Radian girder stiffness in K-ft²

Card 7 INDEX = Case loading designation

Card 8 DEF(I) = All mesh point deflections in ft

The output data are the bridge identification, case loadings, moments in the radial and angular directions, and torques at all mesh points.

Figure 7-13. Detail of Model.

Example Input Data: In order to explain the use of the curved finite difference computer programs, a stiffened plate that was physically tested [32, 33] will be analyzed. The geometry of the curved plate is shown in Figures 7-12 through 7-15. Simple supports were positioned to create an angle of $\theta = 83°$, as shown in Figure 7-15. The model was divided into 88 mesh points, which gave

$$\lambda = 1.5 \text{ in} = 0.125 \text{ ft}$$

$$\theta_0 = 9.23° = 0.1611 \text{ radians}$$

$$R_{\min} = 33.5 \text{ in} = 2.79 \text{ ft}$$

The plate stiffness, as obtained from stiffness tests [32, 33], were equal to

$$H = 27.5 \text{ K-in}^2/\text{in}$$

$$D_\theta = 2631.0 \text{ K-in}^2/\text{in}$$

$$D_r = 0.4775 \text{ K-in}^2/\text{in} = 0.03979 \text{ K-ft}^2/\text{ft}$$

The relative stiffness parameters α and β are, therefore

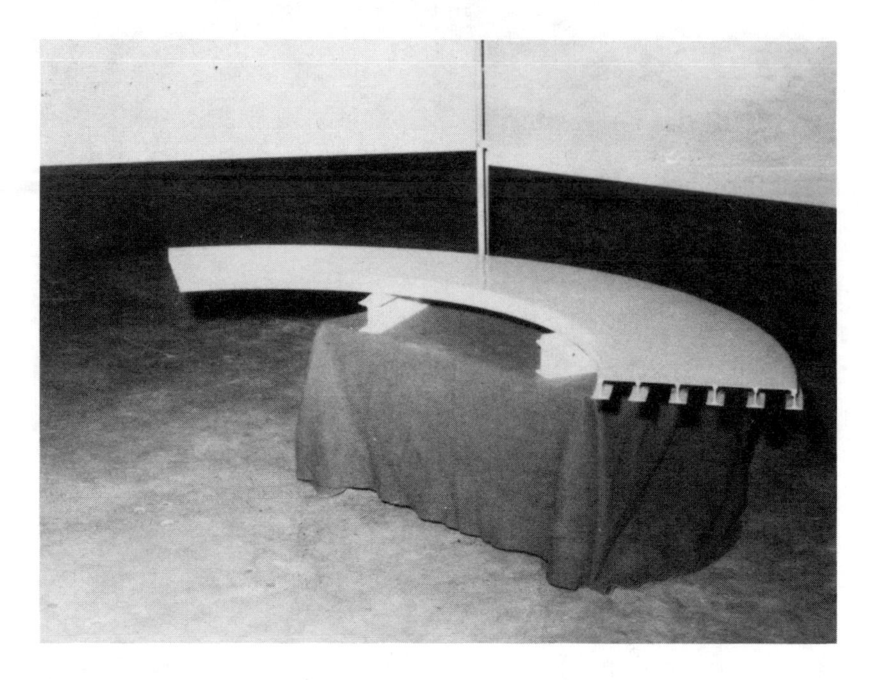

Figure 7-14. Elevation View of Model.

$$\alpha = D_\theta / D_r = 2631.0/0.4775 = 5509.9$$

$$\beta = H / D_r = 27.5/0.4775 = 57.62$$

The girder stiffnesses J and K are equal to zero, as the stiffeners are assumed part of the plate. The loads to be applied for this example, designated as Case 5, consist of 10-lb loads (0.01 Kips) at 24 points. The loads were applied along lines $L3$, $L1$, $R1$, and $R3$ of the model, shown in Figure 7-15.

The required input for this model is shown in Table 7-3. Tables 7-4 and 7-5 are the resulting output data.

Using these deflection data, Table 7-5, and the required plate stiffnesses and parameters for the moment program, the required input data is as shown in Table 7-6. The final plate moments are then computed as given in Table 7-7, by using the computer program of Appendix A-3.

A comparison of these resulting analytical model deflections and moments (stresses) has been made with experiments [32, 33]. Some of these results are shown in Figures 7-16 and 7-17.

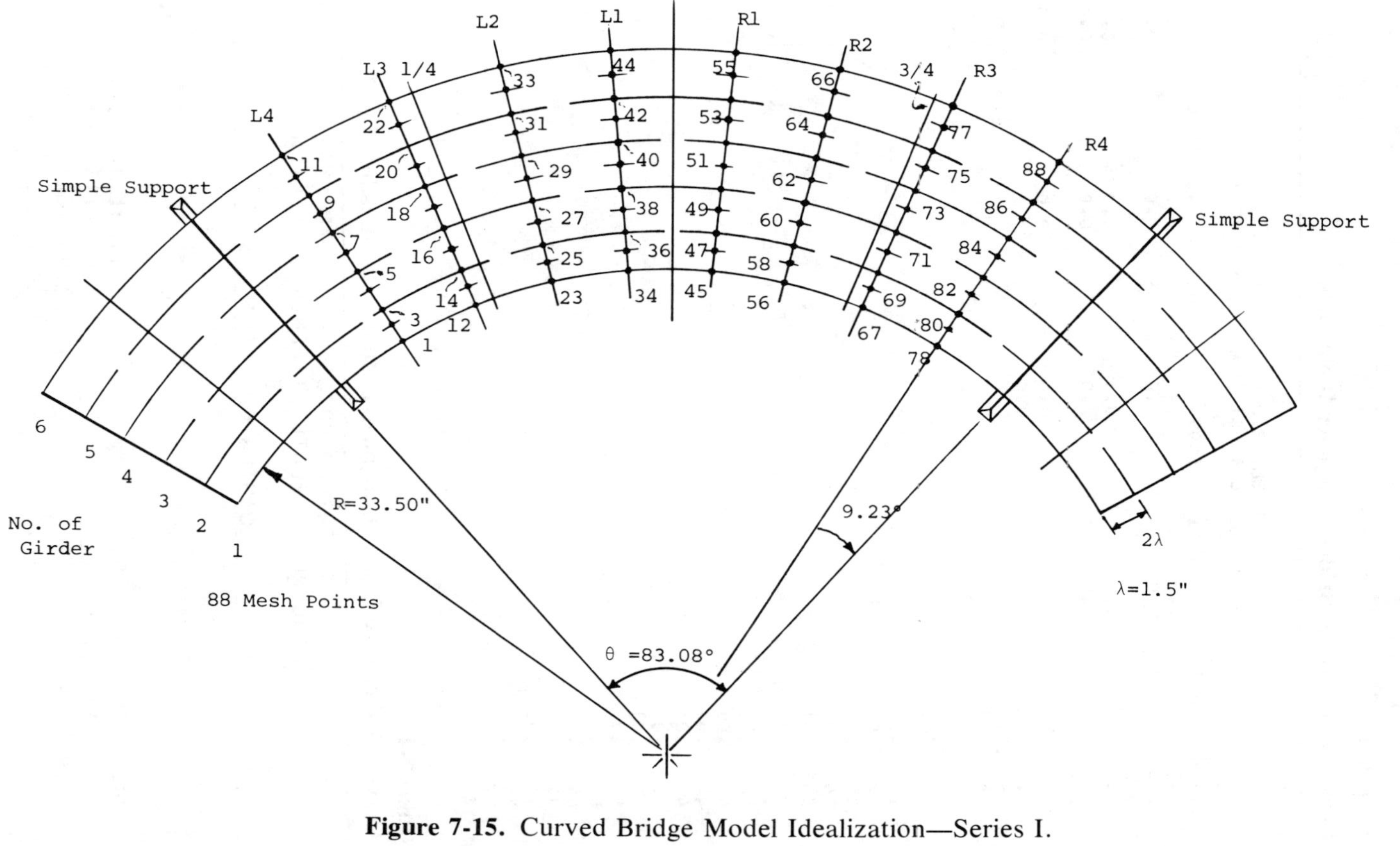

Figure 7-15. Curved Bridge Model Idealization—Series I.

Table 7-3
Curved Orthotropic Bridge Computer Input Data

```
$DATA
1                             CURVED BRIDGE MODEL ANALYSIS
                        ORTHOTROPIC PLATE STIFFNESSES,SERIES I
  88    11      3      1      1      1
 2.79166    0.125    0.1611    5509.      57.62      .0397938
  0.0             0.0             0.0             0.0             0.0
  0.0             0.0             0.0             0.0             0.0
  0.0
  0.0             0.0             0.0             0.0             0.0
  0.0             0.0             0.0             0.0             0.0
  0.0
  0.0             0.0             0.0             0.0             0.0
  0.0             0.0             0.0
   5    24
  12    .01
  14    .01
  16    .01
  18    .01
  20    .01
  22    .01
  34    .01
  36    .01
  38    .01
  40    .01
  42    .01
  44    .01
  45    .01
  47    .01
  49    .01
  51    .01
  53    .01
  55    .01
  67    .01
  69    .01
  71    .01
  73    .01
  75    .01
  77    .01
```

Table 7-4
Curved Orthotropic Bridge Computer Output Data—Loading Coefficients

```
                         CURVED BRIDGE MODEL ANALYSIS
                    ORTHOTROPIC PLATE STIFFNESSES,SERIES I
                                  CASE  5
                          LOADING COEFFICIENTS

0.                  0.                  0.                  0.
0.                  0.                  0.                  0.
0.                  0.                  0.
0.98162275E-03      0.                  0.49081138E-03      0.
0.49081138E-03      0.                  0.49081138E-03      0.
0.49081138E-03      0.                  0.98162275E-03
0.                  0.                  0.                  0.
0.                  0.                  0.                  0.
```

Table 7-4 (cont.)

0.	0.	0.	
0.98162275E-03	0.	0.49081138E-03	0.
0.49081138E-03	0.	0.49081138E-03	0.
0.49081138E-03	0.	0.98162275E-03	
0.98162275E-03	0.	0.49081138E-03	0.
0.49081138E-03	0.	0.49081138E-03	0.
0.49081138E-03	0.	0.98162275E-03	
0.	0.	0.	0.
0.	0.	0.	0.
0.	0.	0.	
0.98162275E-03	0.	0.49081138E-03	0.
0.49081138E-03	0.	0.49081138E-03	0.
0.49081138E-03	0.	0.98162275E-03	
0.	0.	0.	0.
0.	0.	0.	0.
0.	0.	0.	

Table 7-5
Curved Orthotropic Bridge Computer Output Data—Deflections

CURVED BRIDGE MODEL ANALYSIS
ORTHOTROPIC PLATE STIFFNESSES,SERIES I
CASE 5

POINT	DEFLECTION IN INCHES
1	0.54758918E-01
2	0.66143972E-01
3	0.79253701E-01
4	0.94180079E-01
5	0.11132039E 00
6	0.13068675E 00
7	0.15269481E 00
8	0.17737725E 00
9	0.20516904E 00
10	0.23607469E 00
11	0.27012458E 00
12	0.10294445E 00
13	0.12431303E 00
14	0.14897169E 00
15	0.17700399E 00
16	0.20924301E 00
17	0.24561564E 00
18	0.28700776E 00
19	0.33336671E 00
20	0.38563307E 00
21	0.44368914E 00
22	0.50779948E 00
23	0.13863597E 00
24	0.16747522E 00
25	0.20066038E 00
26	0.23846372E 00

Table 7-5 (cont.)

27	0.28185249E 00
28	0.33089890E 00
29	0.38661180E 00
30	0.44911876E 00
31	0.51947075E 00
32	0.59771448E 00
33	0.68383826E 00
34	0.15770337E 00
35	0.19045702E 00
36	0.22822540E 00
37	0.27118406E 00
38	0.32056341E 00
39	0.37630239E 00
40	0.43970259E 00
41	0.51074396E 00
42	0.59080213E 00
43	0.67976382E 00
44	0.77792404E 00
45	0.15770337E 00
46	0.19045702E 00
47	0.22822540E 00
48	0.27118406E 00
49	0.32056341E 00
50	0.37630239E 00
51	0.43970259E 00
52	0.51074396E 00
53	0.59080213E 00
54	0.67976382E 00
55	0.77792404E 00

CURVED BRIDGE MODEL ANALYSIS
ORTHOTROPIC PLATE STIFFNESSES,SERIES I
CASE 5

POINT	DEFLECTION IN INCHES
56	0.13863597E 00
57	0.16747522E 00
58	0.20066038E 00
59	0.23846372E 00
60	0.28185249E 00
61	0.33089890E 00
62	0.38661180E 00
63	0.44911876E 00
64	0.51947075E 00
65	0.59771448E 00
66	0.68383826E 00
67	0.10294445E 00
68	0.12431303E 00
69	0.14097169E 00
70	0.17700399E 00
71	0.20924301E 00
72	0.24561564E 00
73	0.28700776E 00
74	0.33336671E 00
75	0.38563307E 00
76	0.44368914E 00

Table 7-5 (cont.)

77	0.50779948E 00
78	0.54758918E-01
79	0.66143972E-01
80	0.79253701E-01
81	0.94180079E-01
82	0.11132039E 00
83	0.13068675E 00
84	0.15269481E 00
85	0.17737725E 00
86	0.20516904E 00
87	0.23607469E 00
88	0.27012458E 00

```
MM = 3
CHECK = 0.99117503E-08
```

Table 7-6
Curved Orthotropic Bridge Moment Computer Input Data

```
$IBJOB           GO,MAP
1                           CURVED BRIDGE MODEL ANALYSIS
                 ORTHOTROPIC PLATE STIFFNESSES,SERIES I
   88    11  0.27916600E 01  0.12500000E 00  0.16110000E 00     1
 0.2740301E 02 0.2740301E 02 0.2740301E 02 0.2740301E 02 0.2740301E 02
 0.2740301E 02 0.2740301E 02 0.2740301E 02 0.2740301E 02 0.2740301E 02
 0.2740301E 02
 0.2866148E 00 0.2866148E 00 0.2866148E 00 0.2866148E 00 0.2866148E 00
 0.2866148E 00 0.2866148E 00 0.2866148E 00 0.2866148E 00 0.2866148E 00
 0.2866148E 00
 0.4974225E-02 0.4974225E-02 0.4974225E-02 0.4974225E-02 0.4974225E-02
 0.4974225E-02 0.4974225E-02 0.4974225E-02
    5
 0.4563243E-02 0.5511998E-02 0.6604475E-02 0.7848340E-02 0.9276700E-02
 0.1089056E-01 0.1272457E-01 0.1478144E-01 0.1709742E-01 0.1967289E-01
 0.2251038E-01 0.8578704E-02 0.1035942E-01 0.1241431E-01 0.1475033E-01
 0.1743692E-01 0.2046797E-01 0.2391731E-01 0.2778056E-01 0.3213609E-01
 0.3697409E-01 0.4231662E-01 0.1155300E-01 0.1395627E-01 0.1672170E-01
 0.1987198E-01 0.2348771E-01 0.2757491E-01 0.3221765E-01 0.3742656E-01
 0.4328923E-01 0.4980954E-01 0.5698652E-01 0.1314195E-01 0.1587142E-01
 0.1901878E-01 0.2259867E-01 0.2671362E-01 0.3135853E-01 0.3664188E-01
 0.4256200E-01 0.4923351E-01 0.5664699E-01 0.6482700E-01 0.1314195E-01
 0.1587142E-01 0.1901878E-01 0.2259867E-01 0.2671362E-01 0.3135853E-01
 0.3664188E-01 0.4256200E-01 0.4923351E-01 0.5664699E-01 0.6482700E-01
 0.1155300E-01 0.1395627E-01 0.1672170E-01 0.1987198E-01 0.2348771E-01
 0.2757491E-01 0.3221765E-01 0.3742656E-01 0.4328923E-01 0.4980954E-01
 0.5698652E-01 0.8578704E-02 0.1035942E-01 0.1241431E-01 0.1475033E-01
 0.1743692E-01 0.2046797E-01 0.2391731E-01 0.2778056E-01 0.3213609E-01
 0.3697409E-01 0.4231662E-01 0.4563243E-02 0.5511998E-02 0.6604475E-02
 0.7848340E-02 0.9276700E-02 0.1089056E-01 0.1272457E-01 0.1478144E-01
 0.1709742E-01 0.1967289E-01 0.2251038E-01
```

Table 7-7
Curved Orthotropic Bridge Moment Computer Output Data

CURVED BRIDGE MODEL ANALYSIS
ORTHOTROPIC PLATE STIFFNESSES,SERIES I

CASE 5

POINT	ANGULAR MOMENT IN KIP-FT	RADIAL MOMENT IN KIP-FT	TORQUE KIP-FT
1	-.28934115E-03	.00000000E-99	-.35601698E-02
2	.57737320E-02	-.45753956E-04	-.35960333E-02
3	.64947927E-02	-.48194430E-04	-.39429711E-02
4	.71476718E-02	-.58734057E-04	-.43351684E-02
5	.74941224E-02	-.59053999E-04	-.47490895E-02
6	.81597681E-02	-.70084840E-04	-.51891795E-02
7	.85246930E-02	-.70947570E-04	-.56506076E-02
8	.92523031E-02	-.82487772E-04	-.61376021E-02
9	.97945223E-02	-.82608745E-04	-.66391035E-02
10	.11754923E-01	-.83414171E-04	-.71047772E-02
11	.20881519E-01	.00000000E-99	-.64170679E-01
12	.12228458E-02	.00000000E-99	-.29099914E-02
13	.11071510E-01	-.87283402E-04	-.29323945E-02
14	.13233692E-01	-.89497847E-04	-.32144718E-02
15	.13604786E-01	-.11160409E-03	-.35333760E-02
16	.15164783E-01	-.10965897E-03	-.38714153E-02
17	.15549981E-01	-.13316278E-03	-.42295313E-02
18	.17153732E-01	-.13176841E-03	-.46062256E-02
19	.17644655E-01	-.15671753E-03	-.50022952E-02
20	.19590481E-01	-.15359451E-03	-.54096906E-02
21	.22462633E-01	-.16061732E-03	-.57786025E-02
22	.42217847E-01	.00000000E-99	-.73826698E-01
23	-.10348237E-02	.00000000E-99	-.18976839E-02
24	.14482875E-01	-.11529378E-03	-.19138389E-02
25	.16245377E-01	-.12251715E-03	-.20980951E-02
26	.17973938E-01	-.14817619E-03	-.23064328E-02
27	.18774636E-01	-.15009266E-03	-.25269653E-02
28	.20520149E-01	-.17685638E-03	-.27608368E-02
29	.21377259E-01	-.18024044E-03	-.30066680E-02
30	.23262461E-01	-.20812475E-03	-.32655055E-02
31	.24576442E-01	-.20935995E-03	-.35324111E-02
32	.29506438E-01	-.20905115E-03	-.37767859E-02
33	.52152521E-01	.00000000E-99	-.74651955E-01
34	.93398703E-03	.00000000E-99	-.65017558E-03
35	.16845788E-01	-.13303544E-03	-.66362894E-03
36	.19727987E-01	-.13769609E-03	-.72849756E-03
37	.20752281E-01	-.17033656E-03	-.80179790E-03
38	.22658901E-01	-.16871297E-03	-.87767410E-03
39	.23709749E-01	-.20324762E-03	-.95964830E-03
40	.25678431E-01	-.20271598E-03	-.10443820E-02

Table 7-7 (cont.)

41	.26896958E-01	-.23920530E-03	-.11353069E-02
42	.29384733E-01	-.23620644E-03	-.12294223E-02
43	.34218240E-01	-.24402513E-03	-.13261741E-02
44	.63099917E-01	.00000000E-99	-.66460840E-01
45	.93398703E-03	.00000000E-99	.65017558E-03
46	.16845788E-01	-.13303544E-03	.66362894E-03
47	.19727987E-01	-.13769609E-03	.72849756E-03
48	.20752281E-01	-.17033656E-03	.80179790E-03
49	.22658901E-01	-.16871297E-03	.87767410E-03
50	.23709749E-01	-.20324762E-03	.95964830E-03
51	.25678431E-01	-.20271598E-03	.10443820E-02
52	.26896958E-01	-.23920530E-03	.11353069E-02
53	.29384733E-01	-.23620644E-03	.12294223E-02
54	.34218240E-01	-.24402513E-03	.13261741E-02
55	.63099917E-01	.00000000E-99	-.50170696E-01
56	-.10348237E-02	.00000000E-99	.18976839E-02
57	.14482875E-01	-.11529378E-03	.19138389E-02
58	.16245377E-01	-.12251715E-03	.20980951E-02
59	.17973938E-01	-.14817619E-03	.23064328E-02
60	.18774636E-01	-.15009266E-03	.25269653E-02
61	.20520149E-01	-.17685638E-03	.27608368E-02
62	.21377259E-01	-.18024044E-03	.30066680E-02
63	.23262461E-01	-.20812475E-03	.32655055E-02
64	.24576442E-01	-.20935995E-03	.35324111E-02
65	.29506438E-01	-.20905115E-03	.37767859E-02
66	.52152521E-01	.00000000E-99	-.27936997E-01
67	.12228458E-02	.00000000E-99	.29099914E-02
68	.11071510E-01	-.87283402E-04	.29323945E-02
69	.13233692E-01	-.89497847E-04	.32144718E-02
70	.13604786E-01	-.11160409E-03	.35333760E-02
71	.15164783E-01	-.10965897E-03	.38714153E-02
72	.15549981E-01	-.13316278E-03	.42295313E-02
73	.17153732E-01	-.13176841E-03	.46062256E-02
74	.17644655E-01	-.15671753E-03	.50022952E-02
75	.19590481E-01	-.15359451E-03	.54096906E-02
76	.22462633E-01	-.16061732E-03	.57786025E-02
77	.42217847E-01	.00000000E-99	-.22901650E-02
78	-.28934115E-03	.00000000E-99	.35601698E-02
79	.57737320E-02	-.45753956E-04	.35960333E-02
80	.64947927E-02	-.48194430E-04	.39429711E-02
81	.71476718E-02	-.58734057E-04	.43351684E-02
82	.74941224E-02	-.59053999E-04	.47490895E-02
83	.81597681E-02	-.70084840E-04	.51891795E-02
84	.85246930E-02	-.70947570E-04	.56506076E-02
85	.92523031E-02	-.82487772E-04	.61376021E-02
86	.97945223E-02	-.82608745E-04	.66391035E-02
87	.11754923E-01	-.83414171E-04	.71047772E-02
88	.20881519E-01	.00000000E-99	.15621163E-01

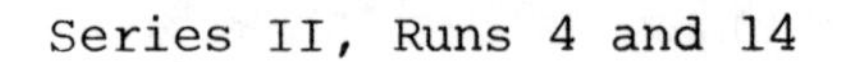
Series II, Runs 4 and 14

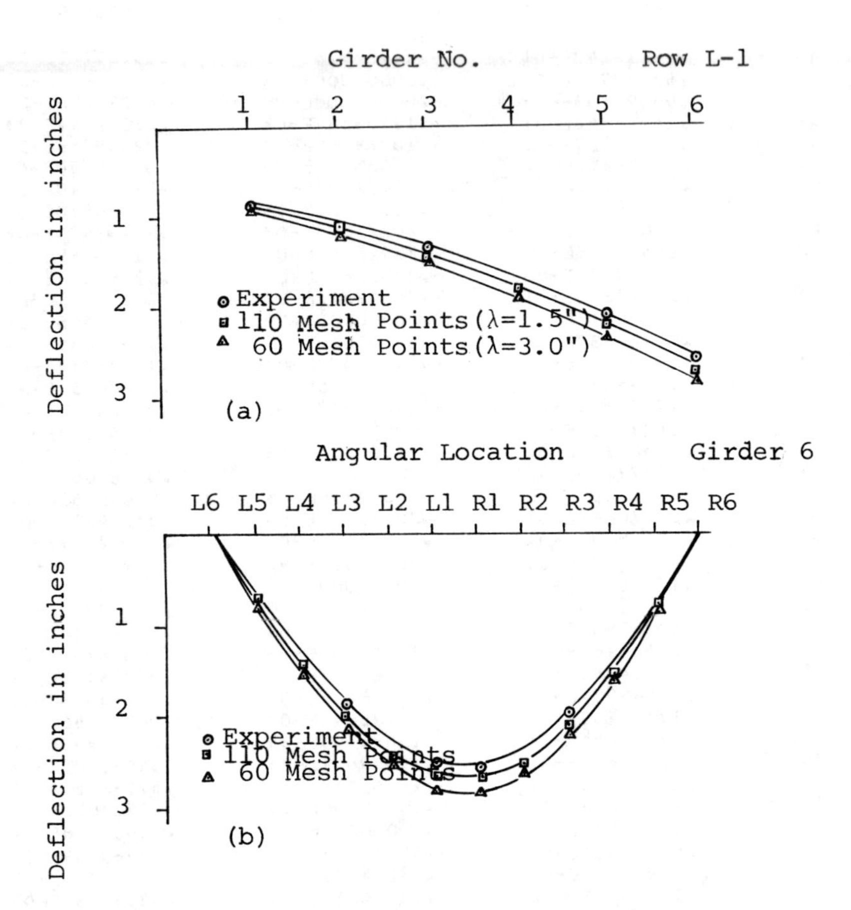

Figure 7-16. Curved Bridge Model Deflection Results.

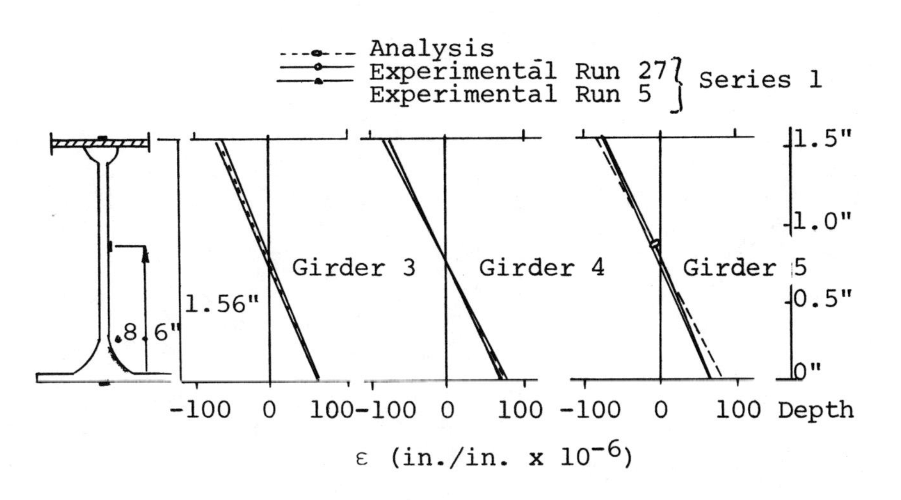

Figure 7-17. Curved Bridge Model Stress Results.

Bridge Data and Examples

8.1 Plate and Girder Stiffnesses

In order to apply the orthotropic plate theory, the plate stiffness parameters D_x, D_y, and H must be known. As described in section 4.6.2, the original development of the plate equations assumed material variations in the definition of terms D_x, D_y, and H. In order to apply these terms relative to engineering structures, in which the material is generally isotropic, but the stiffnesses vary, a new definition will be conceived, i.e.,

$$\begin{aligned} D_x &= EI_x/\text{width} \\ D_y &= EI_y/\text{width} \\ H &= GK_T/\text{width} \end{aligned} \tag{8.1}$$

where E = modulus of elasticity of the material

G = Shear modulus of the material

I_x = Bending stiffness

I_y = Bending stiffness

K_T = Torsional stiffness

With these definitions, a plate that has longitudinal stiffeners and thus a given EI_x, EI_y, and H can be equated to an orthotropic plate. Then the solution of the proper plate equation and structure response can be obtained by use of the various computer programs and theories.

It should be noted that for the various plate classifications, the stiffness parameter $\alpha = H/\sqrt{D_x D_y}$ is equal to, greater than, or less than one.

Various researchers have conducted investigations relative to the determination of the stiffnesses D_x, D_y, and H. The following sections will present these results relative to various structural configurations. The last section will present an experimental procedure that will permit evaluation of the stiffnesses when an analytical technique is questionable or when exact stiffnesses are required.

8.1.1 Uniform Thick Plate ($\alpha = 1.0$)

Isotropic: If the plate is isotropic, then the material properties are equal and thus $D_x = D_y = H = D$, where, as given in section 4.2.5 for a uniform plate:

$$D = \frac{Et^3}{12(1 - \mu^2)} \tag{8.2}$$

Reinforced Concrete Slab: As given in reference [4], a concrete slab that has two-way reinforcement in the x and y directions can be assumed to have the following stiffnesses:

$$\begin{aligned} D_x &= \frac{E_c}{1 - \mu_c^2}[I_{cx} + (n - 1)I_{sx}] \\ D_y &= \frac{E_c}{1 - \mu_c^2}[I_{cy} + (n - 1)I_{sy}] \\ D_{xy} &= \frac{1 - \mu_c}{2}\sqrt{D_x D_y} \end{aligned} \tag{8.3}$$

and

$$H = \sqrt{D_x D_y} \quad \text{or} \quad H/\sqrt{D_x D_y} = 1.0$$

I_{cx} = Moment of Inertia of slab

I_{sx} = Moment of Inertia of reinforcement about neutral axis of section x = constant

I_{cy}, I_{sy} = similar constants for section y = constant

where E_s equals Young's modulus of steel, E_c that of the concrete, μ_c equals Poisson's ratio for concrete, and $n = E_s/E_c$.

8.1.2 Corrugated Sheets

For a plate that forms the shape of a continuous sine wave, Figure 8-1, the properties can be computed as [4]

$$D_x = \left(\frac{L}{S}\right)\frac{Et^3}{12(1 - \mu^2)}$$

$$D_y = EI \tag{8.4}$$

$$H = \left(\frac{S}{L}\right)\frac{Et^3}{12(1 + \mu)}$$

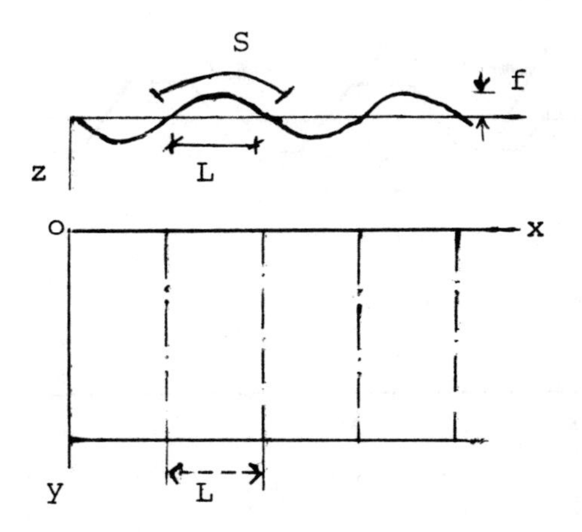

Figure 8-1. Corrugated Plate.

in which

$$S = L\left(1 + \frac{\pi^2 f^2}{4L^2}\right)$$

$$I = \frac{f^2 h}{2}\left[1 - \frac{0.81}{1 + 2.5(f/2L)^2}\right]$$

If the plate has trough type corrugation, as shown in Figure 8-2, recent tests [49] have indicated that the stiffnesses can be computed from a modified version of equation (8.4):

$$D_x = \frac{P}{\bar{S}}\,\frac{Et^3}{12(1 - \mu^2)}$$

$$D_y = EI/\text{width} \tag{8.5}$$

$$H = \frac{\bar{S}}{P}\,\frac{Et^3}{12(1 + \mu)}$$

8.1.3 Open Ribs ($\alpha < 1.0$)

A plate may be reinforced by a series of thin stiffeners in one direction or both directions, symmetrical about the plating. The stiffnesses for such a plate [4], as per Figure 8-3, are:

Plate Reinforced by Equidistant Stiffeners in One Direction: As shown in

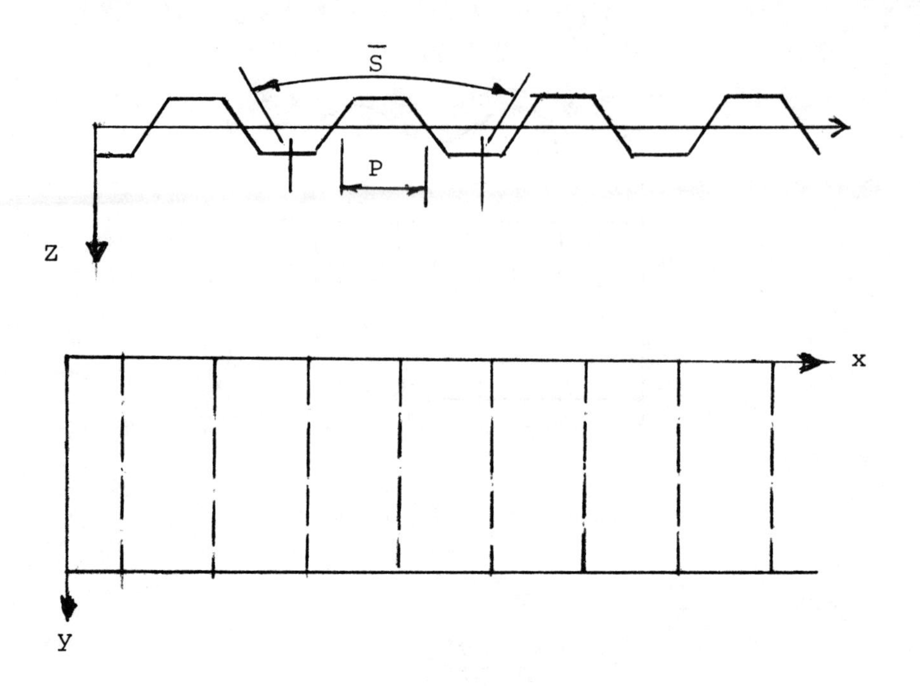

Figure 8-2. Ribbed Plate.

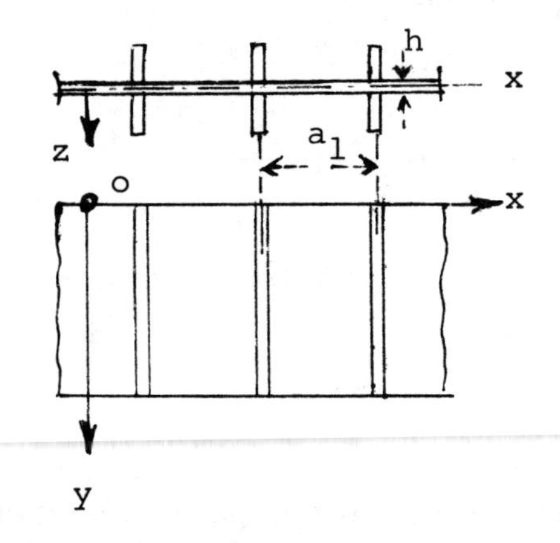

Figure 8-3. Stiffened Plate—Both Sides.

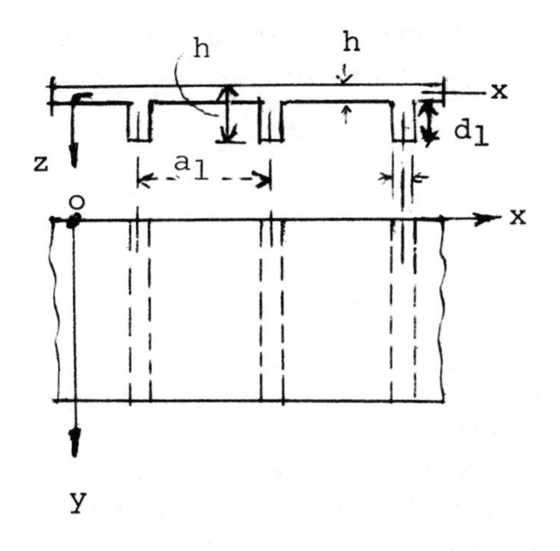

Figure 8-4. Stiffened Plate—One Side.

Figure 8-3, the stiffnesses are

$$D_x = H = \frac{Eh^3}{12(1 - \mu^2)} \tag{8.6}$$

$$D_y = \frac{Eh^3}{12(1 - \mu^2)} + \frac{E'I}{a_1}$$

where E and μ are the elastic constants of the material of the plating, E' Young's modulus, and I the moment of inertia of a stiffener, taken with respect to the middle axis of the cross section of the plate.

Plate Cross-Stiffened by Two Sets of Equidistant Stiffeners: The plate stiffnesses are

$$D_x = \frac{Eh^3}{12(1 - \mu^2)} + \frac{E'I_1}{b_1}$$

$$D_y = \frac{Eh^3}{12(1 - \mu^2)} + \frac{E'I_2}{a_1} \tag{8.7}$$

$$H = \frac{Eh^3}{12(1 - \mu^2)}$$

I_1 being the moment of inertia of one stiffener and b_1 the spacing of the stiffeners in direction x, and I_2 and a_1 being the respective values for the stiffening in direction y.

If the plate has one stiffener attached to one face of the plate, as shown in Figure 8-4, the stiffnesses may be computed as [4]:

$$D_x = \frac{Ea_1h^3}{12(a_1 - t + \alpha^3 t)}$$

$$D_y = EI/a_1 \qquad (8.8)$$

$$H = \frac{Gh^3}{6} + k_1 \frac{G}{2} dt^3/a_1$$

where E is the modulus of the material, I the moment of inertia of a T section of width a_1, and $\alpha = h/H$.

If the plate is also stiffened by stiffeners along the x-axis at a distance a_2 and depth d_2, thickness t_2, the expression given by D_y would then apply to the D_x term. The equations would then be

$$D_x = \frac{EI_x}{a_2}$$

$$D_y = \frac{EI_y}{a_1} \qquad (8.9)$$

$$H = \frac{Gh^3}{6} + \frac{G}{2}\left[\frac{d_1t_1^3}{a_1}k_1 + \frac{d_2t_2^3}{a_2}k_2\right]$$

where k_1 and k_2 in equations (8.8) and (8.9) are dependent on the ratio of the stiffener dimensions (d/t) and are given as follows:

d/t	k	d/t	k
1.0	0.141	2.5	0.249
1.2	0.166	3.0	0.263
1.5	0.196	4.0	0.281
1.75	0.213	5.0	0.291
2.0	0.229	10.0	0.312
2.25	0.240	∞	0.333

In the evaluation of the stiffnesses for a plate with stiffeners, the eccentricity of the stiffener will affect the location of the neutral surface. Thus, the application of the orthotropic equation, which assumes a neutral surface at mid-depth, is in error. In order to account for this discrepancy, Giencke [50‘ rederived the plate equation including the eccentricity effects

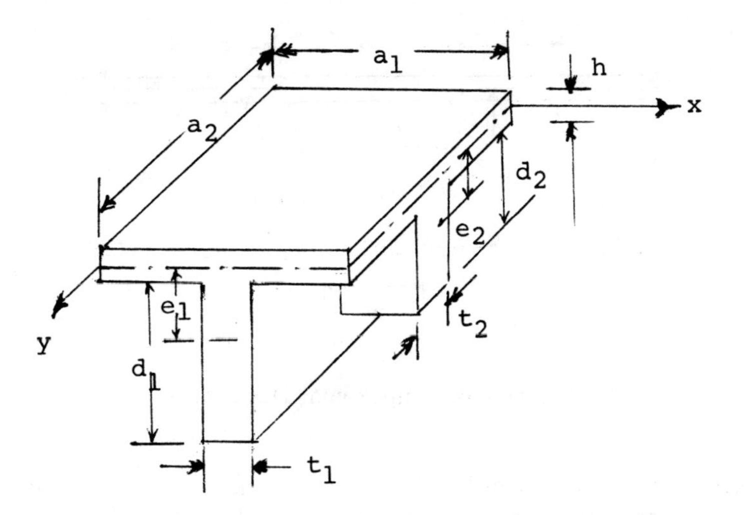

Figure 8-5. T-Ribbed Plate.

from the mid-plate thickness to the centroid of the respective T section, then

$$D_x = EI_x/a_2 \tag{8.10a}$$

$$D_y = EI_y/a_1 \tag{8.10b}$$

$$H = D + \frac{D_{xy}}{2} + \frac{D_{yx}}{2} + \mu e_1^* e_2^* B + (e_1^* + e_2^*)^2 \left(\frac{1-\mu}{4}\right) B \tag{8.10c}$$

where $$e_1^* = \frac{e_1 A_1}{A_1 + h/(1-\mu^2)}$$

$$e_2^* = \frac{e_2 A_2}{A_2 + h/(1-\mu^2)}$$

$$D = \frac{Eh^3}{12(1-\mu^2)}$$

$$B = \frac{Eh}{(1-\mu^2)}$$

A_1, A_2 = Cross-sectional area of stiffeners per unit width

D_{xy}, D_{yx} = Torsional rigidity of ribs of $[G(dt^3/a)k]$ as per notation in equation (8.9)

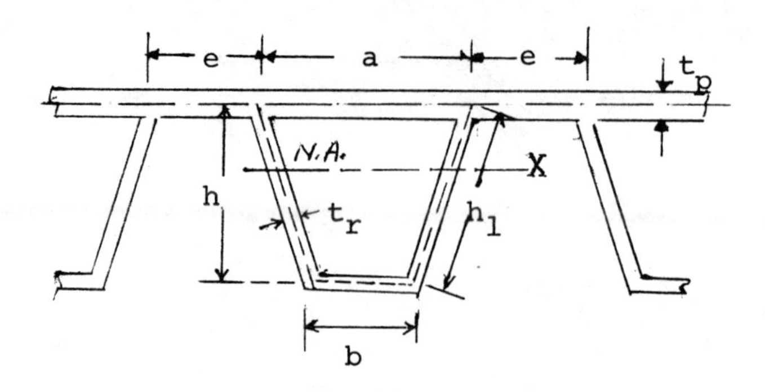

Figure 8-6. Trapezoidal Deck Plate.

Recent work by Cusens [51] has improved the torsional equation H by including a correction term, as verified by tests [52]. The equation (8.10c) becomes

$$H = D + \frac{D_{xy}}{2} + \frac{D_{yx}}{2} + \frac{\mu E t_1 t_2}{2a_1 a_2} d_1 \left\{ A[A - (e_1 + e_2)] + \frac{1}{3} d_1^2 \right\} \quad (8.10d)$$

where $A = (d_1 + h)$, and e_1, e_2 are as shown in Figure 8-5.

8.1.4 Separated Closed Ribs ($\alpha > 1.0$)

In addition to the stiffening of a plate with a single plate unit (open rib), the ribs may be shaped such that a box or closed rib is constructed. Such ribs afford a greater torsional stiffness H, but require more fabrication of the plate.

The equations that will be presented were derived by Pelikan and Esslinger [53], and are presented in detail by Wolchuck [54] and Cusens [51].

Trapezoidal Ribs: For the rib shown in Figure 8-6:

$$D_x = \frac{EI_B}{(a + e)}$$

$$D_y = \frac{Et_p^3}{12(1 - \mu^2)} \quad (8.11)$$

$$H = \frac{RGK_T}{2(a + e)} + \frac{Gt_p^3}{12} + \frac{\mu E t_p^3}{12(1 - \mu^2)}$$

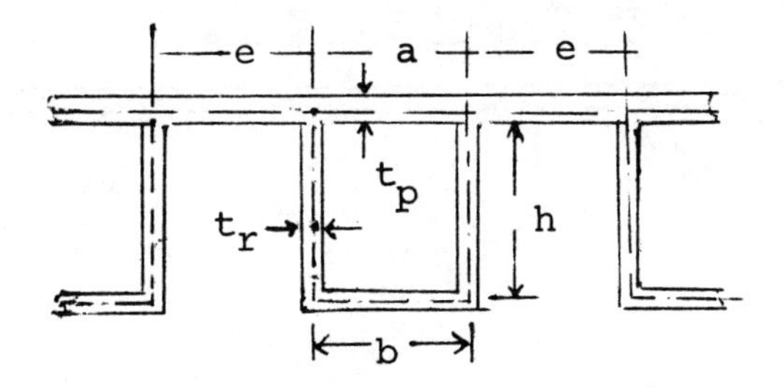

Figure 8-7. Rectangular Deck Plate.

where:

$$K_T = \frac{4\left\{\frac{1}{2}(a+b)h\right\}^2}{\left(\frac{a}{t_p} + \frac{2h_1+b}{t_r}\right)} \tag{8.12}$$

$$\frac{1}{R} = 1 + \frac{GK_T}{EI_p}\,\frac{a^3}{12(a+e)^2}\left(\frac{\pi}{L}\right)^2\left\{\left(\frac{e}{a}\right)^3 + \left(\frac{e-b}{a+b} + \lambda_t\right)^2 + \frac{\lambda_t}{\rho}\left(\frac{b}{a}\right)^3 + \frac{24h_1}{\rho a}(C_1^2 + C_1C_2 + C_2^2/3)\right\} \tag{8.13}$$

and

$$C_1 = \frac{\lambda_t b}{2a}$$

$$C_2 = \frac{\lambda_t(a-b)}{2a} - \left(\frac{a+e}{a+b}\right)\frac{b}{2a}$$

$$\lambda_t = \frac{(2a+b)(a+e)bh_1 - \rho a^3(e-b)}{(a+b)2h_1(a^2+ab+b^2) + b^3 + \rho a^3} \tag{8.14}$$

$$I_p = \frac{t_p^3}{12(1-\mu^2)}$$

$\rho = (t_r/t_p)^3$

L = Simple beam span length of rib; if continuous $L = 0.81\ L_c$

Rectangular Ribs: The stiffnesses of this type of rib, shown in Figure 8-7, is

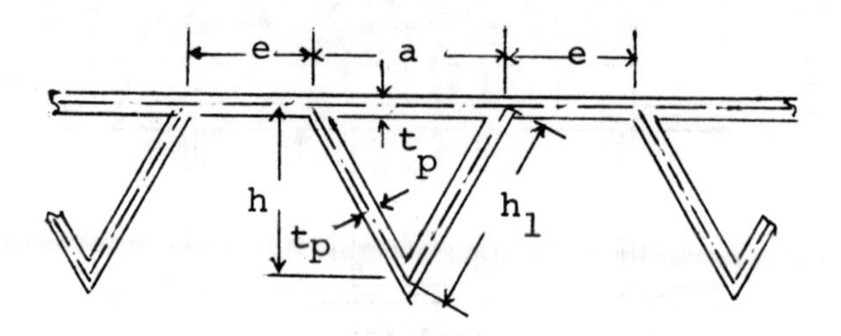

Figure 8-8. Triangular Deck Plate.

obtained from the previous equations by setting $a = b$, $h = h_1$ where similarly D_x, D_y, and H are

$$D_y = \frac{EI_B}{(a + e)}$$

$$D_x = \frac{Et_p^3}{12(1 - \mu^2)} \tag{8.11}$$

$$H = \frac{RGK_T}{2(a + e)} + \frac{Gt_p^3}{12} + \frac{\mu Et_p^3}{12(1 - \mu^2)}$$

and

$$K_T = \frac{4(ah)^2}{\left(\frac{2h + a}{t_r} + \frac{a}{t_p}\right)} \tag{8.15}$$

$$\frac{1}{R} = 1 + \frac{GK_T}{EI_p} \frac{a^3}{12(a + e)^2} \left(\frac{\pi}{L}\right)^2 \left\{\left(\frac{e}{a}\right)^3 + \left(\frac{e - a}{2a} + \lambda_t\right)^2 + \frac{\lambda_t^2}{\rho} + \frac{24h}{\rho a}\left(C_1^2 + C_1C_2 + \frac{C_2^2}{3}\right)\right\} \tag{8.16}$$

$$\lambda_t = \frac{3h(a + e) - \rho a(e - a)}{2a[6h + a(1 + \rho)]}$$

$$C_1 = \frac{\lambda_t}{2} \tag{8.17}$$

$$C_2 = -\left(\frac{a + e}{4a}\right)$$

Triangular Ribs: Shown in Figure 8-8, the stiffnesses D_x, D_y, and H for this rib are as given previously in equation (8.11). The constants R and K_t are

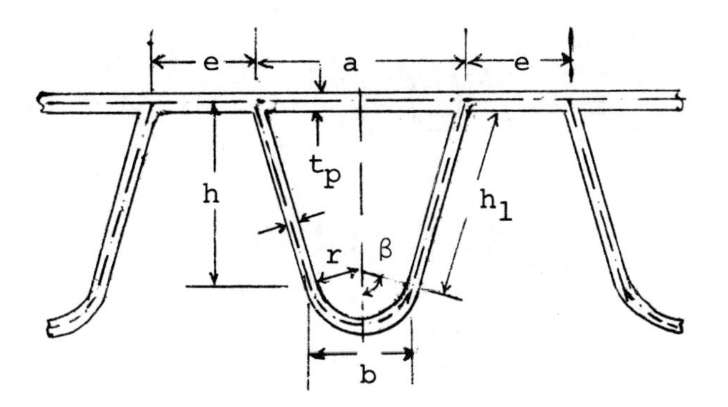

Figure 8-9. Rounded Trapezoidal Deck Plate.

changed and are equal to

$$K_T = \frac{4(ah/2)^2}{(2h_1/t_r + a/t_p)} \tag{8.18}$$

and

$$\frac{1}{R} = 1 + \frac{GK_T}{EI_p} \cdot \frac{a^3}{12(a + e)^2}$$
$$\left(\frac{\pi}{L}\right)^2 \left\{\left(\frac{e}{a}\right)^3 + \left(\frac{e}{2a} + \lambda_t\right)^2 + \frac{2h_1\lambda_t^2}{\rho a}\right\} \tag{8.19}$$

and

$$\lambda_t = \frac{-\rho e}{(2h_1 + \rho a)} \tag{8.20}$$

Rounded Trapezoidal Ribs: This type of deck is described in Figure 8-9. The stiffeners D_x, D_y, and H still conform to the expressions given by equations (8.11). The constants K_T and R, however, are modified as given below.

$$K_T = 4A^2/(2h_1/t_r + 2\beta r/t_r + a/t_p) \tag{8.21}$$

$\beta =$ Angle subtended by rounded portion of stiffener—radians

where A is the area of the enclosed section which is

$$A = h\left(\frac{a + b}{2}\right) + r^2\left(\beta - \frac{b}{2r} \cos \beta\right)$$

also

$$\frac{1}{R} = 1 + \frac{GK}{EI_p}\left(\frac{a^3r^2}{2A^2}\right)\left(\frac{\pi}{L}\right)^2 (A_1 + A_2 + A_3 + A_4) \qquad (8.22)$$

and

$$A_1 = \left[\frac{2A}{r(a + e)}\right]^2 \left(\frac{e}{a}\right)^3 \frac{1}{24}$$

$$A_2 = \frac{1}{24}(\lambda_t - B)^2$$

$$A_3 = \frac{1}{\rho}\left(\frac{r}{a}\right)^3 \left[\frac{\beta^3}{3} - 2(1 + \lambda_t)\left(\frac{b}{2r} - \beta \cos \beta\right) + \frac{(1 + \lambda_t)^2}{2}\left(\beta - \frac{b}{2r} \cos \beta\right)\right]$$

$$A_4 = \frac{1}{\rho}\left(\frac{r^2h_1}{a^3}\right)(C_1^2 + C_1C_2 + C_2^2/3)$$

$$\lambda_t = \frac{B + C}{1 + D}$$

The coefficients B, C, D, C_1 and C_2 associated with these variables are defined as follows:

$$B = \frac{2A}{r(a + e)} - \frac{h}{r} - (1 - \cos \beta) \qquad (8.23)$$

$$C = \frac{24}{\rho}\left(\frac{r}{a}\right)^3 \left[\frac{b}{2r}\left(1 + \frac{1}{2} \cos \beta\right) - \beta\left(\frac{1}{2} + \cos \beta\right) + \frac{h_1(a + b)}{4r^2}\left(\beta - \frac{b}{2r}\right) + \frac{h_1^2(2a + b)}{12r^3}(1 - \cos \beta)\right]$$

$$D = \frac{24}{\rho}\left(\frac{r}{a}\right)^3 \left[\frac{1}{2}\left(\beta - \frac{b}{2r} \cos \beta\right) + \frac{h_1(a^2 + ab + b^2)}{12r^3}\right]$$

$$C_1 = \beta - \frac{b}{2r}(1 + \lambda_t)$$

$$C_2 = \frac{h_1}{r}(1 - \cos \beta) - \lambda_t\left(\frac{a - b}{2r}\right)$$

U Shaped Ribs: This type of deck is shown in Figure 8-10 and noting from Figure 8-9, $a = b = 2r$ and $h = h_1, \beta = \pi/2$. The equations given previously reduce to the following:

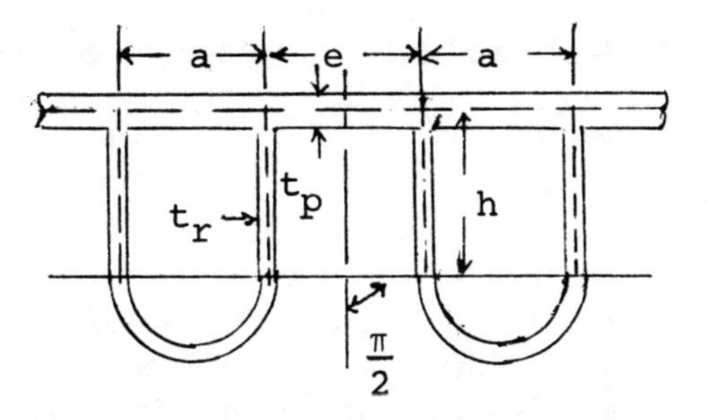

Figure 8-10. U Shaped Deck Plate.

$$K_T = \frac{4\{ah + \pi a^2/8\}^2}{\left(\dfrac{2h + \pi a/2}{t_r} + \dfrac{a}{t_p}\right)} \tag{8.24}$$

and

$$\frac{1}{R} = 1 + \frac{GK_T}{EI_p}\frac{a^5}{8A^2}\left(\frac{\pi}{L}\right)^2 (A_1 + A_2 + A_3 + A_4) \tag{8.25}$$

$$A = ah + \pi a^2/8$$

$$A_1 = \left[\frac{4A}{a(a + e)}\right]^2 \left(\frac{e}{a}\right)^3 \frac{1}{24}$$

$$A_2 = \frac{1}{24}(\lambda_t - B)^2$$

$$A_3 = \frac{1}{8\rho}\left\{\frac{\pi^3}{24} - 2(1 + \lambda_t) + \frac{(1 + \lambda_t)^2}{2}\left(\frac{\pi}{2}\right)\right\}$$

$$A_4 = \frac{h}{4\rho a}(C_1^2 + C_1C_2 + C_2^2/3) \tag{8.26}$$

$$B = \frac{4A}{a(a + e)} - \frac{2h}{a} - 1$$

$$C = \frac{3}{\rho}\left[1 - \frac{\pi}{4} + \frac{2h}{a}\left(\frac{\pi}{2} + 1\right) + 2\left(\frac{h}{a}\right)^2\right]$$

$$D = \frac{3}{\rho}\left(\frac{\pi}{4} + \frac{2h}{a}\right)$$

$$C_1 = \frac{\pi}{2} - (1 + \lambda_t)$$

$$C_2 = \frac{2h}{a}$$

The stiffnesses D_x, D_y, and H are as computed by equation (8.11).

8.1.5 Continuous Closed Ribs

In the previous section the closed ribs were attached to a continuous top plate of uniform thickness. The ribs are, in these cases, generally welded to the top plate along the top of the rib after forming or welding of the boxes. It is possible to form the cells as a continuous unit [55, 56] and then burn weld the cells to the continuous top plate, resulting in the section shown in Figure 8-11. The stiffnesses given previously could be applied by averaging the plate thicknesses; however, more exact equations have been developed by Stroud [57] and are as follows.

$$D_y = E \cdot \bar{I} \tag{8.27}$$

where

$$\begin{aligned}
\bar{I} = \frac{2}{P}\Bigg[& \frac{(a+b)}{12} t_p^3 + \left(\bar{Y} - \frac{t_p}{2}\right)^2 (a+b)t_p + \frac{(a+a_i)}{12} t_c^3 \\
& + (e_1 + r_1)^2 (a + a_i) t_c + t_c r_1^3 \left(\frac{\theta}{2} + \frac{\sin 2\theta}{4}\right) + 2 t_c e_1 r_1^2 \sin\theta \\
& + e_1^2 r_1 \theta t_c + \frac{t_c}{12} a_k^3 \sin^2\theta + \frac{a_k t_c^3}{12} \cos^2\theta \\
& + \left(t_p + \frac{t_c}{2} + a_j + \frac{a_k}{2}\sin\theta - \bar{Y}\right)^2 a_k t_c + t_c r_2^3 \left(\frac{\theta}{2} + \frac{\sin 2\theta}{4}\right) \\
& + 2 t_c e_2 r_2^2 \sin\theta + e_2^2 r_2 \theta t_c + \frac{g t_c^3}{12} + (e_2 + r_2)^2 t_c g \Bigg]
\end{aligned} \tag{8.28}$$

The term $\bar{Y}$ is computed as:

$$\bar{Y} = \frac{\left[\begin{aligned} & \left[\frac{t_p^2}{2}(b+a) + \left(t_p + \frac{t_c}{2}\right)(a+a_i)t_c + \left(t_p + \frac{t_c}{2} + r_1 \right.\right. \\ & \left. - \frac{r_1 - \sin\tau}{\theta}\right) r_1 \theta t_c + \left(t_p + \frac{t_c}{2} + a_j + \frac{a_n - a_j}{2}\right) a_k t_c \\ & + \left(t_p + \frac{t_c}{2} + d - r_2 + \frac{r_2 \sin\theta}{\theta}\right) r_2 \theta t_c + \left.\left(t_p + \frac{t_c}{2} + d\right) g t_c\right] \end{aligned}\right]}{[t_p(b+a) + (a+a_i)t_c + r_1\theta t_c + a_k t_c + r_2 \theta t_c + g t_c]}$$

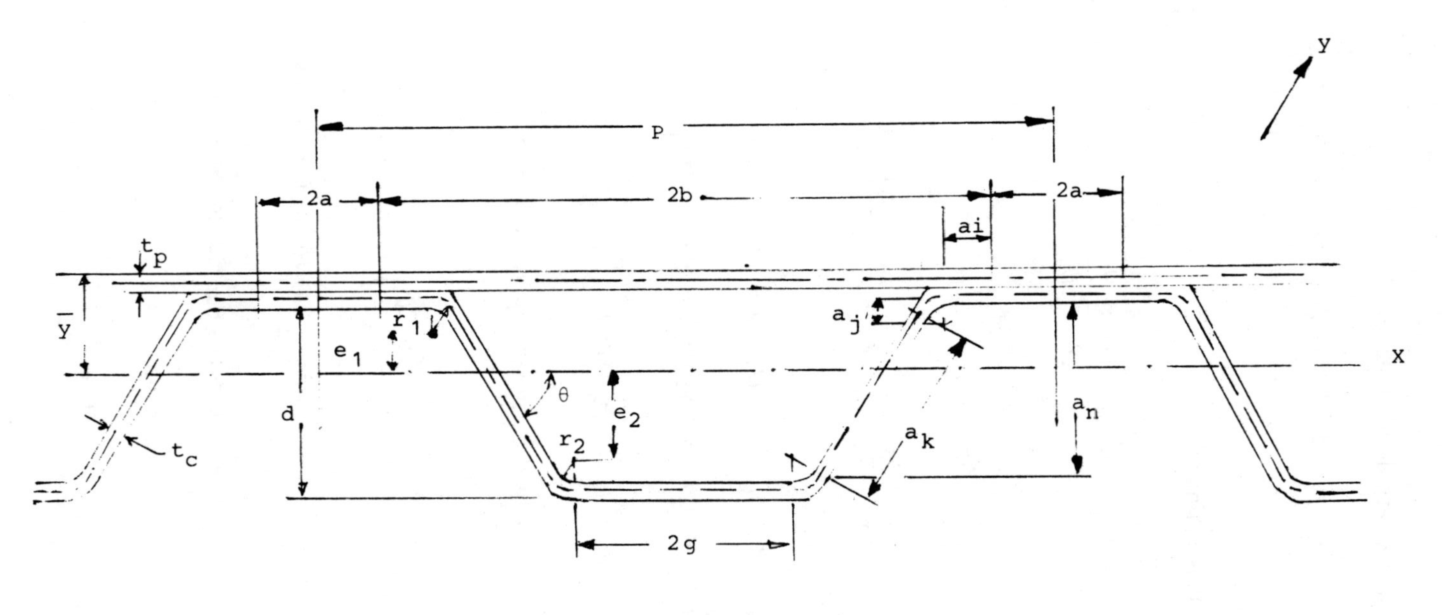

Figure 8-11. Continuous Closed Ribs.

and e_1, e_2, a_j, and a_n are defined as:

$$e_1 = \bar{Y} - (r_1 + t_p + t_c/2)\,; \qquad e_2 = d - (r_1 + r_2 + e_1)$$
$$a_j = r_1(1 - \cos\theta)\,; \qquad a_n = d - r_2(1 - \cos\theta)$$

The stiffness D_x is defined as:

$$D_x = \frac{a + b}{\dfrac{a}{D''_x} + \dfrac{b}{D'_x}} \tag{8.29}$$

where

$$D'_x = \frac{bD_c}{x} + D_p\,; \qquad D_c = \frac{Et_c^3}{12}\,;$$
$$D_p = \frac{Et_p^3}{12}\,; \qquad D''_x = \frac{E(t_c + t_p)^3}{12}$$

also,

$$x = a_i + (r_1 + r_2)\theta + a_k + g - \bar{H}\Big[r_1^2(\theta - \sin\theta) + a_j a_k + \frac{a_k^2}{2}\sin\theta + r_2(a_n\theta - r_2\theta\cos\theta + r_2\sin\theta) + gd\Big]$$

$$\bar{H} = \Big[r_1^2(\theta - \sin\theta) + a_j a_k + \frac{a_k^2}{2}\sin\theta + r_2(a_n\theta - r_2\theta\cos\theta + r_2\sin\theta) + gd\Big]\Big/U1 \tag{8.30}$$

and

$$U1 = r_1^3\left(\frac{3\theta}{2} - 2\sin\theta + \frac{\sin 2\theta}{4}\right) + a_j^2 a_k + a_j a_k^2 \sin\theta + \frac{a_k^3}{3}\sin^2\theta + r_2 a_n^2\theta - 2a_n r_2^2\theta\cos\theta + 2a_n r_2^2\sin\theta + \frac{r_2^3}{2}(3\theta - 3\sin\theta\cos\theta - 2\theta\sin^2\theta) + gd^2$$

The torsional stiffness H is computed as:

$$H = \frac{G \cdot BJ}{2} \tag{8.31}$$

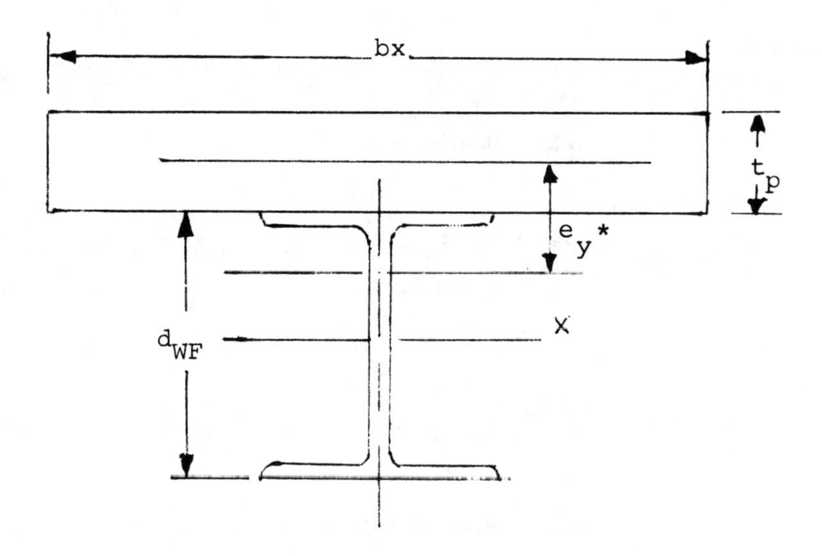

Figure 8-12. Composite Girder.

$$\text{where} \quad BJ = \frac{2}{P \cdot 1K}\left\{b(t_c + t_p) + d\left[b + g - a_i + (r_2 - r_1)\tan\frac{\theta}{2}\right]\right.$$

$$\left. + 2(r_1^2 - r_2^2)\tan\frac{\theta}{2} + (r_2^2 - r_1^2)\theta\right\}^2 \tag{8.32}$$

$$1K = \frac{b}{t_p} + \frac{1}{t_c}[a_i + \theta(r_1 + r_2) + a_k + g]$$

8.1.6 Composite Section

If the structure is composed of a common composite concrete slab supported by interconnected steel girders, it is possible to assume that the system is an equivalent to an orthotropic plate [58, 59]. The general shape of such a girder-slab is shown in Figure 8-12. Following a similar procedure as used by Geincke [50], the properties are computed as:

$$\begin{aligned} D_y &= E_c(\bar{I}_y - e_y^* \bar{S}_y) \\ D_x &= E_c I_p \\ H &= E_c[I_p + (e_y^*)^2 t_p/4] \end{aligned} \tag{8.33}$$

where

$$
\begin{aligned}
A_y &= A_{WF}/bx \\
e_y &= 0.5(d_{WF} + t_p) \\
I_y &= I_{WF}/bx + A_y(e_y)^2 \\
I_p &= t_p^3/12 \\
\bar{A}_y &= t_p + nA_y \\
\bar{I}_y &= I_p + nI_y \\
\bar{S}_y &= A_y e_y n \\
e_y^* &= \bar{S}_y/\bar{A}_y \\
n &= E_s/E_c \\
A_{WF} &= \text{Area of wide flange} \\
E_s, E_c &= \text{Modulus of elasticity of steel and concrete, respectively}
\end{aligned}
\tag{8.34}
$$

8.1.7 Box Cells

If a deck is composed of cellular or monalithic box beams in both directions, as shown in Figure 8-13, the equivalent plate stiffnesses are computed as [60]:

$$
\begin{aligned}
D_x &= EI_x/\text{width} \\
D_y &= EI_y/\text{width} \\
2H &= \left[\frac{GK_{Tx}}{2} + \frac{GK_{Ty}}{2}\right]\Big/\text{width}
\end{aligned}
\tag{8.35}
$$

where the torsional terms K_{Tx} and K_{Ty} are determined as follows:

Several Cells:

$$
K_{Ty} = K_{Tx} = \frac{A^2 t}{b}\left[1 - \frac{2\rho\alpha(\alpha^n - 1)}{n(\alpha - 1)(\alpha - 1)(\alpha^n - 1) + (\alpha^n + \alpha)}\right] \tag{8.36}
$$

where $2b$ = Width of enclosed section (see Figure 8-13)

d_1 = Depth of enclosed section between mid-flange points

A = Area of enclosed section ($A = d_1 \times 2b$)

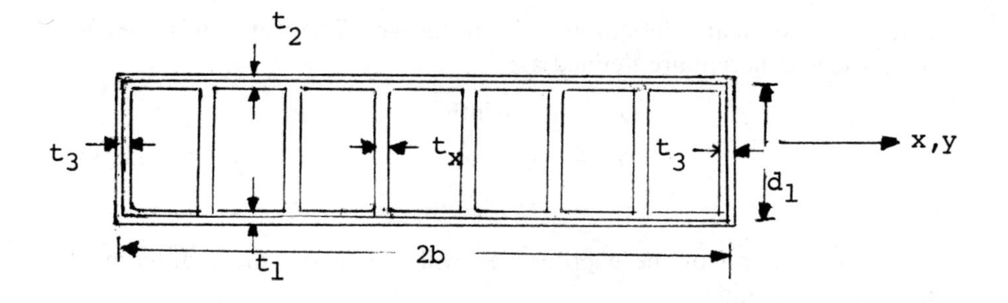

Figure 8-13. Multi-cell Section.

n = Number of cells

t_3 = Thickness of end webs

$t = t_1 = t_2$ = Thickness of top and bottom flanges

t_x = Thickness of internal web

G = Modulus of rigidity

$$r = \frac{2bt_x}{nd_1t_3}$$

$$\alpha = 1 + r + \sqrt{2r + r^2}$$

$$\rho = t_x/t_3$$

Five or More Cells:

$$K_{Tx}, K_{Ty} = \frac{8d_1^2b}{\left(\frac{2b}{t_2} + \frac{2b}{t_1} + \frac{d_1}{t_3} + \frac{d_1}{t_3}\right)} \tag{8.37}$$

These equations apply when the web thickness t_3 and flange thicknesses t_1, and t_2 are small compared to the overall dimensions d_1 and $2b$.

8.2 Experimental Determination of Plate Stiffnesses

If a plate has deformed a configuration or attachments, and it is desired to evaluate as accurately as possible the plate stiffnesses (D_x, D_y, and H), experiments on the plate may be conducted. Various schemes have been proposed to test such plates [32, 57, 61, 62, 63, 64] depending on the exactness desired and definition of the stiffnesses. The plate stiffnesses to be considered herein are defined as:

$$D_x = EI_x/\text{width}$$
$$D_y = EI_y/\text{width} \qquad (8.38)$$
$$H = (GK_T/\text{width})_{\text{avg}}$$

The evaluation of these properties will utilize the finite difference procedure, recalling

$$M_x = -D_x \frac{\partial^2 w}{\partial x^2} = -D_x\left(\frac{w_l - 2w_0 + w_r}{\lambda^2}\right) \qquad (8.39)$$

$$M_y = -D_y \frac{\partial^2 w}{\partial y^2} = -D_y\left(\frac{w_a - 2w_0 + w_b}{n\lambda^2}\right) \qquad (8.40)$$

$$M_{xy} = H \frac{\partial^2 w}{\partial x \partial y} = H\left(\frac{w_{ar} - w_{br} - w_{al} + w_{bl}}{4n\lambda^2}\right) \qquad (8.41)$$

Consider the plate shown in Figure 8-14(a), which has stiffeners in one direction. If the plate is now constructed to have the ribs spanning longitudinally, where the length $3a$ is much greater than the width, as shown in Figure 8-14(b), loads may be applied at 1/3 points. Measurement of the deformations in the pure moment region at spacings of λ will permit evaluation of

$$D_x = EI_x = Pa\lambda^2/[w_l - 2w_0 + w_r] \qquad (8.42)$$

Similarly for the transverse stiffness D_y, as shown in Figure 8-14(c),

$$D_y = EI_y = Pa(n\lambda)^2/(w_a - 2w_0 + w_b) \qquad (8.43)$$

where P is a line load (force/length).

For the torsional stiffness, the plate is constructed with equal lengths, as shown in Figure 8-14(d). Concentrated loads are applied diagonally, with supports on the other two diagonal edges. Meausrement of the deflections gives

$$H = \left(\frac{GK_T}{\text{width}_{\text{avg}}}\right) = \left(\frac{P}{2}a\right)\frac{4n\lambda^2}{a}\Big/(w_{ar} - w_{br} - w_{al} + w_{bl}) \qquad (8.44)$$

8.3 Estimate of Floor Beam Forces

Using the slope-deflection method and the resulting computer program, a study [48] was conducted on the induced forces in floor beams associated with orthotropic bridges. In conducting such a study, certain bridge param-

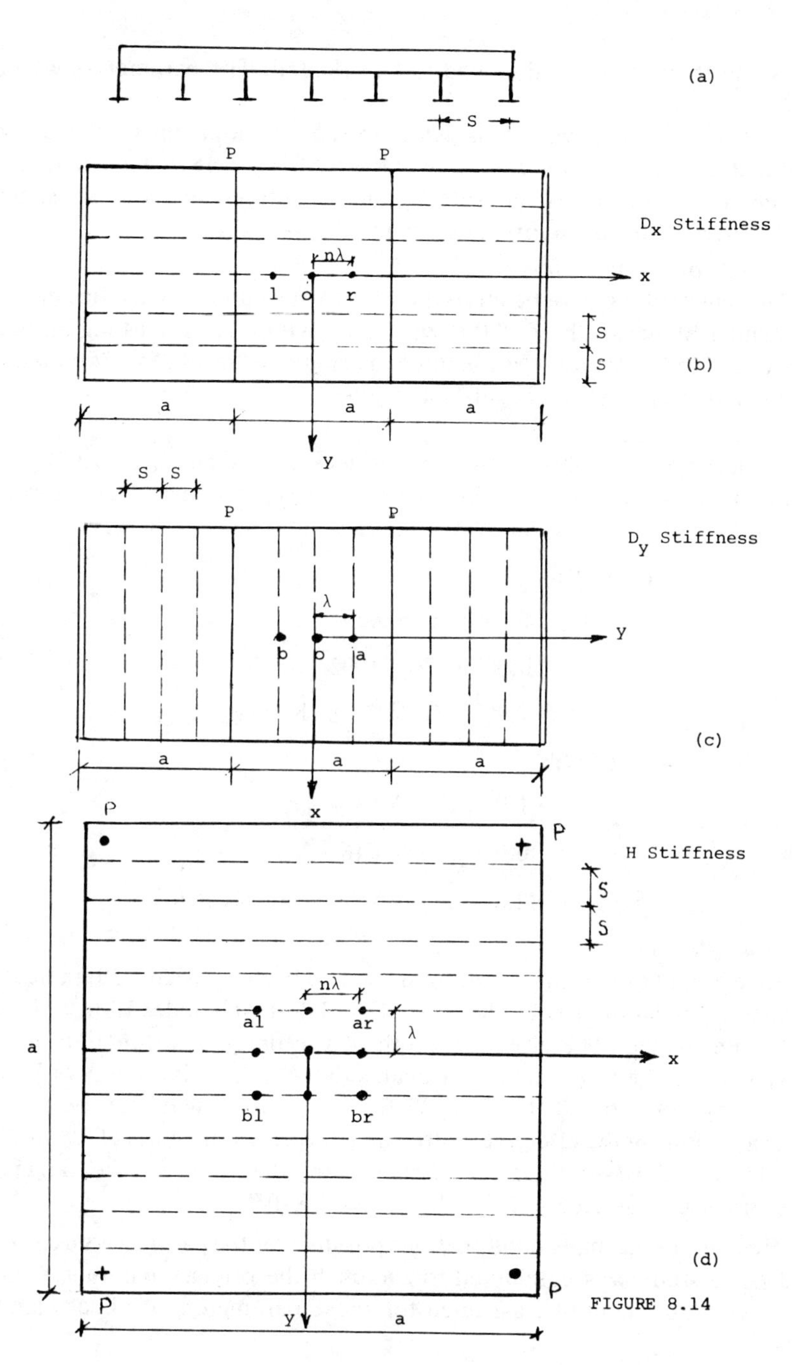

Figure 8-14. Experimental Evaluation of Plate Stiffnesses.

eters and the type of loading had to be selected. The parameters were:

Loading

Only live loading was considered in the development of the design equations. The loading consisted of the AASHTO HS20-44 vehicle. The vehicles were positioned side by side, centered in their lanes, and oriented in the same direction for maximum effect.

Length of Floor Beams

The length of the floor beams is dictated by the number of traffic lanes. A standard lane width of 10.0 ft was used, with a 2.5-ft curbing on both sides of the roadway. This resulted in bridge widths of 25 ft, 35ft, and 45 ft for two-, three-, and four-lane bridges.

Deck Properties

As described previously, the type of deck can be characterized by the parameters α and β. A complete study of typical deck configurations [54, 65, 66] indicates the following practical ranges in the parameters:

Open Rib:

$$0.05 < \alpha < 0.35$$

$$0.002 < \beta < 0.02$$

$$7.98 < D_y < 71.8 \text{ (K-ft}^2\text{/ft)}$$

Closed Rib:

$$1.0 < \alpha < 3.6$$

$$0.02 < \beta < 0.16$$

$$12.8 < D_y < 52.5 \times 10^3 \text{ (K-ft}^2\text{/ft)}$$

Floor Beam

An examination of the literature of typical bridge designs [39] indicates that a typical floor beam has a 36-in × 3/8-in web and a 10-in × 1/2-in bottom flange. The effective width of top flange is a function of the spacing of the transverse floor beams [54, 65]. Considering floor beam spacings of 5 ft, 10 ft, 15 ft, 20 ft, and 25 ft, which then permitted evaluations of the effective width and a top plate thickness of 3/8-in, the effective cross-sectional properties were determined. The range in stiffnesses was $1.65 < EI < 2.18 \times 10^6$ (K-ft^2).

Because of the many combinations possible for the parameters α, β, D_y, and EI, a study was conducted to establish the primary parameters and limitations that could be assumed for those parameters of minor significance.

Open Rib

Examination of the various influences the properties of an open rib ($\alpha <$

1.0) have on the resulting forces in the girder system indicated the following:

1. α does not have great influence, therefore, it was set equal to a minimum of 0.05 and a maximum of 0.35 in conducting the study.
2. The primary parameter is D_y/EI, the y plate stiffness to the floor beam stiffness, which had ranges of $0.003 < D_y/EI < 0.05$.
3. β was held constant at 0.002.

Closed Rib

The effects of a closed rib ($\alpha > 1.0$) on the forces in the system indicated:

1. α has minimal effects, therefore, a minimum of 1.0 and maximum of 5.0 were used.
2. β was held constant at 0.02.
3. The stiffness D_y/EI had a range of 0.005 to 0.05.

In combination with the various stiffnesses and the bridge dimensions and loadings, the resulting girder forces and deformations were computed. It was found that the following relationship could represent the interaction of these parameters [48].

$$\log (f/f^*) = -0.11N - 0.86(1 - \log S) + (-0.0485N + 0.046) \log (D_y/EI) \qquad (8.45)$$

where (f/f^*) represents (δ/δ^*), (M/M^*), (V/V^*) and

δ = Girder system deflection

M = Girder system moment

V = Girder system shear

δ^* = Simple beam deflection

M^* = Simple beam moment

V^* = Simple beam shear

D_y = Primary plate bending stiffness/width

S = Floor beam spacing

EI = Floor beam stiffness

N = Number of traffic lanes

The girder system values f represent the response of the floor beams when interacting as part of an orthotropic bridge. The f^* values are the functions computed when the floor beam is isolated as a simple beam and subjected to a set of single AASHTO HS-20 axles, as determined by the number of lanes.

8.4 Orthotropic Plate Sizes

In reviewing the literature [39] that describes existing orthotropic steel plate bridges, the following proportions of members, as listed in Tables 8-1 and 8-2 were determined. In general, bridges with open cross sections will have the following parameters:

Size of rib	(3/8 in × 8 in) to (1 in × 12 in)
Rib spacing	12 to 16 in
Floor beam spacing	4 to 7 ft

and those bridges with closed cross sections will have the following parameters:

Cell spacing	24 to 28 in
Size of cell	12 × 12 × 5/6 in
Floor beam spacing	4 to 15 ft

8.5 Design Examples

8.5.1 System Analysis

In order to illustrate the application of the slope-deflection analytical method an orthotropic bridge, which was initially designed as described in reference [48], will now be analyzed. The initial properties for this example were based on the variables given in section 8.4, as previously determined from existing orthotropic bridge structures [39]. The general plans of this bridge are described in Figure 8-15. The dimensions of the transverse floor beams, longitudinal girders, and closed cellular deck are also given in detail in Figure 8-15. The stiffness properties of the various members and cell are listed as follows:

Deck

$$D_x = 236.5 \text{ k-in}^2/\text{in}$$

$$D_y = 248869.5 \text{ k-in}^2/\text{in}$$

$$H = 37531.6 \text{ k-in}^2/\text{in}$$

Table 8-1
Closed Cross Sections

Bridge	*Rib Space*	*Rib Type*	*Rib Size*	*Deck Plate*	*s/l*	*Depth Ratio*[a]	*Floor Beam Properties*[b]
Weser	24″	12″, 12″	12″ × 12″ × 1/4″	1/2″ to 3/4″	0.40	1:2.4	
Duisburg	23 1/2″	10″	5/16″	9/16″ to 11/16″	0.133	1:8.4	
Port Mann	24″	12″, 11″	11″ × 12″ × 5/16″	7/16″	0.093	1:4	36″ × 5/16″ 12″ × 5/8″
AISC Exp.	24″	13″, 9.5″, 6″	1/4″	3/8″	0.30	1:3 to 1:4	4′ to 5′ × 5/16″ 10″ × 9/16″
Calif.	24″	12″, 8.5″, 6″	5/16″	3/8″ to 7/16″	0.65	1:3	24″ × 1/2″ 10″ × 7/8″
St. Louis	26″	12″, 10″, 6″		1/2″	0.50	1:4	
Beth. Steel	6″	5.5″, 2.5″, 3″	3/16″	3/16″	1.0	1:3	34.5″ × 0.28″ 11″ × 1″

[a]Ratio of depth of rib to depth of floor beam
[b]Floor beam properties: depth and thickness of web and size of flange

Table 8-2
Open Cross Sections

Bridge	*Rib Space*	*Rib Type*	*Rib Size*	*Deck Plate*	*s/l*	*Depth Ratio*[a]	*Spacing Ratio*[b]	*Thickness Ratio*[c]
Save River	12″	Open	4 3/4″ × 7/16″ 10 1/2″ × 1″	3/8″ to 1″	0.133	1:9 to 1:3	0.025	0.079 0.095
Duesseldorf Neuss	17″	Split I Bm	7″ to 11″ deep	9/16″ to 1 3/32″	0.250	1:2	0.0575	0.080 0.099
Duesseldorf North	16″	Angles	8″ × 4″ × 7/16″	9/16″	0.125	1:4	0.0272	0.070
Severin	12″	Open	9 1/2″ × 5/16″ 11 1/2″ × 3/4″	3/8″	0.105	1:3.5	0.0161	0.040 0.030
Cologne Muelhein	12″	Bulb	7″ × 5/16″	1/2″	0.095	1:4	0.0178	0.0715
AISC Exp.	12″	Open	8 1/2″ × 1/2″	3/8″	0.120	1:3 to 1:4	0.02	0.044

[a]Ratio of depth of rib to depth of floor beam
[b]Ratio of rib spacing to space width l
[c]Ratio of deck plate thickness to depth of plate stiffener

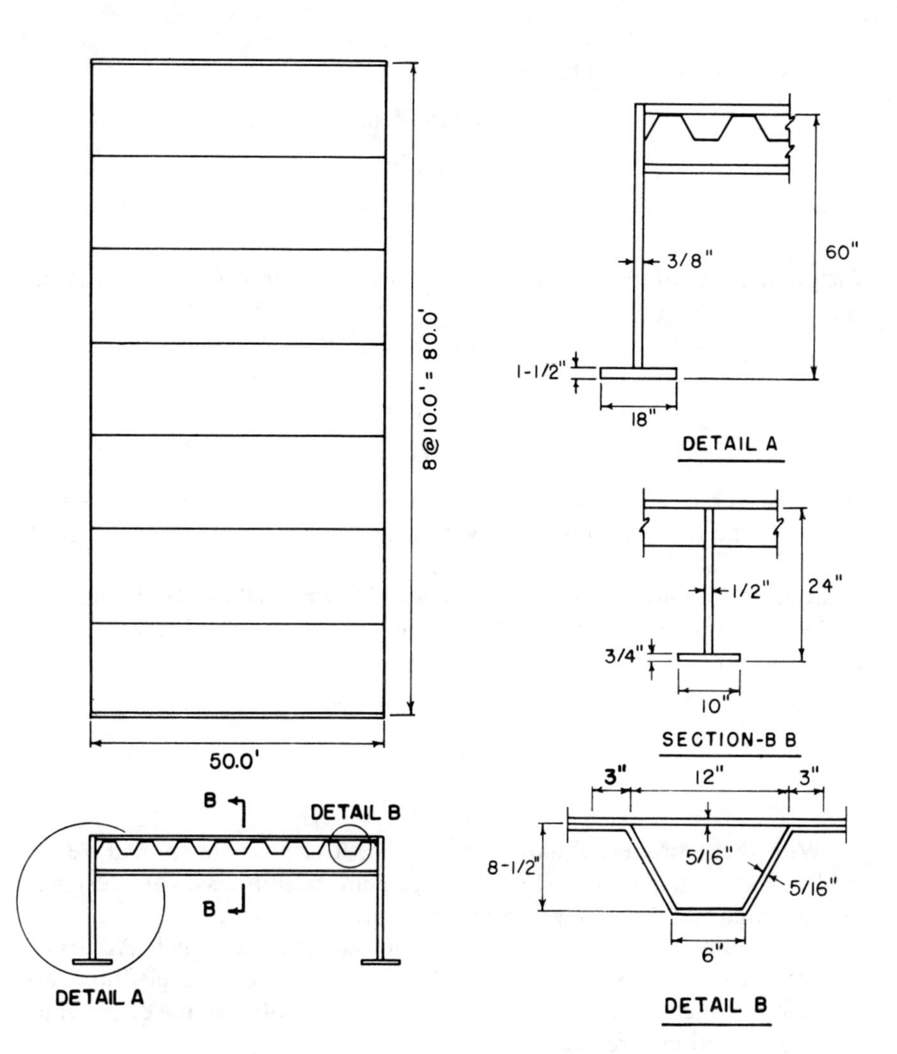

Figure 8-15. Orthotropic Bridge Example.

These properties were determined by equations (8.27) through (8.30) and computer programs [48]. The plate constants are, therefore, equal to:

$$\alpha = H/\sqrt{D_x D_y} = 4.892$$

$$\beta = H/D_y = 0.1508$$

Transverse Floor Beam: The transverse floor beam, neglecting the top plate stiffness, is computed as

$$I_x = 1180.5 \text{ in}^4$$

$$S_t = 72.0 \text{ in}^3$$

$$S_B = 154.0 \text{ in}^3$$

Longitudinal Girders: The longitudinal main girder stiffness, neglecting the top plate effect, is

$$I_y = 73461.17 \text{ in}^4$$

$$S_T = 1530.0 \text{ in}^3$$

$$S_B = 4960.0 \text{ in}^3$$

The transverse and longitudinal girder stiffnesses can be increased by including the top plate effective width.

Transverse Composite Section Properties of Entire Cellular Deck and Two Main Longitudinal Girders: The entire cross-sectional bridge section properties are

$$I_y = 239377.76 \text{ in}^4$$

$$S_T = 24000.0 \text{ in}^3$$

$$S_B = 4770.0 \text{ in}^3$$

With these properties now determined, it is possible to analyze the bridge. Examining Figure 8-16, the bridge plan and stiffnesses are designated, as required by the slope-deflection method.

The loading to be applied to the bridge is a simulated single AASHTO HS-20 truck, as shown in Figure 8-17. Two truck loading positions are considered; one is at the center of the structure, the other at the edge. Other loading conditions are presented elsewhere [48].

With the loadings and bridge properties known, the slope-deflection method was applied, resulting in the deflections and forces in the bridge system.

Recall that for this type of structure, the slope-deflection technique must be analyzed in two parts in order to obtain the final deflections and moments [48]; the first is the plate solution and the second the beam solution.

In addition to the slope-deflection method, the finite difference technique, as outlined in section 6.1.2, was applied [48]. The modeling of the bridge based on the finite difference technique is shown in Figure 8-18.

Part of the results from this study for the floor beams are given in

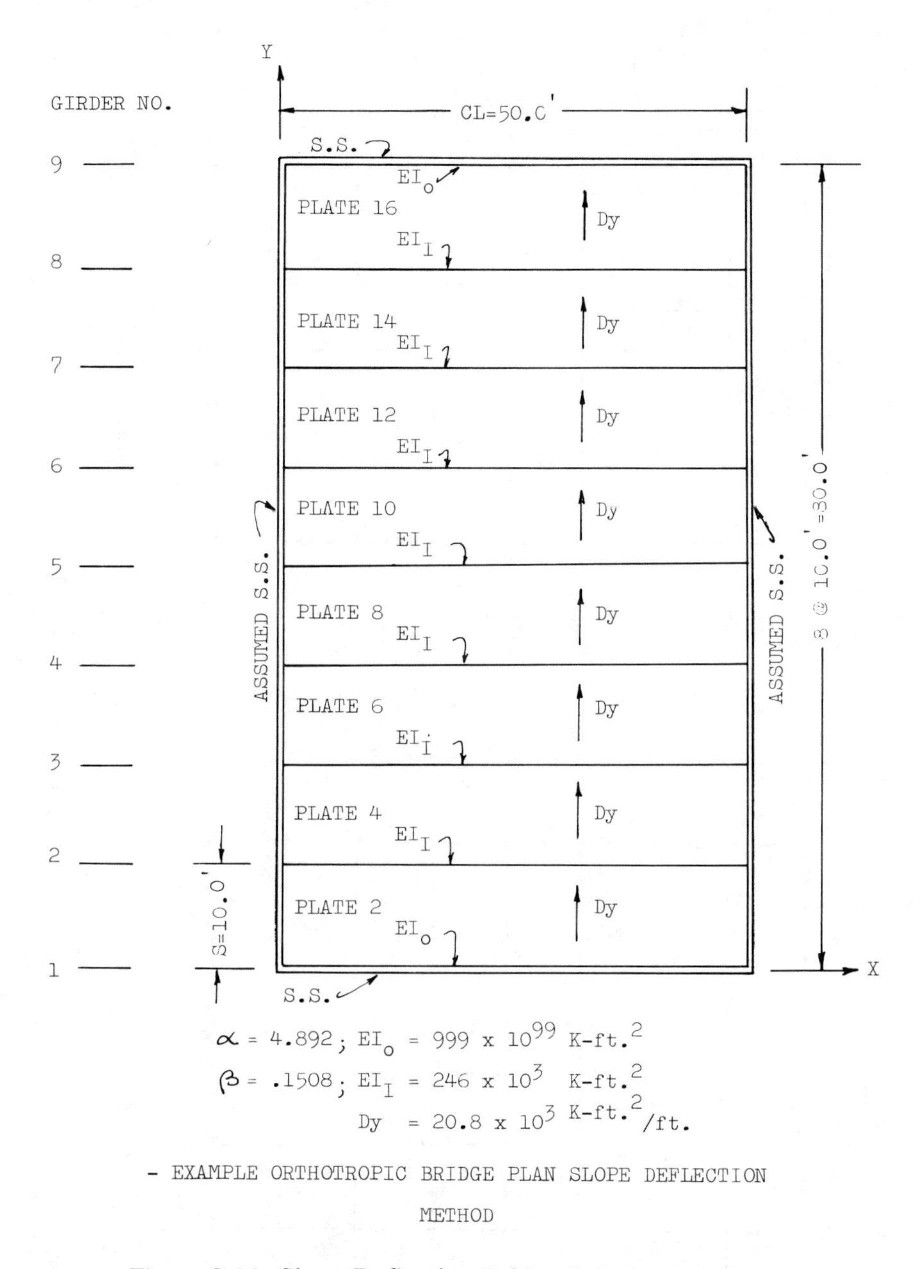

Figure 8-16. Slope Deflection Bridge Solution Parameters.

Figures 8-19 and 8-20. For the maximum moments in the transverse floor beams, Case 1 loading gives the following.

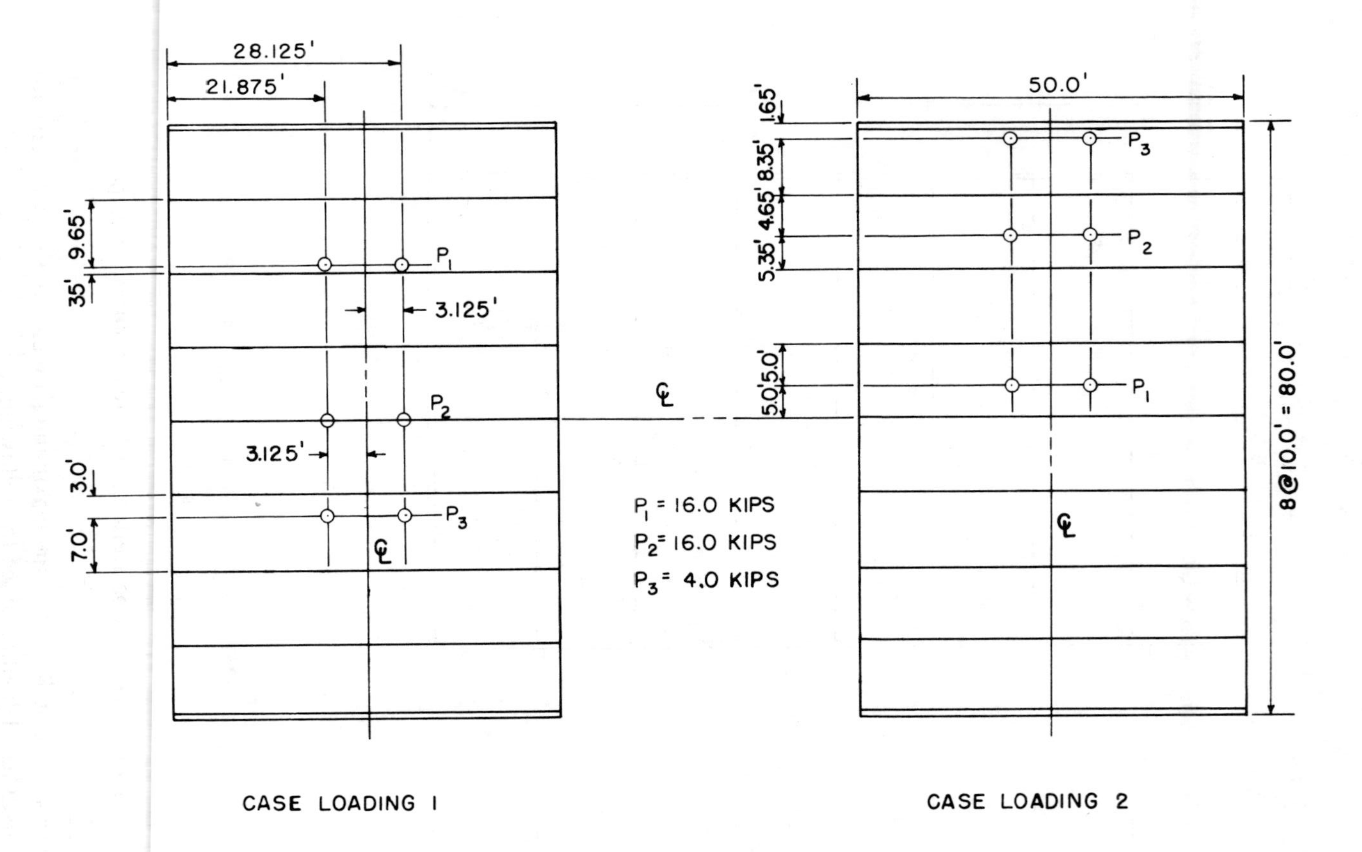

Figure 8-17. Vehicle Loading Patterns.

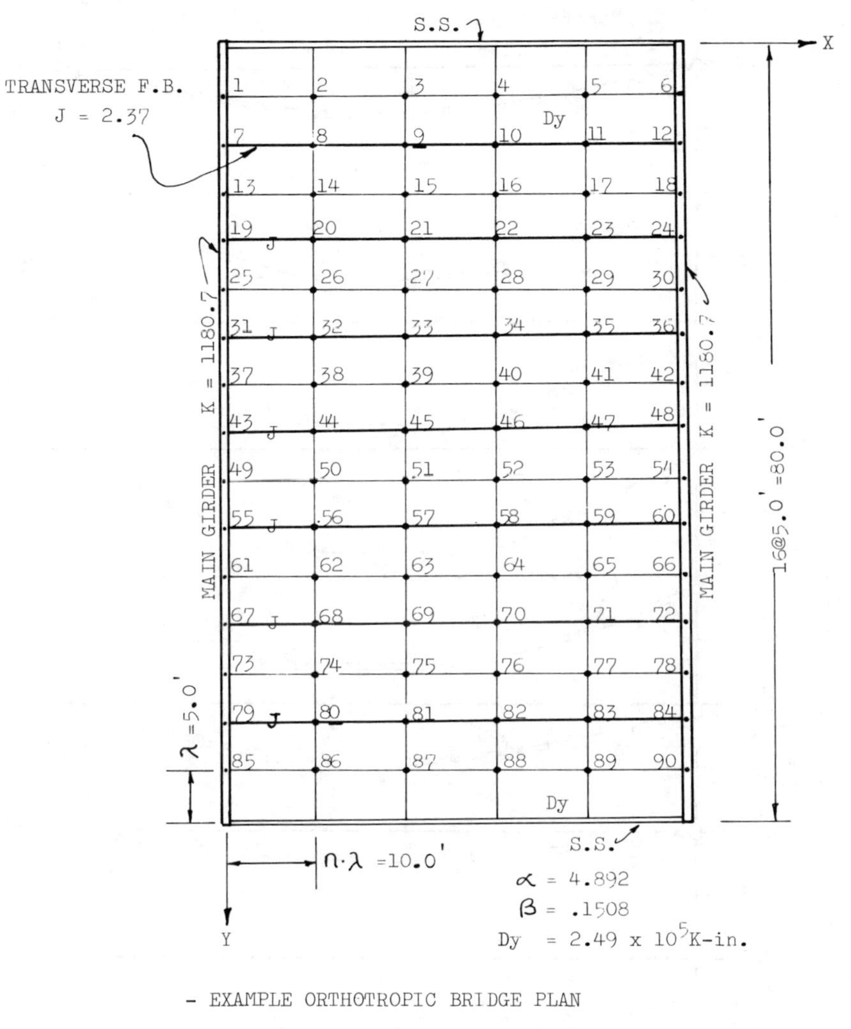

Figure 8-18. Finite Difference Bridge Solution Parameters.

Slope-Deflection: $M_x = 172.0 \times 10^4$ lb-in

Finite-Difference: $M_y = 151.6 \times 10^4$ lb-in

resulting in a difference of 11.6%.

The resulting stresses due to these applied moments may be determined by using the section properties listed previously. Examination of these stresses would indicate a revision in section properties.

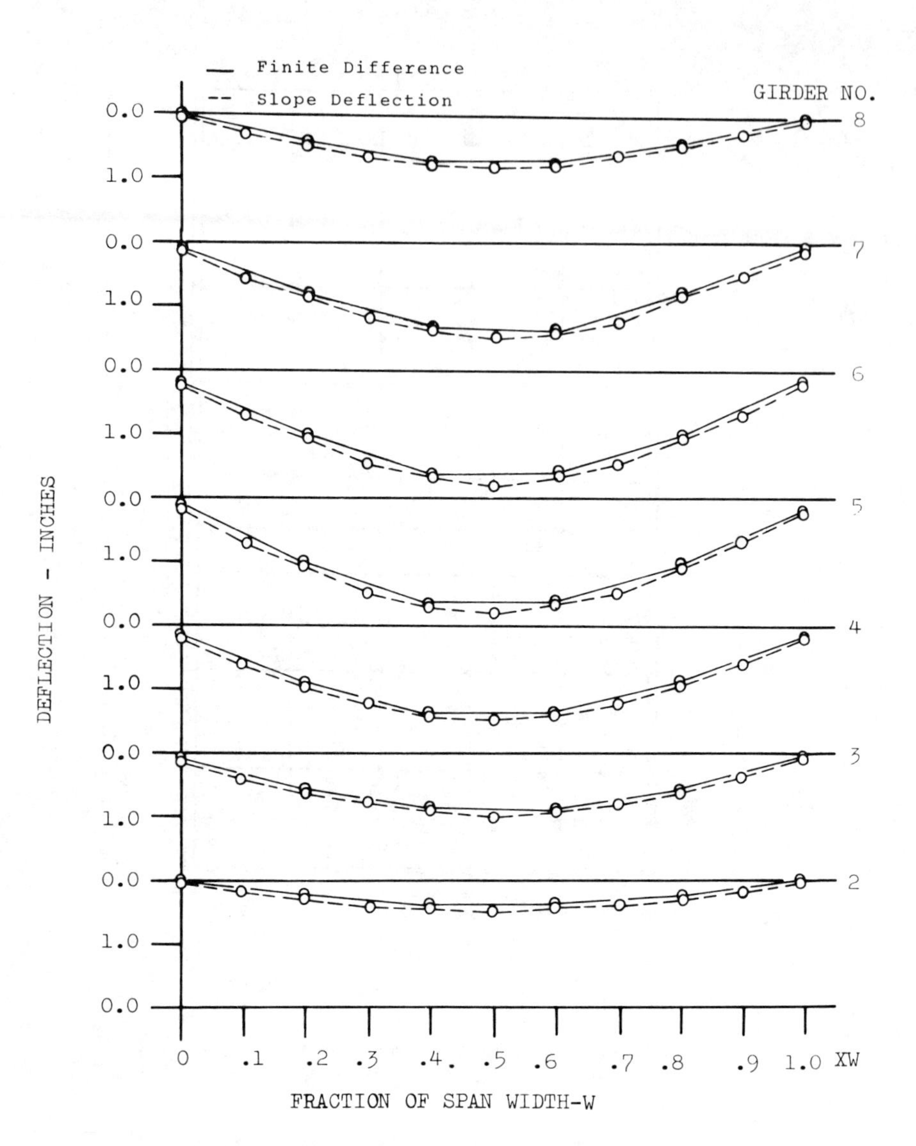

Figure 8-19. Comparison of Results—Floor Beam Deflections Case Load 1.

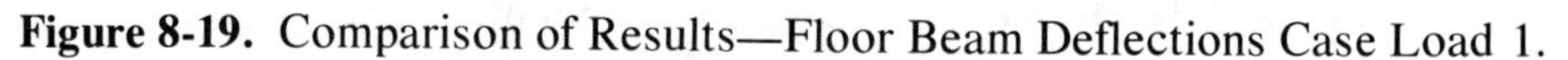

A comparison between resulting deflections by both methods, using Case 1 loading, gives the following for the transverse floor beams.

Slope-Deflection = 1.753 in

Finite-Difference = 1.758 in

resulting in 0.28% difference between methods.

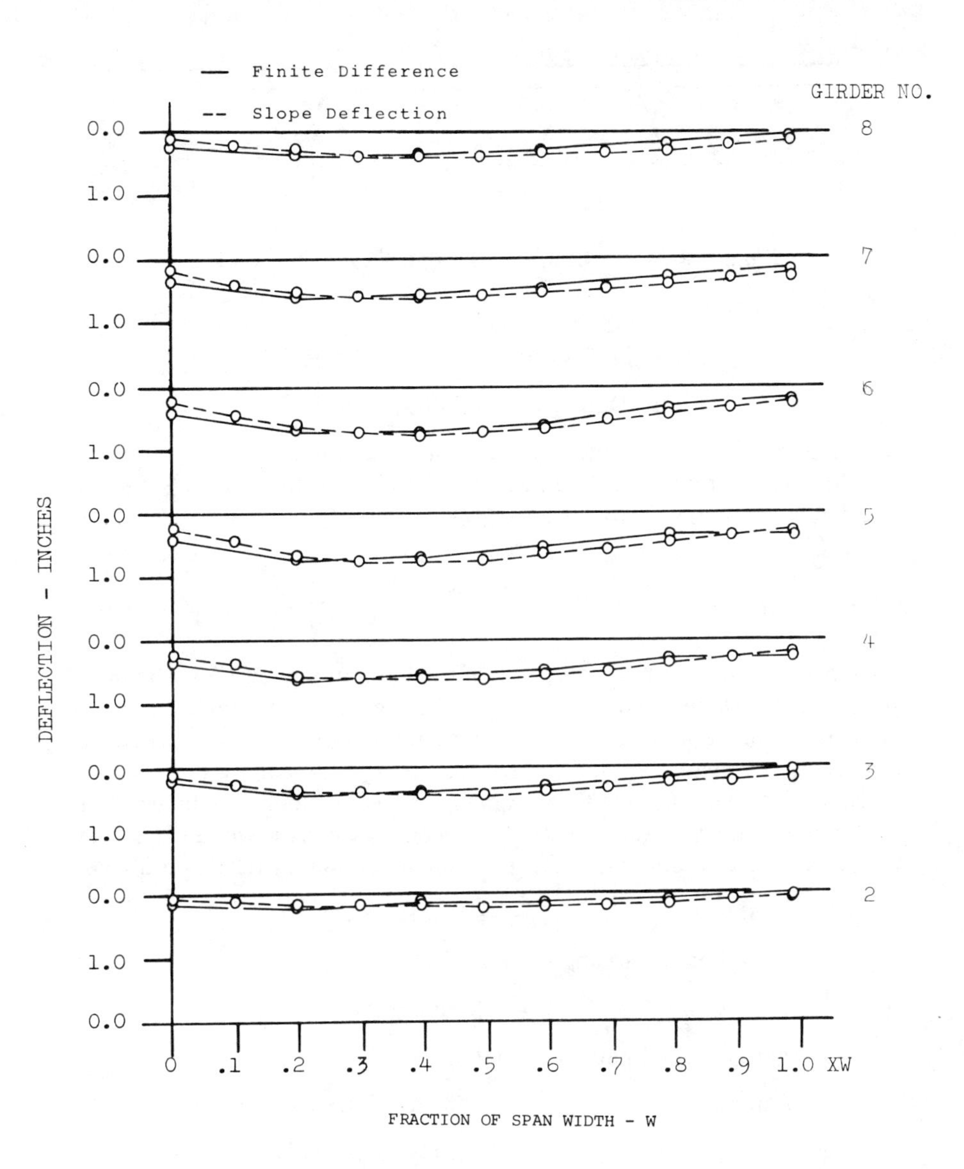

Figure 8-20. Comparison of Results—Floor Beam Deflections Case Load 2.

For the longitudinal girders, Case 1 loading, the deflections are:

$$\text{Slope-Deflection} = 0.1577 \text{ in}$$

$$\text{Finite-Differences} = 0.2137 \text{ in}$$

resulting in a 26.6% difference [48]. The maximum longitudinal girder moments, according to Case 2 loading, are:

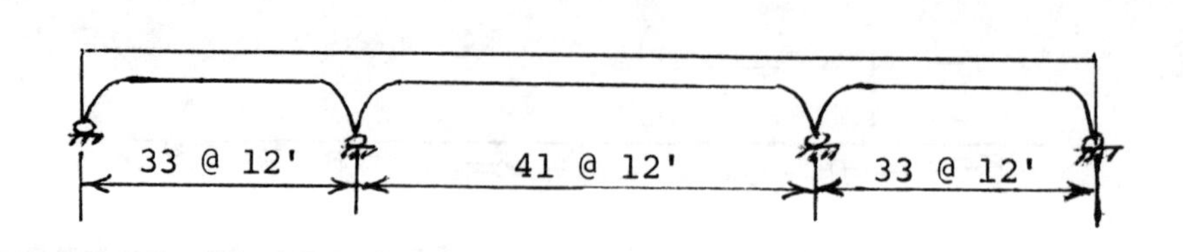

Figure 8-21. Orthotropic Bridge Design Example.

Slope-Deflection $= 127.9 \times 10^5$ lb-in

Finite-Difference $= 118.3 \times 10^5$ lb-in

resulting in a 7.5% difference between methods [39].

This same bridge has been analyzed by the grid system technique (section 4.9), giving similar results [34].

8.5.2 System Design

Comparisons: In order to apply any of the computer programs, the initial girder sizes must be shown. Utilizing the data given in section 8.4 and with the help of the approximate floor beam equation (8.44), initial properties can readily be evaluated and then a computer study can be conducted.

In order to demonstrate the design equation, consider a six-lane bridge, continuous over three spans with floor beams spaced at twelve-foot intervals, as shown in Figure 8-21. The preliminary geometry for this bridge is:

Roadway	— 72 feet (6 12 foot lanes)
Floor Beam Sapcing	— 12 feet
Rib Spacing	— 18 inches
Main Girder Spacing	— 60 feet
Material	— A588, A572, and A36 steels
Wearing Surface	— 2 inches thick asphalt

The AASHTO HS20-44 live loading to be applied is as shown in Figure 8-22.

The bridge will be designed and then analyzed by three procedures:

1. Pelikan-Esslinger Method [53], which is basically the same procedure as the slope-deflection technique, but neglects some plate stiffness parameters
2. Approximate design equation
3. Slope-deflection computer program (section 7.1.1)

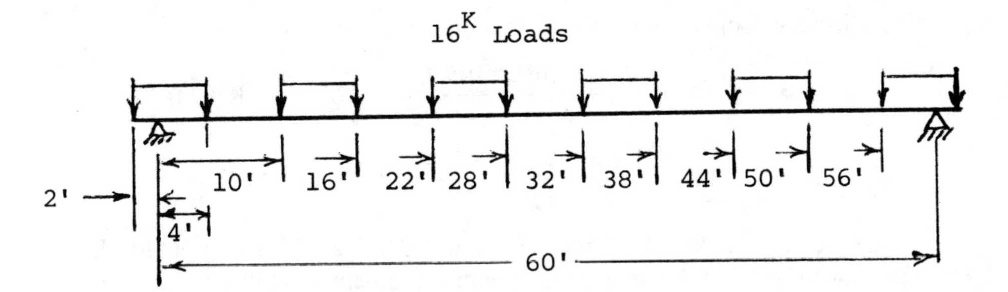

Figure 8-22. Vehicle Loading.

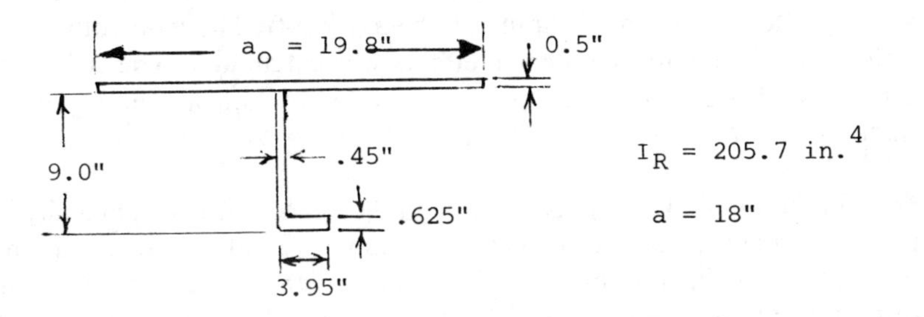

Figure 8-23. Rib Section.

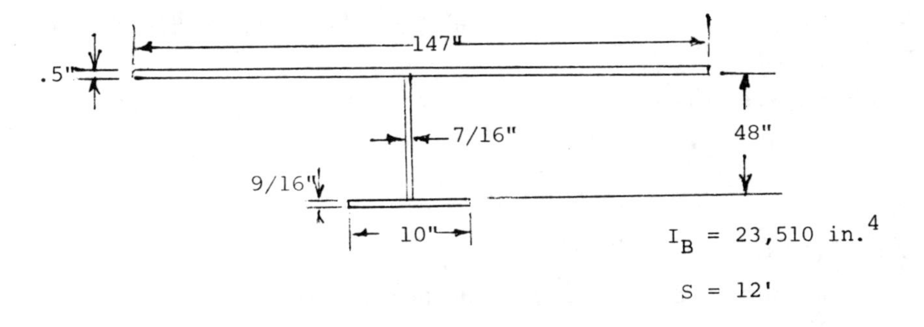

Figure 8-24. Floor Beam Section.

Pelikan-Esslinger Method: As given by Hall [67], the required deck plate thickness based on allowable deflection between ribs is 0.5 inch. Based on this deck plate thickness and the assumed rib and floor beam sizes, as shown in Figures 8-23 and 8-24, the rib and floor beam section properties are

$$I_R = 205.7 \text{ in}^4; \qquad I_B = 23{,}510.0 \text{ in}^4$$

The stiffness properties for these sections are, therefore,

$$D_y = \frac{E \times I_R}{a} = \frac{2.9 \times 10^4 \times 205.7}{18 \times 12} = 2.76 \times 10^4 \text{ K-ft}^2/\text{ft}$$

$$EI = E \times I_B = \frac{2.9 \times 10^4 \times 23{,}510}{144} = 4.73 \times 10^6 \text{ Kip-ft}^2$$

Using the Pelikan-Esslinger method [53], Hall [67] computed the maximum rigid floor beam live load bending moment as 1130 kip-ft with the maximum thirty percent impact factor and with a ninety percent factor for moment reduction with three lanes loaded causing the critical loading condition. The bending moment in the floor beam due to the uniformly distributed dead load was computed to be 337 kip-ft. The moment relief in the floor beams due to their elastic effects was calculated to be 267 kip-ft, again considering a thirty percent impact factor and a ninety percent reduction factor.

Approximate Design Equations: Although the span width is technically six lanes, the wheel loads overhanging the supports tend to reduce the moments caused by the interior loads. The span may therefore be considered a five-lane span for preliminary design. Equation (8.44) may be used to determine M/M^*:

$$\log_{10}(M/M^*) = -0.11N - 0.86(1 - \log_{10}S) + (-0.0485N + 0.046)\log_{10}(D_y/EI) \qquad (8.45)$$

where $N = 5$ lanes

$S = 12$ ft

$D_y = 2.76 \times 10^4$ kip-ft²/ft

$EI = 4.73 \times 10^6$ kip-ft²

Substituting these values into equation (8.45) gives

$$\frac{M}{M^*} = 0.906$$

From Figure 8-25, $M^* = 1192$ kip-ft. The maximum bending moment may now be determined.

$$M = \frac{M}{M^*} \times M^* = 0.906 \times 1192$$

$$M = 1080 \text{ kip-ft}$$

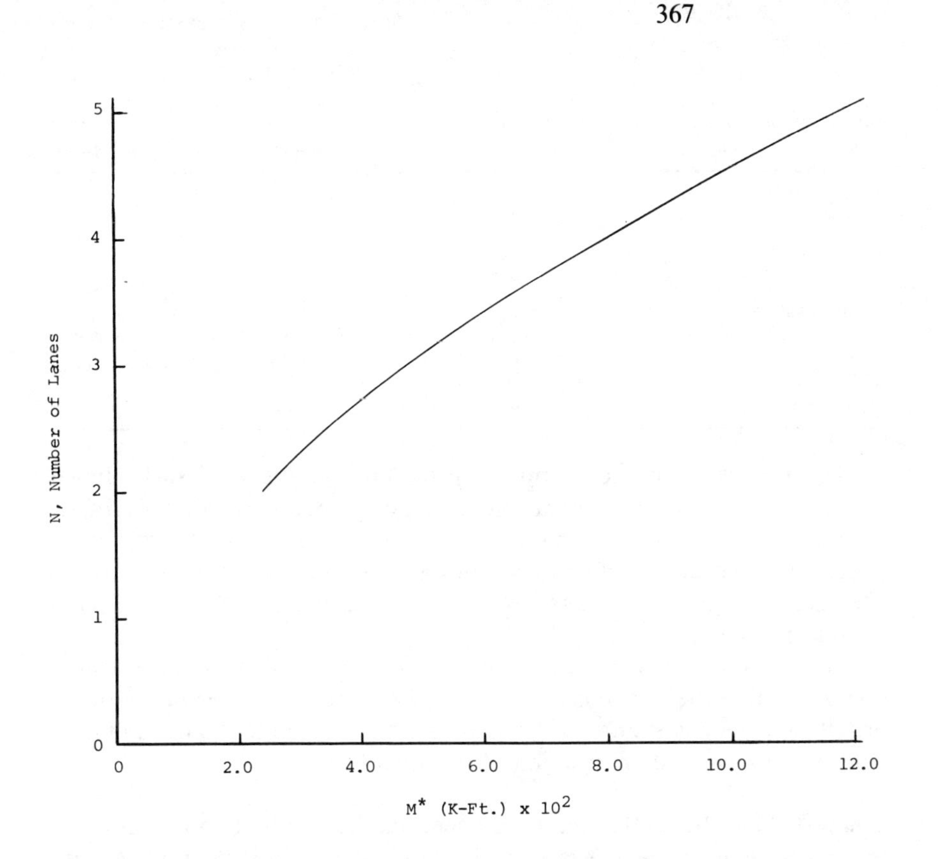

Figure 8-25. Actual Simple Beam Moment versus Number of Lanes.

Slope-Deflection Program: The bridge was analyzed using the computer program described in section 7.1.1. The loading followed the pattern described in Figure 8-22. The maximum floor beam bending moment was computed to be

$$M = 866 \text{ kip-ft}$$

using the equivalent five-lane span.

Comparison of Results: The following table compares the results obtained from the Pelikan-Esslinger technique [53], equation (8.44), and the computer program. The impact factor of 1.27 and the live load moment reduction of ninety percent used by Hall have been eliminated from his results. The dead load moment has been included.

Maximum Floor Beam Bending Moment	*Reference* [53] *Pelikan-Esslinger*	*Design Equation*	*Computer Program*
Rigid System Dead Load	+337	—	—
Rigid System Live Load	+988	—	—
Elastic System Live Load	−234	+1080	+866
Total Live Load	+754	+1080	+866
Total Load	+1091	+1080	+866

The results from the computer program and the Pelikan-Esslinger method are in close agreement, differing by thirteen percent. This difference is in the range by which the actual transverse floor beam deflections exceed the theoretical deflections in bridge tests, as given in reference [35]. The design equation results are virtually identical with the Pelikan-Esslinger results.

The point should be made here that the resulting design is not dependent on the fact that the bridge is continuous (three spans). The behavior of the orthotropic deck is such that the loads are transferred to the main girders within a relatively short fraction of the span length.

Complete Floor Beam Design: In this section, the required floor beam for a three-lane span will be determined. The preliminary design parameters are given as follows:

Span Width — 35 feet (3 lanes)

Floor Beam Spacing — 15 feet

Rib Spacing — 16 inches

Rib Height — 10 inches

Rib Thickness — 3/8 inches

Asphalt Topping Thickness — 2 inches

Deck Plate Design: The deck plate thickness required to limit the deflection between ribs to $a/300$ is [54]:

$$t_p = 0.0065\, a\sqrt[3]{P}$$

where t_p = Plate thickness (in)

a = Rib spacing (in)

P = Wheel pressure for one 12 kip wheel

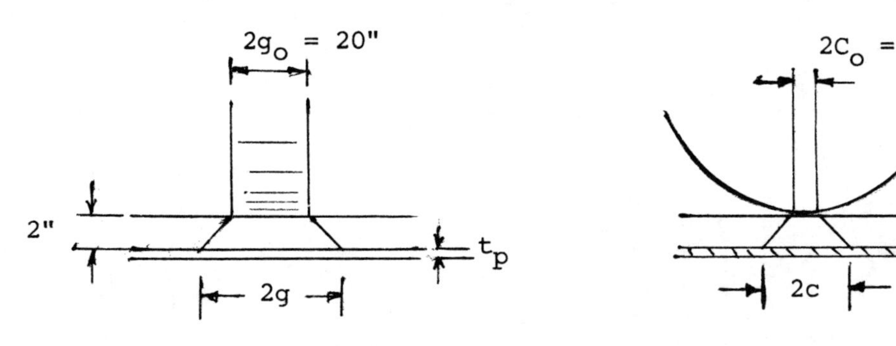

Figure 8-26. Wheel Load.

The contact area of the wheel on the plate through the asphalt topping, as shown in Figure 8-26, is found as:

$$2g = 20 + 2 \times 2 = 24 \text{ in } 2c = 8 + 2 \times 2 = 12 \text{ in}$$

where the contact area = 24 in × 12 in = 288 in².
Therefore,

$$P = \frac{12 \times 1000}{288} = 41.67 \text{ psi}$$

$$t_p = 0.0065 \times 16 \times \sqrt[3]{41.67} = 0.361 \text{ in}$$

Use $t_p = 0.376$ in = 3/8 in

Deck Stiffness, D_y: The deck stiffness per unit of width, D_y, is

$$D_y = \frac{E \times I_R^*}{a}$$

where E = Young's modulus

I_R^* = Moment of inertia of one rib including effective width of deck plate

a = Rib spacing

$$\text{Effective rib span, } l_1 = \infty$$

$$\text{Ideal rib span, } a^* = a = 16 \text{ in}$$

$$\beta = \frac{\pi a^*}{l_1} = 0$$

From Figure 8-27 with $\beta = 0$

$$\lambda = \frac{a_0}{a^*} = 1.10$$

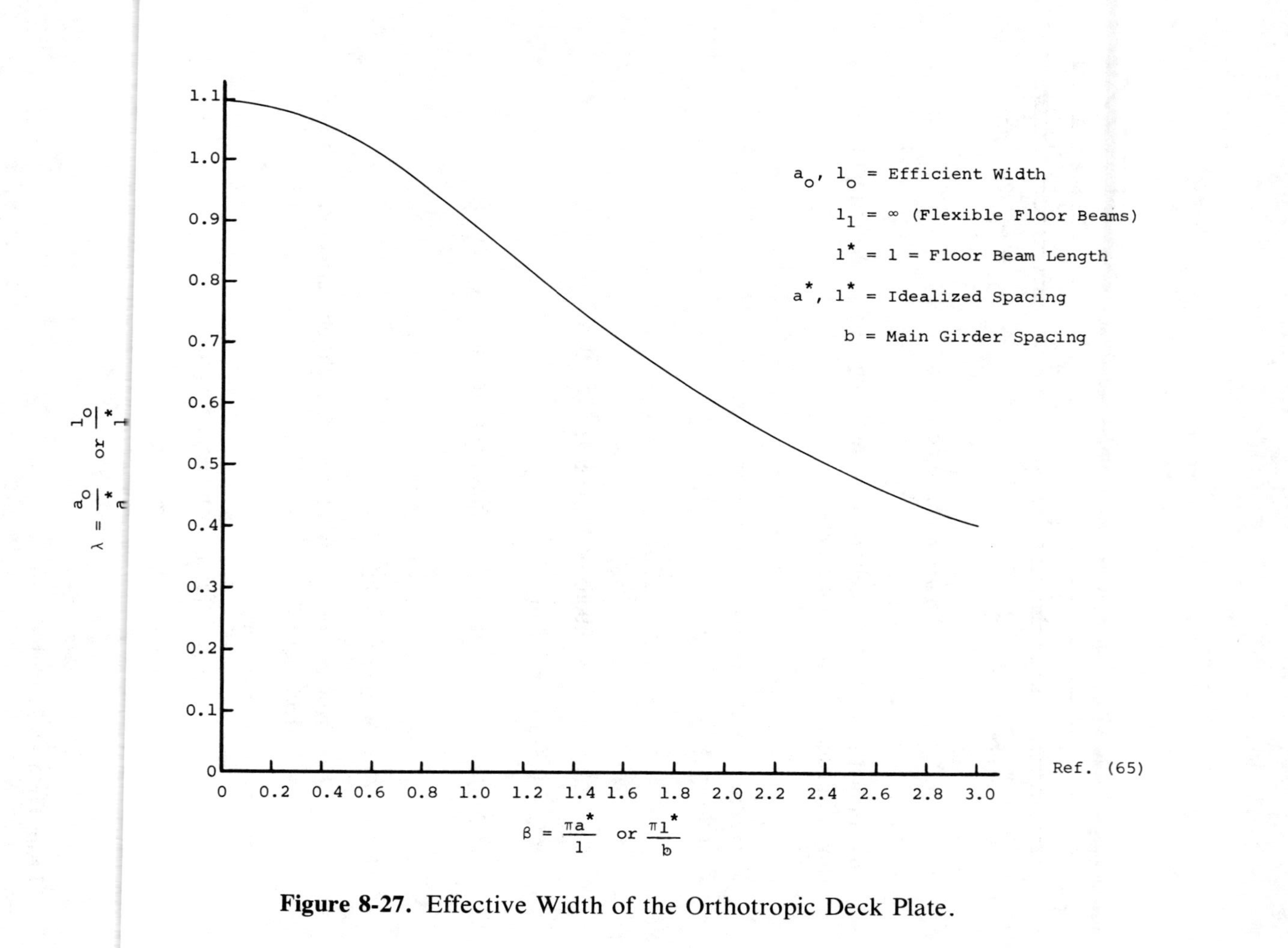

Figure 8-27. Effective Width of the Orthotropic Deck Plate.

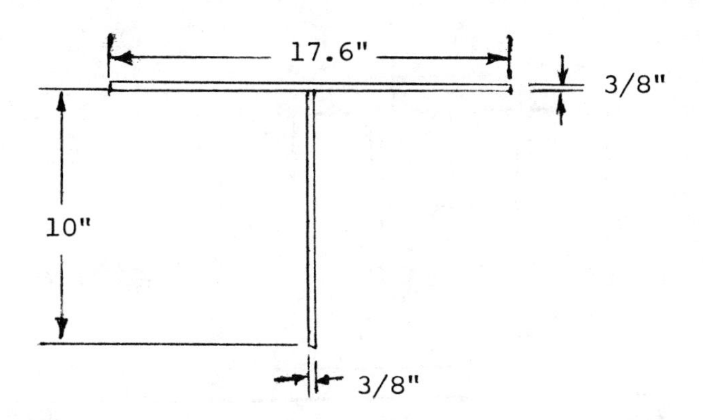

Figure 8-28. Rib Section.

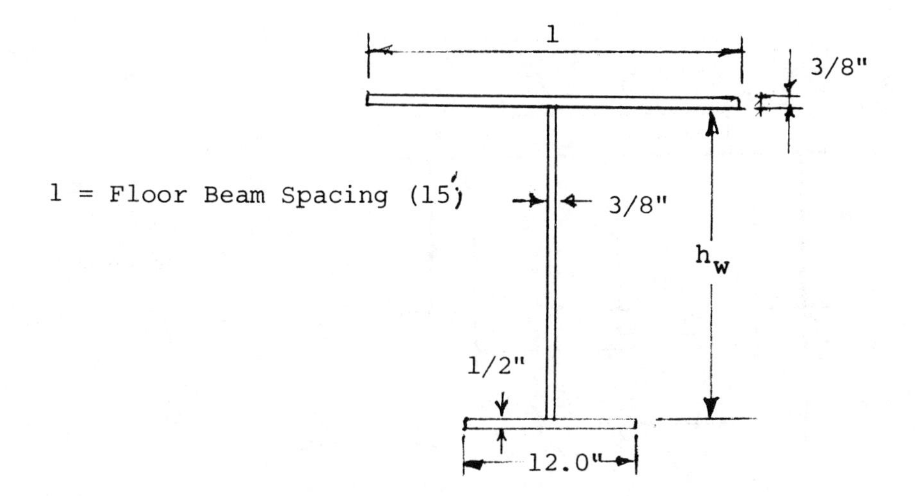

Figure 8-29. Floor Beam Section.

and the effective deck plate width as defined in Figure 8-30, is

$$a_0 = \frac{a_0}{a^*} a^*$$

therefore,

$$a_0 = 1.10 \times 16 = 17.6 \text{ in}$$

The geometry of the deck is shown in Figure 8-28. The properties are computed as

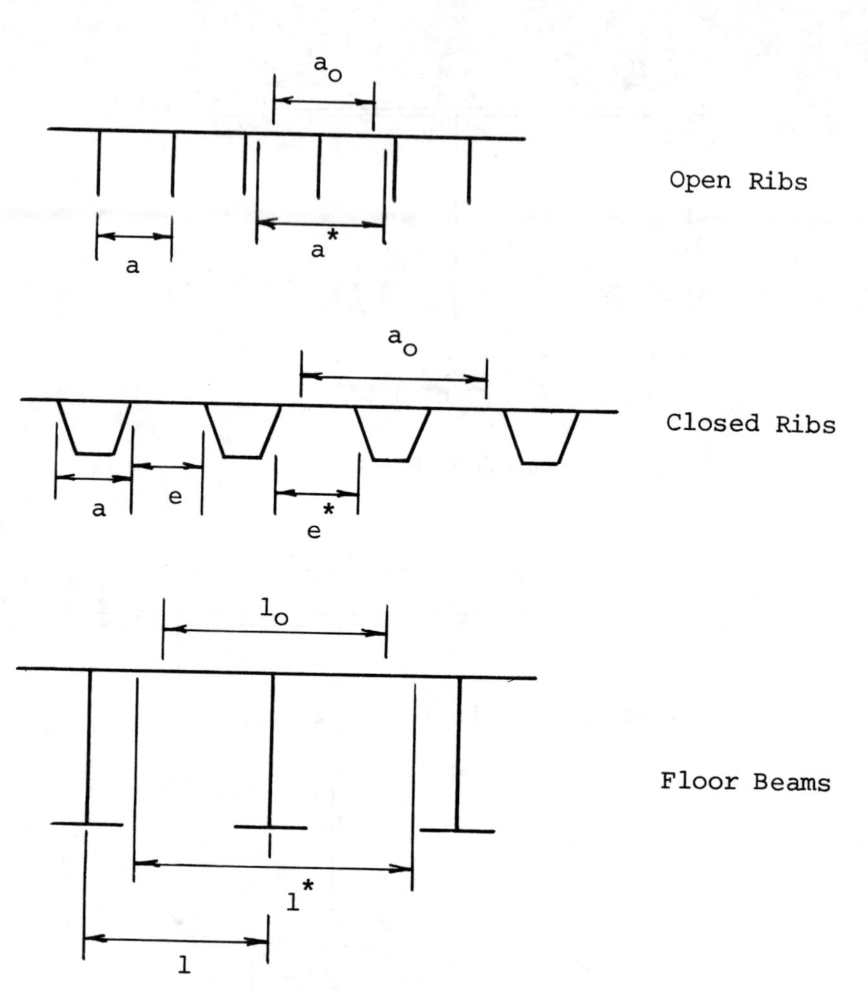

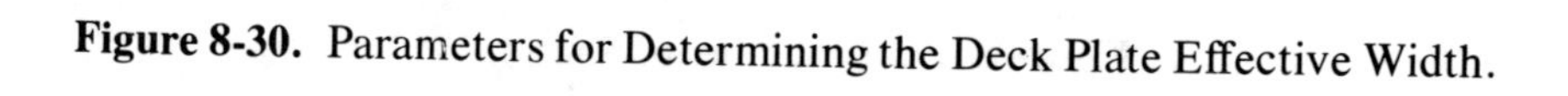

Figure 8-30. Parameters for Determining the Deck Plate Effective Width.

$$\bar{y} = \frac{10 \times 0.375 \times 5 + 17.6 \times 0.375 \times 10.1875}{10 \times 0.375 + 17.6 \times 0.375}$$

$$= \frac{18.75 + 67.15}{3.75 + 6.6} = \frac{85.90}{10.35} = 8.31 \text{ in}$$

$$I_R^* = \frac{1}{12}(0.375 \times 10^3 + 17.6 \times 0.375^3)$$

$$+ 3.75 \times 3.31^2 + 6.6 \times 1.88^2$$

$$I_R^* = \frac{1}{12} \times 375\times + 41.1 + 23.3 = 94.8 \text{ in}^4$$

$$D_y = \frac{2.9 \times 10^4 \times 94.8}{16 \times 12} = 1.432 \times 10^4 \text{ kip-ft}^2/\text{ft}$$

Floor Beam Stiffness, EI: Assuming $h_w = 36.0$ in, $t_w = 3/8$ in and a constant bottom flange of 1/2 in × 12.0 in as shown in Figure 8-29 with an effective top plate width, equal to the floor beam spacing of 15.0 ft, from Figure 8-31:

$$EI = 2.24 \times 10^6 \text{ kip-ft}^2$$

Preliminary Design Equation Solution:

$$\begin{aligned}\log_{10}(M/M^*) &= -0.11N - 0.86(1 - \log_{10}S) \\ &\quad + (-0.0485N + 0.046)\log_{10}(D_y/EI) \\ &= -0.11 \times 3 - 0.86(1 - \log_{10}15) \\ &\quad + (-0.0485 \times 3 + 0.046)\log_{10}\frac{1.432 \times 10^4}{2.24 \times 10^6}\end{aligned}$$

$$\log_{10}(M/M^*) = +0.0408$$

$$\frac{M}{M^*} = 1.1$$

From Figure 8-25,

$$M^* = 472 \text{ kip-ft}$$

$$M = \frac{M}{M^*} \times M^* = 1.1 \times 472$$

$$M = 519 \text{ kip-ft} \times \left(1 + \frac{50}{35 + 125}\right) \times 0.9 = 613 \text{ kip-ft}$$

Dead Load Bending Moment:

2-in Asphalt Topping: (2/12) × 15 × 120	=	300 lb/ft
Deck Plate: 1/12 × 0.375 × 15 × 490	=	230 lb/ft
Web: 0.375 × 36.0 × 1/144 × 490	=	46 lb/ft
Bottom Flange: 0.5 × 12.5 × 1/144 × 490	=	21 lb/ft
Ribs: 3.75 × 15 × 1/144 × 490 × 12/16	=	144 lb/ft
Total		741 lb/ft

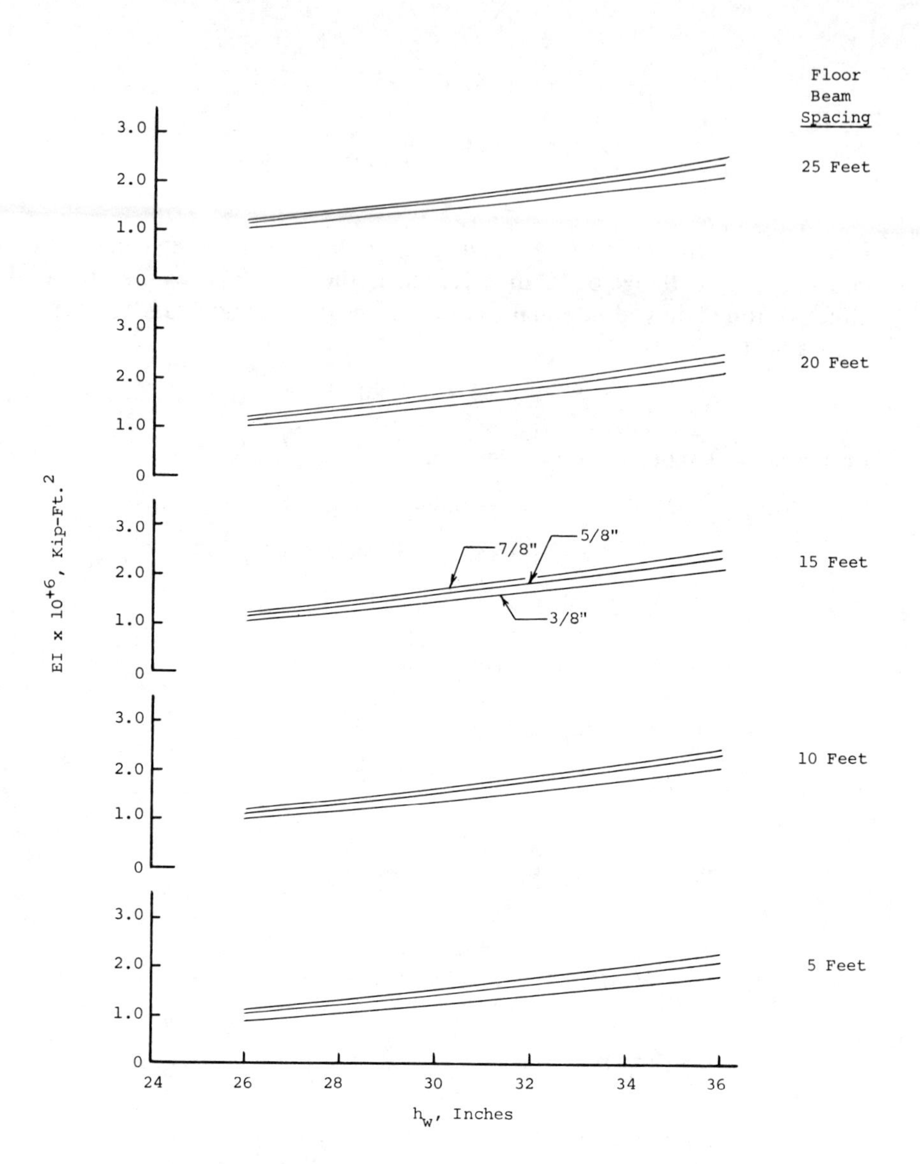

Figure 8-31. Floor Beam Stiffness versus Web Height and Deck Plate Thickness.

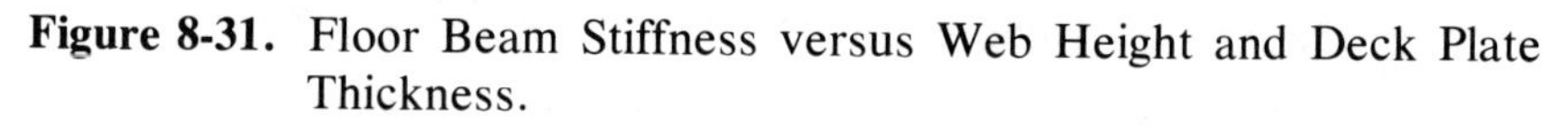

$$M_{DL} = \frac{wl^2}{8} = \frac{741 \times 35^2}{8 \times 100} = 113 \text{ kip-ft}$$

Total Maximum Floor Beam Moment: The total maximum moment is the sum of the dead and live load moment:

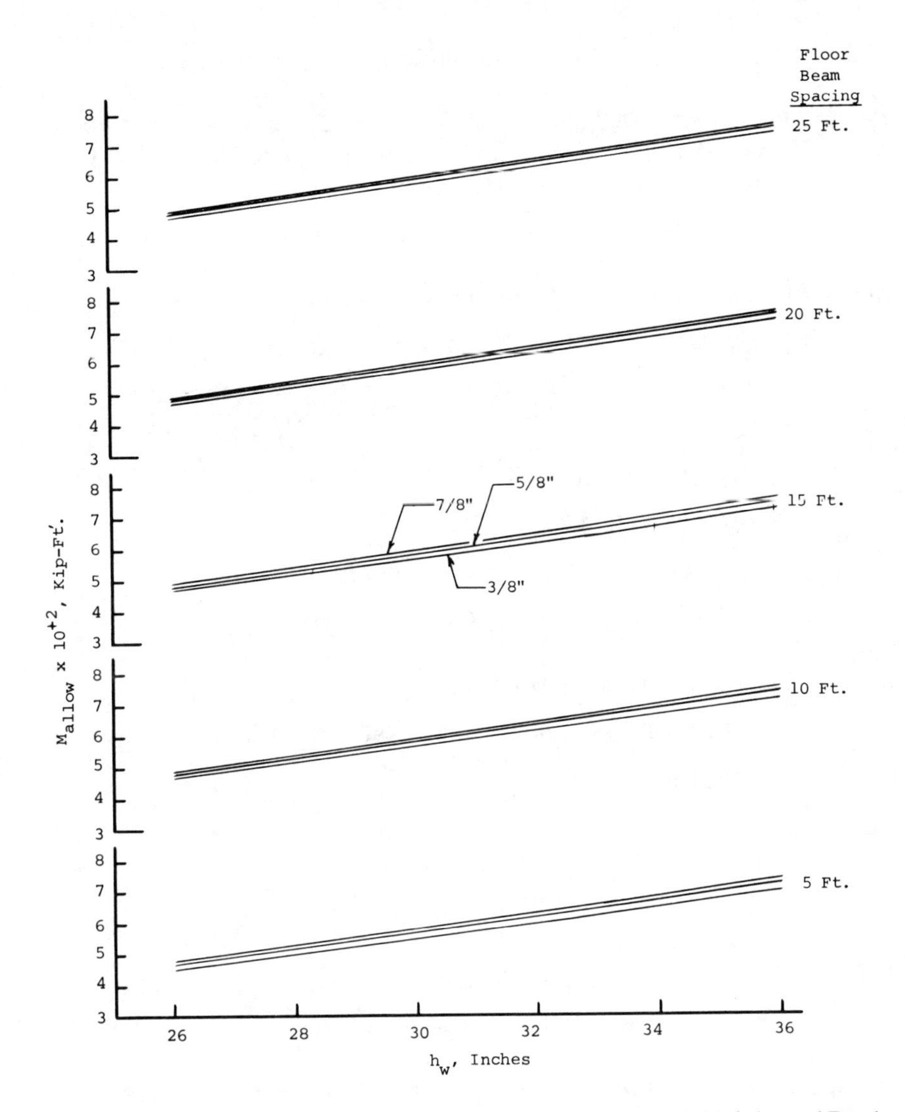

Figure 8-32. Allowable Floor Beam Moment versus Web Height and Deck Plate Thickness

$$M = 613 + 113 = 726 \text{ kip-ft}$$

Allowable Moment: From Figure 8-32, the allowable bending moment has been computed for A36 steel where the allowable stress is

$$F_b = 0.55\ F_y = 20{,}000 \text{ psi}$$
$$M_{\text{allow}} = 700 \text{ kip-ft}$$

Since $M > M_{\text{allow}}$, $h_w = 36$ in is not sufficient. Try $h_w = 42$ in.

Revised Floor Beam Stiffness, EI: With $h_w = 42.0$ in, $t_w = 3/8$ in and a bottom flange of 12 in × 1/2 in, and an effective plate width of 15.0 ft, extending the curve on Figure 8-31, gives:

$$EI = 3.19 \times 10^6 \text{ kip-ft}^2$$

Revised Preliminary Design Equation Solution:

$$\log_{10}(M/M^*) = -0.11 \times 3 - 0.86(1 - \log_{10}15) + (-0.0485 \times 3 + 0.046)\log_{10}\frac{1.432 \times 10^4}{3.19 \times 10^6}$$

$$\log_{10}(M/M^*) = +0.055$$

$$M/M^* = 1.135$$

$$M = 1.135 \times 472 = 536 \text{ kip-ft}$$

$$536 \times 1.3 \times 0.9 = 628 \text{ kip-ft}$$

Revised Dead Load Bending Moment:

Web:	$0.375 \times 6 \times 1/144 \times 490 =$	8 lb/ft
Previous:		741 lb/ft
	Total	749 lb/ft

$$M_{DL} = \frac{749 \times 35^2}{8 \times 1000} = 115 \text{ kip-ft}$$

Revised Maximum Floor Beam Moment:

$$M = 628 + 115 = 743 \text{ kip-ft}$$

Revised Allowable Moment: For $h_w = 42.0$ in extending the curve on Figure 8-32 for $l = 15$ ft:

$$M_{\text{allow}} = 850 \text{ kip-ft}$$

Since $M < M_{\text{allow}}$, 42 inches is acceptable.

Computer Program Analysis: Using the parameters D_y and EI, tabulated in the design example section entitled "Deck Stiffness and Revised Floor Beam Stiffness," and assuming values of $\alpha = 0.35$ and $\beta = 0.002$, the live load bending moments were computed using the computer program described in chapter 7. The live loading scheme is consistent with that described in chapter 7, Figure 7-9. The resulting maximum bending moment, obtained from the computer study [48], is 508 kip-ft, which compares very favorably with the computed static live load moment of 536 kip-ft, which excludes impact and reduction factor.

Large Deflection Theory and Plate Buckling

In chapter 3, the behavior of plates subjected to forces in their plane was examined. Chapter 4 described the response of plates when subjected to external forces with induced in-plane bending and torsional and shear forces. In this chapter the combined effects of these two force systems will be examined.

9.1 Rectangular Plate Action under Lateral and In-Plane Loads

The in-plane normal forces acting on a differential element are shown in Figure 9-1. If this element is now subjected to in-plane bending forces, the element will develop some curvature, as shown in Figures 9-2 and 9-3 with respect to y-axis and x-axis, respectively. This curvature will cause the forces to have components relative to the z-axis. Examining the component effect of the normal forces and summing forces with respect to z gives

y-Axis (Figure 9-2) ΣF_z:

$$\sum F_z = -N_y dx\frac{\partial w}{\partial y} + \left(N_y + \frac{\partial N_y}{\partial y}dy\right)dx\left(\frac{\partial w}{\partial y} + \frac{\partial^2 w}{\partial y^2}dy\right)$$

$$= -N_y\frac{\partial w}{\partial y}dx + N_y\frac{\partial w}{\partial y}dx + N_y\frac{\partial^2 w}{\partial y^2}dy\,dx$$

$$+ \frac{\partial N_y}{\partial y}dy\left(\frac{\partial w}{\partial y} + \frac{\partial^2 w}{\partial y^2}dy\right)dx$$

and neglecting higher order terms:

$$\sum F_z = N_y\frac{\partial^2 w}{\partial y^2}dy\,dx + \frac{\partial N_y}{\partial y}\frac{\partial w}{\partial y}dx\,dy \tag{9.1}$$

x-Axis (Figure 9-3) ΣF_z:

$$\sum F_z = -N_x dy\frac{\partial w}{\partial x} + \left(N_x + \frac{\partial N_x}{\partial x}dx\right)dy\left(\frac{\partial w}{\partial x} + \frac{\partial^2 w}{\partial x^2}dx\right)$$

$$= -N_x dy\frac{\partial w}{\partial x}Q + N_x dy\frac{\partial w}{\partial x} + \frac{\partial N_x}{\partial x}dx\,dy\left(\frac{\partial w}{\partial x} + \frac{\partial^2 w}{\partial x^2}dx\right)$$

$$+ N_x\frac{\partial^2 w}{\partial x^2}dx\,dy$$

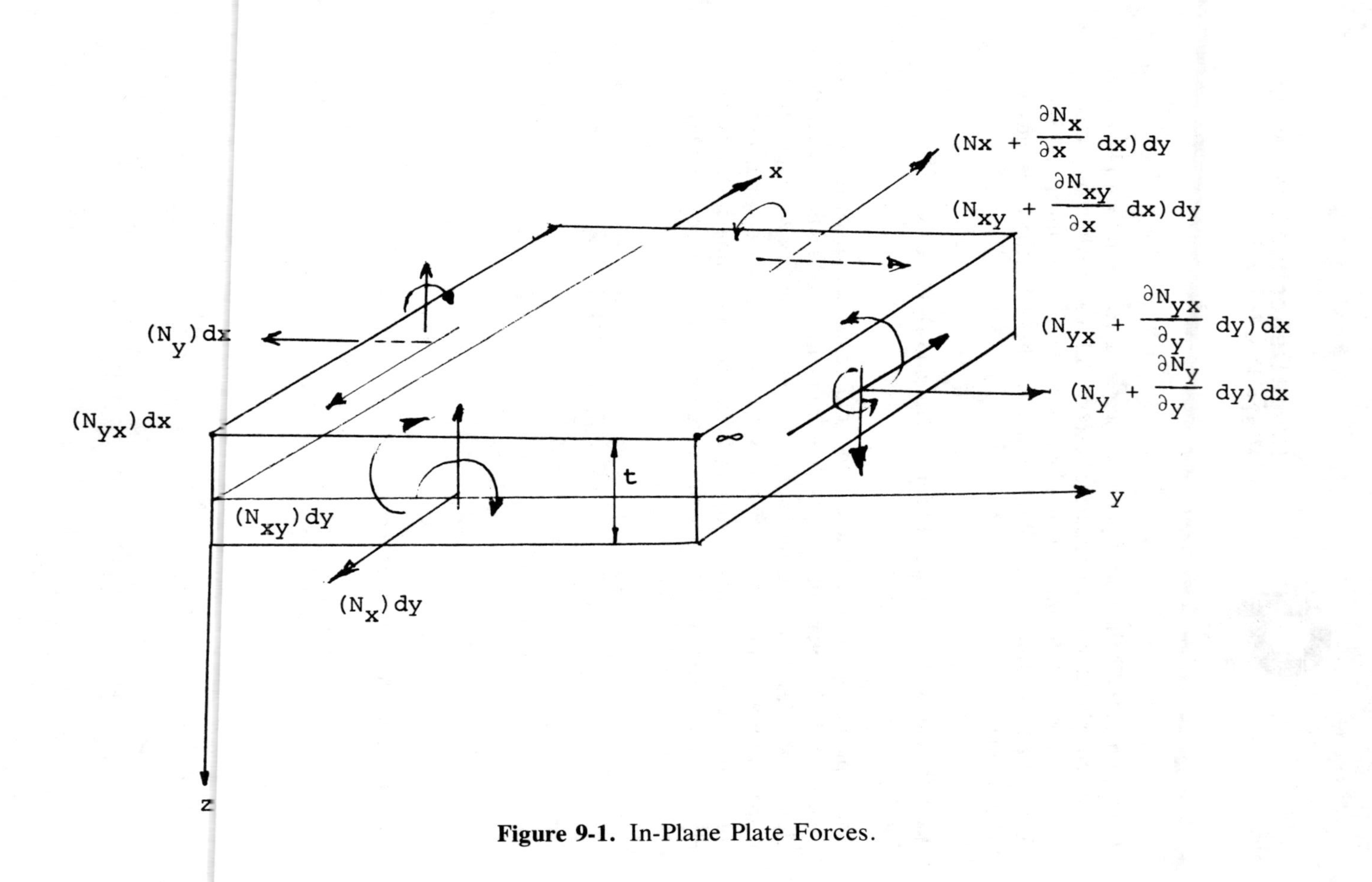

Figure 9-1. In-Plane Plate Forces.

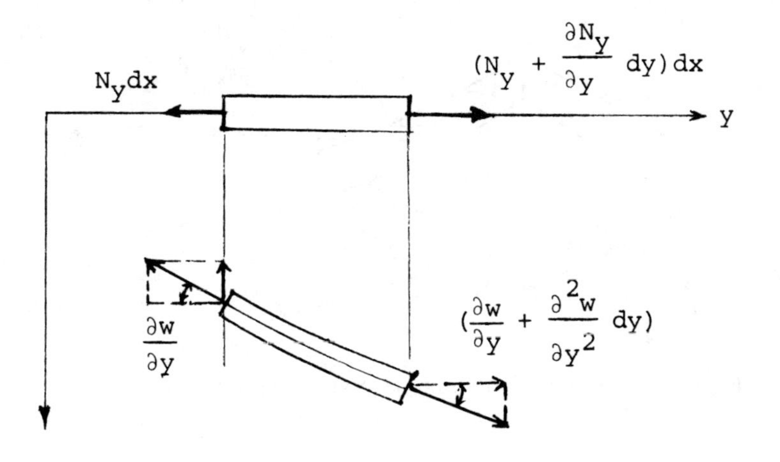

Figure 9-2. Plate Distortion *y*-Axis.

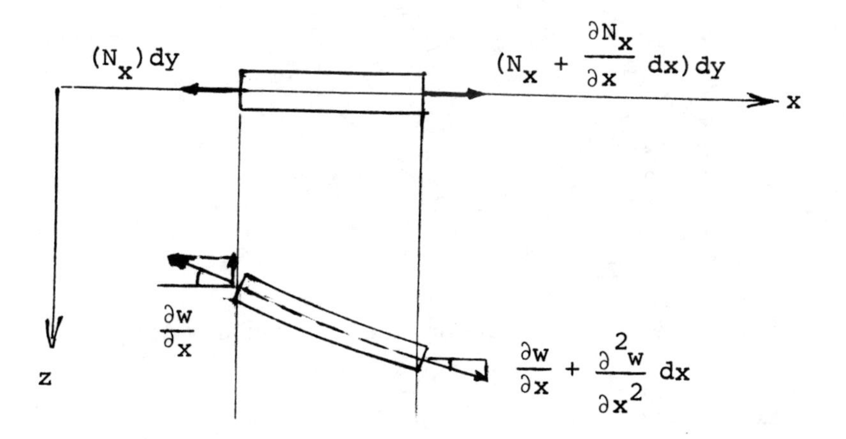

Figure 9-3. Plate Distortion *x*-Axis.

and neglecting higher order terms gives:

$$\sum F_z = N_x \frac{\partial^2 w}{\partial x^2} dx\,dy + \frac{\partial N_x}{\partial x} \frac{\partial w}{\partial x} dx\,dy \tag{9.2}$$

Also due to this bending of the element, the shear forces on the surface of the plate will develop vertical components. Examining the two faces of the element, with respect to the *y*- and *x*-axis, as shown in Figures 9-4 and 9-5, gives:

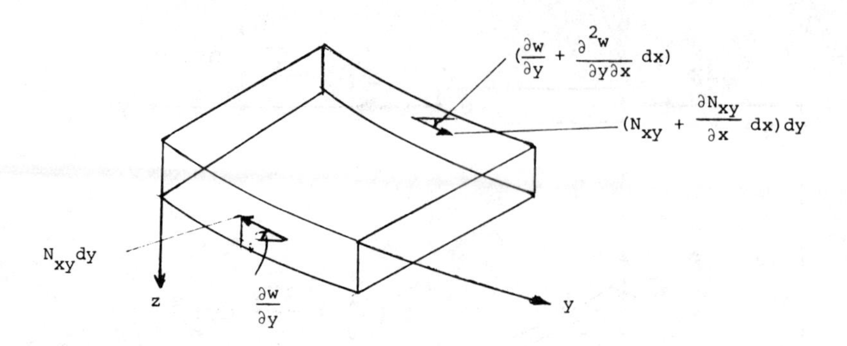

Figure 9-4. Plate Distortion—Slope Variation, y-Axis.

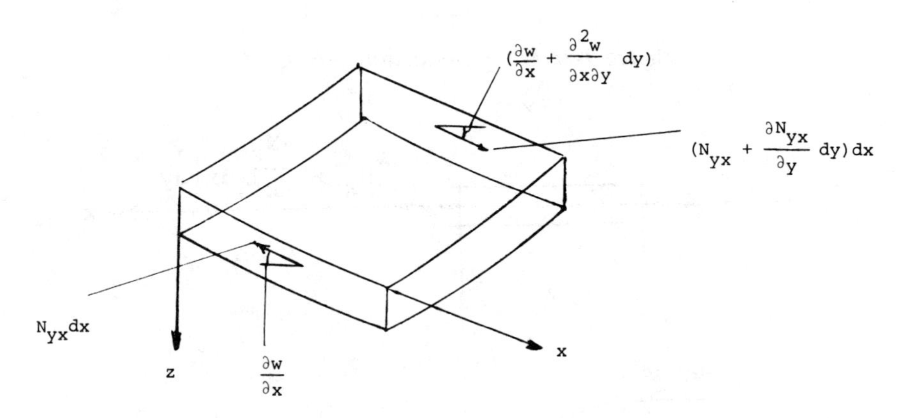

Figure 9-5. Plate Distortion—Slope Variation, x-Axis.

y-Axis (Figure 9-3) ΣF_z:

$$\sum F_z = -N_{xy}\,dy\frac{\partial w}{\partial y} + \left(N_{xy} + \frac{\partial N_{xy}}{\partial x}\,dx\right)dy\left(\frac{\partial w}{\partial y} + \frac{\partial^2 w}{\partial y\,\partial x}\,dx\right)$$

$$= -N_{xy}\frac{\partial w}{\partial y}\,dy + N_{xy}\frac{\partial w}{\partial y}\,dy + N_{xy}\frac{\partial^2 w}{\partial y\,\partial x}\,dx\,dy$$

$$+ \frac{\partial N_{xy}}{\partial x}\,dx\,dy\left(\frac{\partial w}{\partial y} + \frac{\partial^2 w}{\partial y\,\partial x}\right)$$

Reducing this expression gives

$$\sum F_z = N_{xy}\frac{\partial^2 w}{\partial y\,\partial x}\,dx\,dy + \frac{\partial N_{xy}}{\partial x}\,\frac{\partial w}{\partial y}\,dx\,dy \tag{9.3}$$

x-Axis (Figure 9-4) ΣF_z:

$$\sum F_z = -N_{yx}\,dx\frac{\partial w}{\partial x} + \left(N_{yx} + \frac{\partial N_{yx}}{\partial y}\,dy\right)dx\left(\frac{\partial w}{\partial x} + \frac{\partial^2 w}{\partial x\,\partial y}\,dy\right)$$

Collecting terms gives

$$\sum F_z = N_{yx}\frac{\partial^2 w}{\partial\,\partial y}\,dx\,dy + \frac{\partial N_{yx}}{\partial y}\,\frac{\partial w}{\partial x}\,dx\,dy \qquad (9.4)$$

Noting that $N_{xy} = N_{yx}$, as shown in chapter 2, equations (9.3) and (9.4) can be added, giving,

$$\sum F_z = 2N_{xy}\frac{\partial^2 w}{\partial x\,\partial y}\,dx\,dy + \frac{\partial N_{xy}}{\partial x}\,\frac{\partial w}{\partial y}\,dx\,dy + \frac{\partial N_{xy}}{\partial y}\,\frac{\partial w}{\partial x}\,dx\,dy \qquad (9.5)$$

Equations (9.1), (9.2), and (9.5) now represent the contribution the in-plane force has on resisting the total vertical force with respect to the z-axis. In chapter 4, equation (4.6), a relationship between the externally applied load q and internal vertical shears (Figure 4-3) gave the following:

$$\left(\frac{\partial V_x}{\partial x} + \frac{\partial V_y}{\partial y} + q\right)\,dx\,dy = 0 \qquad (4.6)$$

To this equation must now be added the vertical effects of the in-plane forces, as given by equations (9.1), (9.2), and (9.5), or:

$$\frac{\partial V_x}{\partial x} + \frac{\partial V_y}{\partial y} + q + N_y\frac{\partial^2 w}{\partial y^2} + \frac{\partial N_y}{\partial y}\,\frac{\partial w}{\partial y} + N_x\frac{\partial^2 w}{\partial x^2} + \frac{\partial N_x}{\partial x}\,\frac{\partial w}{\partial x}$$

$$+ 2N_{xy}\frac{\partial^2 w}{\partial x\,\partial y} + \frac{\partial N_{xy}}{\partial x}\,\frac{\partial w}{\partial y} + \frac{\partial N_{xy}}{\partial y}\,\frac{\partial w}{\partial x} = 0 \qquad (9.6)$$

where the terms $dx\,dy$ have been cancelled out. This equation can be reduced in its complexity by considering the other in-plane equilibrium equations, i.e., $\Sigma F_x = 0$, $\Sigma F_y = 0$.

$$\sum F_x = 0 = -N_x dy + \left(N_x + \frac{\partial N_x}{\partial x}\,dx\right)dy - N_{yx}\,dx$$

$$+ \left(N_{yx} + \frac{\partial N_{yx}}{\partial y}\,dy\right)dx$$

which gives

$$\frac{\partial N_x}{\partial x} + \frac{\partial N_{yx}}{\partial y} = 0 \qquad (9.7)$$

Also,

$$\sum F_y = 0 = -N_y\,dx + \left(N_y + \frac{\partial N_y}{\partial y}\,dy\right)dx - N_{xy}\,dy + \left(N_{xy} + \frac{\partial N_{xy}}{\partial x}\,dx\right)dy$$

which gives

$$\frac{\partial N_y}{\partial y} + \frac{\partial N_{xy}}{\partial x} = 0 \tag{9.8}$$

Substituting equations (9.7) and (9.8) into equation (9.6) gives

$$\frac{\partial V_x}{\partial x} + \frac{\partial V_y}{\partial y} + q + N_y\frac{\partial^2 w}{\partial y^2} + \left(\frac{\partial N_y}{\partial y} + \frac{\partial N_{xy}}{\partial x}\right)\frac{\partial w}{\partial y} + N_x\frac{\partial^2 w}{\partial x^2} + \left(\frac{\partial N_x}{\partial x} + \frac{\partial N_{xy}}{\partial y}\right)\frac{\partial w}{\partial x} + 2N_{xy}\frac{\partial^2 w}{\partial x\,\partial y} = 0$$

or,

$$\frac{\partial V_x}{\partial x} + \frac{\partial V_y}{\partial y} + \left(q + N_y\frac{\partial^2 w}{\partial y^2} + N_x\frac{\partial^2 w}{\partial x^2} + 2N_{xy}\frac{\partial^2 w}{\partial x\,\partial y}\right) = 0 \tag{9.9}$$

The shear terms, V_x and V_y, given in equation (9.9) can be equated to moments, as described in section 4.2.2, or:

$$\frac{\partial V_x}{\partial x} = \frac{\partial^2 M_x}{\partial x^2} + \frac{\partial^2 M_{xy}}{\partial x\,\partial y}$$
$$\frac{\partial V_y}{\partial y} = \frac{\partial^2 M_y}{\partial y^2} + \frac{\partial^2 M_{xy}}{\partial x\,\partial y} \tag{9.10}$$

Substituting (9.10) into (9.9) gives

$$\frac{\partial^2 M_x}{\partial x^2} + 2\frac{\partial^2 M_{xy}}{\partial x\,\partial y} + \frac{\partial^2 M_y}{\partial y^2} = -\left(q + N_y\frac{\partial^2 w}{\partial y^2} + N_x\frac{\partial^2 w}{\partial x^2} + 2N_{xy}\frac{\partial^2 w}{\partial x\,\partial y}\right) \tag{9.11}$$

The moments are related to curvature, as given by equations (4.31), (4.32), and (4.33) for an orthotropic plate. Substituting the derivatives of these expressions into (9.11), and collecting terms gives

$$D_x \frac{\partial^4 w}{\partial x^4} + 2H \frac{\partial^4 w}{\partial x^2 \partial y^2} + D_y \frac{\partial^4 w}{\partial y^4} =$$

$$\left[q + N_y \frac{\partial^2 w}{\partial y^2} + N_x \frac{\partial^2 w}{\partial x^2} + 2N_{xy} \frac{\partial^2 w}{\partial x \partial y} \right] \tag{9.12}$$

Equation (9.12) represents the general orthotropic plate equation (4.38) with the inclusion of membrane or in-plane forces. The solution of equation (9.12) cannot be directly obtained as in the case when no membrane action was considered ($N_y = N_x = N_{xy} = 0$). Previously the only unknown was the vertical deflection w of the plate; now the equation contains the unknowns of the three forces, N_y, N_x, and N_{xy}. In order, therefore, to solve plate equation (9.12), additional relationships are needed. The following will describe the development of these needed equations and their relationship to equation (9.12).

9.2 Large Deflections of Orthotropic Rectangular Plates

In the previous development it was assumed that in-plane forces, as well as bending forces, occurred simultaneously. Therefore, the in-plane strains in the middle surface are not zero. It was noted in chapter 4 that if the vertical deflections were of the order of less than or equal to 0.3 times the plate thickness, then the in-plane forces would be sufficiently small such that the neutral surface would be unstrained. If, however, the deflections become large, then part of the induced load will be carried by membrane action, i.e., forces N_x, N_y, and N_{xy}, and the neutral surface will be strained.

In section 9.1, the general orthotropic plate equation was modified in order to include these in-plane forces. It is now necessary to develop relationships for these in-plane forces in order to obtain a solution. Consider first the strains induced by the in-plane forces. In chapter 4, the induced strains in an element for in-plane bending resulted in horizontal displacements u and v due to rotation ($\partial w/\partial x$, $\partial w/\partial y$) of the plate, as given in Figure 4-5. Chapter 2 describes the resulting action or displacement of the plate due to no rotation (plane force effects), as shown in Figure 2-7. If the combined displacement occurs, then the resulting displacements are as shown in Figure 9-6. The combined action of these displacements gives the following strains [68].

$$\epsilon_x = \frac{\partial u}{\partial x} + \frac{1}{2}\left(\frac{\partial w}{\partial x}\right)^2 \tag{9.13a}$$

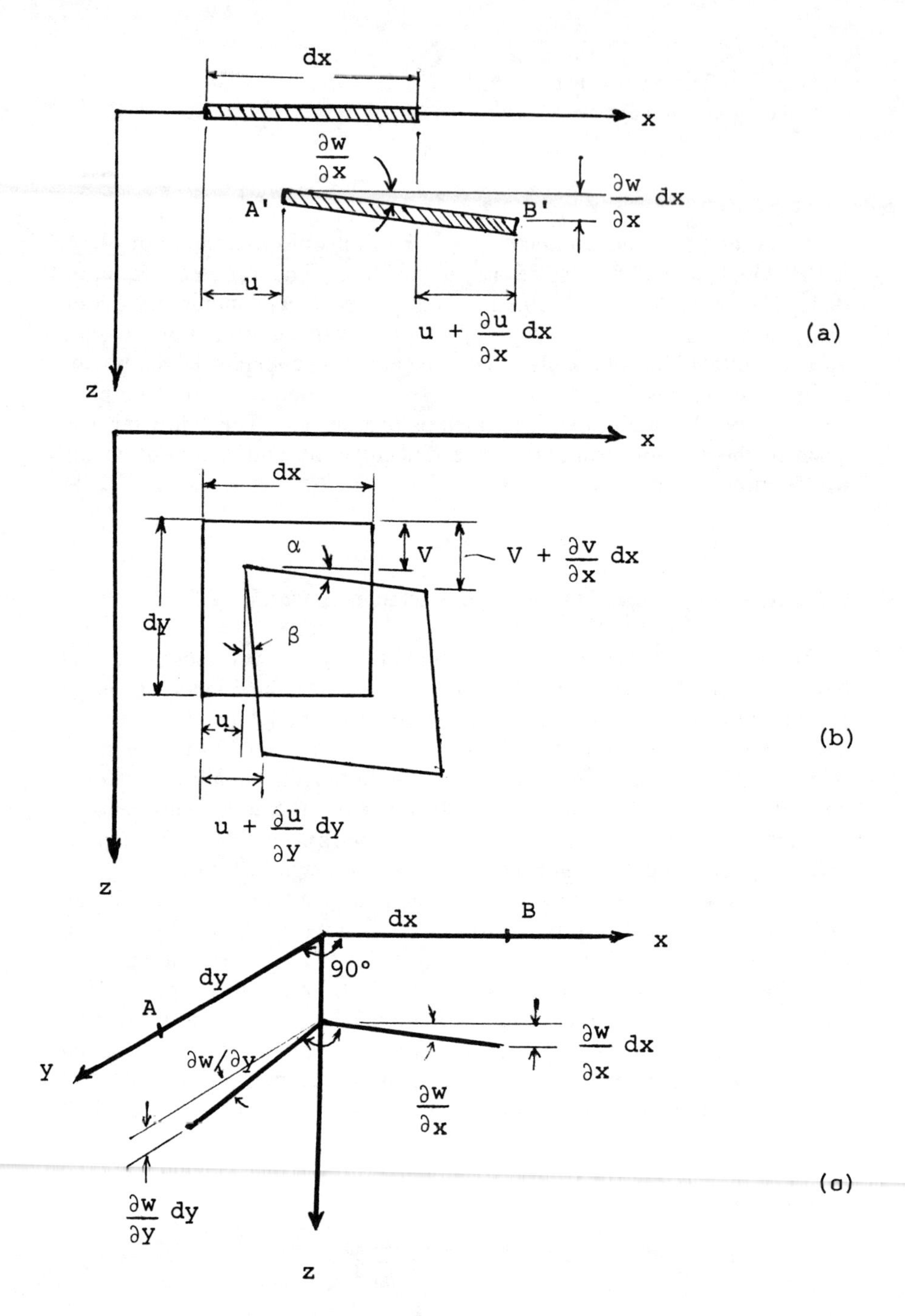

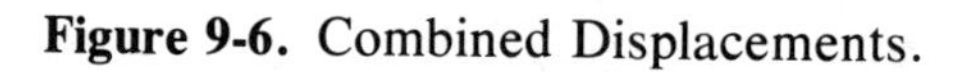

Figure 9-6. Combined Displacements.

$$\epsilon_y = \frac{\partial v}{\partial y} + \frac{1}{2}\left(\frac{\partial w}{\partial y}\right)^2 \tag{9.13b}$$

$$\gamma_{xy} = \frac{\partial u}{\partial y} + \frac{\partial v}{\partial y} + \frac{\partial w}{\partial x}\frac{\partial w}{\partial y} \tag{9.13c}$$

Combining the three equations of (9.13) gives

$$\frac{\partial^2 \epsilon_x}{\partial y^2} + \frac{\partial^2 \epsilon_y}{\partial x^2} - \frac{\partial^2 \gamma_{xy}}{\partial x \partial y} = \left(\frac{\partial^2 w}{\partial x \partial y}\right)^2 - \frac{\partial^2 w}{\partial x^2}\frac{\partial^2 w}{\partial y^2} \tag{9.14}$$

The relationships between strain in the middle plane of the plate of thickness t and the in-plane forces,.as given in part by equations (4.28), are

$$\begin{aligned} \epsilon_x &= \sigma_x/E_x - \mu_y \sigma_y/E_y \\ \epsilon_x &= \frac{N_x}{tE_x} - \mu_y \frac{N_y}{tE_y} \\ \epsilon_y &= \sigma_y/E_y - \mu_x \sigma_x/E_x \\ \epsilon_y &= \frac{N_y}{tE_y} - \mu_x \frac{N_x}{tE_x} \\ \gamma_{xy} &= N_{xy}/tG \end{aligned} \tag{9.15}$$

Equation (9.15) can be substituted into (9.14) to get another independent equation. Thus the solution of four expressions, equation (9.7), (9.8), (9.12), and (9.14) provide enough information to evaluate the four unknowns, N_x, N_y, N_{xy}, and w. However, equations (9.7), (9.8), and in particular (9.14) can be simplified by introducing a stress function. Let

$$\begin{aligned} N_x &= t\frac{\partial^2 F}{\partial y^2} \\ N_y &= t\frac{\partial^2 F}{\partial x^2} \\ N_{xy} &= -t\frac{\partial^2 F}{\partial x \partial y} \end{aligned} \tag{9.16}$$

F similar to ϕ

Substituting (9.16) into (9.15) gives

$$\begin{aligned} \epsilon_x &= \frac{1}{E_x}\frac{\partial^2 F}{\partial y^2} - \frac{\mu_y}{E_y}\frac{\partial^2 F}{\partial x^2} \\ \epsilon_y &= \frac{1}{E_y}\frac{\partial^2 F}{\partial x^2} - \frac{\mu_x}{E_x}\frac{\partial^2 F}{\partial y^2} \\ \gamma_{xy} &= -\frac{1}{G}\frac{\partial^2 F}{\partial x \partial y} \end{aligned} \tag{9.17}$$

and the partials of equations (9.17) are

$$\frac{\partial^2 \epsilon_x}{\partial y^2} = \frac{1}{E_x}\frac{\partial^4 F}{\partial y^4} - \frac{\mu_y}{E_y}\frac{\partial^4 F}{\partial x^2 \partial y^2}$$

$$\frac{\partial^2 \epsilon_y}{\partial x^2} = \frac{1}{E_y}\frac{\partial^4 F}{\partial x^4} - \frac{\mu_x}{E_x}\frac{\partial^4 F}{\partial x^2 \partial y^2} \tag{9.18}$$

$$\frac{\partial^2 \gamma_{xy}}{\partial x \partial y} = -\frac{1}{G}\frac{\partial^4 F}{\partial x^2 \partial y^2}$$

Substituting (9.18) into (9.14) gives

$$\frac{1}{E_x}\frac{\partial^4 F}{\partial y^4} - \frac{\mu_y}{E_y}\frac{\partial^4 F}{\partial x^2 \partial y^2} + \frac{1}{E_y}\frac{\partial^4 F}{\partial x^4} - \frac{\mu_x}{E_x}\frac{\partial^4 F}{\partial x^2 \partial y^2} + \frac{1}{G}\frac{\partial^4 F}{\partial x^2 \partial y^2} = \left(\frac{\partial^2 w}{\partial x \partial y}\right)^2 - \frac{\partial^2 w}{\partial x^2}\frac{\partial^2 w}{\partial y^2}$$

Collecting terms gives

$$\frac{1}{E_x} \cdot \frac{\partial^4 F}{\partial y^4} + \left(\frac{1}{G} - \frac{\mu_y}{E_y} - \frac{\mu_x}{E_x}\right)\frac{\partial^4 F}{\partial x^2 \partial y^2} + \frac{1}{E_y}\frac{\partial^4 F}{\partial x^4} = \left(\frac{\partial^2 w}{\partial x \partial y}\right)^2 - \frac{\partial^2 w}{\partial x^2}\frac{\partial^2 w}{\partial y^2}$$

or

$$\frac{\partial^4 F}{\partial y^4} + \left(\frac{E_x}{G} - \mu_y\frac{E_x}{E_y} - \mu_x\right)\frac{\partial^4 F}{\partial x^2 \partial y^2} + \frac{E_x}{E_y}\frac{\partial^4 F}{\partial x^4} = \left[\left(\frac{\partial^2 w}{\partial x \partial y}\right)^2 - \frac{\partial^2 w}{\partial x^2}\frac{\partial^2 w}{\partial y^2}\right]E_x \tag{9.19}$$

Equation (9.19), in conjunction with equation (9.12), can then be solved simultaneously to evaluate w and F. The resulting stresses can then be determined from curvature and stress function relationships.

9.3 Rectangular Stiffened Plates

If the plate is stiffened, as shown in Figure 9-7, then the basic equation can be suitably modified [68]. Assume the plate has an equivalent thickness:

$$t_x = A_x/b$$

$$t_y = A_y/a$$

$$t = \text{Top plate thickness}$$

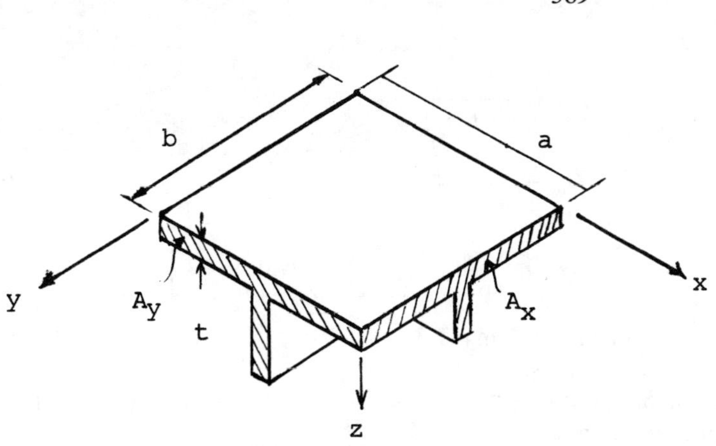

Figure 9-7. Stiffened Plate Geometry.

where A_x, A_y = Area of plate with respect to y- and x-axis

a, b = Effective plate widths

and the material is isotropic, i.e., $E_x = E_y = E$, $\mu_x = \mu_y = \mu$.

The forces N_x, N_y, and N_{xy}, equations (9.16), become

$$N_x = t_x \frac{\partial^2 F}{\partial y^2}$$

$$N_y = t_y \frac{\partial^2 F}{\partial x^2} \qquad (9.20)$$

$$N_{xy} = -t \frac{\partial^2 F}{\partial x \, \partial y}$$

Substituting (9.20) into (9.15) and the resulting derivatives into (9.14) gives

$$\frac{t}{t_y} \frac{\partial^4 F}{\partial x^4} + 2H' \frac{\partial^4 F}{\partial x^2 \, \partial y^2} + \frac{t}{t_x} \frac{\partial^4 F}{\partial y^4} = E\left[\left(\frac{\partial w}{\partial x \, \partial y}\right)^2 - \frac{\partial^2 w}{\partial x^2} \frac{\partial^2 w}{\partial y^2}\right] \qquad (9.21)$$

in which

$$2H' = 2(1 + \mu) - \mu\left(\frac{t}{t_x} + \frac{t}{t_y}\right) \qquad (9.22)$$

Equation (9.12) is the same, except the stiffness terms are defined as follows.

$$D_x = \frac{EI_x}{a(1 - \mu^2)}$$

$$D_y = \frac{EI_y}{b(1 - \mu^2)} \tag{9.23}$$

$$H = \frac{Gt_p^3}{6} + \frac{G}{2}\left[\sum dt^3/\text{width} \times k\right]$$

where k is defined in section 8.1.3 and d = depth of rib, t = thickness of rib, t_p = top plate thickness.

For isotropic-uniform plate, $t_x = t_y = t$ and $D_x = D_y = H = D$, equations (9.21) and (9.12) give

$$\frac{\partial^4 F}{\partial y^4} + 2\frac{\partial^4 F}{\partial x^2\,\partial y^2} + \frac{\partial^4 F}{\partial x^4} = E\left[\left(\frac{\partial^2 w}{\partial x\,\partial y}\right)^2 - \frac{\partial^2 w}{\partial x^2}\frac{\partial^2 w}{\partial y^2}\right] \tag{9.24}$$

Solve by iteration w & F unknown

$$\frac{\partial^4 w}{\partial x^4} + 2\frac{\partial^4 w}{\partial x^2\,\partial y^2} + \frac{\partial^4 w}{\partial y^4} = \frac{t}{D}\left[\frac{q}{t} + \frac{\partial^2 F}{\partial x^2}\frac{\partial^2 w}{\partial y^2} + \frac{\partial^2 F}{\partial y^2}\frac{\partial^2 w}{\partial x^2} - 2\frac{\partial^2 F}{\partial x\,\partial y}\frac{\partial^2 w}{\partial x\,\partial y}\right] \tag{9.25}$$

For a very thin plate, the plate stiffness $D = 0$, so that equations (9.24) and (9.25) reduce to equations of a membrane; or

$$\nabla^4 F = E\left[\left(\frac{\partial^2 w}{\partial x\,\partial y}\right)^2 - \frac{\partial^2 w}{\partial x^2}\frac{\partial^2 w}{\partial y^2}\right]$$

$$\frac{q}{t} + \frac{\partial^2 F}{\partial y^2}\frac{\partial^2 w}{\partial x^2} - 2\frac{\partial^2 F}{\partial x\,\partial y}\frac{\partial^2 w}{\partial x\,\partial y} + \frac{\partial^2 F}{\partial x^2}\frac{\partial^2 w}{\partial y^2} = 0 \tag{9.26}$$

9.4 Solution of Stiffened Rectangular Plates

The solution of the two fourth order differential equations (9.12) and (9.21) can conveniently be obtained by applying the finite difference technique [68]. Assuming a mesh pattern as described in Figure 9-8, the partial derivatives relative to the mesh points can be written as:

$$\frac{\partial^2 w}{\partial x^2} = \frac{1}{h_x^2}(w_1 + w_3 - 2w_0)\,; \qquad \frac{\partial^2 w}{\partial y^2} = \frac{1}{h_y^2}(w_2 + w_4 - 2w_0)$$

$$\frac{\partial^2 w}{\partial x\,\partial y} = \frac{1}{4h_x h_y}(w_5 - w_6 + w_7 - w_8)$$

$$\frac{\partial^4 w}{\partial x^4} = \frac{1}{h_x^4}(6w_0 - 4w_1 - 4w_3 + w_9 + w_{11}) \tag{9.27}$$

$$\frac{\partial^4 w}{\partial y^4} = \frac{1}{h_y^4}(6w_0 - 4w_2 - 4w_4 + w_{10} + w_{12})$$

$$\frac{\partial^4 w}{\partial x^2\,\partial y^2} = \frac{1}{h_x^2 h_y^2}[4w_0 - 2(w_1 + w_2 + w_3 + w_4) + w_5 + w_6 + w_7 + w_8]$$

Derivatives for the function F would be similar to the derivatives of w shown above. Substituting the finite difference equations into equations (9.12) and (9.21) gives

$$w_0\left[\frac{6D_x}{h_x^4} + \frac{8H}{h_x^2 h_y^2} + \frac{6D_y}{h_y^4} + \frac{2t(F_2 + F_4 - 2F_0)}{h_x^2 h_y^2} + \frac{2t(F_1 + F_3 - 2F_0)}{h_x^2 h_y^2}\right]$$
$$+ (w_1 + w_3)\left[-\frac{4D_x}{h_x^4} - \frac{4H}{h_x^2 h_y^2} - \frac{t(F_2 + F_4 - 2F_0)}{h_x^2 h_y^2}\right]$$
$$+ (w_2 + w_4)\left[-\frac{4D_y}{h_y^4} - \frac{4H}{h_x^2 h_y^2} - \frac{t(F_1 + F_3 - 2F_0)}{h_x^2 h_y^2}\right]$$
$$+ (w_5 + w_6 + w_7 + w_8)\left[\frac{2H}{h_x^2 h_y^2}\right]$$
$$+ (w_5 - w_6 + w_7 - w_8)\left[\frac{t}{8h_x^2 h_y^2}(F_5 - F_6 + F_7 - F_8)\right]$$
$$+ (w_9 + w_{11})\left[\frac{D_x}{h_x^4}\right] + (w_{10} + w_{12})\left[\frac{D_y}{h_y^4}\right] - q(x, y) = 0 \qquad (9.28)$$

and

$$F_0\left[6\left(\frac{t}{t_y h_x^4} + \frac{t}{t_x h_y^4}\right) + \frac{8H'}{h_x^2 h_y^2}\right] + (F_1 + F_3)\left[-4\left(\frac{t}{t_y h_x^4} + \frac{H'}{h_x^2}\right)\right]$$
$$+ (F_2 + F_4)\left[-4\left(\frac{t}{t_x h_y^4} - \frac{H'}{h_x^2 h_y^2}\right)\right] + (F_5 + F_6 + F_7 + F_8)\left[\frac{2H'}{h_x^2 h_y^2}\right]$$
$$+ (F_9 + F_{11})\left[\frac{t}{t_y h_x^4}\right] + (F_{10} + F_{12})\left[\frac{t}{t_x h_y^4}\right] = \frac{E}{16h_x^2 h_y^2}(w_5 - w_6$$
$$+ w_7 - w_8)^2 - \frac{E}{h_x^2 h_y^2}(w_1 + w_3 - 2w_0)(w_2 + w_4 - 2w_0) \qquad (9.29)$$

Equations (9.28) and (9.29) are shown in mesh form in Figures 9-9 and 9-10.

These mesh patterns are then modified to accommodate an actual physical plate. The solution of the resulting set of equations is accomplished as follows [68].

1. Solve for the deflections w, assuming all the Fs are zero; i.e., equation (9.28).

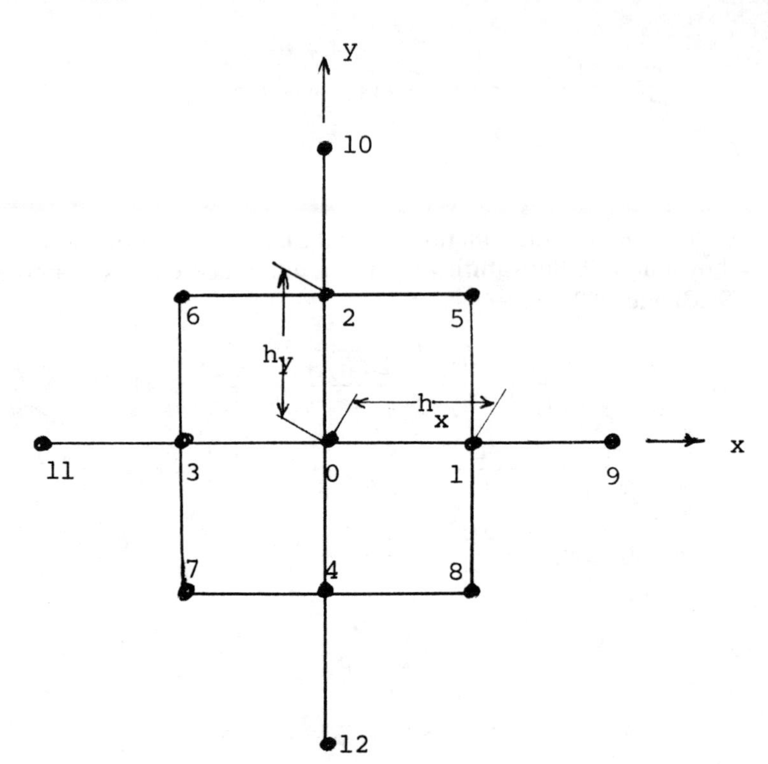

Figure 9-8. Finite Difference Mesh Pattern.

2. Use these resulting *w*s and solve for *F*s, using equation (9.29).
3. Return to equation (9.28), using the previously evaluated *F*s, and solve for *w*s.
4. Repeat process until both *w*s and *F*s converge to within a tolerable limit.
5. Use these resulting *F*s and *w*s to evaluate bending and membrane stresses at each node point.

9.5 Curved Orthotropic Plates

The development of the bending and membrane equations for rectangular orthotropic plates, as given in the previous section, 9.1, etc., can be extended to curved orthotropic plates. The combined action of bending and membrane effects will be examined separately and then combined effects added.

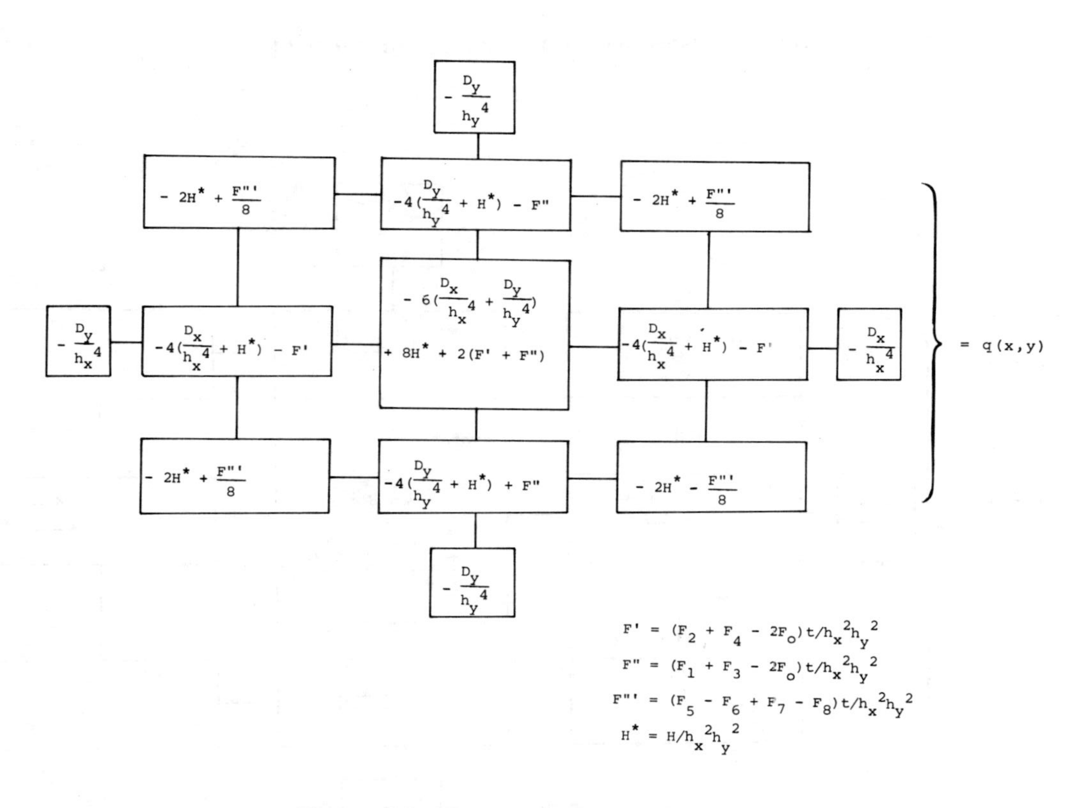

Figure 9-9. General Plate Mesh Equations.

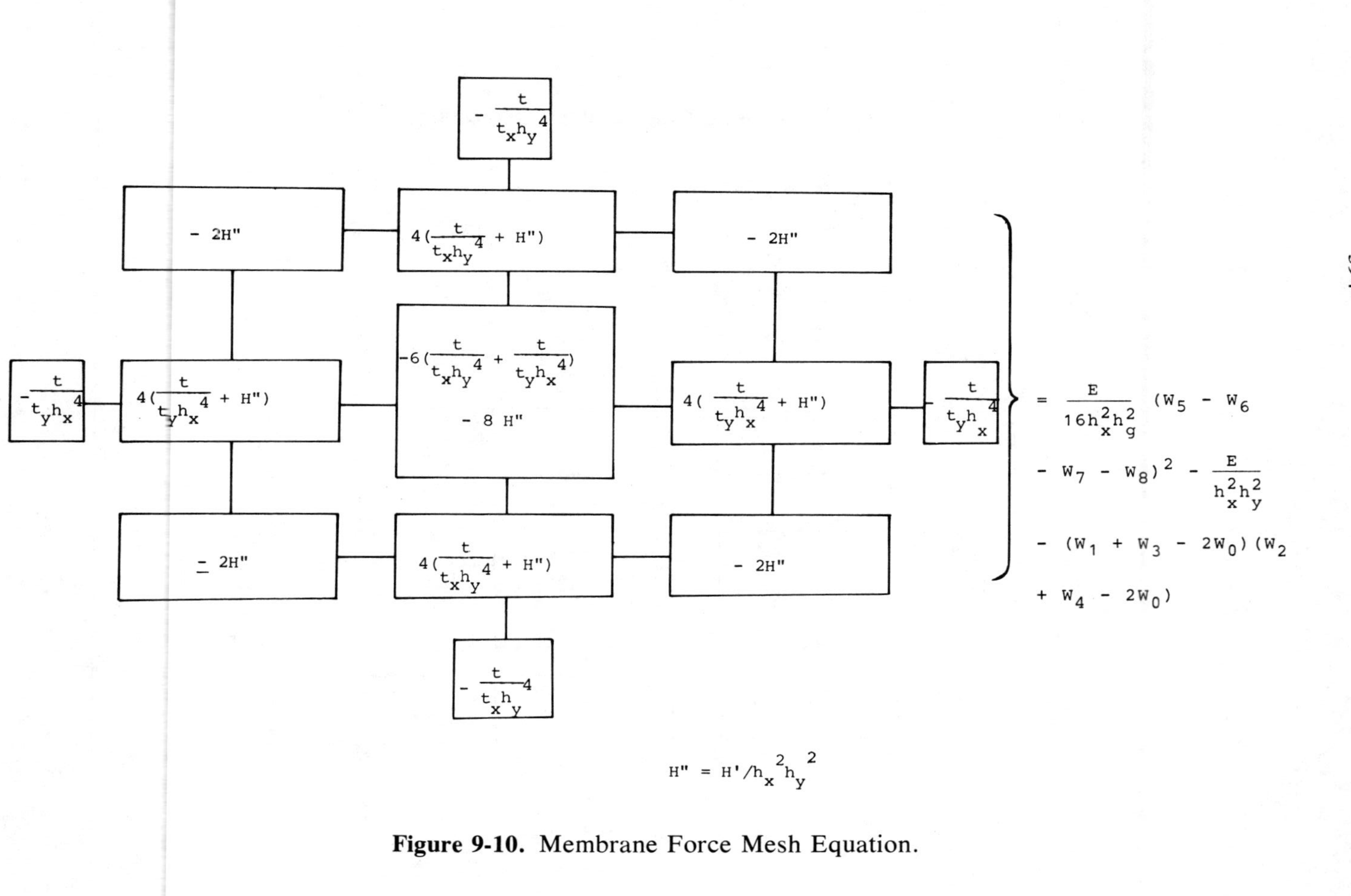

Figure 9-10. Membrane Force Mesh Equation.

9.5.1 *Plane Bending*

The load-deformation relationship [4, 25, 69] of a plate sector described in Figure 9-11 and given in section 5.5 is

$$D_r \frac{\partial^4 w}{\partial r^4} + \frac{(2H + \mu D_r + \mu D_\theta)}{r^2} \frac{\partial^4 w}{\partial r^2 \partial \theta^2} + \frac{D_\theta}{r^4} \frac{\partial^4 w}{\partial \theta^4}$$
$$+ \frac{(2 + \mu)D_r + \mu D_\theta}{r} \frac{\partial^3 w}{\partial r^3} - \frac{2(H + \mu D_r)}{r^3} \frac{\partial^3 w}{\partial r \partial \theta^2}$$
$$+ \frac{2(H + \mu D_r + D_\theta)}{r^4} \frac{\partial^2 w}{\partial \theta^2} - \frac{D_\theta}{r^2} \frac{\partial^2 w}{\partial r^2} + \frac{D_\theta}{r^3} \frac{\partial w}{\partial r} = q_1(r, \theta) \quad (9.30)$$

where w is the deflection of the plate in the z direction and $q_1(r, \theta)$ is a uniformly distributed load.

The plate stiffnesses D_r, D_θ and H are defined as

$$D_r = \frac{Et_r^3}{12(1 - \mu^2)} = \frac{EI_\theta}{2(1 - \mu^2)} \quad (9.31)$$

$$D_\theta = \frac{Et_\theta^3}{12(1 - \mu^2)} = \frac{EI_r}{r\theta_0(1 - \mu^2)} \quad (9.32)$$

$$2H = 2(1 - \mu)D_r \quad (9.33)$$

in which λ and $r\theta_0$ are spacing of the ribs, as described in Figure 9-12, and I_r and I_θ are moments of inertia per unit of width.

9.5.2 *Membrane Action*

Figure 9-13 describes the membrane forces acting on a differential plate element. Examination of the distortions of this element relative to a uniformly distributed load $q_2(r, \theta)$ and these forces, as given in [4, 25, 69], yields

$$N_r \frac{\partial^2 w}{\partial r^2} + N_\theta \left(\frac{1}{r} \frac{\partial w}{\partial r} + \frac{1}{r^2} \frac{\partial^2 w}{\partial \theta^2} \right) + N_{r\theta} \left(\frac{-2}{r^2} \frac{\partial w}{\partial \theta} + \frac{2}{r} \frac{\partial^2 w}{\partial r \partial \theta} \right)$$
$$= -q_2(r, \theta) \quad (9.34)$$

The total differential equation of the neutral surface is then found by combining equations (9.30) and (9.34), by letting $q_1(r, \theta) + q_2(r, \theta) = q(r, \theta)$. The externally applied uniform load gives the following:

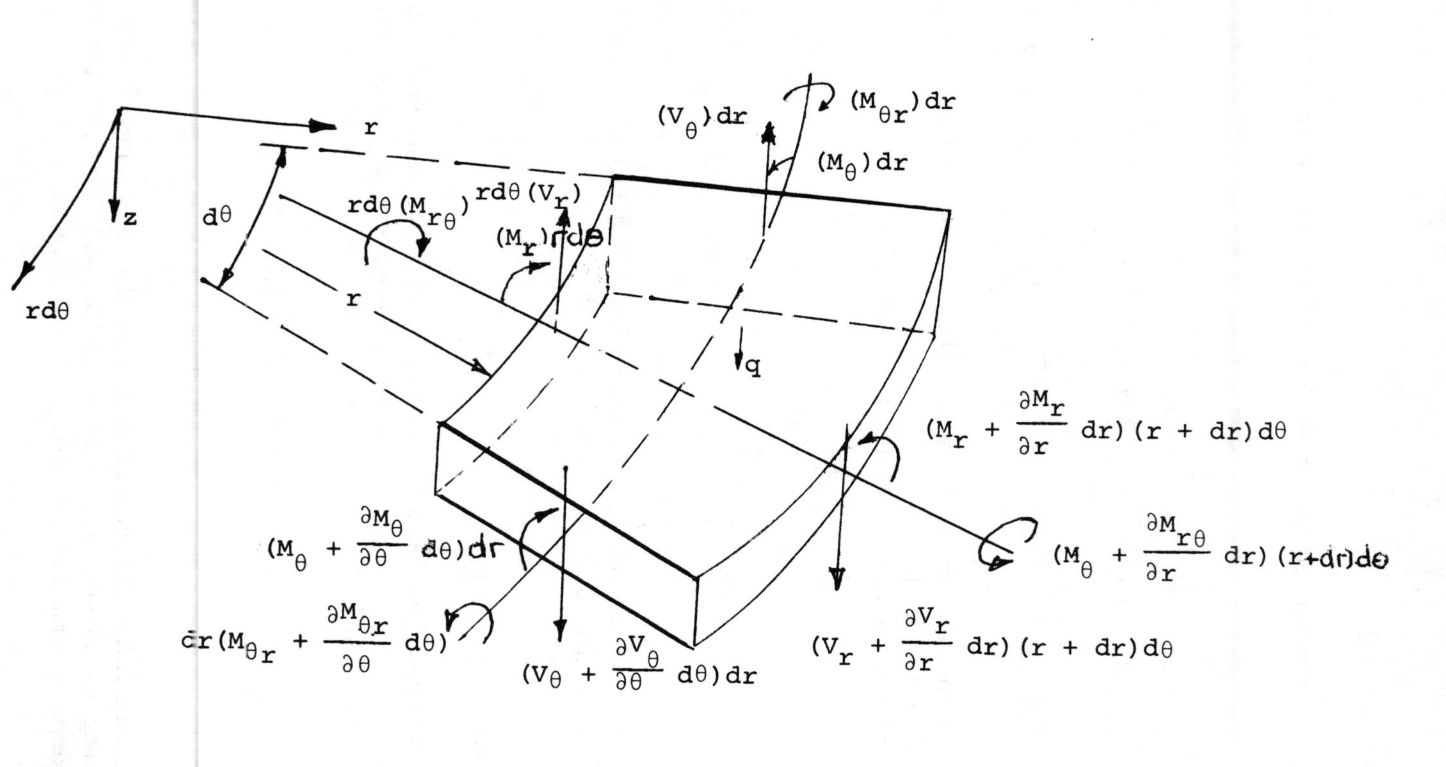

Figure 9-11. Bending Forces on a Curved Plate Element.

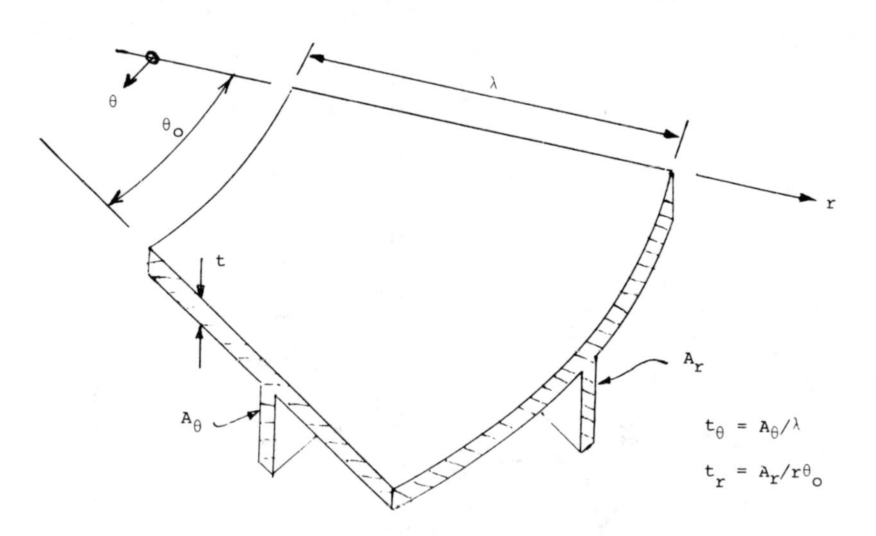

Figure 9-12. Curved Stiffened Plate Element.

$$
\begin{aligned}
D_r \frac{\partial^4 w}{\partial r^4} &+ \frac{(2H + \mu D_r + \mu D_\theta)}{r^2} \frac{\partial^4 w}{\partial r^2 \partial \theta^2} + \frac{D_\theta}{r^4} \frac{\partial^4 w}{\partial \theta^4} \\
&+ \frac{(2 + \mu) D_r + \mu D_\theta}{r} \frac{\partial^3 w}{\partial r^3} - \frac{2(H + \mu D_r)}{r^3} \frac{\partial^3 w}{\partial r \partial \theta^2} \\
&+ \frac{2(H + \mu D_r + D_\theta)}{r^4} \frac{\partial^2 w}{\partial \theta^2} - \frac{D_\theta}{r^2} \frac{\partial^2 w}{\partial r^2} + \frac{D_\theta}{r^3} \frac{\partial w}{\partial r} \\
&= q(r, \theta) + N_r \frac{\partial^2 w}{\partial r^2} + N_\theta \left(\frac{1}{r} \frac{\partial w}{\partial r} + \frac{1}{r^2} \frac{\partial^2 w}{\partial \theta^2} \right) \\
&+ N_{r\theta} \left(-\frac{2}{r^2} \frac{\partial w}{\partial \theta} + \frac{2}{r} \frac{\partial^2 w}{\partial r \partial \theta} \right) \qquad (9.35)
\end{aligned}
$$

9.5.3 Strain in the Neutral Surface

The strain induced in the middle surface of the plate, as caused by large plate deflections and thus membrane forces, are given elsewhere [4, 25, 68] and result in the following compatibility equation:

$$
\begin{aligned}
\frac{\partial^2 \epsilon_\theta}{\partial r^2} &+ \frac{1}{r^2} \frac{\partial^2 \epsilon_r}{\partial \theta^2} + \frac{2}{r} \frac{\partial \epsilon_\theta}{\partial r} - \frac{1}{r} \frac{\partial \epsilon_r}{\partial r} - \frac{1}{r^2} \frac{\partial \nu_{r\theta}}{\partial \theta} - \frac{\partial \nu_{r\theta}}{r \partial r \partial \theta} \\
&= \left[\frac{\partial}{\partial r} \left(\frac{1}{r} \frac{\partial w}{\partial \theta} \right) \right]^2 - \frac{\partial^2 w}{\partial r^2} \left(\frac{1}{r} \frac{\partial w}{\partial r} + \frac{1}{r^2} \frac{\partial^2 w}{\partial \theta^2} \right) \qquad (9.36)
\end{aligned}
$$

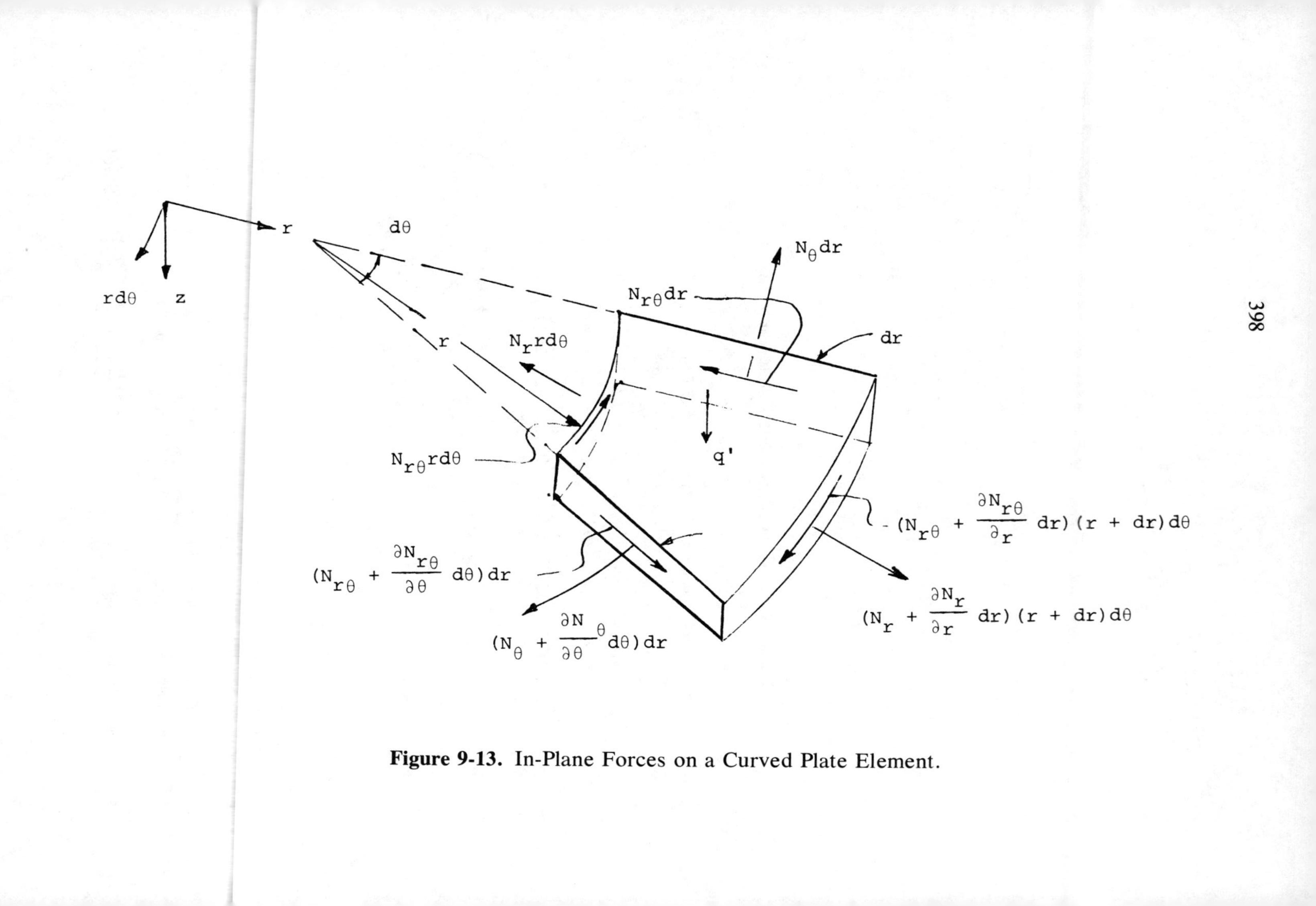

Figure 9-13. In-Plane Forces on a Curved Plate Element.

The relationship between strain in the middle plane and the membrane forces can be given by

$$\epsilon_r = \frac{N_r}{t_r E} - \mu \frac{N_\theta}{t_\theta E} \tag{9.37}$$

$$\epsilon_\theta = \frac{N_\theta}{t_\theta E} - \mu \frac{N_r}{t_r E} \tag{9.38}$$

$$\nu_{r\theta} = \frac{N_{r\theta}}{tG} \tag{9.39}$$

where t_r, t_θ, and t are average plate thicknesses, as shown in Figure 9-12. Using Airy's stress function $F(r, \theta)$, the membrane forces can be defined as

$$N_r = \left(\frac{1}{r}\frac{\partial F}{\partial r} + \frac{1}{r^2}\frac{\partial^2 F}{\partial \theta^2}\right)t \tag{9.40}$$

$$N_\theta = \frac{\partial^2 F}{\partial r^2} t \tag{9.41}$$

$$N_{r\theta} = \left(\frac{1}{r^2}\frac{\partial F}{\partial \theta} - \frac{1}{r}\frac{\partial^2 F}{\partial r \partial \theta}\right)t \tag{9.42}$$

Substituting equations (9.27) through (9.42) into equation (9.36) gives

$$\frac{t}{Et_\theta}\frac{\partial^4 F}{\partial r^4} + \left[\frac{1}{G} - \frac{\mu t}{Et_r} - \frac{\mu t}{Et_\theta}\right]\frac{1}{r^2}\frac{\partial^4 F}{\partial r^2 \mu \theta^2} + \frac{t}{Et_r}\frac{1}{r^4}\frac{\partial^4 F}{\partial \theta^4}$$
$$+ \left[\frac{(2+\mu)t}{Et_\theta} - \frac{\mu t}{Et_r}\right]\frac{1}{r}\frac{\partial^3 F}{\partial r^3} - \left[\frac{1}{G} - \frac{2\mu t}{Et_r}\right]\frac{1}{r^3}\frac{\partial^3 F}{\partial r \partial \theta^2}$$
$$- \frac{t}{Et_r}\frac{1}{r^2}\frac{\partial^2 F}{\partial r^2} + \frac{1}{r^4}\left[\frac{2t}{Et_r} - \frac{2\mu t}{Et_r} + \frac{1}{G}\right]\frac{\partial^2 F}{\partial \theta^2} + \frac{1}{r^3}$$
$$\frac{t}{Et_r}\frac{\partial F}{\partial r} = \left[\frac{\partial}{\partial r}\left(\frac{1}{r}\frac{\partial w}{\partial \theta}\right)\right]^2 - \frac{\partial^2 w}{\partial r^2}\left(\frac{1}{r}\frac{\partial w}{\partial r} + \frac{1}{r^2}\frac{\partial^2 w}{\partial \theta^2}\right) \tag{9.43}$$

Using the stress function relation of equations (9.40) through (9.42), equation (9.35) will reduce to a partial differential equation of two variables w and F. This equation can be solved by considering equation (9.43), and solving equation (9.35) and (9.43) simultaneously. The following will describe the solution of these equations.

9.6 Solution of Stiffened Curved Plates

The solution of the two coupled differential equations (9.35) and (9.43), as

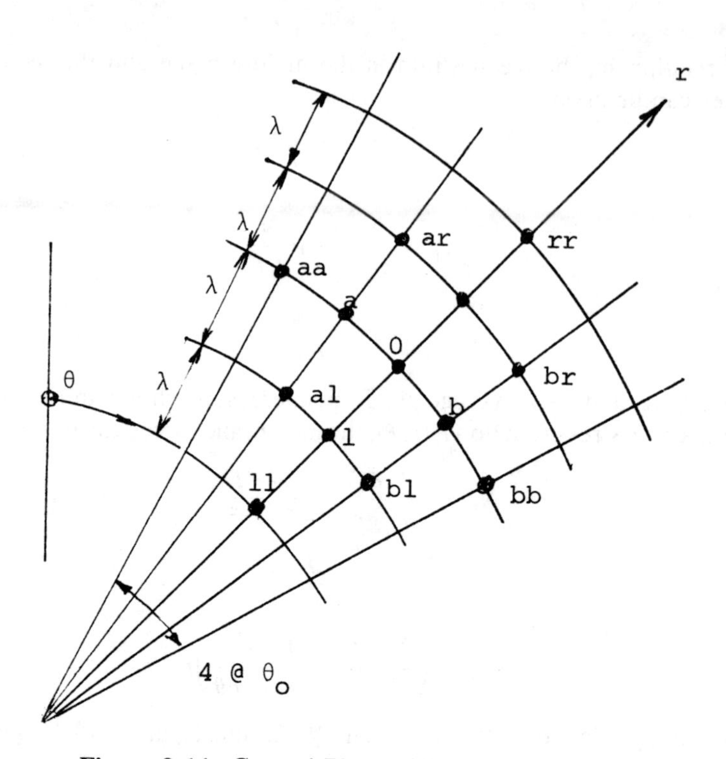

Figure 9-14. Curved Plate Difference Pattern.

functions of w and F, is obtained by applying the central finite difference technique [69]. The mesh pattern associated with the difference equations is as given in Figure 9-14 and as was used in chapter 5. Defining the following constants:

$$
\begin{aligned}
DI &= [\mu(D_r + D_\theta) + 2H]/D_r \\
DJ &= D_\theta/D_r \\
DK &= [(2 + \mu)D_r - \mu D_\theta]/2D_r \\
DL &= (\mu D_r + H)/D_r \\
DM &= 2(\mu D_r + D_\theta + H)/D_r \\
WA &= (t\lambda r^3\theta_0^4)/2D_r \\
WB &= (t\lambda^2 r^2\theta_0^2)/D_r \\
WC &= (t\lambda^2 r\theta_0^2)/4D_r \\
WD &= (t\lambda^4\theta_0^2)/2D_r
\end{aligned}
\tag{9.44}
$$

Equation (9.35) can be written in difference notation giving

$$
\begin{aligned}
& r^4\theta_0^4(w_{rr} - 4w_r + w_0 - 4w_l + w_{ll}) \\
& + \lambda^2 r^2 \theta_0^2 DI(w_{ar} - 2w_r + w_{br} - 2w_a + 4w_0 - 2w_b + w_{al} - 2w_l + w_{bl}) \\
& + \lambda^4 DJ(w_{aa} - 4w_a + 62_0 - 4w_b + w_{bb}) \\
& + \lambda r^3 \theta_0^4 DK(w_{rr} - 2w_r + 2w_l - w_{ll}) \\
& - \lambda^3 r \theta_0^2 DL(w_{ar} - w_{al} - 2w_r + 2w_l + w_{br} - w_{bl}) \\
& - \lambda^2 r^2 \theta_0^4 DJ(w_r - 2w_0 + w_l) \\
& + \lambda^4 \theta_0^2 DM(w_a - 2w_0 + w_b) + \lambda^3 r \theta_0^4 DJ(w_r - w_l)/2 \\
& = (qr^4\theta_0^4\lambda^4)/D_r + WA(w_r - 2w_0 + w_l)(F_r - F_l) \\
& + WB(w_r - 2w_0 + w_l)(F_a - 2F_0 + F_b) \\
& + WA(w_r - w_l)(F_r - 2F_0 + F_l) \\
& + WB(w_a - 2w_0 + w_b)(F_r - 2F_0 + F_l) \\
& - WB(w_{ar} - w_{al} - w_{br} + w_{bl})(F_{ar} - F_{al} - F_{br} + F_{bl})/2 \\
& + WC(w_a - w_b)(F_{ar} - F_{al} - F_{br} + F_{bl}) \\
& + WC(w_{ar} - w_{al} - w_{br} + w_{bl})(F_a - F_b) \\
& - WD(w_a - w_b)(F_a - F_b)
\end{aligned}
\tag{9.45}
$$

Expansion of equation (9.45), and collection of terms associated with each node point of Figure 9-14 will yield a general equation of the form

$$w[\ ] = qr^4\theta_0^4\lambda^4/D_r \tag{9.46}$$

where the matrix [ ] is a function of the constants given by equations (9.44) and λ, r, θ_0, and F, as shown in Figure 9-15. Similarly, equation (9.43) can be written in difference form giving

$$
\begin{aligned}
& r^4\theta_0^4 t/Et_\theta(F_{rr} - 4F_r + 6F_0 - 4F_l + F_{ll}) \\
& + 2H' r^2\lambda^2\theta_0^2(F_{ar} - 2F_r + F_{br} - 2F_a + 4F_0 - 2F_b + F_{al} - 2F_l + F_{bl}) \\
& + \lambda^4 t/Et_r(F_{aa} - 5F_a + 6F_0 - 4F_b + F_{bb}) \\
& + D_1 r^3\lambda\theta_0^4/2(F_{rr} - 2F_r + 2F_l - F_{ll}) \\
& - D_2 r\theta_0^2\lambda^3/2(F_{ar} - F_{al} - 2F_r + 2F_l + F_{br} - F_{bl}) \\
& - r^2\lambda^2\theta_0^4 t/Et_r(F_r - 2F_0 + F_l) + \lambda^4\theta_0^2 D_3(F_a - 2F_0 + F_b) \\
& + r\theta_0^4\lambda^3 t/(2Et_r)(F_r - F_l)
\end{aligned}
$$

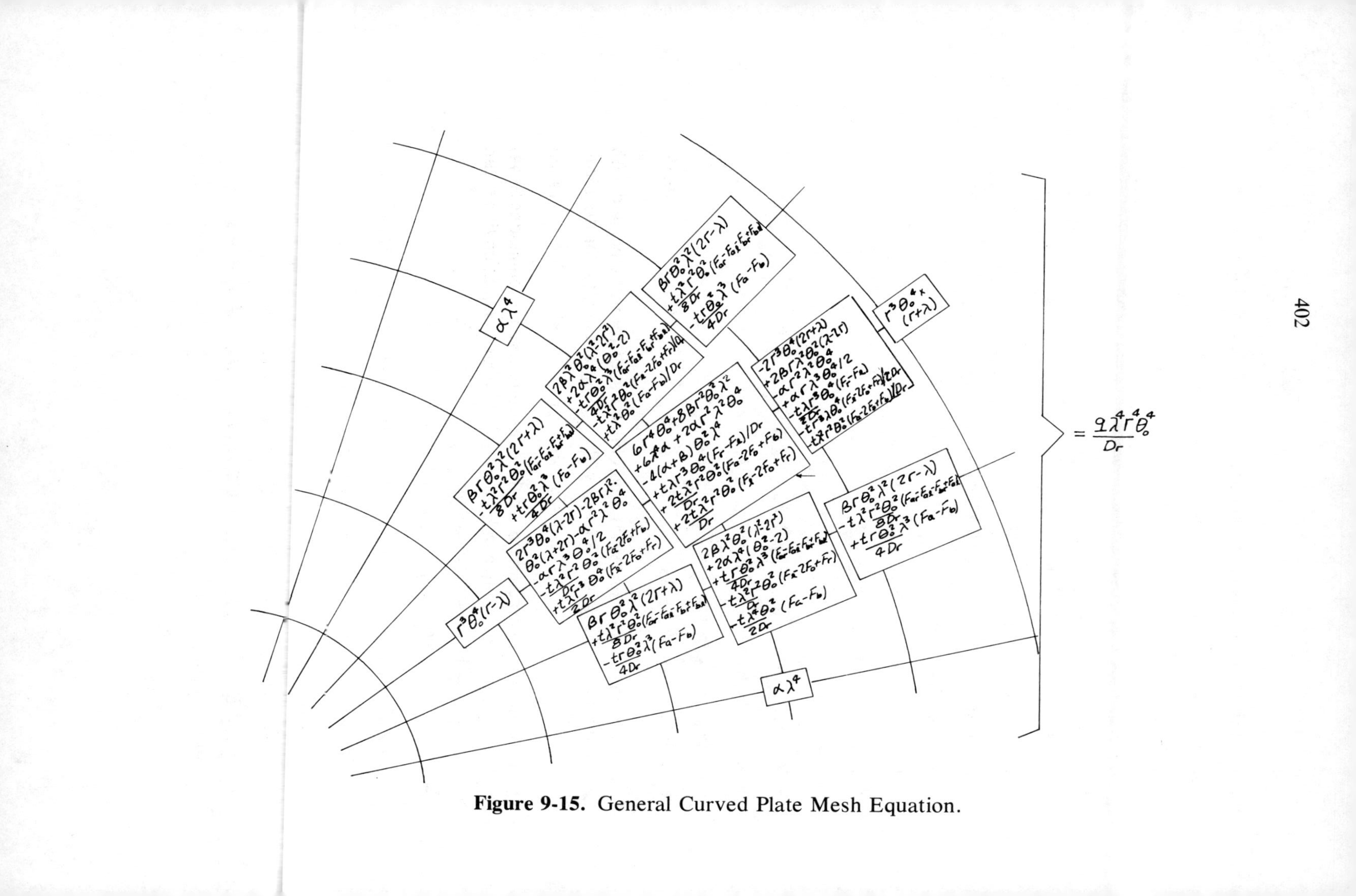

Figure 9-15. General Curved Plate Mesh Equation.

$$= r^2\theta_0^2\lambda^2/16(w_{ar} - w_{al} - w_{br} + w_{bl})^2$$
$$- r\lambda^3\theta_0^2/4(w_a - w_b)(w_{ar} - w_{al} - w_{br} + w_{bl})$$
$$+ \theta_0^2\lambda^4/4(w_a - w_b)^2 - r^3\theta_0^4\lambda/2(w_r - w_l)(w_r - 2w_0 + w_l)$$
$$- r^2\theta_0^2\lambda^2(w_r - 2w_0 + w_l)(w_a - 2w_0 + w_b) \qquad (9.47)$$

where:

$$2H' = \frac{1}{G} - \frac{\mu t}{Et_\theta} - \frac{\mu t}{Et_r}$$

$$D_1 = \frac{(2 + \mu)t}{Et_\theta} - \frac{\mu t}{Et_r}$$

$$D_2 = \frac{1}{G} - \frac{2\mu t}{Et_r}$$

$$D_3 = \frac{2(1 - \mu)t}{Et_r} + \frac{1}{G} \qquad (9.48)$$

Expansion of equation (9.47) and collection of terms relative to each nodal point, as given in Figure 9-16, will result in an expression of the form

$$F[\ ] = BB(I) \qquad (9.49)$$

where the matrix [ ] is a function of the constants given by equation (9.48) and r, θ_0, λ, t, t_t, t_θ, and E. The matrix $BB(I)$ is a function of w at the mesh nodes and of r, θ_0, and λ. The two general finite difference equations (9.46) and (9.49) representing any interior node point are given by Figures 9-15 and 9-16.

9.6.1 *Solution of Equations*

The solution of equations (9.46) and (9.49) utilizing the following boundary conditions:

Radial Simple Supports and

1. Fixed angular edges
2. Free angular edges
3. Elastic supports—angular edges

as shown in Figure 9-17, have been evaluated [70]. The solution of the resulting set of simultaneous equations is obtained by (1) assuming $F = 0$ in equation (9.46) and evaluating w; (2) substituting w into equation (9.49) and

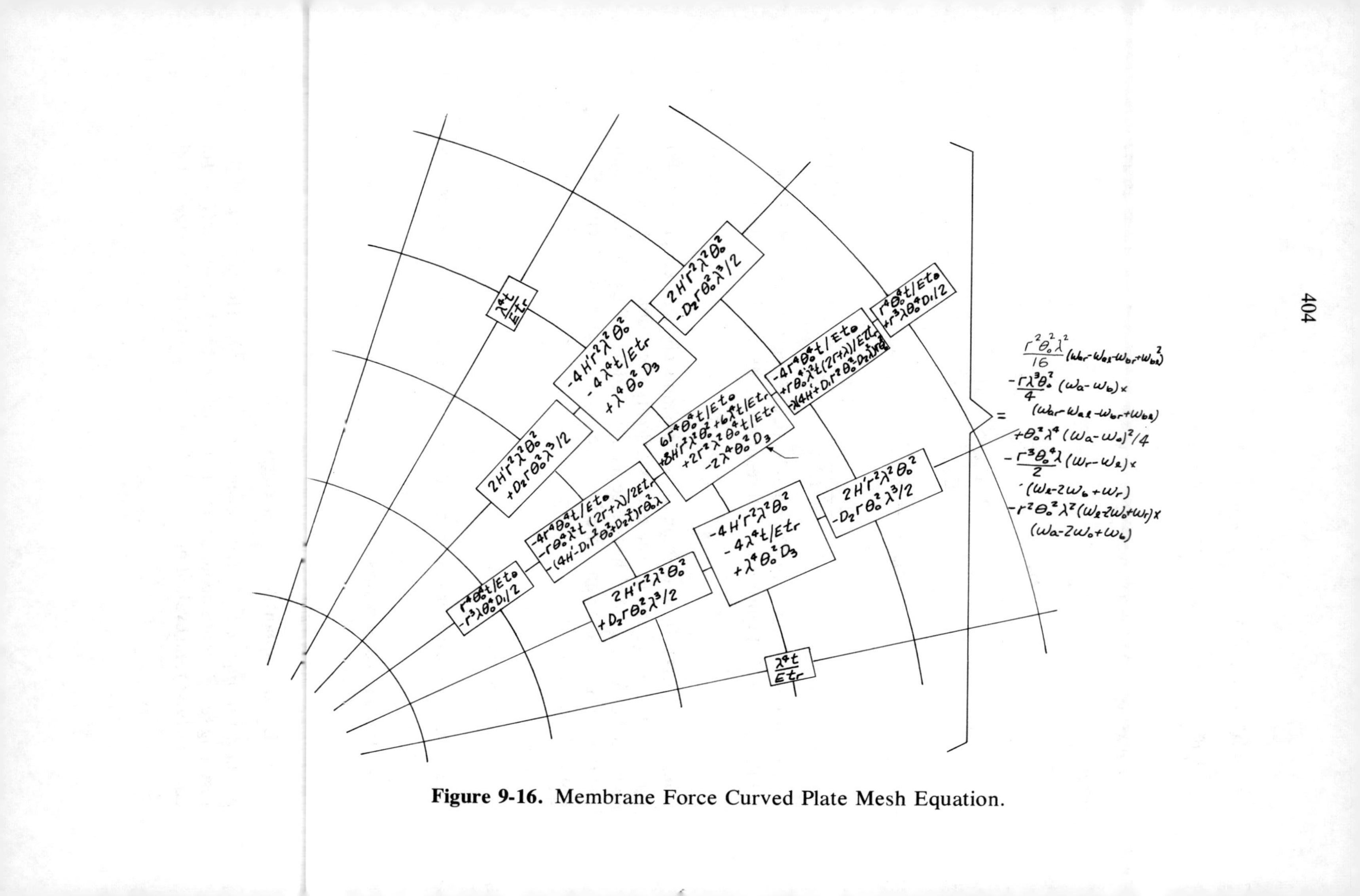

Figure 9-16. Membrane Force Curved Plate Mesh Equation.

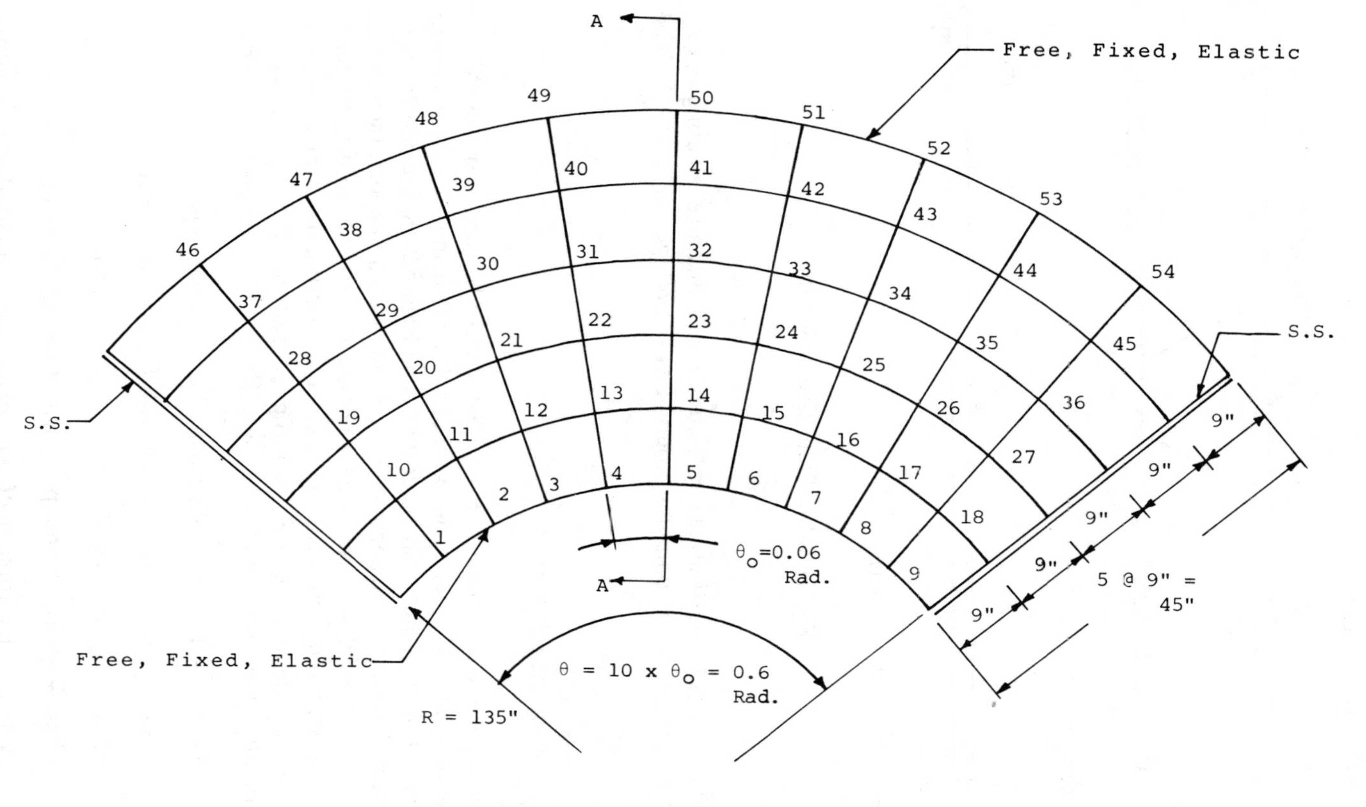

Figure 9-17. Curved Plate Geometry.

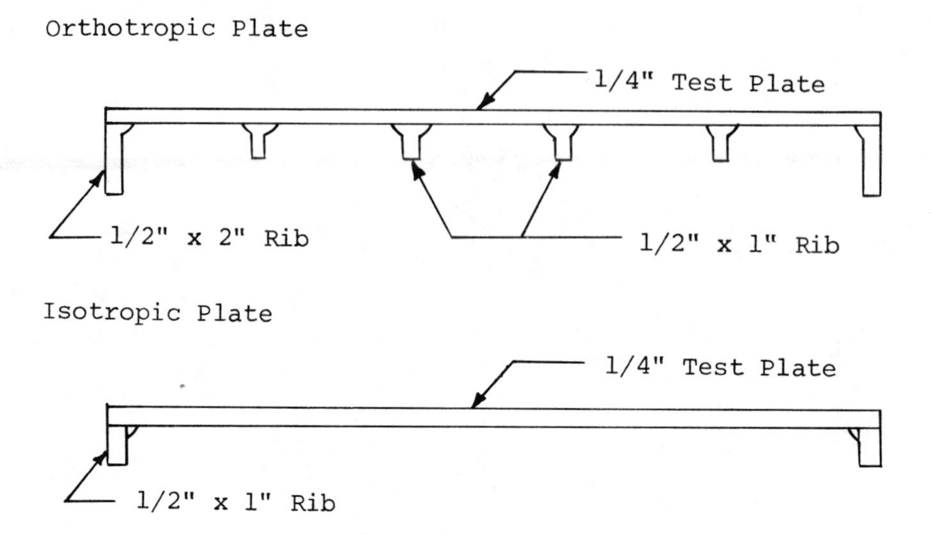

Figure 9-18. Plate Cross Section.

evaluating F; (3) substituting F into equation (9.46) and evaluating w, etc., until convergence. A computer program was developed to handle this task [70].

9.6.2 Experimental Work

A series of six test models, Figure 9-17, were fabricated of ASTM-A36 structural steel. The deck consisted of 1/4-in plate and the stiffeners were 1/2 in × 1 in or 1/2 in × 2 in, as described in Figure 9-18. The models were fabricated initially to accommodate a fixed angular support; then the excess plate material was removed for testing of the free edge support condition. The stiffening ribs were spot welded intermittently to the deck plate. Marine-Tex epoxy glue was then applied between the welds to ensure continuity between the plate and ribs. The six test models were [69]

Plate 1— Isotropic plate with fixed angular edges

Plate 2— Orthotropic plate with fixed angular edges

Plate 3— Isotropic plate with free angular edges

Plate 4— Orthotropic plate with free angular edges

Plate 5— Isotropic plate with elastically supported angular edges

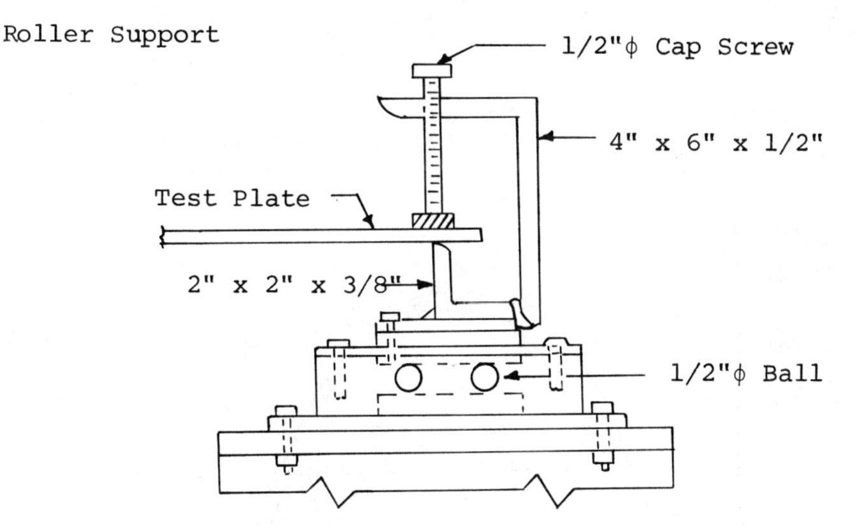

Figure 9-19. Physical Support Construction.

Plate 6— Orthotropic plate with elastically supported angular edges

For all of the test models, the radial supports were considered "simply supported." The physical arrangement required for this support condition is described in Figure 9-19.

Partial results [69] from these tests are given in Figures 9-20 and 9-21, which represent the response of the isotropic and orthotropic curved plates.

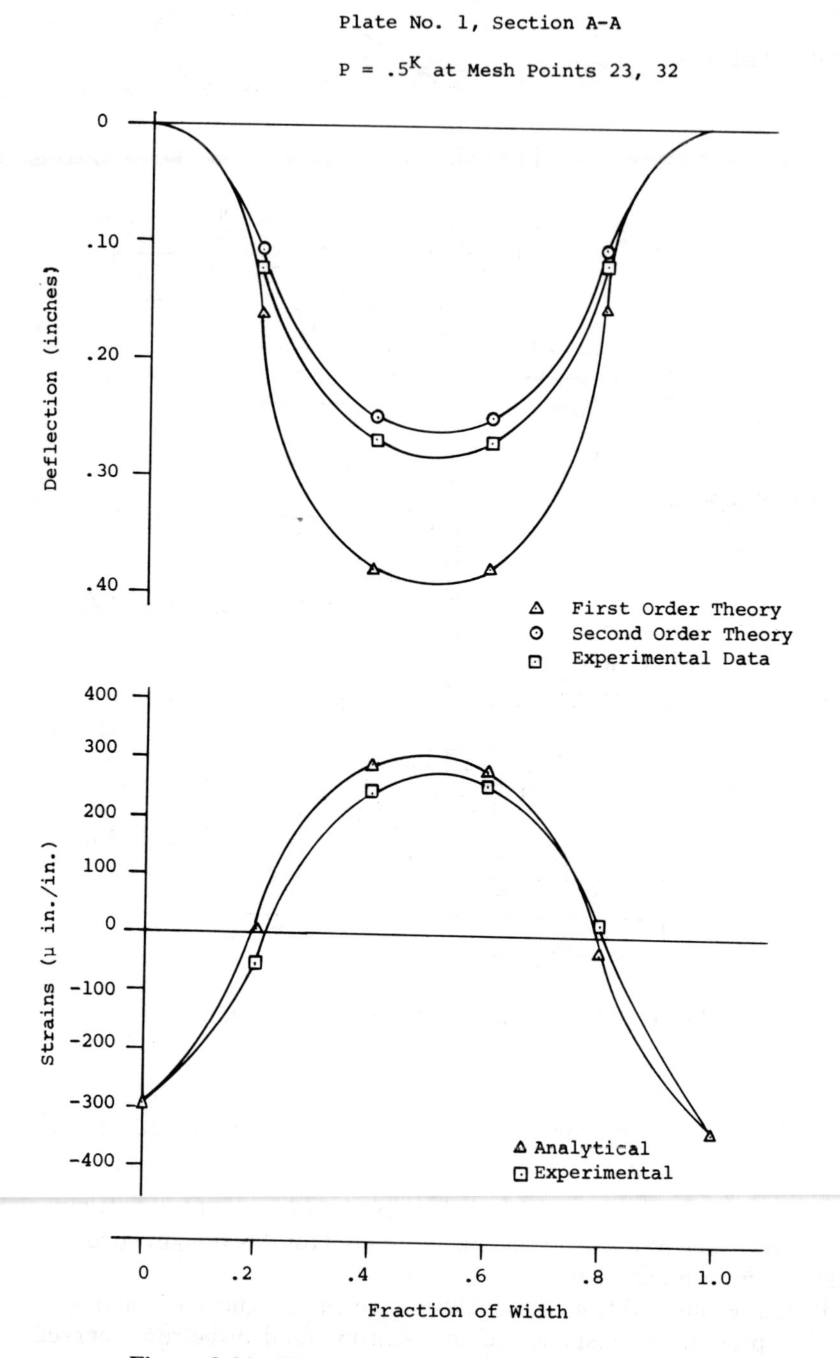

Figure 9-20. Theory versus Test Results—Curved Plate.

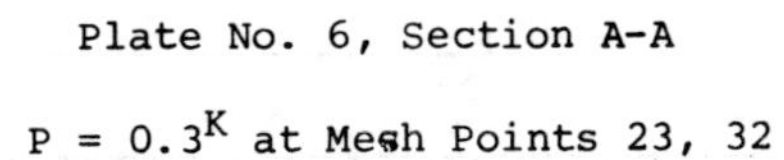

Plate No. 6, Section A-A
P = 0.3K at Mesh Points 23, 32

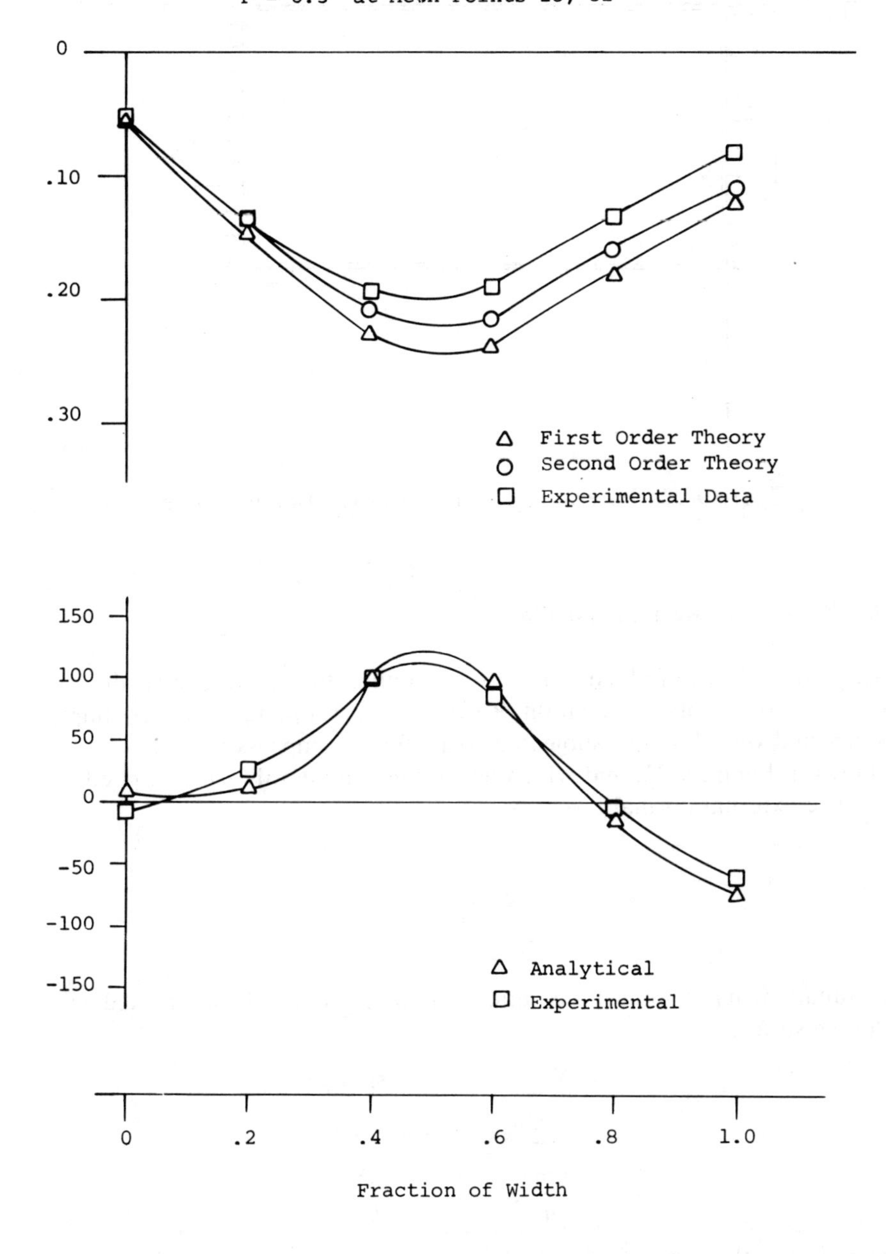

Figure 9-21. Theory versus Test Results—Curved Plate.

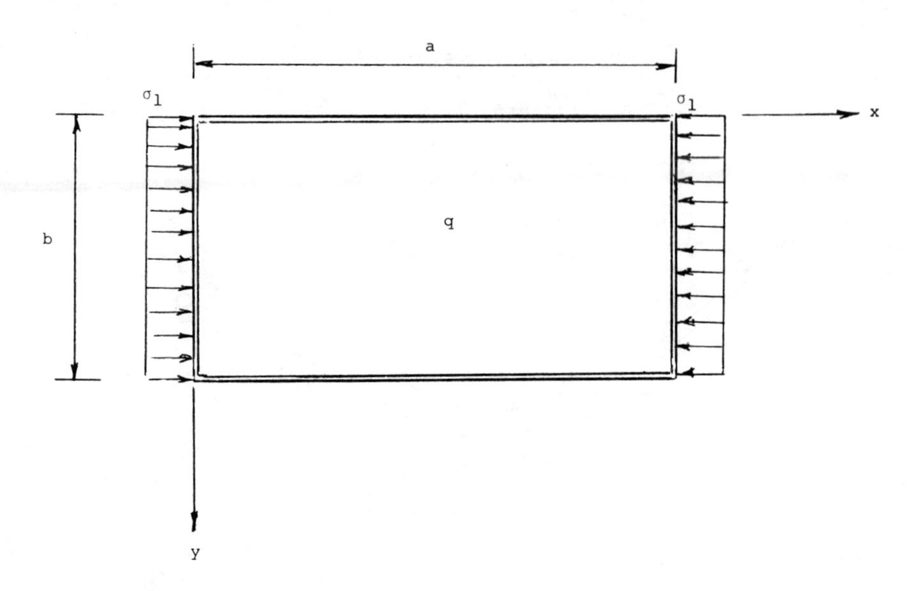

Figure 9-22. Lateral and In-Plane Force Loaded Plate.

9.7 Buckling of Rectangular Plates

The critical buckling load that may be applied to a plate can readily be determined by solving equation (9.12). Consider a plate, which is simply supported on all sides, subjected to uniform compressive stress σ_x as shown in Figure 9-22, and which has a uniform lateral load q applied.

The external normal forces are

$$N_x = -\sigma_1 h$$

$$N_y = 0$$

$$N_{xy} = 0$$

Assuming from chapter 4 the deflection and loads can be expressed as a double series:

$$w = \sum\sum b_{mn} \sin \lambda_m x \sin \alpha_n y$$

$$q = \sum\sum a_{mn} \sin \lambda_m x \sin \alpha_n y$$

$$\lambda_m = \frac{m\pi}{a}; \qquad \alpha_n = \frac{n\pi}{b}$$

Substituting the expressions into the plate equation (9.12) and noting N_x is a compressive force, gives

$$\sum\sum b_{mn}[D_x\lambda_m^4 + 2H\lambda_m^2\alpha_n^2 + D_y\alpha_n^4 + N_x\lambda^2]\sin\lambda_m x \sin\alpha_n y = \sum\sum a_{mn} \sin\lambda_m x \sin\alpha_n y$$

$$m = 1, 2, 3, \ldots; \qquad n = 1, 2, 3, \ldots$$

Solving for the coefficient b_{mn} for integer mn gives

$$b_{mn} = \frac{a_{mn}}{[D_x\lambda_m^4 + 2H\lambda^2\alpha_n^2 + D_y\alpha_n^4 + N_x\lambda_m^2]} \tag{9.50}$$

as given in section 4.5:

$$a_{mn} = \frac{4}{ab}\int_0^a\int_0^b q(x, y)\sin\lambda_m x \sin\alpha_n y \, dx \, dy$$

The deflection is, therefore,

$$w = \frac{b^4}{\pi^4 D_y}\sum\sum \frac{a_{mn}\sin\lambda_m x \sin\alpha_n y}{[(\beta/\alpha)^2(b/a)^4m^4 + m^2n^2(b/a)^2 2\beta + n^4 + N_x/D_y(mb/\pi a)^2 b^2]} \tag{9.51}$$

and for uniform load

$$a_{mn} = 16q_0/\pi^2 mn$$

as given in section 4.5.

The critical buckling load can be determined by noting in equation (9.51) that the deflection w will become infinite as the denominator goes to zero; or

$$[(\beta/\alpha)^2(b/a)^4m^4 + m^2n^2(b/a)^2 2\beta + n^4 + N_x/D_y(mb/\pi a)^2 b^2] = 0$$

The only way this expression can go to zero is if $N_x = 0$, or if

$$\frac{N_x}{D_y}\left(\frac{mb}{\pi a}\right)^2 b^2 = [(\beta/\alpha)^2(b/a)^4m^4 + m^2n^2(b/a)^2 2\beta + n^4]$$

where N_x is a compressive force.

If the plate is isotropic, N_x equals

$$N_x = \frac{\pi^2 D}{b^2}\left[\frac{mb}{a} + \frac{a}{mb}\right]^2 \tag{9.52}$$

In general, this equation can be written as

$$N_x = k\pi^2 D/b^2$$

However,

Table 9-1
Buckling Coefficients

Loading	*Ratio of Bending Stress to Uniform Compression Stress* f_b/f_c	*Minimum Buckling Coefficient*,* k_1	
		Unloaded Edges Simply Supported	*Unloaded Edges Fixed*
$f_2 = -f_1$	∞ (Pure Bending)	23.9	39.6
$f_2 = -\frac{2}{3} f_1$	5.00	15.7	
$f_2 = -\frac{f_1}{3}$	2.00	11.0	
$f_2 = 0$	1.00	7.8	13.6
$f_2 = \frac{f_1}{3}$	0.50	5.8	
$f_2 = f_1$	0.00 (Pure Compression)	4.0	6.97

*Values given are based on plates having loaded edges simply supported and are conservative for plates having loaded edges fixed (ref. [71]).

$$D = Eh^3/12(1 - \mu^2), \qquad N_x = \sigma_1 h$$

$$k = \left[\frac{mb}{a} + \frac{a}{mb}\right]^2$$

Threerefore, the critical stress is

$$\sigma = k \frac{\pi^2 E}{12(1 - \mu^2)(b/h)^2} \tag{9.53}$$

The minimum buckling coefficient k factor for this example is 4.0, and for other boundary conditions it would vary accordingly. Various k factors for different types of boundary conditions and loading are given in Table 9-1 as listed in reference [71].

Appendix A
Computer Programs

Appendix A-1
Slope-Deflection Rectangular Orthotropic Bridge Computer Program

```
C
C      SLOPE DEFLECTION EQUATION SOLUTION
C
C      CASE 1 H*H GREATER THAN DX*DY
C      CASE 2 H*H EQUAL DX*DY
C      CASE 3 H*H LESS THAN DX*DY
C
C      LONGITUDINAL GIRDERS
C
       DOUBLE PRECISION S,GJO,GJI,EII,PI4,VRR,C,RR,R3,VR1
       DOUBLE PRECISION VR4,FERB,VS1,RA,BBD,C1,C3,C6,C8,ABB
       DOUBLE PRECISION H,DY,WIDE,PI,ANT,W,R1,R11,VR2,ALP,CL
       DOUBLE PRECISION FERA,RS,RVV,P1,P,VP,AB,RAD,C4
       DOUBLE PRECISION C7,RAA,BAMD,RADD,EIO,WID,PI2,CNT,BET
       DOUBLE PRECISION CLL,F,VL1,B,R2,R22,CON,VR3,FEMB,RV
       DOUBLE PRECISION S1,P2,VS,VP1,BB,RBD,C2,C5,BBB,G
       DOUBLE PRECISION SL1,DEFL,A,SLOPE,X
       DOUBLE PRECISION RBDD,EIWI,EIWO,U,VL
       DOUBLE PRECISION SINH1,COSH1,SINH2,COSH2,SINV1
       DOUBLE PRECISION COSV1,SINV2,COSV2,SINV3,COSV3,SINV4
       DOUBLE PRECISION COSV4,SINRS,COSRS,SINRV,COSRV,SINVV
       DOUBLE PRECISION COSVV,SINHP,COSHP,SINHT,COSHT,SINHPV
       DOUBLE PRECISION COSHPV,SINPV,COSPV,FEMA,S2
       DOUBLE PRECISION GSHR,DEF1,Q,VR,DSCPT,GMOM
       DOUBLE PRECISION ANDLG,AL,XOL,SUM,BNT
       DIMENSION A(52,52),X(50),ML(50)
       DIMENSION Q(50,16),U(50,16),VL(50,16),VR(50,16)
       DIMENSION SL1(25),DEF1(25)
       DIMENSION SLOPE(11,25),GMOM(11,25),GSHR(11,25),DEFL(11,25)
       DIMENSION DSCPT(14)
       DIMENSION CC(5)
C
C
    1 FORMAT(A1,13A6)
    2 FORMAT(4D12.5,I3)
    5 FORMAT(3D14.7)
    6 FORMAT(I3)
  113 FORMAT(1H0,6(D12.5,5X))
  114 FORMAT(4D14.7)
  400 FORMAT(46H1          SINGLE SPAN ORTHOTROPIC BRIDGE ANALYSIS)
  401 FORMAT(A1,13A6)
  402 FORMAT(14H0BRIDGE LENGTH,6X,D14.7)
  403 FORMAT(13H0BRIDGE WIDTH,7X,D14.7)
  404 FORMAT(18H0NUMBER OF GIRDERS,2X,I3)
  405 FORMAT(15H0GIRDER SPACING,5X,D14.7)
  407 FORMAT(1H0,5X,3HIWI,14X,3HIWO,14X,3HGJI,14X,3HGJO,14X,3HEII,14X,3H
     1EIO)
  408 FORMAT(6H0ALPHA,14X,D14.7)
  409 FORMAT(5H0BETA,15X,D14.7)
  410 FORMAT(3H0DY,17X,D14.7)
  633 FORMAT(1H1,8X,18HPROGRAM OUTPUT FOR,I3,2X,18HFOURIER SERIES SUM)
  634 FORMAT(A1,13A6)
  701 FORMAT(1H0,34X,6HGIRDER,I3)
  702 FORMAT(1H0,1X,3HX/L,5X,10HDEFLECTION,3X,11HSHEAR FORCE,3X,11HBEND
     1MOMENT,7X,5HSLOPE)
  705 FORMAT(1H0,F5.2,4(3X,D11.4))

  976 FORMAT(2I3)
 3999 FORMAT(37X,15HPLATE LOAD DATA)
 4000 FORMAT(10H0     PLATE,5X,7HLOAD NO,6X,9HMAGNITUDE,7X,12HX-COORDINAT
     1E,10X,2HVL,15X,2HVR)
```

```
 4003 FORMAT(6X,I3,8X,I3,3X,4(3X,D14.7))
 4004 FORMAT(17X,I3,3X,4(3X,D14.7))
 6000 FORMAT (1H0,5X,5HDY/EI,9X,7HS*DY/EI,9X,4HD/D*,11X,4HV/V*,11X,4HM/M
     1*)
 6001 FORMAT(1H ,5(E12.6,3X))
C
C
C       READ INPUT DATA
C       READ JOB DESCRIPTION
  301 READ(5,1) (DSCPT(I),I=1,14)
C       READ ALPHA, BETA, PLATE WIDTH, SPAN, 2X NO. GIRDERS
  300 READ(5,2) ALP,BET,S,CL,N
C       READ NUMBER OF DIVSIONS TAKEN ALONG THE GIRDERS
        READ(5,6) NDGJ
C       READ WARPING CONSTANTS
        READ(5,114) EIWO,EIWI,GJO,GJI
C       READ OUTSIDE AND INSIDE GIRDER STIFFNESS
C       READ EIO, EII, DY
        READ(5,5) EIO,EII,DY
C
C       INITIALIZE PLATE LOAD DATA
        DO 5000 M=2,N,2
        ML(M)=0
        DO 5000 J1=1,16
        Q(M,J1)=0.0
        U(M,J1)=0.0
        VL(M,J1)=0.0
        VR(M,J1)=0.0
 5000 CONTINUE
C
C       READ PLATE LOAD DATA
C
C       READ NUMBER OF PLATES LOADED
  977 READ(5,6) NUM
        DO 940 I=1,NUM
C       READ PLATE NUMBER AND NUMBER OF LOADS
        READ(5,976) M,ML(M)
        MK=ML(M)
        DO 950 J1=1,MK
C       READ LOAD MAGNITUDE AND LOCATION
        READ(5,114) Q(M,J1),U(M,J1),VL(M,J1),VR(M,J1)
  950 CONTINUE
  940 CONTINUE
C
C
        NP1=N+1
        NM1=N-1
        NM2=N-2
        NM3=N-3
        NWID=N/2-1
        WIDE=NWID
        WID=WIDE*S
        NUM=N/2
        NDLG=NDGJ+1
        PI=3.1415927
        PI2=PI**2
        PI4=PI2**2
C
C       WRITE OUT INPUT DATA
C
        WRITE(6,400)
        WRITE(6,401) (DSCPT(I),I=1,14)
        WRITE(6,403) CL
        WRITE(6,402) WID
        WRITE(6,404) NUM
        WRITE(6,405) S
        WRITE (6,407)
        WRITE(6,113) EIWI,EIWO,GJI,GJO,EII,EIO
        WRITE(6,408) ALP
```

```
      WRITE(6,409) BET
      WRITE(6,410) DY
C
C     WRITE PLATE LOAD DATA
C
      WRITE(6,3999)
      DO 4001 M=2,NM2,2
      MK=ML(M)
      IF(MK.EQ.0) GO TO 4006
      WRITE(6,4000)
      DO 4005 J1=1,MK
      IF(J1.GT.1) GO TO 4002
      WRITE(6,4003) M,J1,Q(M,J1),U(M,J1),VL(M,J1),VR(M,J1)
      GO TO 4005
 4002 WRITE(6,4004) J1,Q(M,J1),U(M,J1),VL(M,J1),VR(M,J1)
 4005 CONTINUE
 4006 CONTINUE
 4001 CONTINUE
C
C     INITIALIZE COUNTERS FOR SUCESSIVE FOURIER SERIES
C
      ANT=0.
      CNT=0.
C
C     READ NUMBER OF FOURIER SERIES TERMS DESIRED
C
      READ(5,6) NT
      ANT=NT
      ANT=ANT+1.
C
C     INITIALIZE MATRICIES
C
      DO 899 K=1,NUM
      DO 900L=1,NDLG
      SLOPE(L,K)=0.0
      DEFL(L,K)=0.0
      GSHR(L,K)=0.0
      GMOM(L,K)=0.0
  900 CONTINUE
  899 CONTINUE
  110 CNT=CNT+1.
      CLL=CL/CNT
      IF(DABS(CNT-ANT).LT..1D-34) GO TO 301
      DO 120 I=1,52
      DO 120 J=1,52
  120 A(I,J)=0.0
      DO 17 I=1,50
   17 X(I)=0.0
C
C     CHECK FOR NUMBER OF FOURIER SERIES DESIRED
C
C
C     READ GIRDER OR PLATE DATA
C
C     READ GIRDER OR PLATE DATA
      DO 440 M=2,N,2
      MK=ML(M)
      F=0.0
      DO 419 J1=1,MK
C     COMPUTE FORCE
      F=2.*(Q(M,J1)*DSIN(PI*U(M,J1)/CLL))/CL
      IF(VL(M,J1).GT.0.) GO TO 431
      IF(DABS(VR(M,J1)).LT..1D-34) GO TO 432
  432 A(M,NP1)=A(M,NP1)+F
      GO TO 419
  431 IF(ALP-1.) 12,11,10
C
C     COMPUTE CONSTANTS
C
```

```
C       CASE 1 H*H GREATER THAN DX*DY
C
   10 VL1=(S/CLL-VL(M,J1)/CLL)*PI
      VRR=(S/CLL-VR(M,J1)/CLL)*PI
      W=SQRT(BET/ALP)
      B=SQRT(ALP+SQRT(ALP*ALP-1.0))
      C=SQRT(ALP-SQRT(ALP*ALP-1.0))
      R1=W*B
      R2=W*C
      RR=R2**2-R1**2
      R11=R1*S/CLL*PI
      R22=R2*S/CLL*PI
      R3=S/CLL*PI
      SINH1=.5*(EXP(R11)-EXP(-R11))
      COSH1=.5*(EXP(R11)+EXP(-R11))
      SINH2=.5*(EXP(R22)-EXP(-R22))
      COSH2=.5*(EXP(R22)+EXP(-R22))
      CON=2.*R1*R2*(1.-COSH1*COSH2)+(R1*R1+R2*R2)*SINH1*SINH2
      VR1=R1*VL1
      VR2=R2*VL1
      SINV1=.5*(EXP(VR1)-EXP(-VR1))
      COSV1=.5*(EXP(VR1)+EXP(-VR1))
      SINV2=.5*(EXP(VR2)-EXP(-VR2))
      COSV2=.5*(EXP(VR2)+EXP(-VR2))
      FEMA=(-R2/R1*SINV1+SINV2)*(COSH2-COSH1)*R1
      FEMA=(FEMA+(R1*SINH2-R2*SINH1)*(COSV1-COSV2))/(CON*R3)*(-1.0)
      FERA=(R2*SINH2-R1*SINH1)*(R2/R1*SINV1-SINV2)
      VR3=R1*VRR
      FERA=(FERA+R2*(COSH2-COSH1)*(COSV2-COSV1))*R1/CON
      VR4=R2*VRR
      SINV3=.5*(EXP(VR3)-EXP(-VR3))
      COSV3=.5*(EXP(VR3)+EXP(-VR3))
      SINV4=.5*(EXP(VR4)-EXP(-VR4))
      COSV4=.5*(EXP(VR4)+EXP(-VR4))
      FEMB=(-R2/R1*SINV3+SINV4)*(COSH2-COSH1)*R1
      FEMB=(FEMB+(R1*SINH2-R2*SINH1)*(COSV3-COSV4))/(CON*R3)*(-1.0)
      FERB=(R2*SINH2-R1*SINH1)*(R2/R1*SINV3-SINV4)
      FERB=(FERB+R2*(COSH2-COSH1)*(COSV4-COSV3))*R1/CON
      A(M-1,NP1)=A(M-1,NP1)+FEMA*F*S
      A(M+1,NP1)=A(M+1,NP1)-FEMB*F*S
      A(M,NP1)=A(M,NP1)+FERA*F
      A(M+2,NP1)=A(M+2,NP1)+FERB*F
      GO TO 419
C
C       CASE 2 H*H EQUAL DX*DY
C
   11 VL1=S/CLL-VL(M,J1)/CLL
      VR1=S/CLL-VR(M,J1)/CLL
      W=SQRT(BET/ALP)
      W=W*PI
      RS=(W/CLL)*S
      RV=W*VL1
      SINRS=.5*(DEXP(RS)-DEXP(-RS))
      COSRS=.5*(DEXP(RS)+DEXP(-RS))
      SINRV=.5*(DEXP(RV)-DEXP(-RV))
      COSRV=.5*(DEXP(RV)+DEXP(-RV))
      CON=(SINRS**2-RS**2)
      FEMA=(RS*SINRS*SINRV-RV*RS*COSRV*SINRS)
      FEMA=(FEMA-RV*SINRV*(SINRS-RS*COSRS))/(CON*RS)
      FERA=(RS*COSRS+SINRS)*SINRV+RS*SINRS*SINRV*RV
      FERA=(FERA-RV*COSRV*(RS*COSRS+SINRS))/CON
      RVV=W*VR1
      SINVV=.5*(DEXP(RVV)-DEXP(-RVV))
      COSVV=.5*(DEXP(RVV)+DEXP(-RVV))
      FEMB=(RS*SINRS*SINVV-RVV*RS*COSVV*SINRS)
      FEMB=(FEMB-RVV*SINVV*(SINRS-RS*COSRS))/(CON*RS)
      FERB=(RS*COSRS+SINRS)*SINVV+RS*SINRS*SINVV*RVV
      FERB=(FERB-RVV*COSVV*(RS*COSRS+SINRS))/CON
      A(M-1,NP1)=A(M-1,NP1)+FEMA*F*S
```

```
      A(M+1,NP1)=A(M+1,NP1)-FEMB*F*S
      A(M,NP1)=A(M,NP1)+FERA*F
      A(M+2,NP1)=A(M+2,NP1)+FERB*F
      BNT=ANT-1.
      IF(DABS(CNT-BNT).LT..1D-34) GO TO 446
      GO TO 419
  446 WRITE(6,447)  A(M-1,NP1),A(M+1,NP1),A(M,NP1),A(M+2,NP1)
  447 FORMAT(2X,D14.7,2X,D14.7,2X,D14.7,2X,D14.7)
      GO TO 419
C
C     CASE 3 H*H LESS THAN DX*DY
C
   12 VL1=(S/CLL-VL(M,J1)/CLL)*PI
      VR1=(S/CLL-VR(M,J1)/CLL)*PI
      W=SQRT(BET/ALP)
      B=SQRT(.5*(1.+ALP))
      C=SQRT(.5*(1.-ALP))
      P1=W*B
      S1=W*C
      S2=S1*S/CLL*PI
      P=P1*S/CLL*PI
      P2=S/CLL*PI
      SINHP=.5*(EXP(P)-EXP(-P))
      COSHP=.5*(EXP(P)+EXP(-P))
      SINHT=.5*(EXP(2.*P)-EXP(-2.*P))
      COSHT=.5*(EXP(2.*P)+EXP(-2.*P))
      CON=(S1*SINHP)**2-(P1*DSIN(S2))**2
      VP=P1*VL1
      VS=S1*VL1
      SINHPV=.5*(EXP(VP)-EXP(-VP))
      COSHPV=.5*(EXP(VP)+EXP(-VP))
      FEMA=SINHP*DSIN(S2)*(P1*COSHPV*DSIN(VS)-S1*SINHPV*DCOS(VS))
      FEMA=FEMA-SINHPV*DSIN(VS)*(P1*COSHP*DSIN(S2)-S1*SINHP*DCOS(S2))
      FEMA=FEMA/(-CON*P2)
      FERA=(P1*COSHP*DSIN(S2)+S1*SINHP*DCOS(S2))
      FERA=FERA*(-P1*COSHPV*DSIN(VS)+S1*SINHPV*DCOS(VS))
      FERA=FERA+(P1*P1+S1*S1)*SINHP*DSIN(S2)*SINHPV*DSIN(VS)
      FERA=FERA/CON
      VP1=P1*VR1
      VS1=S1*VR1
      SINPV=.5*(EXP(VP1)-EXP(-VP1))
      COSPV=.5*(EXP(VP1)+EXP(-VP1))
      FEMB=SINHP*DSIN(S2)*(P1*COSPV*DSIN(VS1)-S1*SINPV*DCOS(VS1))
      FEMB=FEMB-SINPV*DSIN(VS1)*(P1*COSHP*DSIN(S2)-S1*SINHP*DCOS(S2))
      FEMB=FEMB/(-CON*P2)
      FERB=(P1*COSHP*DSIN(S2)+S1*SINHP*DCOS(S2))
      FERB=FERB*(-P1*COSPV*DSIN(VS1)+S1*SINPV*DCOS(VS1))
      FERB=FERB+(P1*P1+S1*S1)*SINHP*DSIN(S2)*SINPV*DSIN(VS1)
      FERB=FERB/CON
      A(M-1,NP1)=A(M-1,NP1)+FEMA*F*S
      A(M+1,NP1)=A(M+1,NP1)-FEMB*F*S
      A(M,NP1)=A(M,NP1)+FERA*F
      A(M+2,NP1)=A(M+2,NP1)+FERB*F
  419 CONTINUE
  440 CONTINUE
C
C     COMPUTE CONSTANTS FOR MATRIX
C
  170 IF(ALP-1.) 173,172,171
C
C     CASE 1 H*H GREATER THAN DX*DY
C
  171 RA=R1*R2*RR*(COSH2-COSH1)*R3*R3
      RA=RA/CON
      AB=((RR*(R1*COSH1*SINH2-R2*COSH2*SINH1))/(-CON))*R3
      BB=((RR*(R1*SINH2-R2*SINH1))/CON)*R3
      BBD=(R1*R1+R2*R2)*(1.-COSH1*COSH2)+2.*R1*R2*SINH1*SINH2
      BBD=BBD*R1*R2*R3*R3/(-CON)
      RAD=R1*R2*RR*(R2*SINH2-R1*SINH1)*R3**3/CON
```

```
      RBD=R1*R2*RR*(R2*SINH2*COSH1-R1*SINH1*COSH2)*R3**3/CON
      C1=AB*DY/S+PI2*GJO/(CLL**2)-PI4*EIWO/(CLL**4)
      C2=BBD*DY/(S*S)
      C3=BB*DY/S
      C4=RA*DY/(S*S)
      C5=RBD*DY/(S**3)+PI4*EIO/(CLL)**4
      C6=RAD*DY/(S**3)
      C7=AB*DY*2./S+PI2*GJI/(CLL**2)-PI4*EIWI/(CLL**4)
      C8=RBD*DY*2./(S**3)+PI4*EII/(CLL)**4
      GO TO 16
C
C     CASE 2 H*H EQUAL TO DX*DY
C
  172 RA=(2.*RS**3*SINRS)/CON
      AB=(2.*RS*(COSRS*SINRS-RS))/CON
      BB=(2.*RS*(RS*COSRS-SINRS))/CON
      BBD=(((SINRS)**2+RS**2)*RS**2)/CON
      RAD=((RS*COSRS+SINRS)*RS**3)/CON*2.
      RBD=(2.*RS**3*(SINRS*COSRS+RS))/CON
      C1=AB*DY/S+PI2*GJO/(CLL**2)-PI4*EIWO/(CLL**4)
      C2=BBD*DY/(S*S)
      C3=BB*DY/S
      C4=RA*DY/(S*S)
      C5=RBD*DY/(S**3)+PI4*EIO/(CLL)**4
      C6=RAD*DY/(S**3)
      C7=AB*DY*2./S+PI2*GJI/(CLL**2)-PI4*EIWI/(CLL**4)
      C8=RBD*DY*2./(S**3)+PI4*EII/(CLL)**4
      GO TO 16
C
C     CASE 3 H*H LESS THAN DX*DY
C
  173 RAA=(2.*S1*P1*(S1*S1+P1*P1)*SINHP*DSIN(S2)*P2*P2)/CON
      BBB=((P1*S1)*(-S1*SINHT+P1*DSIN(2.*S2)))/(-CON)*P2
      ABB=((2.*P1*S1)*(S1*SINHP*DCOS(S2)-P1*COSHP*DSIN(S2)))/(-CON)*P2
      BAMD=((S1*S1+P1*P1)*(S1*S1*(SINHP)**2+P1*P1*DSIN(S2)**2))/CON
      BAMD=BAMD*P2*P2
      G=P1*(S1*S1*P1*P1+S1**4)
      H=S1*(S1*S1*P1*P1+P1**4)
      RADD=(2.*(G*SINHP*DCOS(S2)+H*COSHP*DSIN(S2))*P2**3)/CON
      RBDD=((G*SINHT+H*DSIN(S2*2.))*P2**3)/CON
      C1=BBB*DY/S+PI2*GJO/(CLL**2)-PI4*EIWO/(CLL**4)
      C2=BAMD*DY/(S*S)
      C3=ABB*DY/S
      C4=RAA*DY/(S*S)
      C5=RBDD*DY/(S**3)+EIO/(CLL)**4*PI4
      C6=RADD*DY/(S**3)
      C7=BBB*DY/S*2.+PI2*GJI/(CLL**2)-PI4*EIWI/(CLL**4)
      C8=RBDD*DY/(S**3)*2.+PI4*EII/(CLL)**4
      GO TO 16
C
C     PUT CONSTANTS IN MATRIX
C
   16 DO 180 I=3,NM3,2
      A(I,I-2)=C3
      A(I,I-1)=C4
      A(I,I)=C7
      A(I,I+2)=C3
  180 A(I,I+3)=-C4
      DO 190 I=4,NM2,2
      A(I,I-3)=-C4
      A(I,I-2)=-C6
      A(I,I)=C8
      A(I,I+1)=C4
  190 A(I,I+2)=-C6
      A(1,1)=C1
      A(1,2)=C2
      A(1,3)=C3
      A(1,4)=-C4
      A(2,1)=C2
```

```
      A(2,2)=C5
      A(2,3)=C4
      A(2,4)=-C6
      A(N-1,N-3)=C3
      A(N-1,N-2)=C4
      A(N-1,N-1)=C1
      A(N-1,N)=-C2
      A(N,N-3)=-C4
      A(N,N-2)=-C6
      A(N,N-1)=-C2
      A(N,N)=C5
C
C     SOLVE SIMULTANEOUS EQUATIONS
C
   20 DO 30 J=2,NP1
   30 A(1,J)=A(1,J)/A(1,1)
      DO 60 J=2,NP1
      DO 60 I=2,N
      SUM=0.0
      IF(I-J)35,50,50
   35 IL=I-1
      DO 40 K=1,IL
   40 SUM=SUM+A(I,K)*A(K,J)
      A(I,J)=(A(I,J)-SUM)/A(I,I)
      GO TO 60
   50 JL=J-1
      DO 55 K=1,JL
   55 SUM=SUM+A(I,K)*A(K,J)
      A(I,J)=A(I,J)-SUM
   60 CONTINUE
      X(N)=A(N,NP1)
      DO 70 J=1,NM1
      I=N-J
      IP=I+1
      SUM=0.0
      DO 65 K=IP,N
   65 SUM=SUM+A(I,K)*X(K)
   70 X(I)=A(I,NP1)-SUM
      DO 450 I=1,N,2
      K=(I+1)/2
      SL1(K)=X(I)
  450 CONTINUE
      DO 460I=2,N,2
      K=I/2
      DEF1(K)=X(I)
  460 CONTINUE
C     COMPUTE THE SLOPES AND DEFLECTIONS ALONG THE GIRDERS
C     COMPUTE THE MOMENTS,SHEARS, AND TORSIONAL MOMENTS ALONG THEGIRDERS
      ANDLG=NDGJ
      NDLG=NDGJ+1
      DO 590  L=1,NDLG
      DO 580 K=1,NUM
      LM1=L-1
      AL=LM1
      AL=AL*CNT
      SLOPE(L,K)=SL1(K)*DSIN(3.1415927*AL/ANDLG)+SLOPE(L,K)
      DEFL(L,K)=DEF1(K)*DSIN(3.1415927*AL/ANDLG)+DEFL(L,K)
      IF(K.EQ.1) GO TO 601
      IF(K.EQ.NUM) GO TO 601
      GMOM(L,K)=GMOM(L,K)+EII*((PI*CNT/CL)**2)*DEF1(K)*
     1DSIN(PI*AL/ANDLG)
      GSHR(L,K)=GSHR(L,K)-EII*((PI*CNT/CL)**3)*DEF1(K)*
     1DCOS(PI*AL/ANDLG)
      GO TO 580
  601 CONTINUE
      GMOM(L,K)=GMOM(L,K)+EIO*((PI*CNT/CL)**2)*DEF1(K)*
     1DSIN(PI*AL/ANDLG)
      GSHR(L,K)=GSHR(L,K)-EIO*((PI*CNT/CL)**3)*DEF1(K)*
     1DCOS(PI*AL/ANDLG)
```

```
580 CONTINUE
590 CONTINUE
    BNT=ANT-1.
    IF(DABS(CNT-BNT).LT..1D-34) GO TO 632
    GO TO 110
632 WRITE(6,633) NT
    WRITE(6,634)  (DSCPT(I) ,I=1,14)
    DO 704 K=1,NUM
    WRITE(6,701) K
    WRITE(6,702)
    DO 704 L=1,NDLG
    LM1=L-1
    AL=LM1
    XOL=AL/ANDLG
704 WRITE(6,705)XOL,DEFL(L,K),GSHR(L,K),GMOM(L,K),SLOPE(L,K)
    GO TO 110
    END
```

Appendix A-2
Finite Difference Deflection Curved Orthotropic Bridge Computer Program

```
C CURVED ORTHOTROPIC BRIDGE DECK ANALYSIS BY THE METHOD OF FINITE
C DIFFERENCES UTILIZING THE IBM 7094 COMPUTER AND FORTRAN IV LANGUAGE.
C THE RESULTING SIMULTANEOUS EQUATIONS ARE SOLVED BY MEANS OF A
C GAUSS ITERATIVE ELIMINATION SUBROUTINE.
      COMMON AA
      DIMENSION AA(121,121) ,BB(121) ,XX(121) ,KCON(11)
      DIMENSION JSTIF(11) , KSTIF(11) , LDCOEF(11) , H(11)
C SET VARIABLES EQUAL TO EITHER INTEGER OR REAL VALUES
      REAL MINRAD,KCON,KSTIF,JSTIF,LDCOEF,KSTFOR,JSTFOR
      REAL  LAM, LAM2, LAM3, LAM4
      INTEGER  ROW, COL
 999  READ (5, 966)
  966 FORMAT(1H1,14X,40H                                        )
      READ (5, 977)
  977 FORMAT ( 15X,  40H                                        )
C READ TOTAL NO. OF POINTS, NO. OF PTS. PER ROW, MAX. NO. OF ITERATIONS
C TO PERFORM, AND NO. OF CASES TO BE CONSIDERED.
C IF QUA EQUALS 1,NO, SYSTEM OF EQUATIONS
C IF QUA EQUALS 2,YES,SYSTEM OF EQUATIONS
C IF IORTHO EQUALS 1,OUTPUT STIFFNESSES ARE ORTHOTROPIC PLATE STIFFNESSES
      READ(5,100)N,NN,MM,ICASE,IQUA,IORTHO
  100 FORMAT (6I5)
C READ MINIMUM RADIUS, RADIUS INCREMENT, ANGLE INCREMENT, STIFFNESS
C CONSTANTS ALPHA AND BETA, AND VALUE D.
      READ(5,182) MINRAD, LAM, OTHETA, ALP, BET, DR
 182  FORMAT ( 6F10.0 )
      READ (5, 990)  (KCON(I), I=1,NN)
      READ (5, 990)  (KSTIF(I), I=1,NN)
      NUMROW = N / NN
      READ (5,990)  (JSTIF(I), I=1,NUMROW)
 990  FORMAT (5F15.0)
      WRITE (7,966)
      WRITE (7,977)
      WRITE(7,300) N,NN,MINRAD,LAM,OTHETA,ICASE
  300 FORMAT(2I5,3E16.8,I5)
      IF(IORTHO.EQ.1)GO TO 1600
      WRITE(7,189) (KSTIF(I),I=1,NN)
      DO 222 I=1,NN
  222 H(I)=KSTIF(I)/KCON(I)
      WRITE (7,189) (H(I),I=1,NN)
      WRITE(7,189) (JSTIF(I),I=1,NUMROW)
      GO TO 1604
 1600 KSTFOR=ALP*DR*LAM
      HORTHO=BET*DR*LAM
      JSTFOR=DR*LAM
 1601 WRITE(7,189)(KSTFOR,I=1,NN)
 1602 WRITE(7,189)(HORTHO,I=1,NN)
 1603 WRITE(7,189) (JSTFOR,I=1,NUMROW)
C CLEAR AA ARRAY TO ZERO - ONLY NON-ZERO COEFFICIENTS NEED BE CALCULATED
 1604 DO 111 I=1,N
      DO 111  J=1,N
 111  AA(I,J) = 0.0
C CALCULATE RECURRING CONSTANTS
      TEMP = NUMROW + 1
      THETA = OTHETA * TEMP
      LAM2 = LAM ** 2
      LAM3 = LAM ** 3
      LAM4 = LAM ** 4
```

```
      OTHET2 = OTHETA ** 2
      OTHET4 = OTHETA ** 4
      DO 301 I=1,NUMROW
  301 JSTIF(I)=JSTIF(I)/DR
      RAD = MINRAD
      DO 59  I=1,NN
      CK = 1.0
      IF(I.EQ.1.OR.I.EQ.NN) CK=2.0
      LDCOEF(I)=CK*LAM3/DR
      KSTIF(I)=KSTIF(I)*CK*LAM3/DR
   59 RAD = RAD + LAM
      ROW = 0
      COL = 0
      RAD = MINRAD
      CAABB = ALP * LAM4
      NTEST=(N/2)*2
      IF(NTEST.NE.N) NCENT=(NUMROW/2)+1
      DO 21  K = 1, N, NN
      M = K + NN - 1
      ROW = ROW + 1
      TEMP = NUMROW + 1 - ROW
      XTHETA = TEMP * OTHETA
C PROCEED POINT BY POINT ON ROW
      DO 20  I = K, M
C GROUP RECURRING TERMS AND SET VALUES FOR COEFFICIENT CALCULATIONS
      RAD3 = (RAD*OTHETA)**3
      C1 = BET * RAD * OTHET2 * LAM2
      C2 = 2.0 * RAD ** 2
      C3 = 2.0 * BET * LAM2 * OTHET2
      C4 = 2.0 * ALP * LAM3
      C5 = LAM * OTHET2
      C6 = RAD * OTHET2
      C7 = RAD * LAM
      C8 = RAD ** 4 * OTHET4
      C9 = LAM * RAD ** 3 * OTHET4
      C10 = BET * RAD * LAM3 * OTHET2
      C11 = BET * RAD ** 2 * LAM2 * OTHET2
      C12 = ALP * RAD * OTHET4 * LAM3
      C13 = ALP * RAD ** 2 * OTHET4 * LAM2
      C14 = RAD ** 4 * OTHET4
      C15 = LAM * RAD **3 * OTHET4
      C16 = 4.0 * LAM2 * OTHET2 * BET
      C17 = 2.0 * RAD ** 2
      C18 = LAM2
      C19 = RAD * LAM
      C20 = ALP * LAM2
      C21 = RAD ** 2 * OTHET4
      C22 = LAM2 * OTHET2
      C23 = LAM * RAD * OTHET2
      C24 = LAM * RAD * OTHET4
C UPDATE COLUMN AND PROCEED WITH CALCULATION OF COEFFICIENTS
      COL = COL + 1
      PHO=(-2.0*OTHET2*(1.0+RAD*(1.0+KCON))+6.0)
      PHAB=PHO-10.0
      IF (ROW .EQ. 1) GO TO 35
      IF (ROW .EQ. 2) GO TO 32
C COEFFICIENT AA
      J1 = I - 2 * NN
      AA(I,J1)=CAABB+KSTIF(COL)
   32 IF (COL .EQ. 1) GO TO 33
C COEFFICIENT AL
      J2 = I - NN - 1
      IF(COL.EQ.NN) AA(I,J2) = 2.0*C1*(2.0*RAD+3.0*LAM)
      IF(COL.NE.NN) AA(I,J2) = C1*(2.0*RAD+LAM)
C COEFFICIENT A
   33 J3 = I - NN
      IF(COL.EQ.NN) AA(I,J3)=C3*(3.0*LAM2-C2-C7)+C4*(2.0*C5-2.0*LAM+C6)
     1+KSTIF(COL)*PHAB
      IF(COL.EQ.1) AA(I,J3)=C3*(3.0*LAM2-C2+C7)+C4*(2.0*C5-2.0*LAM-C6)
```

```
     1+KSTIF(COL)*PHAB
      IF(COL.NE.1.AND.COL.NE.NN) AA(I,J3)=C3*(LAM2-C2)+C4*(C5-2.0*LAM)
     1+KSTIF(COL)*PHAB
      IF (COL .EQ. NN) GO TO 35
C COEFFICIENT AR
      J4 = I - NN + 1
      IF(COL.EQ.1) AA(I,J4)=2.0*C1*(2.0*RAD-3.0*LAM)
      IF(COL.NE.1) AA(I,J4)=C1*(2.0*RAD-LAM)
  35  IF (COL .LE. 2) GO TO 36
C COEFFICIENT LL
      IF(COL.EQ.NN) AA(I,I-2)=2.0*C8+2.0*JSTIF(ROW)*RAD3
      IF(COL.NE.NN) AA(I,I-2)=C8-C9+JSTIF(ROW)*RAD3
  36  IF(COL.EQ.1) GO TO 37
C COEFFICIENT L
      IF(COL.EQ.NN) AA(I,I-1)=-4.0*C8-12.0*C10-8.0*C11-3.0*C12-2.0*C13
     1      -4.0*JSTIF(ROW)*RAD3
      IF(COL.EQ.2) AA(I,I-1)=-2.0*C8-2.0*C10-4.0*C11-0.5*C12-C13
     1      -2.0*JSTIF(ROW)*RAD3
      IF(COL.NE.NN.AND.COL.NE.2) AA(I,I-1)=-4.0*C8+2.0*C9-2.0*C10
     1      -4.0*C11-0.5*C12-C13-4.0*JSTIF(ROW)*RAD3
  37  IF(ROW.EQ.1.OR.ROW.EQ.NUMROW)  GO TO 38
C COEFFICIENT O
      IF(COL.EQ.2) AA(I,I)=5.0*C14+C15+C16*(C17-C18)+2.0*C20*(3.0*C18
     1+C21-2.0*C22)+5.0*JSTIF(ROW)*RAD3+KSTIF(COL)*PHO
      IF(COL.EQ.NN-1) AA(I,I)=5.0*C14-C15+C16*(C17-C18)+2.0*C20*(3.0*C18
     1+C21-2.0*C22)+5.0*JSTIF(ROW)*RAD3+KSTIF(COL)*PHO
      IF(COL.EQ.1) AA(I,I)=2.0*C14+C16*(C17-3.0*C18-C19)+C20*(6.0*C18
     1      +2.0*C21-8.0*C22+4.0*C23-3.0*C24)+2.0*JSTIF(ROW)*RAD3+KSTIF
     2(COL)*PHO
      IF(COL.EQ.NN) AA(I,I)=2.0*C14+C16*(C17-3.0*C18+C19)+C20*(6.0*C18
     1      +2.0*C21-8.0*C22-4.0*C23+3.0*C24)+2.0*JSTIF(ROW)*RAD3+KSTIF
     2 (COL)*PHO
      IF(COL.GT.2.AND.COL.LT.NN-1) AA(I,I)=6.0*C14+C16*(C17-C18)+2.0*C20
     1*(3.0*C18+C21-2.0*C22)+6.0*JSTIF(ROW)*RAD3+KSTIF(COL)*PHO
      GO TO 39
  38  CK = +1.0
      IF (ROW .EQ. NUMROW) CK = 1.0
      IF(COL.EQ.2) AA(I,I)=5.0*C14+C15+C16*(C17-C18)+C20*(5.0*C18+2.*C21
     1-4.0*C22)+5.0*JSTIF(ROW)*RAD3+CK*KSTIF(COL)*(PHO-1.0)
      IF(COL.EQ.NN-1) AA(I,I)=5.0*C14-C15+C16*(C17-C18)+C20*(5.0*C18+2.*
     1C21-4.0*C22)+5.0*JSTIF(ROW)*RAD3+CK*KSTIF(COL)*(PHO-1.0)
      IF(COL.EQ.1) AA(I,I)=2.0*C14+C16*(C17-3.0*C18-C19)+C20*(5.0*C18
     1       +2.0*C21-8.0*C22+4.0*C23-3.0*C24)+2.0*JSTIF(ROW)*RAD3+CK
     2 *KSTIF(COL)*(PHO-1.0)
      IF(COL.EQ.NN) AA(I,I)=2.0*C14+C16*(C17-3.0*C18+C19)+C20*(5.0*C18
     1       +2.0*C21-8.0*C22-4.0*C23+3.0*C24)+2.0*JSTIF(ROW)*RAD3+CK
     2 *KSTIF(COL)*(PHO-1.0)
      IF(COL.GT.2.AND.COL.LT.NN-1) AA(I,I)=6.0*C14+C16*(C17-C18)+2.0*C20
     1*(2.5*C18+C21-2.0*C22)+6.0*JSTIF(ROW)*RAD3+CK*KSTIF(COL)*(PHO-1.0)
  39  IF (COL .EQ. NN) GO TO 40
C COEFFICIENT R
      IF(COL.EQ.1) AA(I,I+1)=-4.0*C8+12.0*C10-8.0*C11+3.0*C12-2.0*C13
     1       -4.0*JSTIF(ROW)*RAD3
      IF(COL.EQ.NN-1) AA(I,I+1)=-2.0*C8+2.0*C10-4.0*C11+0.5*C12-C13
     1       -2.0*JSTIF(ROW)*RAD3
      IF(COL.NE.1.AND.COL.NE.NN-1) AA(I,I+1)=-4.0*C8-2.0*C9+2.0*C10
     1       -4.0*C11+0.5*C12-C13-4.0*JSTIF(ROW)*RAD3
  40  IF (COL .GE. NN-1) GO TO 41
C COEFFICIENT RR
      IF(COL.EQ.1) AA(I,I+2)=2.0*C8+2.0*JSTIF(ROW)*RAD3
      IF(COL.NE.1) AA(I,I+2)=C8+C9+JSTIF(ROW)*RAD3
  41  IF (ROW .EQ. NUMROW) GO TO 45
      IF (COL .EQ. 1) GO TO 444
C COEFFICIENT BL
      J5 = I + NN - 1
      IF(COL.EQ.NN) AA(I,J5)=2.0*C1*(2.0*RAD+3.0*LAM)
      IF(COL.NE.NN) AA(I,J5)=C1*(2.0*RAD+LAM)
C COEFFICIENT B
```

```
 444  J6 = I + NN
      IF(COL.EQ.1) AA(I,J6)=C3*(3.0*LAM2-C2+C7)+C4*(2.0*C5-2.0*LAM-C6)
     1+KSTIF(COL)*PHAB
      IF(COL.EQ.NN) AA(I,J6)=C3*(3.0*LAM2-C2-C7)+C4*(2.0*C5-2.0*LAM+C6)
     1+KSTIF(COL)*PHAB
      IF(COL.NE.1.AND.COL.NE.NN) AA(I,J6)=C3*(LAM2-C2)+C4*(C5-2.0*LAM)
     1+KSTIF(COL)*PHAB
      IF (COL . EQ. NN) GO TO 43
C COEFFICIENT BR
      J7 = I + NN + 1
      IF(COL.EQ.1) AA(I,J7)=2.0*C1*(2.0*RAD-3.0*LAM)
      IF(COL.NE.1) AA(I,J7)=C1*(2.0*RAD-LAM)
  43  IF (ROW .EQ. NUMROW-1) GO TO 45
C COEFFICIENT BB
      J8 = I + 2 * NN
      AA(I,J8)=CAABB+KSTIF(COL)
  45  RAD = RAD + LAM
      IF (COL .EQ. NN)  RAD = MINRAD
  20  IF (COL .EQ. NN)  COL = 0
  21  CONTINUE
      DO 4239 I=1,N,NN
      RAD=MINRAD
      M=I+NN-1
      DO 1239 K=I,M
      DO 1238 L=1,N
 1238 AA(K,L)=AA(K,L)/((RAD*OTHETA)**3)
 1239 RAD=RAD+LAM
   16 FORMAT(/    5(3X,1H(,I3,1H,,I3,1H),2X),/     5F14.4)
   17 FORMAT(/    4(3X,1H(,I3,1H,,I3,1H),2X),/     4F14.4)
   18 FORMAT(/    3(3X,1H(,I3,1H,,I3,1H),2X),/     3F14.4)
    3 CONTINUE
      IF(C1.NE.4.0) GO TO 721
      IF(I.EQ.N) GO TO 721
      C1=1.0
      WRITE(6,966)
      WRITE(6,977)
  721 IF(COL.EQ.NN) COL=0
  720 CONTINUE
C MATRIX OF LOAD POINT COEFFICIENTS HAS NOW BEEN CALCULATED AND
C STORED.  PROCEED WITH READ IN OF LOAD VALUES AND CALCULATION OF
C LOADING COEFFICIENTS.
 1501 DO 220 JK=1,ICASE
C READ CASE NO. AND NO. OF LOADING VALUES TO BE READ IN
  666 READ(5,88) INDEX, LL
   88 FORMAT (2I5)
C CLEAR BB ARRAY - ONLY NON-ZERO LOADING VALUES NEED BE READ IN
      DO 556 I=1,N
  556 BB(I)=0.0
C READ IN LOADING VALUES
      DO 1156 L=1,LL
 1156 READ (5,544) I, BB(I)
  544 FORMAT (I5,F10.0)
      WRITE (7,88) INDEX
C MULTIPLY LOAD VALUES BY COLUMN LOADING VALUES TO OBTAIN LOADING
C COEFFICIENTS.
      COL = 0
      DO 828  I = 1, N, NN
      J = I + NN - 1
      DO 828  K = I, J
      COL = COL + 1
      BB(K) = BB(K) * LDCOEF(COL)
  828 IF (COL .EQ. NN)  COL = 0
C WRITE OUT LOADING COEFFICIENTS BY ROWS.
      WRITE (6,966)
      WRITE (6,977)
      WRITE (6,58) INDEX
   58 FORMAT(31X,4HCASE,I3)
      WRITE (6,1114)
```

```
 1114 FORMAT(24X,20HLOADING COEFFICIENTS//)
      DO 228 I=1,N,NN
      K=I+NN-1
  228 WRITE (6,887) (BB(J),J=I,K)
  887 FORMAT(4E18.8)
C ENTIRE MATRIX HAS NOW BEEN CALCULATED AND STORED.  PROCEED WITH
C SOLUTION BY GAUSS SUBROUTINE TO OBTAIN DEFLECTIONS.
C
      CALL GAUSS(N,AA,BB,XX,MM,CHECK)
C
      WRITE (7,189) (XX(I),I=1,N)
  189 FORMAT(5E14.7)
C WRITE OUT DEFLECTIONS SO THAT NO ROW IS SPLIT BETWEEN TWO PAGES WITH
C THE HEADINGS APPEARING ON BOTH PAGES.
 4239 CONTINUE
      IF(IQUA.EQ.1) GO TO 1501
      IF(IQUA.EQ.2) GO TO 669
  669 WRITE(6,966)
      WRITE(6,977)
      C1=1.0
      ROW=0
      COL=0
      DO 720 K=1,N,NN
      ROW=ROW+1
      M=K+NN-1
      DO 721 I=K,M
      C1=C1+1.0
      COL=COL+1
      J1=I-2*NN
      J2=I-NN-1
      J3=I-NN
      J4=I-NN+1
      J5=I+NN-1
      J6=I+NN
      J7=I+NN+1
      J8=I+2*NN
      IW=I-2
      IX=I-1
      IY=I+1
      IZ=I+2
      WRITE(6,4) I
    4 FORMAT(//22X,23HCOEFFICIENTS FOR POINT ,I3//)
      IF(ROW.LE.2) GO TO 1
      WRITE(6,10) I,J1,AA(I,J1)
    1 IF(ROW.EQ.1) GO TO 2
      IF(COL.EQ.1) WRITE(6,11) I,J3,I,J4,AA(I,J3),AA(I,J4)
      IF(COL.EQ.NN) WRITE(6,12) I,J2,I,J3,AA(I,J2),AA(I,J3)
      IF(COL.NE.1.AND.COL.NE.NN) WRITE(6,13) I,J2,I,J3,I,J4,AA(I,J2),
     1      AA(I,J3),AA(I,J4)
    2 IF(COL.EQ.1) WRITE(6,14)I,I,I,IY,I,IZ,AA(I,I),AA(I,IY),AA(I,IZ)
      IF(COL.EQ.2) WRITE(6,15) I,IX,I,I,I,IY,I,IZ,AA(I,IX),AA(I,I),
     1      AA(I,IY),AA(I,IZ)
      IF(COL.EQ.NN-1) WRITE(6,17)I,IW,I,IX,I,I,I,IY,AA(I,IW),AA(I,IX),
     1      AA(I,I),AA(I,IY)
      IF(COL.EQ.NN) WRITE(6,18)I,IW,I,IX,I,I,AA(I,IW),AA(I,IX),AA(I,I)
      IF(COL.GT.2.AND.COL.LT.NN-1) WRITE(6,16)I,IW,I,IX,I,I,I,IY,I,IZ,
     1      AA(I,IW),AA(I,IX),AA(I,I),AA(I,IY),AA(I,IZ)
      IF(ROW.EQ.NUMROW) GO TO 3
      IF(COL.EQ.1) WRITE(6,11) I,J6,I,J7,AA(I,J6),AA(I,J7)
      IF(COL.EQ.NN) WRITE(6,12) I,J5,I,J6,AA(I,J5),AA(I,J6)
      IF(COL.NE.1.AND.COL.NE.NN) WRITE(6,13) I,J5,I,J6,I,J7,AA(I,J5),
     1      AA(I,J6),AA(I,J7)
      IF(ROW.EQ.NUMROW-1) GO TO 3
      WRITE(6,10)I,J8,AA(I,J8)
   10 FORMAT(/28X,  3X,1H(,I3,1H,,I3,1H),2X ,/28X, F14.4)
   11 FORMAT(/28X,2(3X,1H(,I3,1H,,I3,1H),2X),/28X,2F14.4)
   12 FORMAT(/14X,2(3X,1H(,I3,1H,,I3,1H),2X),/14X,2F14.4)
   13 FORMAT(/14X,3(3X,1H(,I3,1H,,I3,1H),2X),/14X,3F14.4)
```

```
   14 FORMAT(/28X,3(3X,1H(,I3,1H,,I3,1H),2X),/28X,3F14.4)
   15 FORMAT(/14X,4(3X,1H(,I3,1H,,I3,1H),2X),/14X,4F14.4)
      DO 61 I=1,N
  61  XX(I)=XX(I)*12.0
      WRITE (6,966)
      WRITE (6,977)
      WRITE(6,58) INDEX
      WRITE (6,42)
   42 FORMAT (20X,5HPOINT,5X,20HDEFLECTION IN INCHES)
      N56N=(56/NN)*NN
      N56N1=N56N+1
      IF(N56N.GE.N) N56N=N
      DO 55 I=1,N56N
   55 WRITE (6,60) I,XX(I)
   60 FORMAT(20X,I4,8X,E15.8)
      IF(N56N.EQ.N) GO TO 551
      WRITE (6,966)
      WRITE (6,977)
      WRITE(6,58) INDEX
      WRITE (6,42)
      N102N=2*N56N
      N103N=N102N+1
      IF(N102N.GE.N)N102N=N
      DO 550 I=N56N1,N102N
  550 WRITE(6,60) I,XX(I)
      IF(N102N.EQ.N) GO TO 551
      WRITE(6,966)
      WRITE (6,977)
      WRITE (6,58) INDEX
      WRITE (6,42)
      DO 560 I=N103N,N
  560 WRITE (6,60) I,XX(I)
  551 WRITE(6,443) MM, CHECK
  443 FORMAT(/////20X,4HMM =,I2,/20X,7HCHECK =,E15.8)
C RETURN TO CONSIDER A NEW CASE OR TO START FOR A NEW BRIDGE.
  220 CONTINUE
      GO TO 999
      END
$IBFTC GSTT
      SUBROUTINE GAUSS(N,AA,BB,XX,MM,CHECK)
C
C     CALCULATES THE SOLUTION OF AA*XX=BB
C     BY AN ITERATIVE GAUSSIAN ELIMINATION WHERE A FIRST SOLUTION FOR
C     XX IS OBTAINED DIRECTLY. THE RESIDUAL B=BB-AA*XX IS CALCULATED
C     USING ACCUMULATED INNER PRODUCTS. THE RESIDUAL EQUATION
C     AA*X=B IS SOLVED AND THE SOLUTION IS ADDED TO XX.
C     THE PROGRAM RETURNS AFTER MM SELF CORRECTIONS OR IF WE ARE NO
C     LONGER ADDING A BINARY BIT TO THE CORRECTED ANSWER. THIS IS
C     CHECKED IN COMPARING THE RATIO OF INFINITY NORMS OF THE
C     CORRECTIONS AT TWO STEPS. AT RETURN THE FINAL ANSWER IS IN XX, THE
C     NUMBER OF SELF-CORRECTIONS TAKEN IS IN MM AND THE RATIO OF THE
C     INFINITY NORM OF THE FINAL CORRECTION TO THE INFINITY NORM OF THE
C     FINAL ANSWER IS IN CHECK.
C
      COMMON A
      DIMENSION AA(121,121) ,A(121,121) ,B(121) ,BB(121)
      DIMENSION X(121) ,XX(121) ,IN(121)
      DOUBLE PRECISION E,CC
      ITCH=0
      VAL2=1.0E37
C
C     SAVE AA AND BB FOR CALCULATING THE RESIDUAL
C     BRING IN A AND B
C     XX IS DEVELOPED AS A SUM
C
      DO 2 I=1,N
      B(I)=BB(I)
      XX(I)=0
```

```
      DO 2 J=1,N
      A(I,J)=AA(I,J)
    2 CONTINUE
      REWIND 1
      WRITE (1) ((AA(I,J),J=1,N),I=1,N)
      END FILE 1
      REWIND 1
C
C     PERFORM THE TRIANGULARIZATION OF A
C
      NN=N-1
      DO 6 L=1,NN
      IN(L)=0
      KP=0
C     FOR ROW NUMBER OF MAX ELEMENT
      T=0.0
C     FOR THE MAX ELEMENT OF THE L TH COL.
C     LOOK AT THE ELEMENTS ON AND BELOW THE MAIN DIAGONAL
      DO 3 K=L,N
      TT=ABS(A(K,L))
      IF(T.GT.TT) GO TO 3
      T=TT
      KP=K
C     ROW NUMBER TO BE MOVED UP
    3 CONTINUE
      IF(L.EQ.KP) GO TO 5
      IN(L)=KP
      DO 4 J=L,N
      T=A(L,J)
      A(L,J)=A(KP,J)
    4 A(KP,J)=T
    5 LL=L+1
      DO 6 K=LL,N
      R=A(K,L)/A(L,L)
      A(K,L)=R
C     SAVE R FOR REDUCTION OF B
      DO 6 J=LL,N
    6 A(K,J)=A(K,J)-R*A(L,J)
C
      REWIND 2
      WRITE (2) ((A(I,J),J=1,N),I=1,N)
      END FILE 2
      REWIND 2
C     THIS COMPLETES THE REDUCTION OF A
C     CARRY OUT CORRESPONDING EXCHANGES ON B
C
   80 READ  (2) ((A(I,J),J=1,N),I=1,N)
      REWIND 2
    8 DO 10 L=1,NN
      KP=IN(L)
      IF(KP.EQ.0) GO TO 9
      T=B(KP)
      B(KP)=B(L)
      B(L)=T
    9 LL=L+1
      DO 10 K=LL,N
      R=A(K,L)
   10 B(K)=B(K)-R*B(L)
C
C     START BACK SUBSTITUTION
C
      X(N)=B(N)/A(N,N)
      DO 12 L=2,N
      LL=N+1-L
      K=LL+1
      T=0.0
      DO 11 J=K,N
   11 T=T+A(LL,J)*X(J)
   12 X(LL)=(B(LL)-T)/A(LL,LL)
```

```
C
C     CALCULATE THE INFINITY NORM OF THE CORRECTION
C
C     SET VAL1 EQUAL TO THE OLD INFINITY NORM
C
      VAL1=VAL2
      VAL2=0
      DO 75 I=1,N
      T=ABS(X(I))
      IF(T.GT.VAL2) VAL2=T
   75 CONTINUE
C
C     ADD CORRECTION TO THE PREVIOUS ANSWER
C
      DO 13 I=1,N
      XX(I)=XX(I)+X(I)
   13 CONTINUE
C
      READ  (1) ((AA(I,J),J=1,N),I=1,N)
      REWIND 1
C     CHECK FOR CONVERGENCE OF NORMS OR MAXIMUM NUMBER OF ITERATIONS
C     REACHED IF EITHER RETURN TO MAIN
C
      IF(ITCH.EQ.MM) GO TO 15
C
C     IF WE GO TO 15 WE ARE FINISHED. IF NOT WE CALCULATE THE RESIDUAL
C     AND BEGIN THE NEXT CORRECTION.
C
      DO 98 I=1,N
      E=0.0
      DO 97 K=1,N
   97 E=E+AA(I,K)*XX(K)
      CC=BB(I)
      B(I)=CC-F
   98 CONTINUE
C
C     NOW REDUCE THIS RESIDUAL IN THE EXACT SAME WAY AS THE RIGHT SIDE
C     UPDATE ITCH
C
      ITCH=ITCH+1
      GO TO 80
C
C     CALCULATE CHECK AND RETURN
C
   15 CONTINUE
      VV=0
      DO 16 I=1,N
      TT=ABS(XX(I))
      IF(TT.GT.VV) VV=TT
   16 CONTINUE
      CHECK=VAL2/VV
C
C     SET MM EQUAL TO THE NUMBER OF CORRECTIONS MADE AND RETURN
C
      MM=ITCH
      RETURN
      END
```

Appendix A-3
Finite Difference Curved Orthotropic Bridge Moment Computer Program

```
C               CURVED ORTHOTROPIC BRIDGE MOMENT PROGRAM
C               COMPUTER PROGRAM NUMBER 2
C   ANGULAR AND RADIAL MOMENTS FOR CURVED ORTHOTROPIC BRIDGE DECKS
      DIMENSION DEF(140),HIA( 20),CK(20),EIA(20),EIR(20),AM(120),RM(120)
      DIMENSION T(140)
  999 READ 90
   90 FORMAT(10X,40H                                         )
      READ 91
   91 FORMAT(10X,40H                                         )
      READ 1, N,NN,RMIN,AL,THO,ICASE
    1 FORMAT(2I5,3E16.8,I5)
      READ 189, (EIA(I),I=1,NN)
      READ 189,(HIA(I),I=1,NN)
      NROW=N/NN
      READ 189, (EIR(I),I=1,NROW)
      TH2=THO**2
      TH4=THO**4
      J1=NN+1
      J2=N+NN
      J3=N+2*NN
      DO 122 IND=1,ICASE
      READ 1, INDEX
      DO 25 I=1,J3
   25 DEF(I)=0.0
      READ 189, (DEF(I),I=J1,J2)
  189 FORMAT(5E14.7)
      DO 7 I=J1,J2
      T(I)=0.0
      AM(I)=0.0
    7 RM(I)=0.0
      NR=0
      DO 10 I=J1,J2,NN
      NC=1
      R=RMIN
      NR=NR+1
      J=I+NN-1
      DO 11 K=I,J
      Z=R**2
      IF(NC-1)12,12,13
   13 IF(NC-NN)14,15,15
   12 TEMP=2.0*DEF(K+1)-2.0*DEF(K)
C     FOR NC=1
      JWAR = K-NN+1
      JWBR = K+NN+1
      JWA = K-NN
      JWB = K+NN
      IF(NR-1) 70,70,71
   71 IF(NR-NROW) 72,73,73
   70 T(K)=HIA(NC)*(2.*DEF(JWB)-2.*DEF(JWBR))/(4.*THO*R*AL)-(DEF(JWA)-DE
     1F(JWB))/(2.*R**2*THO)*HIA(NC)
      GO TO 16
   72 T(K)=HIA(NC)*(2.*DEF(JWAR)-2.*DEF(JWBR)-2.*DEF(JWA)+2.*DEF(JWB))/(
     14.*THO*R*AL)-(DEF(JWA)-DEF(JWB))/(2.*R**2*THO)*HIA(NC)
      GO TO 16
   73 T(K)=HIA(NC)*(2.*DEF(JWAR)-2.*DEF(JWA))/(4.*THO*R*AL)-(DEF(JWA)-DE
     1F(JWB))/(2.*R**2*THO)*HIA(NC)
      GO TO 16
   14 TEMP=DEF(K+1)-DEF(K-1)
C     FOR NC GREATER THAN 1, LESS THAN NN
```

```
      JWA=K-NN
      JWB=K+NN
      JWAL = K-NN-1
      JWAR = K-NN+1
      JWBL = K+NN-1
      JWBR = K+NN+1
      T(K)=HIA(NC)*(DEF(JWAR)-DEF(JWBR)-DEF(JWAL)+DEF(JWBL))/(4.*THO*R*A
     1L)-(DEF(JWA)-DEF(JWB))/(2.*R**2*THO)*HIA(NC)
      GO TO 16
   15 TEMP=2.0*DEF(K)-2.0*DEF(K-1)
C     FOR NC = NN
      JWBL = K-NN+1
      JWBR = K+NN+1
      JWA = K-NN
      JWB = K+NN
      IF(NR-1) 74,74,75
   75 IF(NR-NROW) 76,77,77
   74 T(K)=HIA(NC)*(2.*DEF(JWBL)-2.*DEF(JWB))/(4.*THO*R*AL)-(DEF(JWA)-DE
     1F(JWB))/(2.*R**2*THO)*HIA(NC)
      GO TO 16
   76 T(K)=HIA(NC)*(2.*DEF(JWBL)-2.*DEF(JWAL)-2.*DEF(JWB)+2.*DEF(JWA))/(
     14.*THO*R*AL)-(DEF(JWA)-DEF(JWB))/(2.*R**2*THO)*HIA(NC)
      GO TO 16
   77 T(K)=HIA(NC)*(2.*DEF(JWA)-2.*DEF(JWAL))/(4.*THO*R*AL)-(DEF(JWA)-DE
     1F(JWB))/(2.*R**2*THO)*HIA(NC)
      GO TO 16
   16 CONTINUE
      IF(NR-1)41,41,43
   43 IF(NR-NROW)44,46,46
   41 K3=K+NN
      AM(K)=-EIA(NC)/(2.0*AL*Z*TH2)*(R*TH2*TEMP+2.0*AL*(DEF(K3)
     1-2.0*DEF(K)))
      GO TO 50
   44 K2=K-NN
      K3=K+NN
      AM(K)=-EIA(NC)/(2.0*AL*Z*TH2)*(R*TH2*TEMP+2.0*AL*(DEF(K2)
     1-2.0*DEF(K)+DEF(K3)))
      GO TO 50
   46 K2=K-NN
      AM(K)=-EIA(NC)/(2.0*AL*Z*TH2)*(R*TH2*TEMP+2.0*AL*(DEF(K2)
     1-2.0*DEF(K)))
   50 NC=NC+1
   11 R=R+AL
   10 CONTINUE
      NR=0
      DO 21 I=J1,J2,NN
      NR=NR+1
      I1=I+1
      I2=I+NN-2
      DO 20 J=I1,I2
   20 RM(J)=-((EIR(NR))/AL**2)*(DEF(J-1)-2.0*DEF(J)+DEF(J+1))
   21 CONTINUE
      PUNCH 90
      PUNCH 91
      PUNCH 92, INDEX
   92 FORMAT(/22X,4HCASE,I3/)
      PUNCH 93
   93 FORMAT(2X,5HPOINT,8X,7HANGULAR,14X,6HRADIAL,14X,6HTORQUE)
      PUNCH 94
   94 FORMAT(11X,16HMOMENT IN KIP-FT,4X,16HMOMENT IN KIP-FT,9X,6HKIP-FT/
     1)
      DO 95 I=J1,J2
      K=I-NN
   95 PUNCH 96, K,AM(I),RM(I),T(I)
   96 FORMAT(3X,I3,5X,E16.8,4X,E16.8,4X,E16.8)
  122 CONTINUE
      GO TO 999
      END
```

References

References

1. G.B. Thomas, *Calculus and Analytic Geometry*, Addison-Wesley, Reading, Mass., 1953.
2. C.T. Wang, *Applied Elasticity*, McGraw-Hill Book Co., New York, 1953.
3. S. Timoshenko, J.N. Goodier, *Theory of Elasticity*, McGraw-Hill Book Co., New York, 1951.
4. S. Timoshenko, S. Woinowsky-Krieger, *Theory of Plates and Shells*, McGraw-Hill Book Co., New York, 1959.
5. R.M. Rivello, *Theory and Analysis of Flight Structures*, McGraw-Hill Book Co., New York, 1969.
6. R. Szillard, *Theory and Analysis of Plates*, Prentice-Hall, Englewood Cliffs, N.J., 1974.
7. K. Girkmann, *Flachentragwerke*, Springer Verlag, Vienna, Austria, 1956.
8. M.F. Borodich, *Theory of Elasticity*, Foreign Language Pub. House, Moscow, U.S.S.R., 1960.
9. F. Dischinger, "Contribution to the Theory of Wall-Like Girders," "International Association for Bridge and Structural Engineers," Vol. 1, 1932, Zurich, Switzerland.
10. "Design of Deep Girders," PCA, St66-R/C 12, Portland Cement Assoc., Chicago, Ill., 1960.
11. F.A. Archer, E.M. Kitchen, "Stresses in Single Span Deep Beams," *Australian Journal of Applied Science*, Vol. 7, No. 4, 1956.
12. H.D. Conway, "Analysis of Deep Beams," *Journal of Applied Mechanics*, No. 18, 1951.
13. J.D. Keiegh, R.M. Richard, "Epoxy Bonded Composite T-Beams for Highway Bridges," Univ. of Arizona, Rpt. No. 12, Tucson, Arizona, Oct. 1966.
14. M.G. Salvadori, M.L. Baron, *Numerical Methods in Engineering*, Prentice-Hall, Englewood Cliffs, N.J., 1952.
15. R.G. Stanton, *Numerical Methods for Science and Engineering*, Prentice-Hall, Englewood Cliffs, N.J., 1961.
16. F. Schleicher, *Taschenbuck fur Bauingenieure*, Springer Verlag, Vienna, Austria, 1955.
17. H.H. Schade, "The Effective Breadth of Stiffened Plating under Bending Loads," *Society of Naval Architects and Marine Engineers*, Vol. 59, 1951.

18. E.E. Sechler, *Elasticity in Engineering*, Dover Pub. Co., New York, 1952.

19. F. Levi, "European Concrete Code," *Proceedings of the American Concrete Institute*, Vol. 57, 1960, pp. 1049-1054.

20. H.M. Fan, C.P. Heins, "Effective Slap Width of Simple Span Steel I-Beam Composite Bridges at Ultimate Load," C.E. Rept. No. 57, University of Maryland, College Park, Md., June 1974.

21. I. Holand, K. Bell, *Finite Element Methods,* Technical University of Norway, Trondheim, Norway, 1970.

22. C.S. Desai, J.F. Abel, *Introduction to the Finite Element Method*, Van Nostrand Reinhold Company, New York, N.Y., 1972.

23. O.C. Zienkiewicz, *Finite Element Method*, McGraw-Hill Book Co., New York, 1967.

24. J.S. Przenieniecki, *Theory of Matrix Structural Analysis*, McGraw-Hill Book Co., New York, 1968.

25. S.G. Lekhnitskii, *Anisotropic Plates*, Gordon Breach, New York, 1968.

26. L.G. Jaeger, *Elementary Theory of Elastic Plates*, Pergamon Press, New York, 1964.

27. D.E. McFarland, B.L. Smith, W.D. Bernhart, *Analysis of Plates*, Spartan Books, New York 1972.

28. E.H. Mansfield, *The Bending and Stretching of Plates*, Macmillan Co., New York, 1964.

29. C.P Heins, *Bending and Torsional Design in Structural Members*, D.C. Heath Co., Lexington Books, Lexington, Mass., 1975.

30. A. Parme, "Solutions of Difficult Structural Problems by Finite Differences," *American Concrete Institute*, Vol. 22, No. 3, Chicago, Ill., Nov. 1950.

31. C.P. Heins, C.T.G. Looney, "The Analysis of Curved Orthotropic Highway Bridges by the Finite Difference Technique," CE Rpt. No. 13, University of Maryland, College Park, Md., Jan. 1967.

32. R.L. Hails, C.P. Heins, "The Study of a Stiffened Curved Plate Model Using the Finite Difference Techniques," CE Rpt. No. 22, University of Maryland, College Park, Md., June 1968.

33. C.P. Heins, R.L. Hails, "Behavior of a Stiffened Curved Plate Model," *American Society of Civil Engineers Structural Division Journal*, Vol. 95, No. ST11, Nov. 1969.

34. C.P. Heins, C.H. Yoo, "Grid Analysis of Orthotropic Bridges," *International Association for Bridge and Structural Engineers*, Vol. 30-I, Zurich, Switzerland, 1970.

35. C.P. Heins, C.T.G. Looney, "Bridge Tests Predicted by Finite Difference Plate Theory," *American Society of Civil Engineers Structural Division Journal*, Vol. 95, No. ST2, Feb. 1969.
36. A.D. Sartwell, C.P. Heins, C.T.G. Looney, "Analytical and Experimental Behavior of a Simple Span Girder Bridge," *Highway Research Record*, No. 295, Highway Research Board, Jan. 1969.
37. C.P. Heins, "Building Grid Analysis," *Building Science*, Vol. 7, pp. 265-269, Pergamon Press, New York, 1972.
38. C.P. Heins, C.T.G. Looney, "Bridge Analysis Using Orthotropic Plate Theory," *American Society of Civil Engineers Structural Journal*, Vol. 93, No. ST2, Feb. 1968.
39. C.P. Heins, and C.T.G. Looney, "The Solution of Continuous Orthotropic Plates on Flexible Supports as Applied to Bridge Structures," Progress Report on the Study of the Effect of New Vehicle Weight Laws on Structures, Civil Eng. Department, University of Maryland, March 1966, Vol. I.
40. J.S. Kinney, *Indeterminate Structural Analysis*, Addison Wesley, Reading, Mass., 1957.
41. L.C. Bell, C.P. Heins, "Analysis of Curved Girder Bridges," *American Society of Civil Engineers Structural Division Journal*, Vol. 96, No. ST8, Aug. 1970.
42. C.P. Heins, L.C. Bell, "Curved Girder Bridge Analysis," *Journal of Computers and Structures*, Vol. 2, pp. 785-797, 1972, Pergamon Press, England.
43. C.P. Heins, J. Siminou, "Preliminary Design of Curved Girder Bridges," *American Institute of Steel Construction Engineering Journal*, Vol. No. 2, Ap. 1970.
44. C.P. Heins, "Curved I Girder Design," *Highway Research Record*, No. 547, Highway Research Board, Jan. 1975.
45. C.P. Heins, B. Bonakdarpour, L.C. Bell, "Multi-cell Curved Girder Model Studies," *American Society of Civil Engineers Structural Journal*, Vol. 98, No. ST4, Ap. 1972.
46. C.P. Heins, B. Bonakdarpour, "Behavior of Curved Bridge Models," *Public Roads Journal*, Vol. 36, No. 11, Dec. 1971.
47. C.H. Yoo, J. Buchanan, C.P. Heins, W.L. Armstrong, "An Analysis of a Continuous Curved Box Girder Bridge," *Highway Research Record*, Highway Research Board, Jan. 1975.
48. P.G. Perry, C.P. Heins, "Rapid Design of Orthotropic Bridge Floor Beams," *American Society of Civil Engineers Structural Journal*, Vol. 98, No. ST11, Nov. 1972.

49. C.P. Heins, "Distribution Factors for Bridge Plate Decks," Civil Engineering Study, University of Maryland, College Park, Md., June 1974.

50. F. Gienoko, "Die Grundlelchungen tur die Orthotropische Platten mit Steifen Exzentrischen," *Der Stahlbau,* 24, pp. 128-139, 1955.

51. A.R. Cusens, R.P. Pama, *Bridge Deck Analyses*, John Wiley & Sons, New York, 1975.

52. A.R. Cusens, M.A. Zeidan, and R.P. Pama, "Elastic Rigidities of Ribbed Plates," *Building Science*, 7, pp. 23-32, 1972.

53. W. Pelikan, and H. Esslinger, "Die Stahlfahrbahn Berechnung und Construktion," (*M.A.N. Forschungsheft*), 1957.

54. R. Wolchuk, *Design Manual for Orthotropic Steel Plate Deck Bridges*, American Institute of Steel Construction, 1963.

55. G.W. Zuurbier, "Testing of A Steel Deck Bridge," *Highway Research Record*, Highway Research Board, Washington, D.C., Jan. 1968.

56. G.W. Zuurbier, "Modular Steel Deck Bridge," Homer Research Lab., Bethlehem Steel Corp., Bethlehem, Pa., May 1967.

57. W.J. Stroud, "Elastic Constants for Bending and Twisting of Corrugation Stiffened Panels," NASA TR T-166, Government Printing Office, Washington, D.C., 1963.

58. K.H. Chu, G. Krishnamoorthy, "Use of Orthotropic Plate Theory in Bridge Design," *American Society of Civil Engineers*, Vol. 88, No. ST3, June 1962.

59. V. Vitols, F.J. Clifton, T. Au, "Analysis of Composite Beam Bridges by Orthotropic Plate Theory," *American Society of Civil Engineers*, Vol. 89, No. ST4, Aug. 1963.

60. W.H. Wittrick, "Torsion of a Multi-Webbed Rectangular Tube," *Aircraft Engg.*, 25, p. 372, 1963.

61. R.E. Beckett, R.J. Dohrmann, K.D. Ives, "An Experimental Method for Determining the Elastic Constants of Orthogonally Stiffened Plates," *Experimental Mechanics, Proc. of First International Congress*, Pergamon Press, New York, 1958.

62. R.K. Witt, W.H. Hoppmann, R.S. Buxbaum, "Determination of Elastic Constants of Orthotropic Materials with Special Reference to Laminates," *American Society of Testing Material Bulletin 194*, Dec. 1953.

63. W.H. Hoppmann, "Bending of Orthogonally Stiffened Plates," *Journal of Applied Mechanics, American Society of Mechanical Engineers*, Vol. 77, 1955.

64. R.F.S. Hearmon, E.H. Adams, "The Bending and Twisting of Amsotropic Plates," *British Journal of Applied Physics*, Vol. 3, 1952.

65. *Orthotropic Bridge Decks Using Bethlehem Standard Ribs*, Bridge Design Aids, Bethlehem Steel Corporation, Bethlehem, Pa., 1968.
66. M.S. Troitsky, *Orthotropic Bridges—Theory and Design,* The James F. Lincoln Arc Welding Foundation, Cleveland, 1967.
67. D.H. Hall, *Orthotropic Bridge Examples*, PD-202-2, Bethlehem Steel Corporation, Bethlehem, Pa., March 1971.
68. G.D. Adotte, "Second Order Theory in Orthotropic Plates," *American Society of Civil Engineers Structural Journal*, Vol. 93, No. ST5, Oct. 1967.
69. H.W.L. Lee, C.P. Heins, "Large Deflections of Curved Plates," *American Society of Civil Engineers Structural Journal*, Vol. 97, No. ST4, April 1971.
70. H.W.L. Lee, C.P. Heins, "Local Loading Influence on Curved Decks," *CE Rpt. No. 35*, University of Maryland, College Park, Md., March 1970.
71. R.L. Brockenbrough, B.G. Johnston, *Steel Design Manual*, U.S. Steel Corp., Pittsburgh, Pa., Nov. 1968.

References 72-78 have not been sited herein, but may be of interest to the readers.

72. R.H. Wood, *Plastic and Elastic Design of Slabs and Plates*, Ronald Press, New York, 1961.
73. R. Bases, C. Massonnet, *Le Calcul des Grillages de Poutres et Dalles Orthotropes*, Dunod, Paris, France, 1966.
74. L.S.D. Morley, *Skew Plates and Structures*, Pergamon Press, New York, 1963.
75. D.O. Brush, B.O. Almroth, *Buckling of Bars, Plates and Shells*, McGraw-Hill Book Co., New York, 1975.
76. J.R. Vinson, *The Behavior of Plates and Shells*, John Wiley & Sons, New York, 1974.
77. J. Hahn, *Structural Analysis of Beams and Slabs*, F. Ungar Pub. Co., New York, 1968.
78. R. Bares, *Tables for the Analysis of Plates, Slabs and Diaphragms Based on the Elasticity Theory*, Bauverlag Gimbitt, Berlin, Germany, 1964.

Indexes

Name Index

Subject Index

About the Author

C.P. Heins is Professor of Civil Engineering at The University of Maryland. He received the B.S.C.E. at Drexel Institute of Technology, the M.S.C.E. at Lehigh University and the Ph.D. at the University of Maryland. He is a member of the ASCE, ASEE, ACI, SESA and IABSE and is a committee member of the ASCE Long Span Bridges, ASCE Flexural Members and ASCE Safety and Reliability.

Dr. Heins has participated in the development of various bridge design specifications and was associated with orthotropic bridges, curved bridges and Load Factor Design. He has authored over 60 referenced journal articles in national international journals, and has participated in conferences throughout the world. Dr. Heins has been consultant to various government and state agencies and private consultants and is a registered professional in several states.